T0358937

SINTERING

FROM EMPIRICAL OBSERVATIONS TO SCIENTIFIC PRINCIPLES

RANDALL M. GERMAN

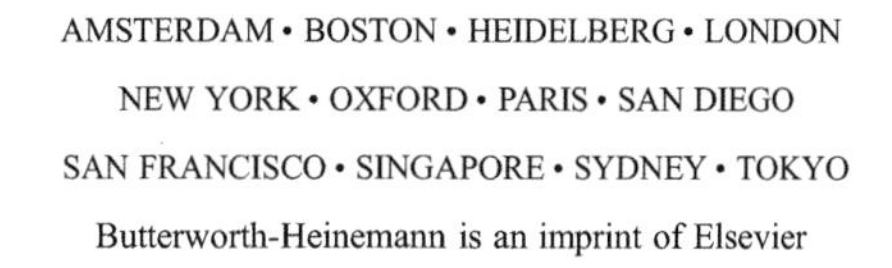
AMSTERDAM • BOSTON • HEIDELBERG • LONDON
NEW YORK • OXFORD • PARIS • SAN DIEGO
SAN FRANCISCO • SINGAPORE • SYDNEY • TOKYO
Butterworth-Heinemann is an imprint of Elsevier

Butterworth-Heinemann is an imprint of Elsevier
225 Wyman Street, Waltham, MA 02451, USA
The Boulevard, Langford Lane, Kidlington, Oxford, OX5 1 GB, UK

First edition 2014

Notices
Knowledge and best practice in this field are constantly changing. As new research and experience broaden our understanding, changes in research methods or professional practices, may become necessary.

Practitioners and researchers must always rely on their own experience and knowledge in evaluating and using any information or methods described herein. In using such information or methods they should be mindful of their own safety and the safety of others, including parties for whom they have a professional responsibility.

To the fullest extent of the law, neither the Publisher nor the authors, contributors, or editors, assume any liability for any injury and/or damage to persons or property as a matter of products liability, negligence or otherwise, or from any use or operation of any methods, products, instructions, or ideas contained in the material herein.

Library of Congress Cataloging-in-Publication Data
A catalog record for this book is available from the Library of Congress

British Library Cataloguing-in-Publication Data
A catalogue record for this book is available from the British Library.

ISBN: 978-0-12-401682-8

For information on all Butterworth-Heinemann publications
visit our web site at books.elsevier.com

Printed and bound in the United States of America

14 15 16 17 18 10 9 8 7 6 5 4 3 2 1

Dedicated to Animesh Bose

CONTENTS

PREFACE

Exciting new designs are enabled by the sintering process. Although practiced for over 28,000 years, recent discoveries are moving sintering into some bold new applications. New energy systems, ranging from solar cells to nuclear reactors, are critically contingent on sintered structures. Another example is the fabrication of porous tissue scaffolds for biomedical implants, in which the device is custom laser sintered to match strength and elastic modulus of the patient. In the same manner, dental crown and bridge constructions are produced overnight using additive computer driven sintering routes. An enormous effort is pushing forward the sintering of thin printed electronic structures, such as small radio frequency identification circuits to be embedded in consumer products, allowing information transfer when activated by near-field cellular telephones. Related efforts are taking place in replacement interconnections for solar cells and a host of capacitor, energy storage, and magnet devices. The field of superabrasives is producing sintered diamond bonded onto sintered cemented carbides substrates to make long-lasting oil and gas drilling tools. Another growth area is that of sintered thermoelectric junctions to convert waste heat into electricity, including waste heat from automobile engines.

This modern era of sintering traces to the early 1800s when the first platinum crucibles were made for melting glass. Significant progress came in the early 1900s with the production of incandescent lamp filaments, but theoretical explanations awaited the development of atomic theory and the atomic motion concepts that emerged in the 1940s. Once atomic theory was melded with sintering observations, quantitative conceptualizations arose. In turn, that effort matured to produce computer simulations. Now those simulations are approaching the accuracy that is demanded by manufacturing. Soon extraterrestrial sintering will use lunar soils and solar energy to construct buildings on the moon. The prospects for expanded applications are outstanding.

This book explains the basics of sintering. The concepts are equally applicable to the fabrication of electronic capacitors, automotive transmission gears, high intensity lights, jet engine control linkages, or high speed end mills. The approach starts with historical concepts, mixes history and science, and outlines the theoretical evolution. The scientific underpinnings arose from simple questions that form the key points covered by this book:

What is sintering?
How do we observe sintering?
What are some of the key parameters?
How can we improve sintering?
Where did sintering theory come from?
Where do we stand on modeling sintering?

Included are chapters on emerging topics, such as the role of rapid heating, and introductions to sintering tools.

The effort of writing this book started in the preparation of a Plenary Seminar for *Sintering 2011* and a Keynote Lecture for the *2012 Materials Science and Technology Conference*. My thanks go to Suk-Joong Kang, Eugene Olevsky, and Khalid Morsi for their early support. Several students helped, notably Wei Li, Timothy Young, Michael Brooks, and Shuang Qiao. Kenneth Brookes provided background information on sintered carbides, while Zak Fang, Animesh Bose, and Donald Heaney organized relevant reviews. Louis Rector and Howard Glicksman provided information on sintered electronic applications. Lanny Pease donated a missing book to provide insights. Other missing information was obtained from the Metal Powder Industries Federation and American Ceramic Society. I am thankful to a host of other individuals for their efforts and kind words, and to San Diego State University for giving semesters without teaching to complete this project.

This book is dedicated to Animesh Bose; he is a testimonial to what can be done by applying sintering. And the scary thing is his children are smarter and more motivated.

Randall M. German

CHAPTER ONE

Introduction

CONTEXT

Sintering is a thermal process used to bond contacting particles into a solid object. Students in school learn to work wet clay to shape a pot, and then heat or fire that body to create a strong pot. That firing process is sintering. In the same manner, newly fallen snow bonds, to harden and eventually form ice—this is a colder version of sintering. For industrial components, sintering is a means to strengthen shaped particles to form useful objects such as electronic capacitors, automotive transmission gears, metal cutting tools, watch cases, heart pacemaker housings, and oil-less bearings.

As a thermal treatment, sintering is crucial to the success of several engineering products; including most ceramics and cemented carbides, several metals, and some polymers. Powder shaping prior to sintering is done by die compaction for simpler shapes, such as automotive transmission gears. For complicated three-dimensional shapes, such as watch cases, injection molding is the favored shaping process. Long, thin objects, such as catalytic converter substrates, are shaped by extrusion through a die, in the same manner as graphite is extruded to form refills for mechanical pencils. There are technologies for shaping flat structures such as ceramic electronic substrates (tape casting), hollow bodies such as porcelain statuary (slip casting), and one of a kind metal prototypes (laser forming). Following each of these forming steps is a sintering treatment—defined by a heating cycle to a peak temperature. The hold time at the sintering temperature ranges from a few minutes to a few hours. Although the shaped body is weak prior to sintering, after the firing cycle it is very strong, competitive in properties with that attained via other manufacturing routes such as casting, machining, grinding, or forging.

This book addresses sintering by describing why it occurs, how it is measured, and the key control parameters. Property changes during sintering are outlined. The prediction of optimal cycles to generate desired properties is a key goal for sintering science. In this book, we see how sintering theory evolved from its empirical base—going to the lab to "see what happens." New materials and detailed phenomenological observations emerged long before atomic structure conceptualizations. Predicative sintering theory awaited an understanding of atoms and atomic motion.

Sintering: From Empirical Observations to Scientific Principles
DOI: http://dx.doi.org/10.1016/B978-0-12-401682-8.00001-X

Diffusion concepts engaged leading scientist in the 1940s. Once atomic motion was understood, the platform was in place to allow rapid progress on sintering theory. Accordingly, a burst of applications, literature, patents, and materials emerged and grew from the late 1940s through the 1960s. The expansion continues today to accommodate the increased composition complexity and more complex designs.

This book details the developments that converged to give today's sintering theory. Optimism abounds in the sintering community as we continue to push forward with improved understanding of a complicated process.

PERSPECTIVES

Archeological findings date sintered objects back 26,000 years. Early fired earthenware structures are found in China, India, Egypt, Japan, Turkey, Korea, Central America, and Southern Europe. About 3000 years ago firing to improve strength was practiced in many locations. By a few hundred years ago sintered products were manufactured under controlled conditions in Spain, China, Korea, Japan, Germany, England, and Russia.

English geologists used the term "sintering" in 1780 to describe the bonding of mineral particles and the formation of crusted stones in Iceland. This was in reference to the way silicates formed hardened crusts around hot geyser vents. The English borrowed the term "cinder" from German to describe the agglomeration or hardening of mineral particles. By 1854 the concept was used to describe the thermal bonding of coal particles and in the 1860s to describe the thermal hardening of iron ore, a process also known as induration.

The United States patent literature shows the first use of the term "sintering" in 1865 with respect to thermal cycles applied to mineral calcination. The agglomeration of flue dust, iron ores, and other minerals were early sintered products. Subsequently the term sintering was widely used to describe agglomeration with an emphasis on sinter plants for iron ore agglomeration.

By the 1880s the term "sintering" was applied to describe gold and silver purification, platinum bonding, iron powder consolidation, and the fabrication of platinum jewelry. In 1913 Coolidge refers to his heating process to form tungsten lamp filaments as involving "... filaments are still further treated to free then from all easily vaporizable components and to sinter together the refractory residue into a coherent conductor ...." [1] In ceramics the term sintering was reserved for describing the agglomeration of refractory, abrasive, or insulator powders. However, in 1939 White and Shremp [2] used "sintering" to describe ceramic particle bonding with reference

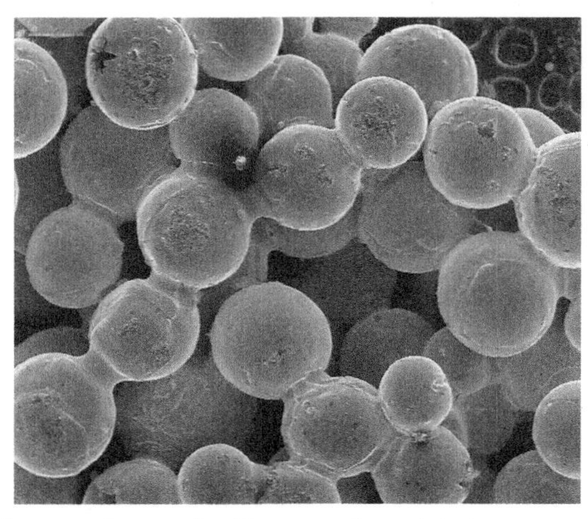

Figure 1.1 Scanning electron micrograph of bronze sphere sintering.

to properties of beryllia heated under different conditions. By World War II the importance of sintering jumped due to its military applications. In 1943, the US Library of Congress published a survey of the field, citing 700 publications and 600 patents [3]. After that time "sintering" was commonly used to describe thermally induced particle bonding [4–7].

Sintering is a thermal treatment to bond particles, leading to improved strength. This is evident in microscopic images, such as that shown in Figure 1.1. These spherical particles were initially poured into a crucible. During heating, bonds grew at the particle contacts. This occurred through atomic motion.

To explain the changes induced by heat, sintering theory emerged to provide a mathematical collection of key parameters such as particle size, heating rate, hold temperature, and hold time. The material is also important since it determines the surface energy, atomic size, activation energy for diffusion, and crystal structure. Consequently, several parameters enter into sintering models, so much background knowledge needed to be developed as a foundation for the models. For example, although liquid surface energy has long been an accepted concept, solid-vapor surface energy only was accepted in the 1940s. Solid surface energy is a necessary concept to explain the stress acting at particle contacts to produce sintering shrinkage.

Thus, the approach used here is as follows:

Assemble an outline of sintering concepts and models

→ trace back to find the important building blocks

→ determine how the building block concepts intersected with sintering

→ isolate early critical events via first publications

→ identify pivotal people and concepts.

This backward tracing is built from many prior assessments [7–43]. These reports were complimented by patent searches and on-line databases. Conflicts were vetted to correct errors in spellings, references, years, and incorrect citations. An example was some early work on spark sintering, which was attributed to the patent agent Arthur Bloxam instead of the inventor Johann Lux.

A master spreadsheet was created to understand the evolution of sintering concepts and the enabling infrastructure. It identified critical steps and individuals. Sometimes the priority was unclear. For example, silicon nitride was formed in 1896, reaction bonding was developed in the 1930s, hot pressing was developed in the 1960s, and pressureless liquid phase sintering emerged in the 1970s [31,44,45]. Since the first densification was by hot pressing, this date was used to tag the emergence of sintered silicon nitride.

By early 2013 over one million articles were indexed under the terms "sinter" and "sintering." This literature is the basis for this book, condensing the sintering concept from a large, constantly expanding body of knowledge.

DEFINITIONS

Sintering as a term arose in the 1800s and became more common in the middle 1900s. Although variants exist, the following definition captures both the historical and modern usage [19,29,46–51]:

Sintering is a thermal treatment for bonding particles into a coherent, predominantly solid structure via mass transport events that often occur on the atomic scale. The bonding leads to improved strength and lower system energy.

A few other terms are important to understanding the overt impact of sintering. Density is the mass per unit volume so it has units of g/cm^3 or kg/m^3. Density depends on the material and changes during most sintering treatments, so it is a common measure of the degree of sintering. Theoretical density corresponds to the pore-free solid density. Fractional or percentage density is useful for comparing the behavior of powder systems without the confusion over differing theoretical densities. For this book the preferred expression for sintered density will be fractional or percentage density based on the ratio of the measured density to the theoretical density. Green density is the density prior to sintering and green strength corresponds to the strength prior to sintering.

Porosity is the unfilled space in a powder compact. Prior to sintering it is called the green porosity. Since there is no mass associated with porosity, it is simply treated as a fraction or percentage of the body. Thus, a sintered component that is 80% dense has 20% porosity. The fractional density and fractional porosity sum to unity. In cases involving liquid phases during sintering, three phases are present—solid, liquid, and porosity. The fractional density in those cases is the sum of the solid and liquid portions.

Particles are discrete solids, generally smaller than 1 mm in size, but larger than an atom. Powders are collections of particles, usually with a range of sizes and shapes. Powders do not fill space efficiently. For example, monosized spheres pour to fill a container at approximately 60% density. After vibration these same spheres will reach a maximum packing density of 64%. Higher densities come from changes to the particle shape, particle size distribution, or by the application of pressure.

Green bodies prior to sintering are termed compacts. They are usually prepared by mixing a binder or lubricant with the powder (wax-type molecule) and applying pressure to the powder to increase density and shape the powder. The pressure ranges from gravity to thousands of atmospheres of applied pressure. A green compact is usually weak; vitamin pills are examples of pressed powders. Hot consolidation relies on the application of pressure during sintering.

Several monitors for sintering appear in this book—surface area, neck size ratio, shrinkage, swelling, and densification. Surface area is the solid-vapor area and is usually captured in terms of area per unit mass, such as m^2/g. Neck size is the diameter of the sinter bond between two particles, and the neck size ratio is the ratio of the neck size divided by the particle size (dimensionless). Shrinkage refers to the decrease in linear dimensions, while swelling refers to an increase in dimensions. They are both linear dimensional changes, where the change is size is divided by the size prior to sintering. Measures such as density and shrinkage are easy to perform, and provide insight into the changes during sintering. Densification is the change in porosity with sintering divided by the starting porosity. If all pore space is eliminated during sintering, then the densification is 100%. It is a useful concept when comparing systems of differing theoretical densities or initial porosities. Densification, final density, neck size, surface area, and shrinkage are related measures of sintering.

Mixed powders of differing compositions are a common basis for sintering. The convention is to list the major component first. Thus, the term WC-Co implies that the bulk of the material is composed of tungsten carbide (WC), with cobalt (Co) being the minor component. When a number is embedded in the formula, such as Fe-8Ni, this designates 8 wt.% nickel powder has been added to iron powder. Composition is given on a mass basis unless explicitly stated otherwise. The atomic composition is given by a chemical formula showing the stoichiometry—for example $MoSi_2$ indicates two silicon atoms for each molybdenum atom. When the powder is pre-compounded, it is designated by the common name where possible, such as stainless steel (Fe-18Cr-8Ni), bronze (Cu-10Sn), or spinel (Al_2O_3-MgO).

SINTERING TECHNIQUES

Sintering theory is most accurate for the case of single phase powders sintered by solid-state diffusion. Unfortunately, this is a small portion of sintering practice. More common are sintering techniques involving multiple phases and liquids. Figure 1.2 organizes the sintering techniques into general categorizations. Pressure is the first differentiation. Most industrial sintering is performed without an external pressure. Pressure-assisted sintering techniques include hot isostatic pressing, hot pressing, and spark sintering. These produce high fractional densities by applying temperature and pressure simultaneously. Pressures range from 0.1 MPa up to 6 GPa.

For pressureless sintering, one major distinction is between solid-state and liquid phase processes. Single phase, solid-state sintering is the best understood form of sintering. Among the solid-state processes, there are options involving mixed phases, such as those to form composites or alloys. Compact homogenization occurs when sintering mixed powders that are soluble in each other and produce an alloy. Activated sintering is a special treatment involving small quantities of insoluble species that segregate to the grain boundaries to accelerate sintering. Mixed phase sintering is often employed to form composites, where one phase is dispersed in a matrix phase. Another variant occurs when a material is intentionally sintered in a two phase field, such as when steel is sintered at a temperature where both body-centered cubic and face-centered cubic phases coexist.

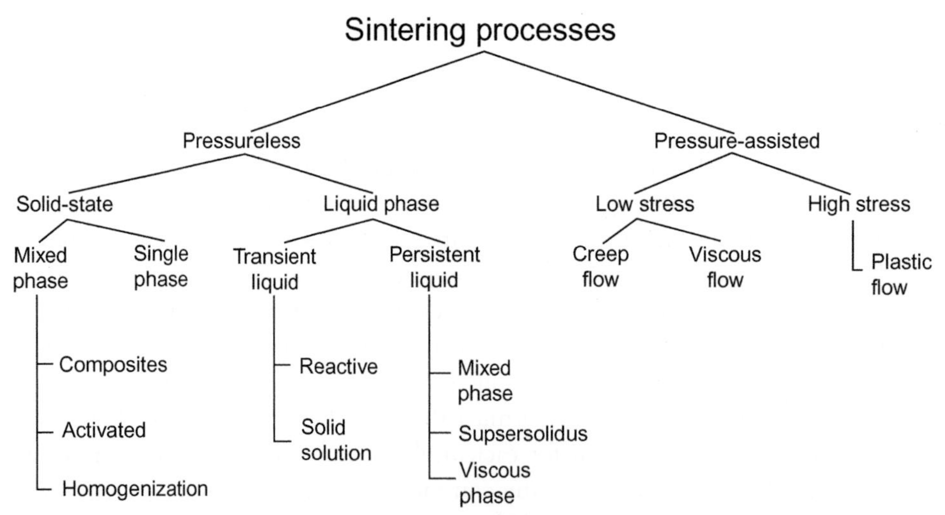

Figure 1.2 The taxonomy of sintering, showing process differentiation by various branches, starting with the application of pressure-assisted versus pressureless sintering.

Commonly, sintering involves a liquid phase that improves the sintering rate. Most industrial sintering involves forming a liquid phase, accounting for nearly 90% of the value of all sintered products. The two forms involve persistent or transient liquids. Persistent liquid phases exist throughout the high temperature portion of the sintering cycle and can be formed using prealloyed powder (supersolidus liquid phase sintering) or from a mixture of powders. Transient liquid phase sintering produces a liquid during heating, but that liquid subsequently dissolves into the solid. In some cases an exothermic heat release occurs, leading to reactive liquid phase sintering when a compound forms.

Although the roadmap shown in Figure 1.2 is schematic, it helps to tie the various chapters together into an overall technological landscape.

KNOWLEDGE

More than a million publications exist on sintering. Of those, almost half are conference proceedings and the other half are a mixture of archival publications and patents. Figure 1.3 plots the cumulative number of archival articles dealing with sintering from 1900 to 2013. To appreciate the acceleration in knowledge, Figure 1.4 plots the data on a log-log basis. By 1900 there were 135 publications on sintering,

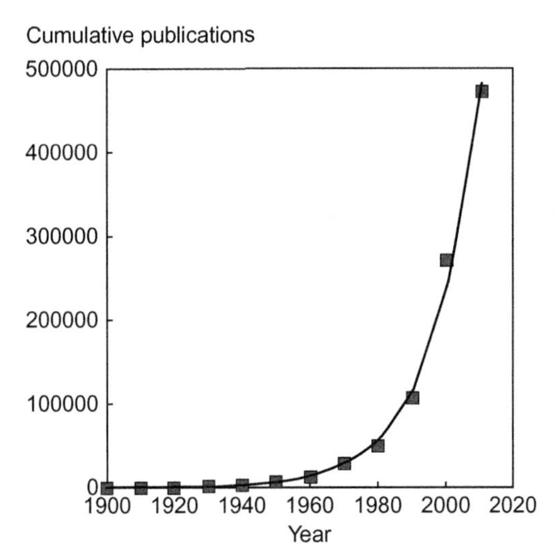

Figure 1.3 Cumulative number of archival journal publications on sintering, showing a surge in recent years. Almost as many conference publications exist, giving over one million total publications by 2012.

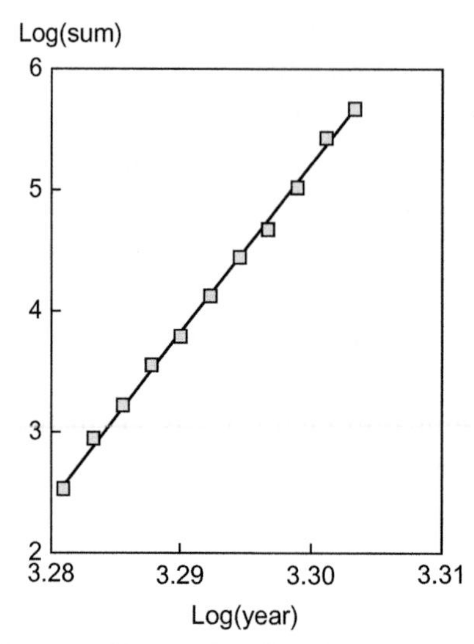

Figure 1.4 A log-log plot of the cumulative sum of the journal publications versus the publication year, producing a very significant regression line.

many of which concerned iron ore hardening. By 2013 the total was 600,000 publications. Regression analysis gives the following relation:

$$\log_{10}\text{(publication sum)} = -461 + 141\log_{10}\text{(year)} \tag{1.1}$$

This fit is highly significant with a correlation coefficient of 0.9979.

Quite possibly the accelerating knowledge generation will continue, as more materials, applications, and techniques emerge. Five nations lead the publication activity—China, Japan, USA, Korea, and Germany, in that order, followed by India, France, United Kingdom, Taiwan, and Spain. In terms of citations, sintering papers from the USA, UK, and Germany are the most frequently cited. With regard to the material treated, the highest impact papers deal with compositions based on alumina, iron, copper, and tungsten. These are followed by cemented carbides (WC-Co), platinum, silica, glass, aluminum, silver, and silicon carbide.

Generally, scientific developments lead patent activity. The first US patent to mention sintering was granted to MacFarlane of Canada in 1865 [52]. Subsequently the US patent literature grew to a position where about seven patents were issued each day in 2012. To appreciate the rise in activity, Figure 1.5 plots the cumulative US patent history, passing 100,000 issued patents in 2011. The rate of patent activity remains high and this indicates much future commercial activity involving sintering.

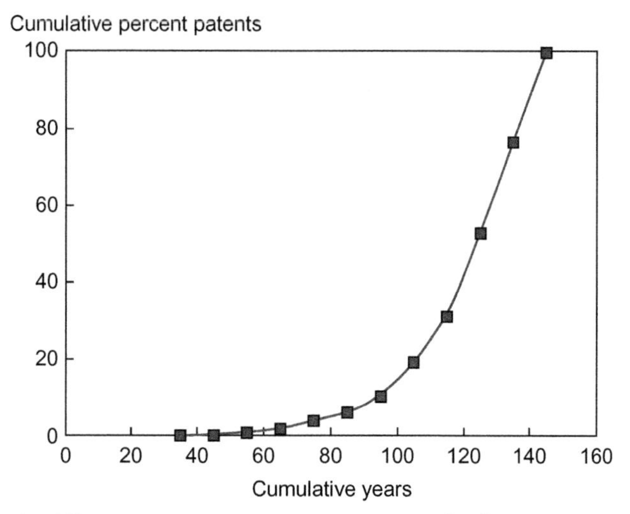

Figure 1.5 Cumulative US patent issuance versus years since the first patent to mention sintering in 1865.

KEY RESOURCES

Information on sintering and sintered products is available in a host of journals. A few journals are very popular, including ceramic and powder metallurgy journals such as these:

Acta Materialia (Acta Metallurgica, Acta Metalluirgica et Materialia)
Ceramic Bulletin (Bulletin of the American Ceramic Society)
Ceramics International
International Journal of Powder Metallurgy
International Journal of Refractory Metals and Hard Materials (formerly Planseeberichte fuer Pulvermetallurgie)
Journal of Applied Physics
Journal of the Korean Powder Metallurgy Institute
Journal of Materials Research
Journal of Materials Science
Journal of the American Ceramic Society
Journal of the Ceramic Society of Japan
Journal of the European Ceramic Society
Journal of the Japan Society of Powder and Powder Metallurgy
Materials Science and Engineering

Materials Transactions

Metallurgical and Materials Transactions (formerly Metallurgical Transactions, Transactions TMS-AIME)

Powder Metallurgy

Powder Technology

Science of Sintering (formerly Physics of Sintering)

Conferences with high sintering content include the following:

International Conference on Sintering—held every four years in South Bend, Indiana; Vancouver, British Columbia; Tokyo, Japan; State College, Pennsylvania; Grenoble, France; JeJu, Korea; Dresden, Germany.

Materials Science and Technology Conference—organized by several societies and, held every fall in cities such as Columbus, Ohio; Cincinnati, Ohio; Pittsburgh, Pennsylvania.

World Congress on Powder Metallurgy—held every two years, rotating between Europe, North America, and Asia; recent meetings were held in Yokohama, Japan; Florence, Italy; Washington, DC; Vienna, Austria; Busan, Korea; Orlando, Florida; Grenada, Spain; Kyoto, Japan.

Plansee Seminar—held every four years in Reutte, Austria—premier conference on sintered hard materials, refractory metals, particulate composites, and high temperature systems.

REFERENCES

[1] W.D. Coolidge, Production of Refractory Conductors; U. S. Patent 1,077,674, issued 5 November 1913.

[2] H.E. White, R.M. Shremp, Beryllium oxide: I, J. Am. Ceram. Soc. 22 (1939) 185–189.

[3] C.G. Goetzel, Treatise on Powder Metallurgy, vol. III, Interscience Publishers, New York, NY, 1952.

[4] B.F. Klugh, The microstructure of sintered iron bearing materials, Trans. TMS-AIME 45 (1913) 330–345.

[5] F.A. Vogel, Sintering and briquetting of flue dust, Trans. TMS-AIME 43 (1912) 381–386.

[6] J. Gayley, The sintering of fine iron bearing material, Trans. TMS-AIME 42 (1912) 180–190.

[7] J.E. Burke, A history of the development of a science of sintering, in: W.D. Kingery (Ed.), Ceramics and Civilization, Ancient Technology to Modern Science, vol. 1, Amer. Ceramic Society, Columbus, OH, 1985, pp. 315–332.

[8] W.D. Jones, Principles of Powder Metallurgy with an Account of Industrial Practice, Edward Arnold, London, UK, 1937.

[9] E.G. Ferguson, Bergman, Klaproth, Vauquelin, Wollaston, J. Chem. Edu. 18 (1941) 3–7.

[10] C.S. Smith, The early development of powder metallurgy, in: J. Wulff (Ed.), Powder Metallurgy, Amer. Society for Metals, Cleveland, OH, 1942, pp. 4–17.

[11] P.E. Wretblad, J. Wulff, Sintering, in: J. Wulff (Ed.), Powder Metallurgy, Amer. Society for Metals, Cleveland, OH, 1942, pp. 36–59.

[12] G.F. Huttig, Die Frittungsvorange innerhalb von Pulvern, weiche aus einer einzigen Komponente bestehen—Ein Beitrag zur Aufklarung der Prozesse der Metall-Kermik und Oxyd-Keramik, Kolloid Z. 98 (1942) 6–33.

[13] C.G. Goetzel, Treatise on Powder Metallurgy, vol. I, Interscience Publishers, New York, NY, 1949, pp. 259–312.

[14] W.D. Jones, Fundamental Principles of Powder Metallurgy, Edward Arnold Publishers, London, UK, 1960.
[15] S.Y. Plotkin, Development of powder metallurgy in the USSR during 50 years of soviet rule, Powder Metall. Metal Ceram. 6 (1967) 844–853.
[16] F.N. Rhines, R.T. DeHoff, R.A. Rummel, Rate of densification in the sintering of uncompacted metal powders, in: W.A. Knepper (Ed.), Agglomeration, Interscience, New York, NY, 1962, pp. 351–369.
[17] V.A. Ivensen, Densification of Metal Powders during Sintering, Consultants Bureau, New York, 1973.
[18] S.Y. Plotkin, G.L. Fridman, History of powder metallurgy and its literature, Powder Metall. Metal Ceram 13 (1974) 1026–1029.
[19] M.M. Ristic, Science of Sintering and Its Future, International Team for Science of Sintering, Beograd, Yugoslavia, 1975.
[20] C.G. Johnson, W.R. Weeks, Powder metallurgy, J.G. Anderson, Metallurgy, (revision), fifth ed., Amer. Technical Publishers, Homewood, IL, 1977, pp. 329–346.
[21] H.E. Exner, Physical and chemical nature of cemented carbides, Inter. Met. Rev. 24 (1979) 149–173.
[22] F.V. Lenel, Powder Metallurgy Principles and Applications, Metal Powder Industries Federation, Princeton, NJ, 1980.
[23] C.A. Handwerker, J.E. Blendell, R.L. Coble, Sintering of ceramics, in: D.P. Uskokovic, H. Palmour, R.M. Spriggs (Eds.), Science of Sintering, Plenum Press, New York, NY, 1980, pp. 3–37.
[24] A. Prince, J. Jones, Tungsten and high density alloys, Historical Metall 19 (1985) 72–84.
[25] W.D. Kingery, Sintering from prehistoric times to the present, in: A.C.D. Chaklader, J.A. Lund (Eds.), Sintering '91, Trans. Tech. Publications, Brookfield, VT, 1992, pp. 1–10.
[26] H. Kolaska, The dawn of the hardmetal age, Powder Metall. Inter. 24 (5) (1992) 311–314.
[27] K.J.A. Brookes, Half a century of hardmetals, Metal Powder Rept. 50 (12) (1995) 22–28.
[28] M.M. Ristic, Frenkel's theory of sintering (1945–1995), Sci. Sintering. 28 (1996) 1–4.
[29] R.M. German, Sintering Theory and Practice, Wiley-Interscience, New York, NY, 1996.
[30] G.H. Haertling, Ferroelectric ceramics: history and technology, J. Amer. Ceram. Soc. 82 (1999) 797–818.
[31] F.L. Riley, Silicon nitride and related materials, J. Amer. Ceram. Soc. 83 (2000) 245–265.
[32] J. Konstanty, Powder Metallurgy Diamond Tools, Elsevier, Amsterdam, Netherlands, 2005.
[33] C.M. Peret, J.A. Gregolin, L.I.L. Faria, V.C. Pandolfelli, Patent generation and the technological development of refractories and steelmaking, Refractories Applic. News 12 (1) (2007) 10–14.
[34] M. Noguez, R. Garcia, G. Salas, T. Robert, J. Ramirez, About the Pre-Hispanic Au-Pt 'Sintering' technique, Inter. J. Powder Metall. 43 (1) (2007) 27–33.
[35] S.J.L. Kang, Sintering Densification, Grain Growth, and Microstructure, Elsevier Butterworth-Heinemann, Oxford, United Kingdom, 2005.
[36] P.K. Johnson, Tungsten filaments—the first modern PM product, Inter. J. Powder Metall. 44 (4) (2008) 43–48.
[37] R.M. German, P. Suri, S.J. Park, Review: liquid phase sintering, J. Mater. Sci. 44 (2009) 1–39.
[38] P. Schade, 100 years of doped tungsten wire, in: P. Rodhammer (Ed.), Proceedings of the Seventeenth Plansee Seminar, vol. 1, Plansee Group, Reutte, Austria, 2009, pp. RM49.1–RM49.12.
[39] R.M. German, Coarsening in sintering: grain shape distribution, grain size distribution, and grain growth kinetics in solid-pore systems, Crit. Rev. Solid State Mater. Sci. 35 (2010) 263–305.
[40] J.F. Garay, Current activated, pressure assisted densification of materials, Ann. Rev. Mater. Res. 40 (2010) 445–468.
[41] K. Morsi, The diversity of combustion synthesis processing: a review, J. Mater. Sci. 47 (2012) 68–92.
[42] Z.A. Munir, D.V. Quach, M. Ohyanagi, Electric current activation of sintering: a review of the pulsed electric current sintering process, J. Am. Ceram. Soc. 94 (2011) 1–19.
[43] R.M. German, History of sintering: empirical phase, Powder Metall. 6 (2) (2013) 117–123.

[44] G.G. Deeley, J.M. Herbert, N.C. Moore, Dense silicon nitride, Powder Metall. 4 (1961) 145–151.
[45] G.R. Terwilliger, F.F. Lange, Pressureless Sintering of Si_3N_4, J. Mater. Sci. 10 (1975) 1169–1174.
[46] R.F. Walker, Mechanism of material transport during sintering, J. Amer. Ceram. Soc. 38 (1955) 187–197.
[47] H.H. Hausner, Discussion on the definition of the term 'Sintering', in: M.M. Ristic (Ed.), Sintering-New Developments, Elsevier Scientific, New York, NY, 1979, pp. 3–7.
[48] R.G. Bernard, Processes involved in sintering, Powder Metall. 2 (1959) 86–103.
[49] M.H. Tikkanen, The application of the sintering theory in practice, Phys. Sintering 5 (2) (1973) 441–453.
[50] A. Mohan, N.C. Soni, V.K. Moorthy, Definition of the term sintering, Sci. Sintering 15 (1983) 139–140.
[51] R.M. German, Sintering, Encyclopedia of Materials Science and Technology, Elsevier Scientific, London, UK, 2002, pp. 8640–8643.
[52] T. Macfarlane, Improved Process of Preparing Chlorine, Bleaching Powder, Carbonate of Soda, and Other Products; U. S. Patent 49,597, issued 22 August 1865.

CHAPTER TWO

History of Sintering

HISTORICAL MILESTONES

Sintering probably started with the observation that clay and ceramic pottery improved in strength as a result of firing in a wood or charcoal fire. There is no record of this discovery, but archeological artifacts show that the early use of sintering probably dates to about 24,000 BC. Only piecemeal remains tell the story. However, between 1700 and 1800 written records emerged on sintering experiments, and starting in the early 1800s several empirical developments are well documented.

Based on reconstructions from various records, a chronological survey of the key segments are as follows [1,2]:

- **archeological artifacts, generally before 1700**
  - retrieved evidence of early sintering successes, most detail is lost
  - examples include early earthenware and simple metals
- **trial and error sintering, starting about 1700**
  - records exist for the key actors with some quantitative process details
  - examples are porcelain, iron, platinum, iron ore
- **qualitative sintering models, starting about 1900**
  - many discoveries, observations, early conjectures, and publications
  - examples are copper, tungsten, cemented carbides, oxide ceramics, and bronze bearings
- **quantitative sintering theory, starting about 1945**
  - mathematical models for neck size, shrinkage, densification, surface area, density, and properties; models emerged to include temperature, time, particle size, heating rate, and atmosphere effects.

Many significant discoveries took place via trial and error efforts without any scientific underpinning. Note the theoretical concepts are relatively recent and occurred in just the last 0.3% of sintering's history. This chapter focuses on the developments up to about 1945 before there was a science of sintering.

Sintering: From Empirical Observations to Scientific Principles
DOI: http://dx.doi.org/10.1016/B978-0-12-401682-8.00002-1

EARLY SINTERED PRODUCTS

Several examples of sintering practice predate written records. Archeological studies on pottery, casting ceramics, early iron, copper, and precious metal structures provide a sense of the early applications. There are surprising process similarities in spite of the very different materials and applications.

Clay Ceramics

Archeological findings show shaped clay ceramic bodies were fired in open fire pits as early as 24,000 BC, in what is now the Czech Republic. This was followed by periodic advances as outlined in Figure 2.1.

As seen in Figure 2.2, early sintered pots were not sintered at high temperatures, so they were weak and rarely survived. They leaked liquids because of their porosity. By approximately 10,000 BC, fired clay vessels were used for water storage, indicating techniques for sealing surface pores had been mastered. Recoveries in China, Egypt, and throughout the Middle East document several subsequent examples of fired beads, amulets, figurines, pots, and earthenware vessels by 6000 BC [1–3].

Glazes were developed about 3500 BC in the Eastern Mediterranean regions, often imitating the blue of lapis lazuli. These are glassy phases which crystallized on cooling. Often the glaze contained lead, especially in Babylon. Impervious coatings often required multiple layers. The lead glazes were supplemented by tin oxide glazes by 700 BC in Persia. Glazed tiles emerged from this base and remain in production today.

Porcelain

Porcelain was a valuable sintered product that arose first in China. A key to porcelain production is in attaining a high firing temperature. Progressive advances in kiln

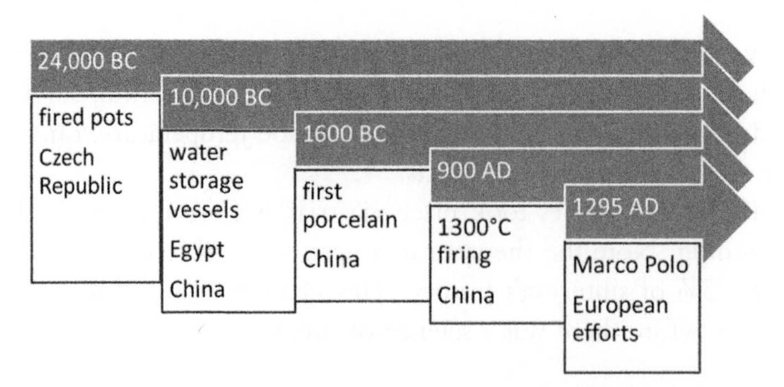

Figure 2.1 An approximate archeological timeline for sintered ceramics, starting with early fired pots found in the Czech Republic and progressing to significant advances in porcelain production.

design generated higher firing temperatures and greater strength. Accordingly, porcelain firing advanced to give leak-free structures by about 1600 BC.

Firing temperatures of 1300°C (1573 K) allowed quartz to partially dissolve into silicate glass which then gained strength by crystal precipitation on cooling. The success of Chinese porcelain was widely recognized, and became a target for considerable trade. An example of the highly valued Tsing dynasty porcelain is shown in Figure 2.3.

About the same time, fired casting molds for bronze were fabricated from sintered ceramics. The idea of firing ceramic molds for additional strength had also emerged

Figure 2.2 Examples of early sintered pottery, where the material was neither strong nor fluid-tight.

Figure 2.3 A Tsing dish fabricated with a high temperature sintering process.

Figure 2.4 This bronze headdress was cast by the Hittites using a sintered ceramic mold.

Figure 2.5 Early Chinese porcelain production excelled in reaching high sintering temperatures by use of sloped dragon kiln design, sketched in this early illustration.

by 1000 BC and spread along the trade routes; for example, Figure 2.4 is an example bronze casting made by the Hittites in Turkey. Trade routes not only carried the products, but also spread the technologies.

By 900 AD porcelain production was an important industry in China, Korea, and Japan. The Italian explorer Marco Polo brought porcelain back from China in about 1295 AD, generating much interest throughout Europe. By 1580 AD an inferior porcelain sintering practice existed in Florence, but it was not competitive with porcelain from China. The secret of the Chinese porcelain was in the furnace design. A novel dragon kiln sketched in Figure 2.5 produced the needed high firing temperatures.

Figure 2.6 European porcelain eventually found composition formulations and sintering techniques capable of forming high value products, initially near Dresden.

Porcelain is based on mixtures that include quartz (SiO_2), feldspar ($KAlSi_3O_8$-$NaAlSi_3O_8$-$CaAl_2Si_2O_8$), and kaolinite ($Al_2Si_2O_5(OH)_4$), the latter being a platelet-shaped clay particle. This mixture was heated in a multiple step cycle reaching about 1400°C (1673 K), giving a final product consisting of quartz, mullite ($Al_6Si_2O_{13}$), and glass. Bulk compositions show 60% SiO_2, 32% Al_2O_3, 4% K_2O, 2% Na_2O, and oxides of iron, titanium, calcium, and magnesium. German porcelain gained traction based on the empirical work by Tschimhaus and Böttger. In the absence of phase diagrams, Böttger's success came from his apothecary training, where he examined a broad range of local minerals, in a systematic array of compositions relying on special sintering enclosures to reach higher firing temperatures.

Böttger became a prisoner because he pretended to be an alchemist. He made the unfortunate mistake of demonstrating his alchemy, most likely by secretly swapping gold for silver overnight. His alchemist trick resulted in his imprisonment to supposedly protect his nonexistent secret. While held in prison, Böttger had outside success with porcelain. Along with von Tschimhaus, their porcelain resulted in a factory that ramped up production between 1708 and 1725. An example of fired porcelain is shown in Figure 2.6. The fabrication of such objects required the isolation of many variables, and today those familiar with sintering would likewise isolate and study the same variables.

Iron, Copper, Silver

As outlined in Figure 2.7, the sintering of metallic objects was developed through discoveries made around the world. Early examples included gold, silver, copper, iron,

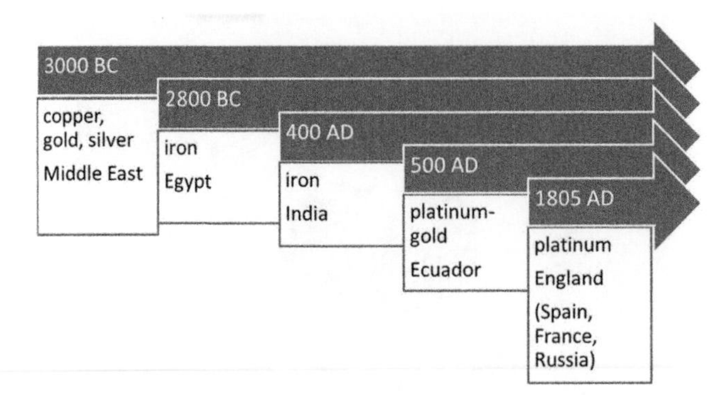

Figure 2.7 An approximate timeline for the evolution of sintered metals, with early examples arising in the Middle East, with apparently independent discoveries in India and Ecuador, and finally via systematic study the Wollaston process was developed in 1805.

and bronze. The first of these are from approximately 3000 BC [4,5]. Most of these easily reduced metals were sintered long ago.

The first sintered ferrous shapes were derived from meteors, which was softer since meteors contain nickel, and then subsequently from smelted iron. Sinter-forged iron artifacts were discovered in Tutankhamen's tomb—one of the few Egyptian tombs not raided in ancient times. Most likely the artifacts were fabricated by the Hittites in Turkey using a recarburization process. In this approach, iron oxide is heated in a reducing charcoal fire, and while hot the sintered agglomerates were hammered to densify the sponge into a relatively simple shape. When reheated, the iron absorbed carbon to become steel as required for swords and shields. Variants of this process were discovered in other parts of the world, including Bulgaria, China, Greece, and India. Chemical analysis indicates these were independent activities.

Gold, silver, and iron sintering were established in India by about 400 AD. The most notable example is the 7.2 m tall Delhi Iron Pillar weighing about 6000 kg [6,7]. A photograph is shown in Figure 2.8. This pillar contains 0.25% phosphorous with 0.15% carbon and traces of nickel, copper, silicon, and manganese. Such a composition forms a passive film to provide excellent corrosion resistance.

To fabricate the pillar, iron granules or lumps were formed using clay crucibles charged with iron oxide, bamboo charcoal, and plant leaves. The reduced iron lump was reheated in charcoal. While hot, the charcoal was swept away and the sintered lumps were hot-hammered together to form ingots. A combination of hot forging, as an additive process, and cold chiseling, as a subtractive process, produced the desired shape.

Subsequent developments in England turned to coke for oxide reduction. Carbon additions to the reduced iron plus quenching and tempering produced exceptional strength. Layers of high and low carbon sponge resulted in a hard but tough laminate structure. Widely heralded armorers formed sword blades using this approach.

Figure 2.8 A photograph of the Delhi Iron Pillar fabricated about 400 AD using sintered lumps of iron.

Platinum Crucibles

Platinum has a melting point of 1769°C (2042 K), much higher than the flame temperature attained by wood, charcoal, and other common combustibles. Since platinum powder is found in nature, early efforts relied on compaction and sintering of collected powder to fabricate platinum objects.

The Inca sintered gold-platinum jewelry as early as 300 BC [8]. The peak sintering temperature was in the 1100°C (1373 K) range, sufficient to melt gold. Indeed, a liquid phase sintering process with gold-silver additives provided a variety of platinum alloy objects—needles, spoons, fish hooks, forceps, nose rings, and safety pins. Such fabrication involved compacting, sintering, hammering, and annealing, with the latter steps being repeated until the desired geometry was attained. Both yellow and platinum colors were formed with these formulations; either high gold (12% platinum) or high platinum (60 to 85%) with traces of copper and silver. An examination of the Au-Pt binary phase diagram, shown in Figure 2.9 reveals the two compositions are on the sides of the two phase field. Figure 2.10 is a photograph of an archeological find which shows the two colors formed in a sintered creation.

After the discovery voyages of Columbus, sintered articles arrived in Spain, resulting in jewelry formed from mixed platinum and gold powders [9]. A rash of platinum sintering efforts followed across Europe, some relying on lead, arsenic, or mercury

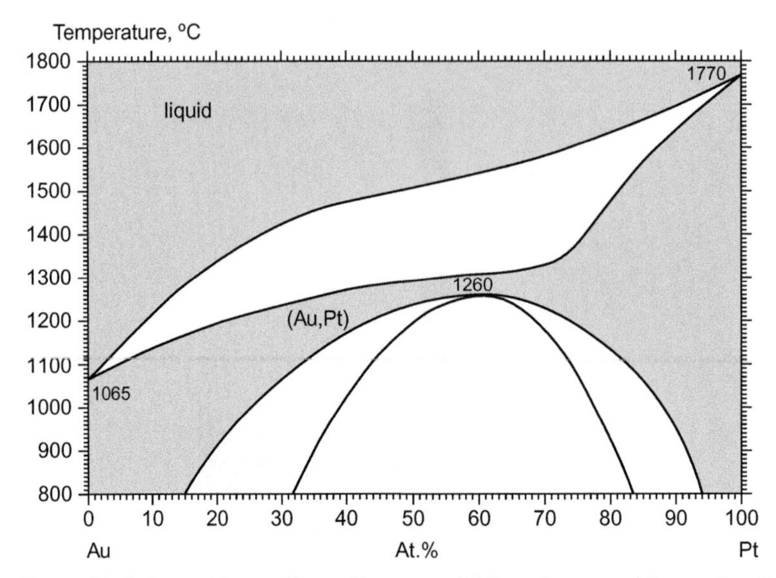

Figure 2.9 The gold-platinum binary phase diagram exhibits a lower melting region in the gold-rich side and a two phase region near the center. The empirical findings out of Ecuador resulted in sintered gold-platinum objects either rich in gold or rich in platinum to avoid the miscibility gap.

Figure 2.10 Photograph of an early sintered gold-platinum decorative medallion.

additives [10–14]. Various forms of sintered platinum appeared throughout Europe by 1750. A dense product required repeated heating and deformation cycles, up to 30 times. Charcoal fires provided sufficient heat if arsenic was added to induce liquid phase sintering. After densification the arsenic was evaporated. Later, mercury served a similar role and was taken in to production. However, the most successful route avoided toxic additions, possibly because those using toxic additions did not survive.

In 1805 Wollaston developed a process where platinum powder was precipitated into discrete small particles, pressed, heated, and hot worked to full density, without mercury or arsenic additions [5,14]. His process was kept confidential until his death in 1828 [15–18]. Between 1805 and 1828 Wollaston became quite wealthy by selling custom-fabricated platinum crucibles, highly regarded for making glass windows. This was the first significant case where public records document the individuals and their contributions.

The Wollaston approach scaled to large, defect-free pieces. Sponge was precipitated from ammonium platinum chloride. The precipitate was heated to remove water and pulverized and sieved to form discrete particles with a high packing density. Compaction was done in a horizontal press, as illustrated in Figure 2.11. The green compact was pre-sintered to increase handling strength and heated to a "white heat," after which it was

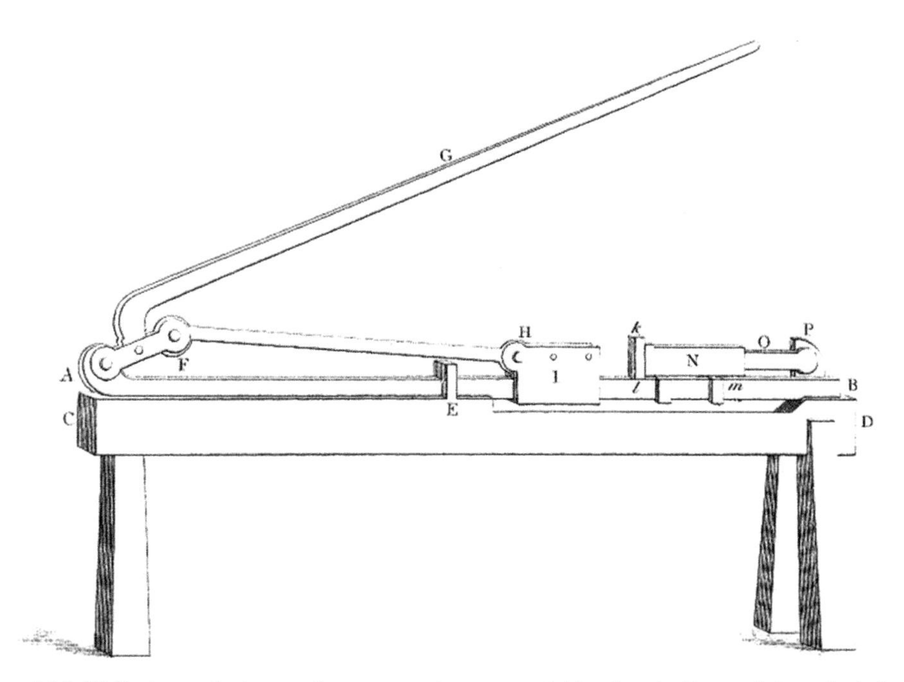

Figure 2.11 Wollaston relied on a lever press to compact his chemically precipitated platinum powder prior to sintering.

hot-forged to shape, probably at around 1000°C (≈1300 K). Note that there was no means to measure high temperatures, so the descriptions were subjective.

Competitive offerings of the time suffered impurity effects, most evident as blisters. By 1809 Wollaston's efforts were producing 13 kg crucibles in his London laboratory, a process subsequently licensed by Johnson, Matthey and Company. Then, as early as 1820 sintered platinum was adapted for incandescent lamps [19].

The Wollaston approach was subsequently adopted for copper in 1830, by von Welsbach for osmium in 1870, and by Coolidge for tungsten in 1910. It is no surprise then that the idea of compaction, sintering, and hot working, initially applied to iron and then platinum, spread to higher temperature materials over the next century.

In Russia, platinum studies were started by Musin-Puskin at the Mining Cadet Corps of St. Petersburg [11]. Like earlier approaches, he used mercury amalgams to form the powder with subsequent distillation to remove the mercury. The resulting sponge powder sintered to give a malleable product, but it failed to move into production.

In about 1825, efforts were restarted in St. Petersburg because of the discovery of new platinum deposits in the Urals. Conversion of platinum powder into coins provided the monarch with a significant source of revenue. Sobolevskii was hired at the same Mining Cadet Corps and quickly succeeded in 1826 to form platinum in a manner similar to Wollaston [5,10,12,13]. The duplication was not initially recognized, since Wollaston waited until 1828 to disclose his process. Between 1828 and 1845, Russian coinage production using sintered platinum totaled about 14,000 kg.

Many important discoveries took place in the quest to produce platinum by sintering. Early recipes were qualitative and imprecise. For example, "red hotness" is a subjective temperature specification; high temperature gas thermometers did not emerge until 1828. Although not alchemy, platinum sintering in the 1800s showed no appreciation for the underlying atomic events. This was evident as late as 1923, when Smith [20] conjectured that platinum sintering was caused by a melting point depression or a crystallization event. Consideration of diffusion events came in the 1940s.

Other metals were produced by related approaches, including copper, silver, and lead. In the Osann sintered copper approach, powder was precipitated from copper carbonate and reduced using charcoal heating [5,14]. Coins were fabricated by sintering compressed powder in containers sealed to avoid oxidation. By 1841, articles of copper, silver, and lead were in production using an approach similar to modern powder metallurgy. The sintered metals developed in a sintering technology resulted in a diversity of approaches as outlined in Figure 2.12. Up to this time the thermal bonding of powder was still not yet termed sintering.

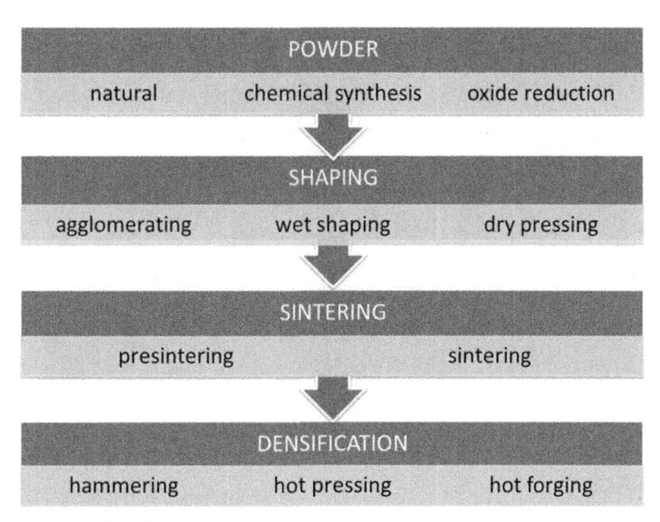

Figure 2.12 Much early effort in sintered metals tended to converge to a few process variants that fundamentally relied on powder shaping, sintering, and hot densification. Modern powder metallurgy likewise follows a similar combination of steps.

Iron Ore Induration

The term induration describes the hardening of a powdery substance. For example, in steel production, iron ore pellets are fed into melt furnaces. To avoid dusting and loss of ore, small oxide particles are agglomerated by sintering. Although initially applied to iron ore, soon the briquette agglomeration concept spread to a variety of materials [21].

Percy [22] describes iron ore agglomeration in 1864 and notes how oxide inclusions are detrimental. By the early 1900s large scale sintering agglomeration systems were in use [23–26]. Figure 2.13 shows one such plant that helps demonstrate the large scale application of sintering by 1912. In the 1930s and 1940s, ore sintering included zinc, lead, lead sulfide, carbonates, chlorides, and precious metals.

Today iron ore agglomeration is the largest tonnage application for sintering, with plants operating at up 20,000 metric tonnes per day. For example, Figure 2.14 is a picture of a modern agglomeration facility which incorporates off-gas capture to reduce environmental damage. In such a facility, the iron ore fines are mixed with fluxes, carbon fuel, and water. The mixture is continuously fed onto trays or belts. As the conveyor moves through the sintering furnace, the mixture is heated to ignite the fuel and sinter the powder. Reaction waste consists of carbon dioxide, carbon monoxide, as well as nitrous oxides and sulfur oxides.

Tungsten Lamp Filaments

The quest for an electric light was significant in the history of sintering. As electricity became better understood, a major application opportunity arose in lighting. Prior to

Figure 2.13 This picture is of an early iron ore sintering plant used to agglomerate powder prior to melting to make steel.

Figure 2.14 A modern iron ore sintering plant, where production rates of sintered material reach upwards of 20,000 tonnes per day.

the electric light, lighting was obtained from gas fired mantles and candles. Davy invented an electric lamp in 1809. In the US, a glass bulb with a carbonized filament was invented in 1854 by Goebel. Early carbon filament light bulbs lasted just hours and were inefficient, with an output of 1.4 lumen per watt. Nernst [Nobel Prize 1920] created an early lamp using mixed magnesium, calcium, and rare earth oxides, delivering 5 lumen per watt.

By 1874, Woodword and Evans of Toronto invented a bulb design, but were unable to secure financing to commercialize the idea and sold it to Edison. Soon after, carbon fiber filaments were reaching lifetimes of 13 hours. Edison patented his version of the light bulb in 1879, relying on a carbon filament in an evacuated glass envelope. It had a filament life of 45 hours, which eventually reached upwards of 1000 hours of life using direct current. There were many inventions and the race to electrification contracts for whole cities drove the search for increaingly durable, low-cost filaments.

By the early 1900s, extended life came from filaments which were based on refractory metals such as osmium and tantalum, produced with difficulty from sintered powders [5]. The formulations mixed metal oxide powders with rubber or sugar for extrusion, relying on binder burnout and hydrogen sintering to reduce the oxide to form a filament.

Weisbach set the stage for tungsten at Osram (named after osmium-wolfram). He initially sintered osmium filaments, but subsequently shifted to tantalum and tungsten. Early tungsten filaments were also formed using chemical vapor deposition of sintered tungsten-nickel compositions. The Russian inventor Lodygin showed an incandescent tungsten filament lamp at the 1900 Paris World Exhibition in Paris. These early tungsten filaments were brittle, but the merits of tungsten as a filament material were well recognized.

To reach the sintering temperature of tungsten, self-heating using intense electric currents, a process termed spark sintering, was necessary. The idea was borrowed from Acheson and Moissan. For tungsten, electric spark sintering was patented by Voelker in 1900 [27]. Tungsten paste was extruded and sintered to form large diameter filaments for high wattage lamps [28–30]. The process patented by Lux relied on electric current pulses of 1 s at $10A/mm^2$ in vacuum (Bloxam acted as his patent agent in London and is often improperly cited as the inventor). This direct current sintering concept became the mainstay process.

It was 30 years from Edison's light bulb patent until a long-lasting ductile tungsten filament was developed for use with alternating current. To work on the problem, Edison hired Whitney, a former student of Nobel laureate Wilhelm Ostwald (1853–1932). Whitney left Massachusetts Institute of Technology to become the research director for General Electric in Schenectady. In turn, Whitney recruited Coolidge from MIT and Langmuir from Stevens Institute of Technology. This was a most successful collaboration [31].

Coolidge developed a sintered filament using pressed tungsten powder sintered by direct electric current [32,33]. The sintered ingot was hot worked and drawn into a wire of the desired diameter. The process was similar to that used earlier in shaping iron, copper, and platinum powders. Unfortunately, critical details of Coolidge's discovery were not understood and his patent was subsequently disallowed, reportedly since it was similar to Wollaston's process for sintering platinum. Years later, researchers were still studying the tungsten filaments to understand their ductility [34,35].

In 1932, Langmuir became the first industrial chemist to win the Nobel Prize. The recognition came from his contributions which detailed how gases influence

Figure 2.15 This photograph shows Whitney, Coolidge, and Langmuir in discussions at the General Electric research laboratories in Schenectady.

evaporation. This complimented Coolidge's discovery of sintered tungsten that retained ductility when hot worked. Figure 2.15 is a photograph of the three pioneers. All three rose to great distinction through their efforts on tungsten sintering, effectively creating a large industrial use for sintering.

In the Coolidge approach, tungsten powder compacts were heated at about 1000°C (1273 K) in a hydrogen-nitrogen atmosphere. Final sintering densification was achieved by direct electric current discharge. Peak temperatures were probably near 2200°C (about 2500 K). The sintered ingots were hot swaged and drawn into wires using diamond dies. When incorporated into Edison's vacuum bulb design, the tungsten filament gave 10 lumens per watt and operated successfully in alternating current for up to hundreds of hours. Of course Tesla was responsible for developing alternating current, much to the chagrin of Edison. Langmuir determined that the addition of nitrogen or argon to the evacuated light bulb retarded tungsten evaporation, enabling longer filament life and higher light intensity at higher operating temperatures. Further, he determined the way that coiling the filament, as illustrated in Figure 2.16, improved life and reduced discoloration of the bulb.

Building an explanation for Coolidge's discovery, in 1917 Pacz determined that intentional potassium additions gave a desirable interlocking grain structure, and in 1922 Smithells identified a range of alkali oxides as most effective, since they

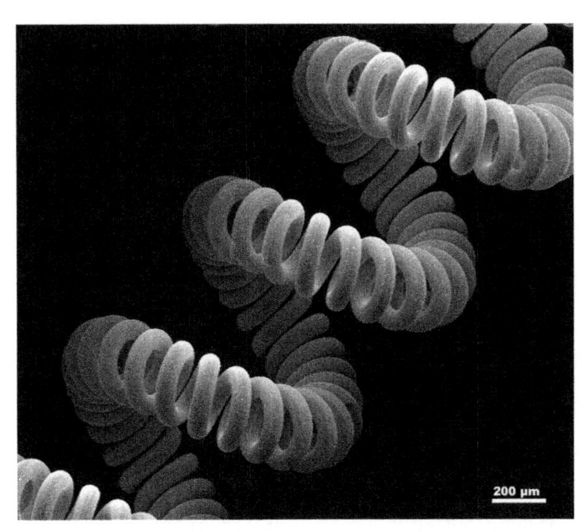

Figure 2.16 This is a scanning electron micrograph of a coiled tungsten lamp filament.

withstood reduction by hydrogen during sintering. Thus, the important ductile-brittle transition temperature effect, inherent to Coolidge's success, was linked to two critical factors:

- the low impurity (oxygen) level that comes from sintering at high temperatures in a reducing hydrogen atmosphere
- the small grain size that comes from grain boundary pinning using potassium-sodium silicate and alumina dispersoids.

High temperature creep resistance is critical to filament life. The combination of microstructure pinning agents and hot deformation produced elongated grains that delay creep failure. Dopants were inadvertently introduced in Coolidge's filaments, reportedly introduced by the crucibles used in the first 1000°C (1273 K) firing cycle, but it took years to understand their critical role [36].

Today 165,000 light bulb 40 W filaments are formed from a kilogram of tungsten powder. However, the incandescent bulb is behind newer lighting technologies. Figure 2.17 plots the performance gains in lighting, in roughly 20 year steps. For comparison, fluorescent bulbs produce at least twice the optical output per watt and these are surpassed at least two-fold by mercury and sodium vapor lamps. The latter are enabled by sintered translucent magnesia doped alumina, and deliver more than 100 lumen per watt. Light emitting diodes give even higher efficiency, and now exceed 200 lumen per watt.

Soon after the commercialization of tungsten incandescent lamps, several related refractory metals were manufactured using similar press-sinter-deform routes. For tungsten wire production, tooling was expensive, since only natural diamonds were

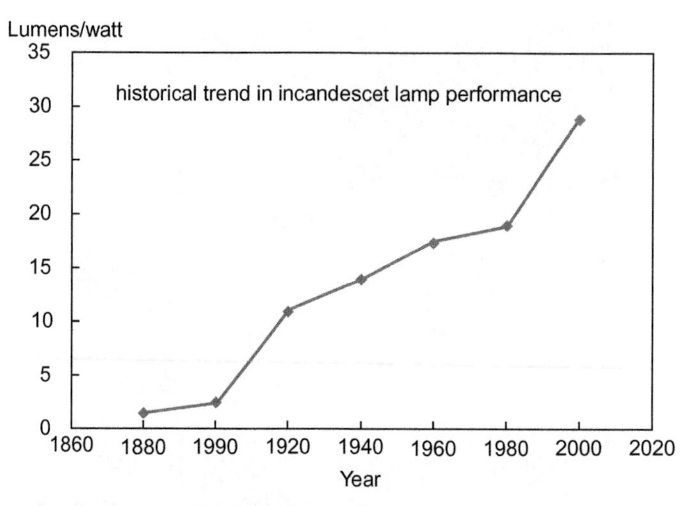

Figure 2.17 Incandescent lamp output in lumens per watt in approximate 20 year steps, indicating the technological advances from early discoveries. The filament life would likewise show considerable gains from the early carbon filaments.

then available as drawing dies. This lead to a most notable sintered product; liquid phase sintered cemented carbides.

Cemented Carbide Dies

Tungsten is one of the hardest metals, so the drawing of lamp filaments quickly wore out the expensive diamond drawing dies. Prior to 1900, carbon steel was the most common metal forming tool material. This was then displaced by tool steels (high speed steels) with higher carbon contents. By 1909, the cobalt-chromium composites quickly became the leading forming tool material. Through the use of improved microscopy and analysis techniques, the hardness of these alloys was traced to precipitated metal carbides; and the newly discovered WC (tungsten carbide) became a target for possible use as a wire drawing die material.

The direct synthesis of tungsten carbide took place using an electric furnace. Moissan [Nobel Prize in 1906] generated much interest with the publication, in 1897, of his book *The Electric Furnace*. He described how a furnace, such as that sketched in Figure 2.18, enabled high temperature synthesis of new compounds. Most likely Moissan discovered a mixture of WC and W_2C (phase identification by X-ray diffraction was as yet unknown) and it was another two years before Williams synthesized pure WC. Reports that WC was as hard as diamond generated much interest. Early carbide drawing dies were formed by casting WC, but they were brittle. By 1914, Voightlander and Lohmann sintered a mixture of WC and Mo_2C at 2200°C (2473 K), but this was also brittle. Meanwhile,

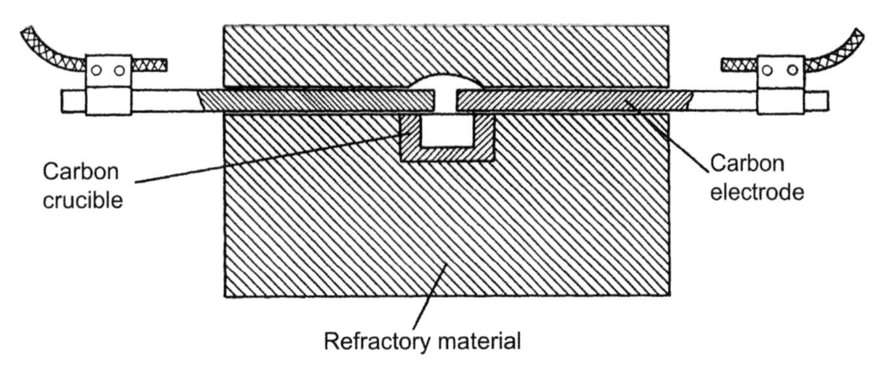

Figure 2.18 Moisson's idea for an electric furnace, initially used to form new compounds and subsequently employed to sinter the compounds.

Schröter, who worked in the chemical department of the German gas lamp firm DGA, transferred in 1908 to the electrical department, where lighting was under study using osmium filaments. Several options were explored on how to make low-cost wire drawing dies, including infiltrating porous carbide bodies with iron, or casting mixtures of carbides and transition metals [5,37].

The 1919 merger of DGA, Siemens Lamp, and AEG resulted in Osram, with Skaupy as director of research on lamp materials, involving Schröter, Baumhauer, and Tammann. The latter was a consultant who was simultaneously developing early sintering concepts. They formed WC by using a slow heating cycle to react tungsten powder with graphite. Initially the carbide was infiltrated, but soon sintering took the forefront. The WC powder was mixed and milled with nickel, cobalt, or iron to enable liquid phase sintering at about 1500°C (1773 K) to form a dense WC-Co composite. Testing verified the success of the cemented carbide as a die for drawing tungsten lamp filaments. The drawing die patent was filed in Germany, the United Kingdom, and the United States, under the names of Schröter and Jenssen [38]. This idea was valuable to several firms; Krupp started Widia (to imply "with diamond properties") and General Electric started Carboloy. The initial application was for tungsten filament wire drawing.

A succession of inventions arose around the initial WC-Co composition, as outlined in Table 2.1 [39–42]. Most were driven by a desire to lower cost and improve properties, generally leading to a 6 to 12% transition metal (Co, Ni, or Fe) content. Schwarzkopf [40,41] relied on solid solutions of WC, TaC, and TiC, and this became a part of the very successful Plansee company. The sintered $W_2Ti_2C_4$ variant was patented by McKenna [42]. Even though the compound was not confirmed, the mixed carbide concept became the basis for the successful Kennametal firm. Subsequently, many carbide combinations were explored prior to World War II [43]. By this time it

Table 2.1 Approximate Evolution of Cemented Carbides to 1960

Approximate Year	Main Ingredients (carbides – matrix)
1900	cast tool steels
1909	cobalt-chromium alloys
1914	cast tungsten carbides
1922	WC – Ni, Co, Fe
1929	WC + TiC – Co
1930	WC + TaC – Co
	WC + VC – Co
	WC + NbC – Co
1931	TaC – Ni
	TiC + TaC – Co
1938	WC + Cr_3C_2 – Co
	TiC + VC – Ni, Fe
1944	TiC + NbC – Ni, Co
1950	TiC + NbC + VC + Mo_2C + TaC – Ni, Co
1951	WC – Ni
	TiC – tool steel
1956	WC + TiC + TaC + NbC + Cr_3C_2 – Co
1959	WC + TiC + HfC – Co

was realized that the liquid phase sintered microstructure consisted of a fully connected solid skeleton. A sintered cemented carbide microstructure is shown in Figure 2.19, consisting of angular carbide grains sintered to form a solid skeleton with matrix phase (solidified WC-Co liquid) between the grains. Both phases are connected, giving an interlaced three-dimensional composite. Thus, contrary to myth, the strength of the cemented carbide is not due to the cobalt phase.

Technological developments included the use of pressure during sintering to remove final porosity. This is performed without a container and is termed sinter-HIP (hot isostatic pressing). Vapor deposited hard coatings of alumina, titanium nitride, diamond, or other phases are added to lengthen service life.

A wide variety of alternatives to WC-Co are available, but the WC-Co composition is the baseline in the field. Annual production of liquid phase sintered hard materials is now (in 2013) valued at $25 billion. Applications range from metal cutting tips, abrasive spray nozzles, mining tools, ball point pen tips, die compaction tools, metal shears, and wire drawing dies. Drawing dies remain an important product for forming electrical wires, and other wires used in a range of products, such as radial tires, surgical suturing, champagne cork retainers, orthodontics bracket tensioning, welding wires, and even student spiral notebooks. Additionally, sintered tungsten carbides have moved into consumer luxury products, such as the carbide wedding band shown in Figure 2.20.

Figure 2.19 The liquid phase sintered microstructure of cemented carbide, consisting of hard and angular grains of WC interlaced with a solidified WC-Co alloy matrix.

Figure 2.20 A recent example of sintered tungsten carbide jewelry in the form of a wedding ring.

Tungsten Heavy Alloy Radiation Shields

Mixing tungsten with other metals to improve sintering was explored during the race to develop tungsten lamp filaments, albeit without success. By the late 1920s, radiation damage was first recognized, leading to a search for shielding materials. Price and Smithells formed radio-opaque materials via liquid phase sintering, which produced

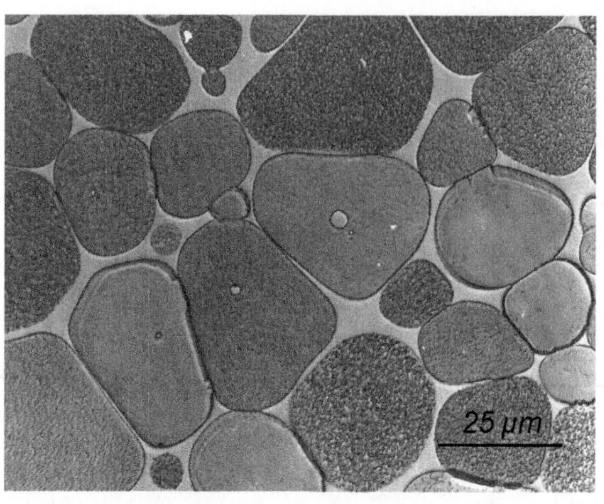

Figure 2.21 Microstructure for liquid phase sintered tungsten heavy alloy, consisting of nearly pure tungsten grains bonded with a matrix of solidified liquid (W-Ni-Fe in this case).

what are now known as tungsten heavy alloys [44]. The initial composition consisted of mixed tungsten, nickel, and copper powders. Applications were in radium radiation therapy where the concentrated mass was useful in containment [45,46]. Higher performance compositions for military use subsequently included cobalt and iron [47,48].

Surprisingly, these mixed powders sintered to full density at relatively low temperatures. Full density was attained with peak temperatures of near 1500°C (1773 K). Figure 2.21 captures the unique sintered microstructure, which consists of large tungsten grains bonded to form a skeletal network with solidified liquid alloy between the grains.

Subsequently, several tungsten heavy alloys were developed around additives of Ni, Cu, Fe, Mn, Co, Mo, Ta, and Re to customize cost and properties. For example, heat dissipation in microelectronics embraced tungsten-copper and tungsten-copper-cobalt alloys [49]. In an alloy consisting of 93 W-5Ni-2Fe, the sintered density is 17.6 g/cm^3, and this gives a tensile strength of 900 MPa with 20% elongation to fracture. Such alloys are used in products ranging from armor piercing projectiles to golf club inserts to adjust the center of gravity. Heavy alloys remain a mainstay for radiation containment, the initial application envisioned when they were discovered. Now these same alloys are applied as collimators in advanced gamma radiation surgical tools.

Bronze Bearings

As the industrial age emerged, the need for reduced friction in motors and engines invoked the search for long-life bearings. In 1870, Gwynn patented the precursor to the sintered self-lubricating bearing [5,50]. He relied on mixtures of zinc, bronze, copper, and tin, and in some cases included natural rubber and intentional porosity. The early approach relied on stirring the powder mixture while heating and then pressing the hot powder in a mold to form the desired shape. His heat and then mold approach is now reversed to mold and then heat.

In 1902, Lowendahl used copper coated graphite to form a bearing, and in 1909 he patented bearings sintered from copper coated with tin, mixed with graphite [51]. Ammonium nitrate was added as a pore former to control pore size and porosity. After sintering in the 300 to 540°C (573 to 813 K) range, the pores were filled with oil, wax, or other lubricant. As the bearing heats during operation, the lubricant expands to provide a hydrodynamic film.

Recognition of the commercial opportunity was evident in a patent for internal combustion engine bearings in 1916 [52]. It relied on copper, tin, lead, and graphite, sintered to form a porous self-lubricating bearing. Mass production of sintered bearings followed, with the first applications being in refrigerators and automobiles. Long hold times were used initially to stabilize the pores [53]. Subsequent efforts converged on mixtures of copper, tin, graphite, and other metals, that sintered to form porous bronze with good strength as is widely used today [54]. The microstructure is about 80% dense bronze with customized pores for lubricant storage [55]. However, empirical discovery was far ahead of theory; it took many years to understand that transient liquid phase sintering was the process required to form a successful bearing. Today, sintered self-lubricating bearings are widely used for applications in computers, home appliances, industrial tools, motors, automobiles, with production rates of hundreds of millions per day.

Automotive Components

The production of complicated shapes by sintering avoids the cost of machining, and this has great traction in automotive component production. A first example was in the production of ceramic insulators for spark plugs. The intent is to form large numbers of complex-shaped components at low cost. Besides automotive applications, several other areas have similar concerns—lawn and garden equipment, home appliances, hardware and hand tools, industrial fittings, and heavy duty equipment. In most cases, the press-sinter route is substantially lower in cost; often just 20% of the cost of machining. A tool needs to be designed to press the powder shape, but otherwise the manufacturing results in a component that averages just $5 per kg.

Early efforts trace to 1864, when Percy [22] detailed the use of iron ores to produce sintered iron. Subsequent work examined the microstructure of the sintered material [23]. Commercial interest intensified during World War II. Germany led the way by sintering rotating bands for artillery shells [56]. Soon other applications emerged, such as oil pump gears for automobiles [5,35,57–59]. Significant gains came

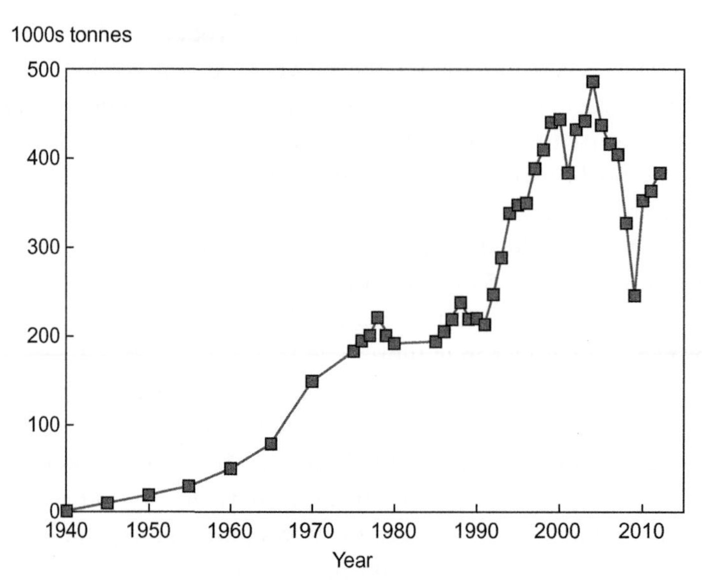

Figure 2.22 Ferrous powder consumption in North America since the first uses of sintered steel in 1940. The consumption of sintered steel is tightly coupled to automotive production.

with the mastery of carbon content during sintering, as required in steels [60,61]. Squire [62] described how graphite could be mixed with iron powder to tailor dissolution during sintering to deliver both high strength and high ductility. These early efforts relied on sponge iron powder, where internal porosity in the particles degraded sintered properties. Similar to earlier ideas, repressing was used for densification [63]. By 1948, sintered nickel alloy steels had reached outstanding strength levels of 1 GPa [64]. By 1950, liquid phase sintered steel, based on mixed iron, copper, and graphite, entered production. Besides structural components, sintered iron powder was applied to an emerging range of magnetic components [65].

Significant early steel component production arose in firms already pressing graphite to make brushes for electric motors. Notable champions were the Stackpole Battery Company and Keystone Carbon. To illustrate this growth, Figure 2.22 plots iron powder consumption in North America after 1940, most of which was used in automotive applications.

Polytetrafluroethylene

Polytetrafluroethylene (PTFE), also known as Teflon®, was discovered in 1938 by Plunkett during research on fluorine refrigerants. PTFE is synthesized as nanoscale particles. The slippery character of sintered PTFE leads to its widespread use in bearings, seals, clothing, shoes, and other situations where chemical resistance is desired.

The polymer melts at temperatures over 320°C (593 K), but can withstand short transients up to 500°C (773 K). It is not a thermoplastic, so the standard approach to forming PTFE objects is by sintering or hot pressing the powder [66]. Early applications arose during World War II in the Manhattan Project, but notable success came subsequently in coating cookware to prevent food from sticking. By 1944, sintered rods were sliced to form tapes as is widely used in plumbing. Generally the compaction pressure was low, in the 30 MPa range, and the sintering was close to the melting temperature, resulting in a mixture of amorphous and crystalline phases [67].

Sintered Abrasives

The history of abrasives dates back to early examples from around 25,000 BC. Initially, hard particles were mixed with pliable substances, but more durable materials were formed using sintering. Grinding wheels formed by using sintering date from 1873. By 1883, Gay was forming mixtures of hard materials in a metal powder matrix [58]. The 1891 discovery of artificial silicon carbide by Acheson opened new possibilities for abrasives, and soon silicon carbide and diamond abrasives were in production.

Since diamond is unstable at high temperatures, in the 1920s and 1930s, short spark sintering cycles were used to fabricate early diamond abrasives [68,69]. One application was in grinding sintered tungsten carbide dies. Cobalt was a favorite bond for diamond, similar to its use in cemented carbides.

In 1955, artificial diamonds were produced based on the discovery of Bundy et al. [70]. Man-made diamonds depended on the high pressure concepts discovered by Bridgman [Nobel Prize in 1931]. Pure diamond is sintered using temperatures of 1500°C and pressures in the 6 GPa range [71], giving a product known as polycrystalline diamond. It is used in oil and gas drilling applications.

For super-abrasives, diamond powder is mixed with tungsten, tungsten carbide, cobalt, iron, copper, or titanium, and either sintered, hot pressed, or spark sintered to full density. Besides diamond "super-abrasives," other hard compounds rely on sintered metal bonds, including titanium carbide, silicon carbide, and cubic boron nitride [71–73].

INTERDEPENDENT DEVELOPMENTS

Advances in sintering, sintered materials and designs have consistently been a part of modern technology. Early developments in platinum subsequently seeded the way tungsten was sintered. This enabled the production of lamp filaments. Improved wire drawing dies for tungsten lamp filament production required new hard materials, leading to the discovery of cobalt bonded tungsten carbide. Then sintered diamond composites were needed to finish the hard carbides dies. When high pressure processing was realized, artificial diamonds become available, but the forming press relied on the strength and stiffness only possible with cemented carbide anvils. In turn, abrasive finishing and wire drawing machinery requires low friction bearings, which lead to sintered bronze bearings. Sensibly a cascade of interdependent developments occurred touched by sintering.

After World War II, sintering entered into the manufacture of electronic, dental, and medical devices. Electrical contacts needed in power switching relied on mixed powders to form the composites of an electrically conductive phase (copper or silver) and an arc erosion resistant phase (tungsten, molybdenum, tungsten carbide) [5,37]. The empirical learning in one area spread and the applications for sintering grew to include a diverse range of products. Meanwhile, non-thermoplastic polymers such as polytetrafluroethylene emerged which required sintering, and with that, research on sintering fundamentals spread to the polymer field.

In the early 1990s, sintering entered the rapid prototyping field as a means of building three-dimensional objects without machining [74]. Desktop manufacturing includes a diversity of additive approaches that rely on sintering. Tool cavities, medical devices, and custom dental restorations are early successes of this approach.

Each of these developments seeded questions about the fundamentals; although commercial products consistently emerged before theory. A need arose to better understand sintering in order to improve productivity, sensibly specify raw materials, establish processing quality controls, and optimize the products. These efforts required a scientific understanding of sintering.

The advent of nanoscale powders opens new applications. These include ferromagnetic composites, titanium tissue scaffolds, circuit interconnections, flexible electronic circuits, diamond composites, electrical circuits, and new materials for catalysis, optical, wear, or thermal applications [75–83]. Later chapters provide more details about these opportunities.

KEY LESSONS FROM SINTERING HISTORY

A lesson from sintering history is that observation leads theory. The major products fabricated from sintering had their origins in empirical discoveries and grew to widespread use without a fundamental understanding. Sintering science gained traction in the late 1940s, but several important products were already in production. The patent literature proved more important than the journal articles.

Today a similar situation persists. Journal publications report on new processes such as spark sintering, while companies work hard to develop high value products for sintered microelectronics, capacitors, high performance magnets, superconductors, and super-abrasives—indeed today sintering has entered an era of "SUPER" as it is widely employed in superconductors, super-abrasives, super-alloys, super-capacitors, and other leading applications. So the key lessons from sintering history are that commercial gains drive the major efforts. Sintering science lags behind sintering practice, but much is to be gained by understanding the basic principles as presented in this book.

REFERENCES

[1] W.D. Kingery, Sintering from prehistoric times to the present, in: A.C.D. Chaklader, J.A. Lund (Eds.), Sintering '91, Trans Tech Publ., Brookfield, VT, 1992, pp. 1–10.

[2] J.E. Burke, A history of the development of a science of sintering, in: W.D. Kingery (Ed.), Ceramics and Civilization, Ancient Technology to Modern Science, vol. 1, Amer. Ceramic Society, Columbus, OH, 1985, pp. 315–332.

[3] T. Ring, Fundamentals of Ceramic Powder Processing and Synthesis, Academic Press, San Diego, CA, 1996.

[4] R.F. Mehl, The historical development of physical metallurgy, in: R.W. Cahn (Ed.), Physical Metallurgy, North Holland Publishing, Amsterdam, Netherlands, 1965, pp. 1–31.

[5] C.G. Goetzel, Treatise on Powder Metallurgy, vol. 1, Interscience Publishers, New York, NY, 1949, pp. 259–312.

[6] R.K. Dube, Further literary and documentary evidence for powder technology in ancient and medieval India, Powder Metall. vol. 36 (1993) 113–131.

[7] R. Balasubramaniam, Novel phosphoric irons based on study of the Delhi Iron Pillar, in: J.V. Kumar (Ed.), Frontiers of Metallurgy and Materials Technology, BS Publications, Hyderabad, India, 2011, pp. 482–501.

[8] M. Noguez, R. Garcia, G. Salas, T. Robert, J. Ramirez, About the Pre-Hispanic Au-Pt 'sintering' technique, Inter. J. Powder Metall. 43 (1) (2007) 27–33.

[9] J.A.P. Elorz, J.I. Verdja-Gonzalez, J.P. Sancho-Martinez, N. Vilela, Melting and sintering platinum in the 18th century: the secret of the Spanish, J. Metals. 51 (10) (1999) 9–12, 41.

[10] M.Y. Balshin, Effect of P.G. Sobolevskii's ideas on the development of powder metallurgy and related branches of technology, Powder Metall. Metal Ceram. 16 (1977) 252–254.

[11] B.N. Menschutkin, Discovery and early history of platinum in Russia, J. Chem. Edu. 11 (1934) 226–229.

[12] S.Y. Plotkin, Petr Grigorevich Sobolevskii, Powder Metall. Metal Ceram. 5 (1966) 993–995.

[13] S.Y. Plotkin, Development of powder metallurgy in the USSR during 50 years of soviet rule, Powder Metall. Metal Ceram 6 (1967) 844–853.

[14] C.S. Smith, The early development of powder metallurgy, in: J. Wulff (Ed.), Powder Metal., American Society for Metals, Cleveland, OH, 1942, pp. 4–17.

[15] W.H. Wollaston, On a method of rendering platina malleable, Phil. Trans. Royal Soc. London 119 (1829) 1–8.

[16] J.A. Chaldecott, Wollaston's platinum thermometer, Platinum Met. Rev. 16 (1972) 57–58.

[17] P.T. Hinde, William Hyde Wollaston; the man and his equivalents, J. Chem. Edu. 43 (1966) 673–676.

[18] E.G. Ferguson, Bergman, Klaproth, Vauquelin, Wollaston, J. Chem. Edu. 18 (1941) 3–7.

[19] K.J. Anderson, Materials for incandescent lighting: 110 years for the light bulb, Mater. Bull. January (1990) 52–53.

[20] R.C. Smith, Sintering – its nature and causes, J. Chem. Soc. Trans. 123 (1923) 2088–2094.

[21] J.R. Wynnyckyj, T.Z. Fahidy, Solid state sintering in the induration of iron ore pellets, Metall. Trans. 5 (1974) 991–1000.
[22] J. Percy, Metallurgy – The Art of Extracting Metals from their Ores, and Adapting Them to Various Purposes of Manufacture, John Murray, London, UK, 1864.
[23] B.G. Klugh, The microstructure of sintered iron bearing materials, Trans. TMS-AIME 45 (1913) 330–345.
[24] B.G. Klugh, The sintering of fine iron bearing materials by the Dwight and Lloyd process, Trans. TMS-AIME 42 (1912) 364–375.
[25] F.A. Vogel, Sintering and briquetting of flue dust, Trans. TMS-AIME 43 (1912) 381–386.
[26] J. Gayley, The sintering of fine iron bearing material, Trans. TMS-AIME 42 (1912) 180–190.
[27] W.L. Voelker, Improvements in the Manufacture of Filaments for Incandescing Electric Lamps, and in Means Applicable for Use in Such Manufacturer, GB Patent 6149, issued 10 February 1900.
[28] S.W.H. Yih, C.T. Wang, Tungsten Sources, Metallurgy, Properties, and Applications, Plenum Press, New York, NY, 1979.
[29] E. Pink, L. Bartha, The Metallurgy of Doped Non-Sag Tungsten, Elsevier Applied Science, London, UK, 1989.
[30] J. Lux, Improved Manufacture of Electric Incandescent Lamp Filaments from Tungsten or Molybdenum or an Alloy Thereof, GB Patent 27,002, issued 13 December 1906.
[31] P.K. Johnson, Tungsten filaments – the first modern PM product, Inter. J. Powder Metall. 44 (4) (2008) 43–48.
[32] W.D. Coolidge, Production of Refractory Conductors, U. S. Patent 1,077,674, issued 5 November 1913.
[33] W.D. Coolidge, Ductile tungsten, Trans. Amer. Inst. Elect. Eng. 29 (1910) 961–965.
[34] C.L. Briant, Potassium bubbles in tungsten wire, Metall. Trans. 24A (1993) 1073–1084.
[35] J.L. Walter, C. Briant, Tungsten wire for incandescent lamps, J. Mater. Res. 5 (1990) 2005–2022.
[36] R. Bergman, L. Bigio, J. Ranish, Filament Lamps, Report 98 CRD 027, General Electric Research and Development Center, Schenectady, NY, 1998.
[37] C.G. Goetzel, Treatise on Powder Metallurgy, vol. III, Interscience Publishers, New York, NY, 1952.
[38] K. Schroter, W. Jenssen, Tool and Die, U. S. Patent 1,551,333, issued 25 Aug 1925.
[39] H.E. Exner, Physical and chemical nature of cemented carbides, Inter. Met. Rev. 24 (1979) 149–173.
[40] P. Schwarzkopf, R. Kieffer, Refractory Hard Metals: Borides, Carbides, Nitrides, and Silicides, Macmillian, New York, NY, 1953.
[41] P. Schwarzkopf, Powder Metallurgy Its Physics and Production, Macmillan, New York, NY, 1947.
[42] P.M. McKenna, Tool materials (cemented carbides), in: J. Wulff (Ed.), Powder Metallurgy, American Society for Metals, Cleveland, OH, 1942, pp. 454–469.
[43] K.J.A. Brookes, Hardmetals and Other Hard Materials, third ed., International Carbide Data, Hertsfordshire, United Kingdom, 1998.
[44] G.H.S. Price, C.J. Smithells, S.V. Williams, Sintered alloys. Part I – copper-nickel-tungsten alloys sintered with a liquid phase present, J. Inst. Metals 62 (1938) 239–264.
[45] J.C. McLennan, C.J. Smithells, A new alloy specially suitable for use in radium beam therapy, J. Sci. Instr. 12 (1935) 159–160.
[46] C.J. Smithells, A new alloy of high density, Nature 139 (1937) 490–491.
[47] R. Cury, Bibliographical survey on the development of tungsten heavy alloys, Proceedings International Conference on Refractory Metals and Hard Materials, 18th Plansee Seminar, Reutte, Austria, 2013, pp. RM19.1–RM19,11.
[48] A. Bose, R. Sadangi, R.M. German. A review on alloying in tungsten heavy alloys. Materials Processing and Interfaces, vol. 1, Proceedings 141st Meeting the Minerals, Metals, and Materials Society, Warrendale, PA, 2012, pp. 455–465.
[49] Y.S. Kwon, S.T. Chung, S. Lee, S.J. Park, R.M. German, Development of thermal management material: nano tungsten coated copper and carbon nanotube reinforced copper, in: P. Rodhammer

(Ed.), Proceedings of the Seventeenth Plansee Seminar, vol. 1, Plansee Group, Reutte. Austria, 2009, pp. RM3.1–RM3.8.
[50] S. Gwynn, Improved Composition of Matter Called Metaline for Journal Bearings Etc., U. S. Patent 101,866, issued 12 April 1870.
[51] V. Lowendahl, Process of Manufacturing Porous Metal Blocks, U. S. Patent 1,051,814, issued 28 Jan 1913.
[52] E.G. Gilson, Bearing Material Suitable for Internal Combustion Engines, U. S. Patent 1,177,407, issued 28 March 1916.
[53] H.M. Williams, A.L. Boegehold, Alloy Structure, U. S. Patent 1,642,349, issued 13 September 1927.
[54] L.F. Pease, W.G. West, Fundamentals of Powder Metallurgy, Metal Powder Industries Federation, Princeton, NJ, 2002.
[55] H.E. Hall, Sintering of copper and tin powders, Met. Alloys 14 (1939) 297–299.
[56] F. Sauerwald, Uber de elementarvorgange beim fritten and sintern von metallpulvern mit besonderer berusksichtigung der realstruktur ihere oberflachen, Kolloid Z. 104 (1943) 144–160.
[57] F.V. Lenel, Oil pump gears, in: J. Wulff (Ed.), Powder *Metallurgy*, American Society for Metals, Cleveland, OH, 1942, pp. 502–511.
[58] W.D. Jones, Fundamental Principles of Powder Metallurgy, Edward Arnold Publishers, London, UK, 1960.
[59] F.V. Lenel, Iron Article and Method of Making Same, U. S. Patent 2,226,520, issued 24 December 1940.
[60] F.C. Kelley, Effect of time, temperature, and pressure upon the density of sintered metal powders, in: J. Wulff (Ed.), Powder Metallurgy, Amer. Society for Metals, Cleveland, OH, 1942, pp. 60–66.
[61] R.P. Smith, Equilibrium of iron-carbon alloys with mixtures of CO-CO_2 and CH_4-H_2, J. Amer. Chem. Soc. 68 (1946) 1163–1175.
[62] A. Squire, Iron-graphite powder compacts, Trans. TMS-AIME 171 (1947) 473–484.
[63] J.F. Kuzmick, Evaluation of the molding, coining, and sintering properties of iron powder, Trans. TMS-AIME 175 (1947) 813–833.
[64] L. Delisle, W.V. Knopp, Nickel steels by powder metallurgy, Trans. TMS-AIME 175 (1948) 791–812.
[65] R. Steinitz, Magnetic properties of iron powder compacts, Trans. TMS-AIME 175 (1948) 834–847.
[66] J.F. Lontz, Sintering of polymer materials, in: L.J. Bonis, H.H. Hausner (Eds.), Fundamental Phenomena in the Materials Sciences, vol. 1, Plenum Press, New York, NY, 1964, pp. 25–47.
[67] J. Yang, R. Williams, K. Peterson, P.H. Geil, T.C. Long, P. Xu, Morphology evolution in polytetrafluroethylene as a function of melt time and temperature. Part III. Effect of prior deformation, Polymer 46 (2005) 8723–8733.
[68] E. Gauthier, Diamond Lap, U. S. Patent 1,625,463, issued 19 April 1927.
[69] J. Konstanty, Powder Metallurgy Diamond Tools, Elsevier, Amsterdam, Netherlands, 2005.
[70] F.P. Bundy, H.T. Hall, H.M. Strong, R.H. Wentorf, Man made diamonds, Nature 176 (1955) 51–55. Errata 1993, vol. 365, p. 19.
[71] R.H. Wentorf, R.C. Devries, F.P. Bundy, Sintered superhard materials, Science 208 (1980) 873–880.
[72] H. Blumenthal, R. Silverman, Infiltration of TiC skeletons, Trans. TMS-AIME 206 (1956) 977–981.
[73] R.A. Alliegro, L.B. Coffin, J.R. Tinklepaugh, Pressure sintered silicon carbide, J. Amer. Ceram. Soc. 39 (1956) 386–389.
[74] C.R. Deckard, Method and Apparatus for Producing Parts by Selective Sintering, U. S. Patent 5,316,580, issued 31 May 1994.
[75] T. Sekino, T. Nakajima, S. Ueda, K. Niihara, Reduction and sintering of a nickel-dispersed-alumina composite and its properties, J. Amer. Ceram. Soc. 80 (1998) 1139–1148.

[76] K. Keskinbora, T.S. Suzuki, I.O. Ozer, Y. Sakka, E. Suvaci, Hybrid processing and anisotropic sintering shrinkage in textured ZnO ceramics, Sci. Tech. Adv. Mater. 11 (2010), article 065006, 11 pages.
[77] B. Twomey, A. Breen, G. Byrne, A. Hynes, D.P. Dowling, Comparison of thermal and microwave assisted plasma sintering of nickel-diamond composites, Powder Metall. 53 (2010) 188–190.
[78] Z. Pan, H. Sun, Y. Zhang, C. Chen, Harder than diamond: superior indentation strength of wurtzite BN and lonsdaleite, Phys. Rev. Lett., 102, (2009), paper 055503.
[79] B. Levine, A new era in porous metals: applications in orthopaedics, Adv. Eng. Mater. 10 (2008) 788–792.
[80] L. Tuchinskiy, Novel fabrication technology for metal foams, J. Adv. Mater. 37 (3) (2005) 60–65.
[81] D. Bera, S.C. Kuiry, S. Seal, Synthesis of nanostructured materials using template-assisted electrodeposition, J. Metals 56 (1) (2004) 49–53.
[82] S. Tada, Z.M. Sun, H. Hashimoto, T. Abe, Fabrication of a dense long rod through pulse discharge sintering assisted by traveling zone heating, Mater. Trans. 44 (2003) 1667–1670.
[83] S. Landwehr, G. Hilmas, T. Huang, A. Griffo, B. White, Functionally designed cemented carbides, Advances in Powder Metallurgy and Particulate Materials – 2003, Part 6, Metal Powder Industries Federation, Princeton, NJ, 2003, pp. 163–171.
[84] W.W. Engle, Cemented carbides, in: J. Wulff (Ed.), Powder Metallurgy, American Society for Metals, Cleveland, OH, 1942, pp. 436–453.
[85] K.J.A. Brookes, Half a century of hardmetals, Metal Powder Rept. 50 (12) (1995) 22–28.

CHAPTER THREE

Infrastructure Developments

QUALITATIVE SINTERING THEORY

Many commercial sintered products arose from the empirical studies outlined in the previous chapter. As with many material phenomena, up to about 1940 the infrastructure did not exist to formulate a theoretical basis. Parallel developments were required in atomic theory, surface energy, microstructure, atomic motion, and tools for quantification of sintering.

Early Conjectures

Several practical sintering technologies progressed without a predictive basis. Qualitative descriptions of sintering arose in the early 1900s, giving several conceptualizations as listed below [1]:

- Sintering has an onset temperature and does not occur below this temperature, similar to the way a glass has a softening transition temperature.
- The sintering temperature for brittle solids corresponds to a dramatic increase in plasticity.
- Only materials that are plastic exhibit sintering.
- Sintering is delayed until surface contamination films evaporate.
- Only metals that undergo polymorphic phase transitions sinter, such as Co and Fe.
- Only mixed powders with a chemical affinity sinter.
- Sintering occurs by melting temperature depression due to surface curvature or external pressure.
- Recrystallization of compacted particles is necessary to give sintering.
- Sintering occurs because surface energy increases on heating.

One early technique for determining the sintering temperature relied on a stirred chamber filled with powder. The sintering temperature was detected when a motor stirring the powder was unable to continue rotating. As an indication of the test accuracy, consider iron in vacuum "sintered" at 750°C (1023 K), but in air iron "sintered" at 150°C (423 K). The lower temperature was due to oxidation, but such reports confused efforts to understand sintering, leading to many effects remaining unexplained [2]. There was even disagreement on the inherent strength of the metallic sinter bond versus that of cast material.

Sintering: From Empirical Observations to Scientific Principles
DOI: http://dx.doi.org/10.1016/B978-0-12-401682-8.00003-3

Rhines and Lenel Stipulations

As observations accumulated, qualitative sintering concepts arose. In 1942 Rhines [3] outlined the key elements:

1. The neck material is the same as the parent material in terms bonding and properties.
2. Pressure increases particle bonding and dislodges interfering films between particles.
3. Initial neck growth occurs by lateral expansion of the neck under surface energy; atomic flow is biased to remove curvature gradients, and surface diffusion is expected to initially dominate sintering, with secondary effects from plastic flow. Both contributions diminish with time.
4. Swelling is associated with gas evolution and gas trapped in pores; compact growth is due to increasing gas pressure inducing plastic flow, late stage swelling is taken as evidence of plastic flow.
5. Recovery, recrystallization, and grain growth are typical aspects of sintering and often initiate at contact points between particles. The grain growth rate increases in proportion to the rate of sintering.
6. Late stage sintering involves the deposition of atoms into pores, causing pore closure, especially for small spherical pores near the compact surface which fill earlier in sintering.
7. Changes in properties due to sintering are simply reflections of structural changes and do not have any unique aspects attributed to sintering.

These observations were followed by similar ideas from Lenel about liquid phase sintering [4]. He noted the two variants were a persistent (WC-Co, W-Ni-Cu) and transient (Cu-Sn) liquid. About 20 different compositions were studied to show that the events during liquid phase sintering commonly involve several steps, starting with compaction, establishing a protective atmosphere, and heating the compact to a temperature where a melt forms. Most systems exhibit solubility of the solid in the liquid, so small grains dissolve and large grains grow by a solution-reprecipitation process. During grain growth, the grains reshape to allow liquid to fill voids, giving rapid densification. On cooling, the liquid freezes to provide a matrix phase between the solid grains. In transient liquid phase sintering, the liquid dissolves into the solid over time and the rapid densification process comes to a halt, so obviously diffusion in the liquid is an important event in sintering.

Frenkel—Surface Energy and Viscous Flow

Surface tension was an accepted concept for liquids, but for solids the idea of a surface energy was controversial up to the 1940s. Surface energy provides the attractive stress important to sintering [5]. It depends on crystal orientation, contacting vapor or liquid, defect structure (for example screw dislocation population), and impurity contamination. The combination of surface energy and atomic motion is the basis for the explanation of sintering.

Frenkel relied on an analogy between atomic diffusion and viscous flow in his sintering model [6–8]. Under the action of surface energy, two viscous spherical particles in point contact coalesce to form a large sphere, as sketched in Figure 3.1. The final sphere is 1.26 times the diameter of the initial spheres. In this model, sintering is tracked by neck growth. Surface energy pulls the spheres together, similarly to the way that soap bubbles coalesce. Surface area and surface energy decline as the bond grows. Mass conservation dictates that the sphere centers move together to give shrinkage.

In this two sphere model crystal structure is ignored, so there is no interparticle grain boundary or dihedral angle. Accordingly, the Frenkel model works for amorphous materials such as glass, but is not applicable to crystalline materials. The success of the Frenkel model is in showing the surface energy decrease during sintering neck growth. In retrospect it can be seen that the viscosity-diffusivity exchange underlying Frenkel's model is an idea promoted from 1905 by the work of Einstein, Stokes, Smoluchowski, and Sutherland.

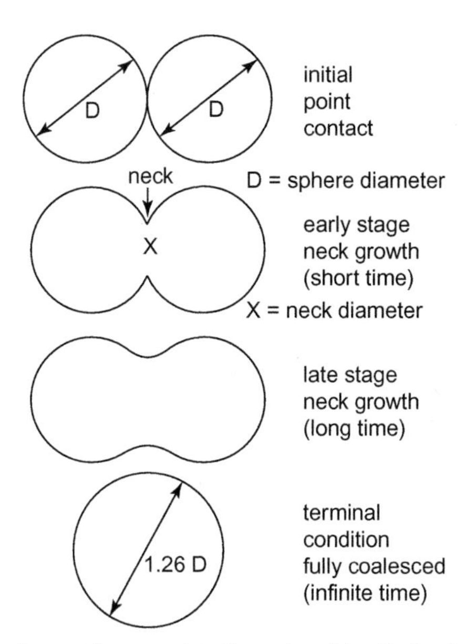

Figure 3.1 The two sphere viscous flow model allows two identical spheres of diameter D to make initial point contact, then by viscous flow the spheres form a neck of diameter X and eventually coalesce into a single sphere.

Frenkel assumed that the surface energy decline due to neck growth was expended on densification. Within ten years, several models were built on his two particle framework.

EMERGENCE OF QUANTITATIVE SINTERING CONCEPTS

A quantitative sintering theory awaited infrastructure gains with respect to understanding atoms. Also, observational tools had to advance, so a quantitative understanding of sintering awaited several enabling developments.

In solid-state sintering, the transition from qualitative to quantitative treatment came with the two sphere neck growth concept of Frenkel [6] and Kuczynski [9], followed in 1955 by the shrinkage model of Kingery and Berg [10]. For the liquid phase sintering problem, quantitative models were developed later [11,12].

In 1948, Kuczynski was working at Sylvania, and was trying to unlock Coolidge's discovery on sintered tungsten lamp filaments [13]. He offered a neck growth model based on diffusion to explain results from experiments in which spherical copper particles bonded to a copper disk [9]. Unfortunately he was unable to explain densification, but his conjecture of a diffusion process was a gain for sintering theory. Just a few years earlier atomic diffusion was not an accepted concept. Thus, infrastructure developments were required to transition from qualitative to quantitative sintering theory.

The goal of sintering theory is to guide practice; but commercial developments remained far ahead of theory. Sintering science is driven by two technical motivations:

- **Design**—how to fabricate a component of specific size and shape by specifying the process variables such as sintering temperature, particle size, or hold time.
- **Properties**—how to adjust process parameters to produce a given property combination, such as strength, hardness, or thermal conductivity, while minimizing cost.

Sintering theory consists of a body of relations that enable solutions to be derived for questions such as those outlined in Figure 3.2. Once mathematical relations are established, optimization is possible. Optimization implies that the relations can be solved for either a maximum (say strength) or minimum (say cost). Relations that link processing variables to the outcome enable optimization.

Figure 3.3 provides a qualitative outline of the parameters captured in a generic sintering theory:

$$F = G\, M\, K \tag{3.1}$$

where F is a measure of the degree of sintering, such as sintered strength, density, component size variation, or cost. The independent input parameters are:

- **G = geometric parameters**—particle size, particle shape, surface area, grain size, component density, pore size ...

Sintering Theory

dF/dY = 0, max or min

F = G M K

F = object function – size, shape, cost, degree of sintering

Y = adjustable parameter

- *G* = function of geometric parameters
- *M* = function of material parameters
- *K* = function of processing parameters

Figure 3.2 A simplified view of how sintering theory should be constructed, where the object function F is optimized with respect to a sintering parameter Y to either maximize or minimize factors such as density or cost. Sintering theory consists of geometric, material, and processing parameters that are often interlinked.

Sintering Models

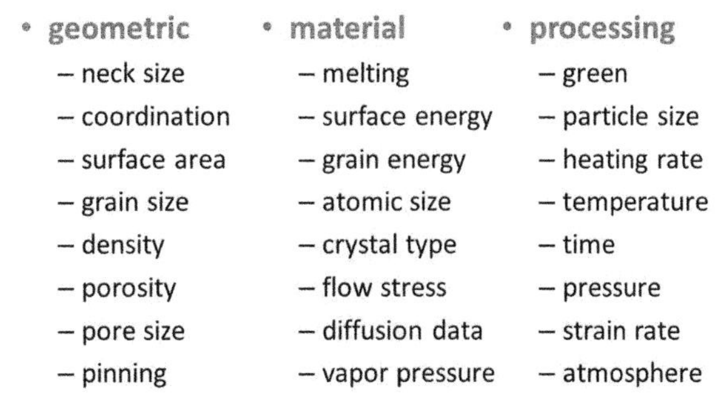

• geometric
– neck size
– coordination
– surface area
– grain size
– density
– porosity
– pore size
– pinning
• material
– melting
– surface energy
– grain energy
– atomic size
– crystal type
– flow stress
– diffusion data
– vapor pressure
• processing
– green
– particle size
– heating rate
– temperature
– time
– pressure
– strain rate
– atmosphere

Figure 3.3 Models for sintering require details on several parameters, divided into geometric, materials, and processing data. This figure lists a few examples for each.

- ***M*** **= material parameters** – melting range, surface energy and its variation with crystal orientation, grain boundary energy, atomic size, crystal type, temperature dependent yield strength, diffusion and creep factors, vapor pressure ...
- ***K*** **= processing parameters** – green density, heating and cooling cycles, peak temperature, hold time, applied pressure, process atmosphere ...

Each has interlaced dependences on the overt processing parameter such as temperature. For example, grain size changes during sintering at a rate that depends on temperature, surface energy depends on temperature, diffusion rates are very sensitive to temperature as is vapor pressure. These dependencies are a part of the scientific infrastructure.

INFRASTRUCTURE DEVELOPMENT

To transcend from sintering discovery to sintering model required an infrastructure. Things we take for granted now, such as crystal structure, are relatively recent ideas. This scientific underpinning was still emerging in the 1960s. For example, final stage sintering required creep concepts developed in 1963, a grain shape model from 1953, and an instability model from 1879 [14–17].

Atomic Theory

Atomic theory was proposed in the late 1600s by Hooke. Cells were used to describe the building blocks of matter. In the early 1800s, Dalton documented five points about atomic structure in the formation of compounds such as ammonia. An example of his atomic concept is shown in Figure 3.4, which is quite accurate considering the speculative nature of his model of the atom. Avogadro followed a few years later, with his demonstration that gas volume did not depend on mass, and in 1869 Mendeleev

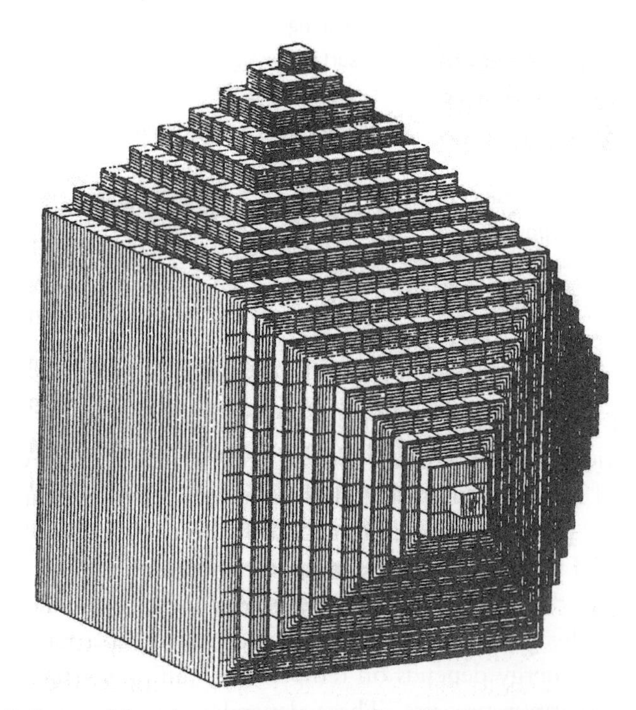

Figure 3.4 John Dalton envisioned crystals made up of small unit atomic masses, what we term as atoms. This is a sketch of how he envisioned assembly of the atoms to form a crystal.

proposed the periodic table to account for the way that different elements have similar properties.

About this time Mitscherlich developed the concept of polymorphs. Multiple crystal types are now widely accepted in several materials, such as diamond-graphite. Such concepts were proposed prior to the emergence of atomic theory.

In spite of the discovery of electrons by Thomson [Nobel Prize 1906], the atomic concept struggled until the Brownian motion theory was developed by Einstein in 1905 [18]. He showed that the erratic motion of small particles can be explained by a transfer of kinetic energies from the surrounding atoms. Measurements by Perrin confirmed the atomic structure and both men won Nobel Prizes. What followed were the atomic models of Rutherford [Nobel Prize 1908] and Bohr [Nobel Prize 1922].

Other developments make a convincing story out of atomic theory; one was that of von Laue. He used X-rays to examine atomic structures [1914 Nobel Prize]. Soon the father-son team of William Henry Bragg and William Lawrence Bragg confirmed crystallinity [1915 Nobel Prize]. Thus, research in the early 1900s finally established atomic structure. However, it was not until 1951 that Muller finally visualized an atom using field ion microscopy. He also confirmed dislocations and vacancies in these images.

A further major step came with Pauling's 1939 book *The Nature of the Chemical Bond* [Nobel Prize 1954] and his linkage of atomic bonding to crystal structure. These atomic theory concepts were a required platform for constructing sintering theory.

Surface Energy

Since the early 1800s, surface tension was a recognized property of liquids. For solids, the concept of surface energy was greatly disputed. Today, we recognize that solid surface energy arises from unsatisfied intermolecular bonds on a surface, so it has units of energy per unit area.

Deformation tests at high temperatures give a means to measure surface energy, so most data are taken near the sintering temperature. Surface energy provides a means for explaining sintering densification [2]. By the early 1940s, solid surface energy was linked to sintering.

In 1942, Wretblad and Wulff [19] calculated that surface energy might be sufficient to induce plastic flow during densification. For the geometry shown in Figure 3.5, the outer neck profile is approximated by the circle of radius R. On this saddle surface, the surface energy γ_{SV} acts to generate a capillary stress σ that pulls the spheres together. There are two curvatures. A convex radius (solid bowed outward) occurs inside the neck, with a curvature perpendicular to the viewing plane of radius of $X/2$. The second curvature occurs in the viewing plane, represented by the concave radius

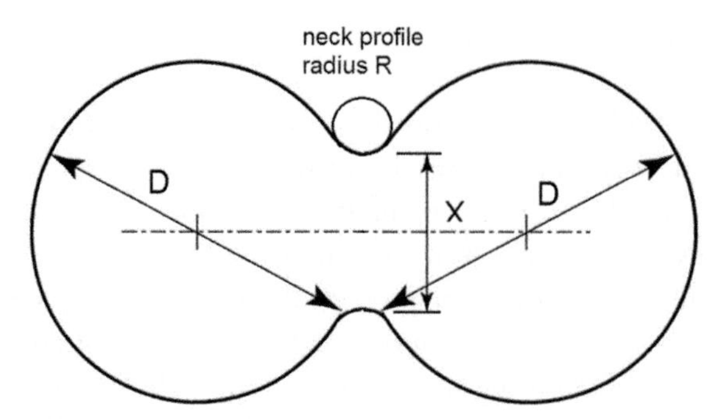

Figure 3.5 The two sphere sintering model with particle diameter D and neck diameter X. The circle approximation to the neck profile is set at a radius R and it is approximately $X^2/4D$. The saddle surface neck has two curvatures, with R located outside as a concave curvature that pulls the spheres together. It is partially offset by a convex circle of radius X/2.

(solid bowed inward) outside the neck with a curvature radius, of opposite sign, equal to $R \approx X^2/4D$. The net capillary stress pulling the two spheres together is estimated as:

$$\sigma = \gamma_{SV}\left[\frac{2}{X} - \frac{4D}{X^2}\right] \tag{3.2}$$

where X is the diameter of the sinter neck and D is the diameter of the particle.

Experimental demonstration of the sintering stress came from Rhines and Coworkers [20,21]. They employed zero creep techniques to stop densification, and showed that the required stress equaled the stress calculated using surface energy, pore size, and porosity.

Subsequently, nonuniform surface energy was demonstrated by particle rotation during sintering, allowing identification of the surface energy cusps associated with preferred crystal misorientation [22].

Two parts of the required infrastructure for sintering theory were established by the early 1950s:

- atomic theory
- surface energy driven mass flow.

Atomic Motion

If surface energy exerts a force on the particles to induce sintering, then how do the atoms move? After the postulation of dislocations in 1934 [23], sintering was assumed to be a case of high temperature plastic flow. Then Kuczynski [9] suggested diffusion, and the thinking switched to diffusion models. Subsequently, sintering concepts oscillated between dislocations and diffusion, before the recognition that both acted simultaneously [24–27]. Experiments by Lenel et al. in 1961 demonstrated that the sintering stress was sufficient to induce plastic flow [28], as validated by subsequent observations [26,29]. Acceptance of plastic flow seemed to prevent acceptance of diffusion and vice versa; early multiple mechanism sintering concepts ignored plastic flow [30–32]. However, conclusive evidence arose to show dislocation motion in the sintering of metals and ionic compounds [33–36]. This led to confusion, until the evidence was rationalized to show cooperative and not exclusive contributions to sintering by the two mechanisms.

Dislocation populations decline over time due to annealing, so neck growth by plastic flow is usually associated with the early portions of sintering. On the other hand, atomic diffusion is dominant after the dislocation population declines.

With respect to diffusion, important gradient laws were developed by Fick in 1855 to explain concentration changes with position and time. His first law is:

$$J = -D\frac{dc}{dx} \tag{3.3}$$

where J is the atomic flux (atoms per area per unit time), D is the diffusion coefficient (area per unit time), dc is the concentration change over distance dx. In a simple sense, this law says that water flows downhill and the steeper the slope the faster it flows. These ideas are used extensively in sintering models.

Ideas about atomic motion in sintering date to 1888, when Boltzmann and Stefan formulated an energy distribution for an ensemble of atoms. Arrhenius used this idea to create a kinetic model based on activation energy [Nobel Prize in 1903].

In 1896, Roberts-Austin demonstrated diffusion in metals, far ahead of atomic theory. His observations were ignored for several years. However, Langmuir [Nobel Prize 1932] worked on describing the surface states of solids, and essentially assumed atomic diffusion. These early conceptualizations were needed to create a quantitative sintering theory.

The critical step came in 1942 [37,38]. Then, Huntington and Seitz accurately predicted the activation energy for diffusion via vacancy migration. These ideas led to the famous vacancy diffusion demonstrations by Kirkendall [39]. Darken modified Fick's law to include thermodynamic chemical potential gradients and allowed for differing diffusion coefficients for each species in binary solutions [40]. Pines [41] conjectured that sintering was a process of pore evaporation. Assuming pores to be collections of atomic vacancies, densification depends on the rate of "evaporation."

Eventually, grain boundary diffusion was accepted as a transport mechanism [42]. It operates much faster than volume diffusion and explained densification in structures with small grain sizes [43]. Wilson and Shewmon [44] extended the idea to include multiple diffusion mechanisms, and determined that grain orientation was a factor in sintering. By 1966, atomic diffusion concepts dominated sintering models, diverting attention from plastic flow models. Later, Schatt et al. [45] brought balance to the treatment with their concept of volume diffusion assisted dislocation climb. This combination of diffusion and plastic flow provides a good explanation for short time sintering behavior. Once the dislocation structure is annihilated, then sintering continues by diffusion. This idea was embraced by Ashby in his sinter maps—which he first presented in the 1970s and fully developed in the 1980s [46–50].

In parallel to atomic diffusion concepts, progress took place using the inverse of diffusion, that being viscosity. Kelvin initially proposed that metals behaved in a viscous manner at higher temperatures. For sintering, surface energy provides the stress, so possibly diffusion provides viscous flow in response to that stress. Today, viscosity ideas are used in computer simulations of sintering, since the implementation of finite element analysis is easier than diffusion solutions. There is a loss of "fundamentals" this way, but predictions are possible for size, shape, and distortion. The models accept that viscosity follows a complex nonlinear behavior which evolves during sintering: Herring conjectured high temperature surface energy induces the viscous flow of crystalline solids [51], an idea embraced by Frenkel in his two sphere sintering model [6]. Such an approach ignores grain boundaries and grain size, so it is simplistic except for amorphous materials.

Mackenzie and Shuttleworth [52] proposed that sintering densification was a viscous flow process. Efforts were made to show that diffusion or plastic flow gives an effective viscosity during sintering. They relied on surface energy as a driving force and connected the sintering stress to strain to explain densification by viscous flow. It was not clear how mass moved, but their idea worked as a phenomenological model. For example, the estimated viscosity for copper powder during sintering was 0.22 GPa · s, about the same as measured in recent experiments [53]. The viscosity model made it possible to realize that an external stress would increase the rate of sintering—a concept now often used in pressure-assisted sintering. However, a yield strength was required to match experimental data, suggesting a Bingham viscosity behavior, such as that observed in plastic flow [54]. One of the difficulties was in predicting negative surface energies for some materials [55]. Thus, by the early 1950s the conceptual linkage between viscosity and densification was established. In this regard, the idea of viscous flow for two particles developed by Frenkel, and that of densification developed by Mackenzie and Shuttleworth proved useful, if not accurate [56].

Microstructure

Buried within sintering theory is a sense of microstructure evolution. Small spherical particles transform to become large polygonal grains. Each grain equals the volume of hundreds of initial particles. Development of a quantitative description of microstructure was a necessary part of sintering theory.

In 1727, Hales described cell structures and the filling of space with polyhedral grains with an average coordination number near 12. Subsequent investigators observed biological cells, etched rocks, and observed crystals or grains. In 1864, Percy published *Metallurgy – Iron and Steel*, and formally started the observation of microstructure. Thirty two years later, Ostwald [Nobel Prize 1909] provided a first model for microstructural coarsening. This introduced the concept of time-dependent grain enlargement. The time-dependent grain size concept was important for understanding the evolution of microstructure in sintering.

Ostwald describes the progressive coarsening during which the larger grains grow at the expense of the smaller grains, leading to a progressive decrease in number of grains, and an increase in characteristic size [57]. It occurs when diffusion takes place through the medium separating the grains. It is convenient to treat the grain boundary region as a separate phase, with a width of five to ten atoms. Coalescence between contacting grains further enhances grain growth, but coalescence involves two grains rotating into a near perfect crystallographic alignment to annihilate the grain boundary.

Ewing and Rosenhain [58] conjectured that solids consisted of crystalline grains in 1900. They showed that additives changed the structure by linking them to grain growth. In recent times the idea of seeded grain growth, termed templated grain growth, has been applied to sintering to form anisotropic structures.

Smith defined the three-dimensional grain structure, giving a quantitative model for grain shape and grain size in polycrystalline solids [59]. This work confirmed the 12 and 14 sided grain shape models from compressed lead sphere structures. The 14-sided tetrakaidecahedron grain shape is illustrated in Figure 3.6. Subsequently, this model was applied to intermediate and final stage sintering models [14]. A transition from intermediate to final stage sintering was based on the Rayleigh instability, where long tubular pores on the grain edges pinch into discrete spherical pores on the grain corners [60]. Grain growth during sintering causes the pores to become unstable, and to pinch into discrete spherical pores. This is the onset of final stage sintering with spherical pores on the grain corners.

At the same time, Rhines reported observations on pore structure evolution during sintering. He and DeHoff collected detailed observations on microstructure trajectories during sintering [61,62]. As an example, Figure 3.7 illustrates the way that surface area declines with densification during the sintering of copper, giving an early and simple means for rationalizing sintering trajectories.

A sintered microstructure consists of a mixture of grain shapes. Aboav and Langdon measured grain shape and grain size distributions for sintered magnesia, including dense and porous pressed variants [63–65], for over 10,000 grain measurements. Up to 1969 the grain structure was not documented in a statistically significant manner. These MgO data enabled subsequent fitting to an exponential (Weibull)

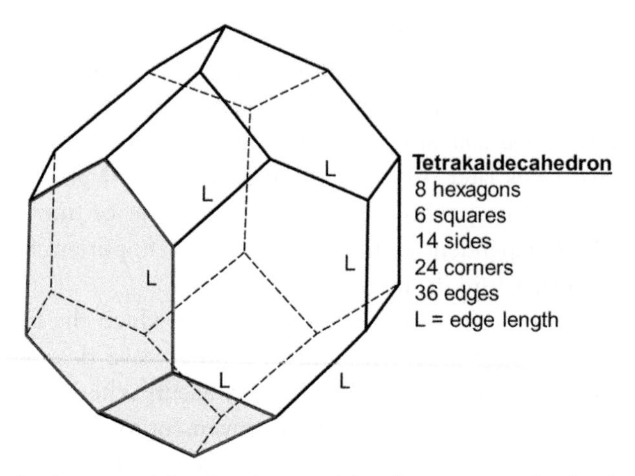

Figure 3.6 The grain shape used for sintering models relies on a tetrakaidecahedron, where each of the edges is of equal length L. It consists of eight hexagons and six squares, giving 14 sides, 24 corners, and 36 edges.

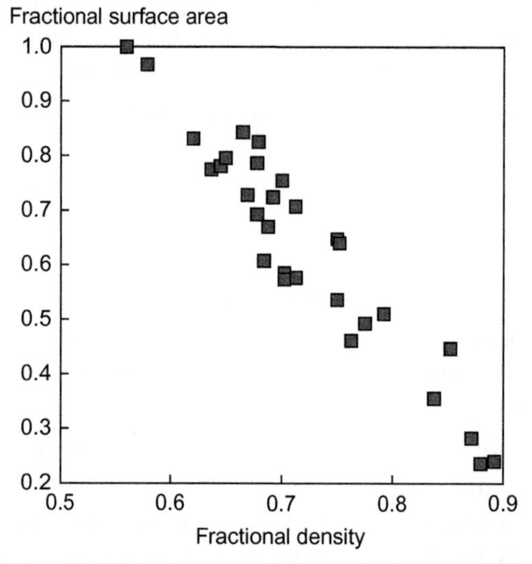

Figure 3.7 A composite of several sintering experiments on copper, plotting the fractional surface area (measured surface area as ratio to initial surface area) versus fractional density [61]. Note the seemingly similar microstructure trajectory.

distribution. Concepts in microstructure coarsening predicted the emergence of self-similar grain size and grain shape distributions [57,66]. Self-similar distributions arise when the shape of the distribution is always the same, even though the median size changes over time. For grain size distributions, the cumulative size distribution is identical if the grain size is normalized to the median grain size. Figure 3.8 plots the cumulative grain size distribution measured in two dimensions (symbols) and Weibull distribution fit (solid line). The plot is for the cumulative grain size distribution, based on grain population versus measured grain size normalized to the median grain size.

Refinements in the grain shape models with the inclusion of a dihedral angle and second phase content came from Beere [67] and Wray [68]. Figure 3.9 is a map of grain corner shape versus a dihedral angle and amount of liquid (effectively the same as the amount of porosity). The map for equilibrium solid grains versus the dihedral angle enables the assessment of possible cell arrangements, where the size and shape of grains are correlated factors [69].

Gas trapped or formed in pores during sintering caused swelling. Markworth [70–73] explained this by balancing the sintering stress, pore pressure, and pore coarsening events during sintering. Pores undergo coarsening, and since gas pressure is reduced by pore enlargement, the material expands. Watanabe and Masuda [74] provided data which showed a seemingly natural trajectory for the pores and grains during sintering, as shown in Figure 3.10 for iron, nickel, and copper. Since the grains coarsen as the porosity declines, the pore-grain interactions are related. Thus, the microstructure evolution during sintering becomes predictable, tracking a similar pathway.

Sintering started with grain shape models that were the same as biological cell structures and isolated relations between parameters such as density, grain size, pore size, and

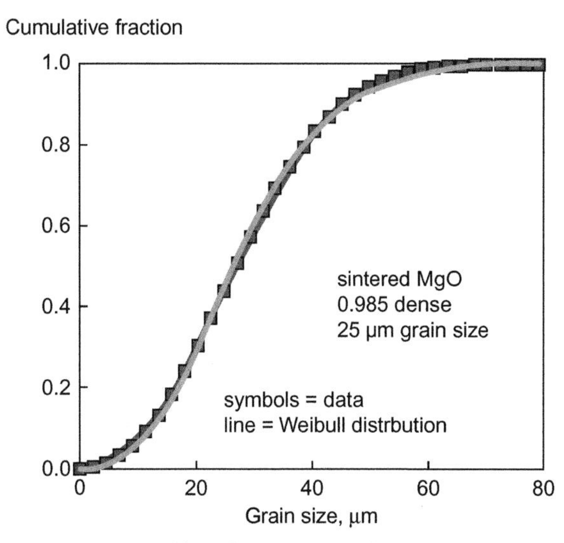

Figure 3.8 Sintering converges to a self-similar grain size distribution as illustrated here by data on sintered magnesia with a 25 μm median grain size [63]. The square symbols are the measured distribution and the solid line represents a Weibull distribution.

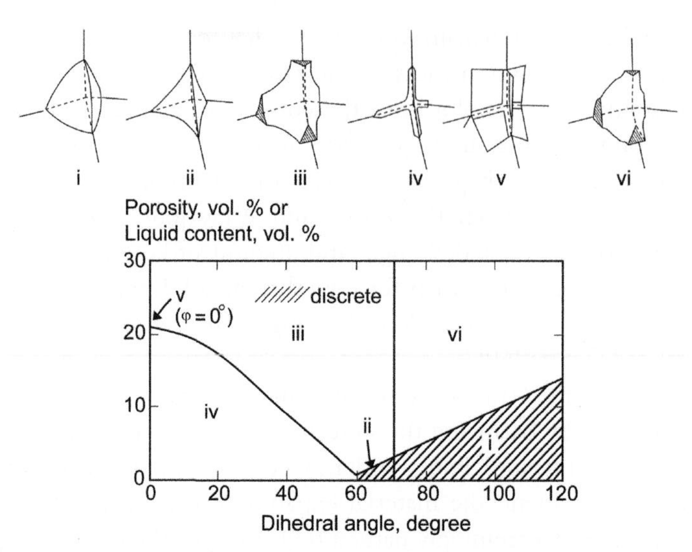

Figure 3.9 Microstructure variants for a second phase (liquid or vapor) and solid grain depend on the dihedral angle and the content of second phase, either porosity or liquid.

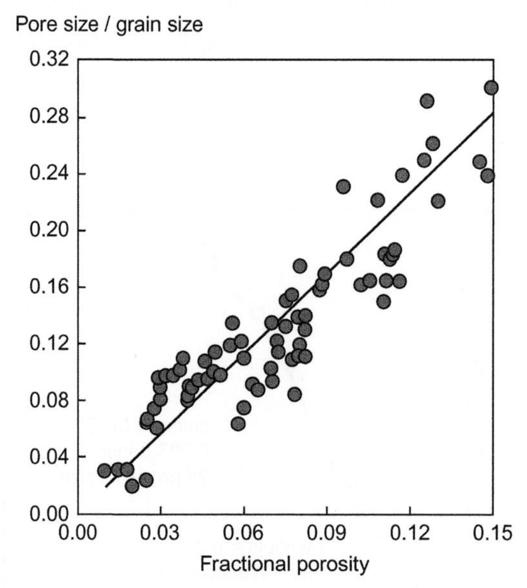

Figure 3.10 Data on the pore size divided by the grain size during sintering densification for Ni, Cu, and Fe [74].

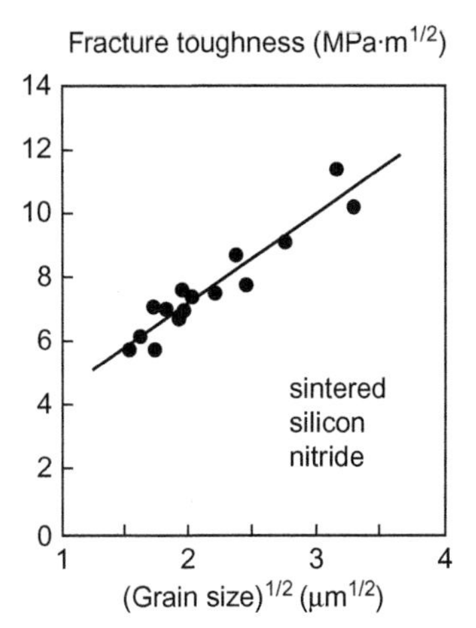

Figure 3.11 Example property linkage to sintered microstructure, in this case the fracture toughness for sintered silicon nitride, is plotted against the square root of the grain size [75].

surface area, to construct models for microstructure evolution. In turn, several studies quantified the microstructure-property links. Figure 3.11 is just one illustration of such a link, in this case showing fracture toughness of sintered silicon nitride versus the square root of the grain size [75]. Such correlations simplified property predictions, since models for microstructure evolution during sintering suggest property variations.

Models emerged to simplify property-microstructure relations, which often sacrificed accuracy. For example the data in Figure 3.12 are taken from a compilation on steel sintering, showing elastic modulus versus fractional density [76]. This gives evidence that other factors beyond density are important in determining the elastic modulus during sintering.

Microstructure relations and an understanding of how microstructure relates to sintering was another ingredient required to build sintering theory. Quantitative microstructure conceptualizations matured by the 1980s, giving the required framework for a predictive sintering theory. Another required part of the infrastructure came with the advent of applicable measurement tools.

EXPERIMENTAL TOOLS

Quantitative measurements enabled a transition from qualitative observations in the 1800s to quantified property measurements in the 1900s. Chapter 4 details the current arsenal of measurement tools, while here a few of the important developments are mentioned.

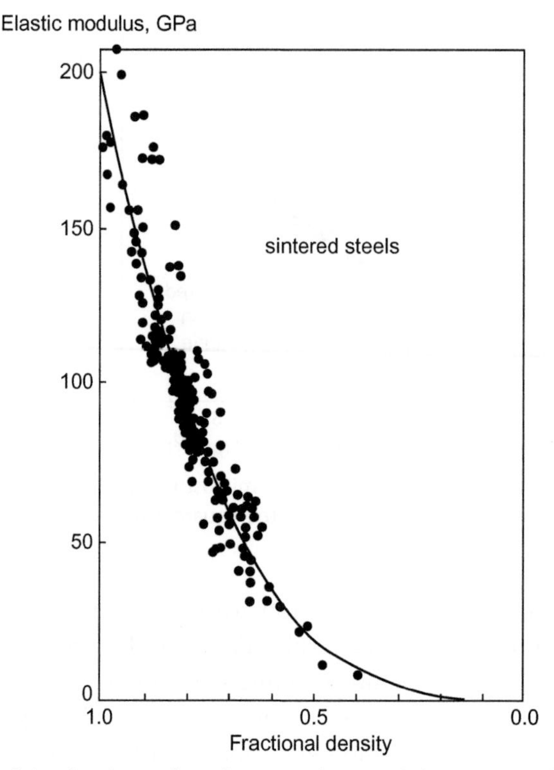

Figure 3.12 Elastic modulus for sintered steels versus fractional density [76]. The solid curve shows a typical relation, but the data are highly scattered reflecting how factors such as pore shape and grain size add to the behavior.

Temperature Measurement

Temperature measurement is fundamental to sintering. The Fahrenheit temperature scale was introduced in 1724 by Daniel Gabriel Fahrenheit (1686–1736) based on an idea from Ole Romer. This scale operated between ice-water-ammonium chloride equilibrium (0°F), ice-water equilibrium (32°F), and body temperature (98°F)—about the coldest and hottest temperatures he observed in a year, divided into 100 units. The Centigrade temperature scale arose by dividing the melting and boiling temperature of water into 100 degrees. The Celsius temperature scale, developed by Anders Celsius (1701–1744), is also defined over the water working range using the triple point for 0°C. Finally, the absolute zero temperature scale was proposed in 1848 by William Thompson, later known as Lord Kelvin. Even so, adoption of a standardized temperature scale did not happen until 1954.

Temperature measurement in the range typical of sintering was delayed until the middle 1800s. The first gas thermometer for high temperature measurement was developed in 1828. About this same time, Seebeck observed the thermoelectric effect in the junction of dissimilar metals. This was the precursor for today's thermocouple. In 1830, Nobill and Melloni combined thermocouples with galvanometers to create a quantitative measurement device. Previous to this development, temperature measurements were subjective. Although the thermocouple is widely applied today, it was not available until the latter portion of the 1800s. Today, we accept that temperature can be defined by standard scales, measured accurately, and these measurements repeated with precision.

Furnaces

The vacuum sintering furnace is a recent development. Vacuum was known in the form of suction pumps in the 1200s, and von Guericke designed the first vacuum pump in 1654. The history of early furnace construction is lost, but charcoal fires were the starting point. As combustible gases became common, flames from methane, acetylene, and hydrogen provided increasingly higher temperatures. Experiments in the 1700s relied on concentrated solar heating, but temperatures were limited. When electric power emerged, high temperature electric discharge heating was developed in parallel. Vacuum furnaces seemed to simply arise as one of several options.

Acheson used an electric arc approach to combine silica and coke to form silicon carbide. In 1896, Moissan [Nobel Prize 1906] developed the electric furnace. He was able to discharge though the powder, generating a wide variety of carbides, borides, nitrides, and silicides, including tungsten carbide. This was the precursor to spark sintering, which was subsequently used by several researchers, including Lux and Coolidge in sintering lamp filaments. This was out of necessity since high temperature furnaces were rare. By 1955, Lenel had applied spark sintering to the consolidation of several powders, including titanium, steel, and zirconium [77]. He reported the benefits of pulsed current, a technique widely used today.

First we needed a temperature scale, then means to measure temperature, and finally a means to generate high temperatures. Other than spark sintering, access to high temperatures was delayed up to the 1940s. Until then electric discharge heating was the most common way to circumvent furnace temperature limits. High vacuum and high temperature combinations relied on the development of the diffusion vacuum pump. A mercury diffusion pump was invented by Gaede in 1915. Silicone oil diffusion pumps arose from the work of Burch in 1928. Finally, Brew started producing high temperature diffusion pumped vacuum furnaces in 1946.

Property Quantification

Another important tool for sintering studies was the development of hardness testing. Mohs developed a scratch test in 1812, useful for minerals, but not useful for steels.

The concept of an indentation test for hardness, relevant to measuring the degree of sintering, was conceived in the early 1900s, and led to several variants. Brinell is credited with establishing the first standardized hardness test. It left a large impression and worked only for softer materials, so various improvements followed.

Although taken for granted now, this was the beginning of the quantification of material properties using tensile, compression, bending, impact, and similar tests. To improve on the Brinell test, new scales emerged. Rockwell hardness (Hugh M. Rockwell and Stanley P. Rockwell) and Vickers microhardness (Smith and Sandland in 1922) followed the Brinell scale, in 1915 and 1924, respectively.

Hardness is probably second to density as a means to track sintering, but was not a quantified parameter up to the early 1900s. About this time, toughness testing arose via the Izod (1903) and Charpy (1904) tests. Now it is routine to use a hardness specification to quantify the degree of sintering and to rely on strength and impact tests during product qualification.

Electrical behavior is inherent to sintered products, such as electrical contacts, light filaments, heating elements, X-ray tubes, and electronic circuits. Goetzel [78] examined the electrical conductivity of copper compacts after sintering in his effort to understand mechanical properties. In 1941, Rhines and Colton [79] went beyond post-sintering tests to use conductivity as a sintering monitor. Several observations are embedded in this 1942 study, including a demonstration of the benefit from smaller particles, an Arrhenius behavior with respect to temperature, and evidence of grain growth. In the early 1950s conductivity data was linked to diffusion rates [80,81].

Surface Area

Surface area as a characterization tool emerged in 1938, based on the BET (Brunauer, Emmett, and Teller) gas absorption concept [82]. Other tools emerged for measuring surface area, but gas absorption is a favorite with smaller powders. An alternative relies on gas permeability to measure surface area prior to pore closure [83]. The flow resistance (measured by the flow rate as a function of the pressure drop according to Poiseuille's law) gives permeability data linked to the open pore surface area. The alternative is to rely on quantitative microscopy, a more tedious approach, but one that is often employed on lower surface area samples [62].

Pressure Generation

Hot pressing was an early approach for the consolidation of hard materials, initially tungsten carbides and related carbides. Early successes arose with hot pressing cemented carbides during the 1920s [84]. Added heat and pressure were effective in densifying powder, and when coupled to spark sintering, this gave a means to consolidate high temperature materials.

Goetzel took up hot pressing in his doctoral research, and became a proponent of the approach in 1942, applying the approach to copper, bronze, mixed copper-tin powders, and iron and brass [84–87]. The approach was then extended to ceramics [88] and carbides [89].

High pressure options resulted in the synthesis of diamond [90] and cubic boron nitride [91], and eventually diamond sintering [92,93]. These latter events emerged from the high pressure research of Bridgman [Nobel Prize 1946].

A lower pressure gas pressure furnace, invented in 1955, was termed hot isostatic pressing. It was applied to perform the pressure cladding of nuclear fuel pellets. By the middle 1960s, hot isostatic pressing (HIP) arose as a means to form fully dense high performance materials including titanium, superalloys, and tool steels. The sinter-HIP approach followed, in which the initial sintering is performed under vacuum, this is followed by pressurization late in the cycle to seal residual pores, and all steps take place in a single furnace.

Newer Tools

We often take the current array of experimental tools for granted. Yet, many significant products were developed without the current infrastructure. In recent years new tools have emerged to increase our ability for monitoring sintering. Although not discussed in detail here, mention is given to some newer tools:

advanced synchrotron radiation sources [94]
combined microwave and plasma heating [95]
computer tomography [96,97]
gas chromatography of evolved decomposition products [98]
gas discharge plasma heating [99]
hot stage electron microscopy [100]
infrared spectroscopy of process gas [101]
in situ eddy current tests for density [102]
in situ strength testing [103]
in situ video imaging [104]
induction heating [105]
mass spectroscopy of sintering atmospheres [106]
microwave heating [107]
molecular dynamic simulations [108]
surface chemistry analysis [109]
thermal or electrical conductivity testing [110]
thermogravimetric analysis [111].

The success of sintering theory is measured by its ability to predict component size, properties, microstructure, and cost. As concepts emerged, a first focus was the prediction of neck size versus time, temperature, and particle size. Then, measurements matured to include density, shrinkage, surface area, and grain size. Later, property measurements and their evolution in sintering emerged using hardness, elastic modulus, strength, ductility, toughness, and conductivity tests. In later years new computational tools linked processing, microstructure, composition, and properties.

ORGANIZATIONAL ADVANCES

The impact of sintering is evident in a wide array of commercial products. Organizations formed to support technical developments. These professional and trade organizations provide forums for exchanging ideas, benchmarking technology, and sounding out scientific principles.

The International Institute for the Science of Sintering, headed by Momcilo Ristic in Beograd, Serbia, is the only organization focused solely on sintering; it arranges conferences and has published *Science of Sintering* (previously known as *Physics of Sintering*) since 1969. Membership of the International Team is decided by election, and is generally limited to about 100 individuals.

The ceramic societies (American, Japanese, and European) publish research on sintering in their journals, offering about 100 papers per year dealing with sintering and sintered materials.

The powder metallurgy community consists of the same three main geographic regions (America, Japan, and Europe, but with rapid growth in China) and each has a journal that reports on sintered products. The Metal Powder Industries Federation has a long-running journal that formerly carried theoretical papers, but now it is focused on market development for sintered steels.

INTEGRATION

Accurate predictions of sintering behavior require detailed information on the powder, chemistry, green body, and thermal cycle. Small errors in the material properties, especially activation energies, result in large variations in the predicted behavior. As the accuracy of sintering theory improves, our ability to simulate important attributes has likewise improved. A good demonstration is the simulation of neck growth for copper [112], as illustrated by the comparison in Figure 3.13, using experimental data for 127 μm copper spheres at 1020°C (1293 K) for upwards of 300 hours hold. The simulation is accurate when volume and surface diffusion are combined.

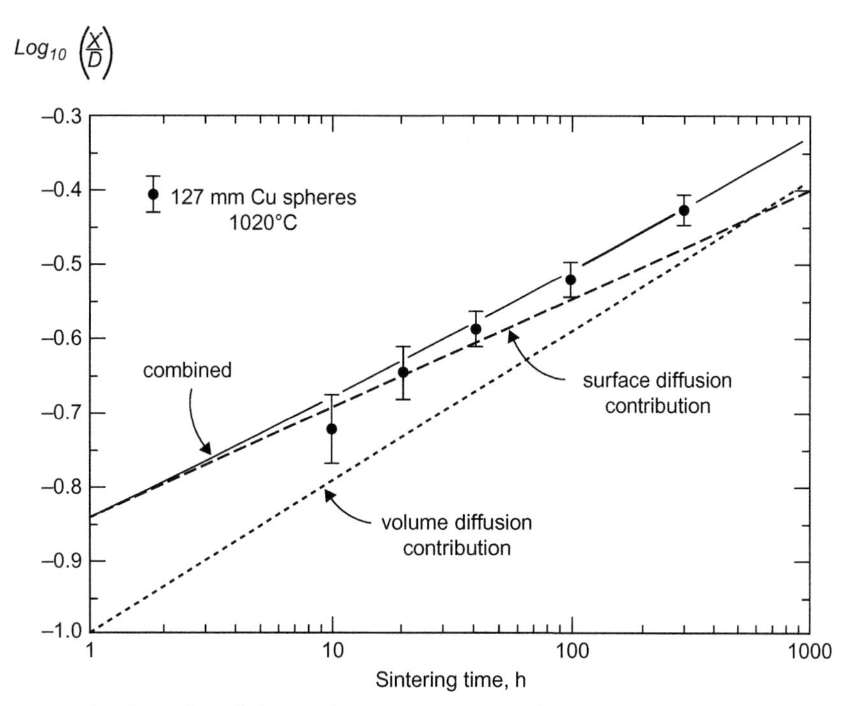

Figure 3.13 A log-log plot of the neck size ratio X/D and sintering time for sintering copper spheres at 1020°C (1293 K) [10]. Three sintering model fits are shown, corresponding to pure volume diffusion, pure surface diffusion, and combined volume and surface diffusion, providing evidence of multiple mechanism sintering [112].

Other studies have demonstrated an ability to predict neck size versus the processing parameters of isothermal hold time, peak temperature, and particle size. More recent efforts moved to density and grain size predictions with the addition of applied pressure, wide particle size distributions, and green density as input parameters [47,48,50,113].

Sintering maps subsequently expanded to cover hot isostatic pressing, giving sintered density versus processing parameters as illustrated in Figure 3.14. This map is for 6 μm tungsten powder starting at 62% density, showing the densification stages, controlling mechanisms, and the density versus hold temperature for hold times of one and ten hours. Subsequent efforts have added nonspherical particles and heating events [114]. These maps are most useful in making trade-off decisions between adjustable factors.

As a newer tool, molecular dynamic models have validated the two sphere sintering models without the need for extensive material data. Attention is focused on atomic level events, treating each atom as a vibrating species. Supercomputers let large

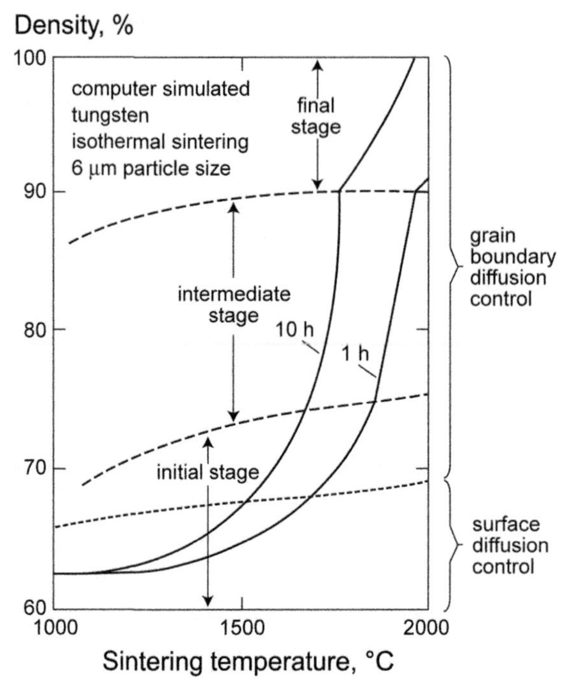

Figure 3.14 A sintering map constructed by computer simulation for 6 μm tungsten powder, plotting the sintered density versus hold temperature for hold times of 1 and 10 h. The stages of sintering and dominant mechanisms are marked. Full density would require about 10 h hold at 1960°C (2233 K) for this powder.

ensembles of atoms interact. Since only an interatomic potential energy curve is required, adjustments are only made to match bulk known properties; the elastic modulus, melting temperature, and thermal expansion coefficient. As illustrated in Figure 3.15, the resulting atomic-level sintering simulations are reminiscent of early experiments. Each sphere represents 50,000 atoms. For short time intervals, the atomic displacement vectors are shown on the spheres. Most of the neck growth results from near-surface atomic motion with some grain boundary diffusion.

For the component fabricator, such simulations are of limited appeal. Instead, these workers need three-dimensional predictions of component size, shape, and properties. This requires additional data on the forming process, since gradients introduced in the green body carry over into the sintered component size. In turn, models exist to correlate processing conditions, raw material, and component features to estimate cost [115–117]. The challenge is to apply the simulations to optimization objectives, such as determination of the time-temperature-heating path that delivers target density, strength, and minimized distortion at the lowest cost [118,119].

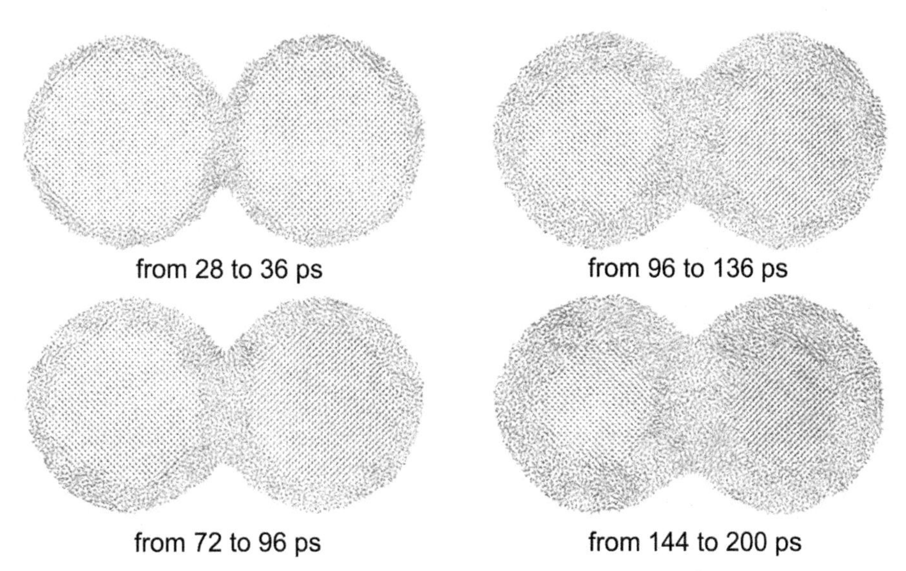

Figure 3.15 Atomic-level sintering simulations for particles consisting of 50,000 atoms, showing neck growth over time. Highlighted on each particle pair are the atomic motion vectors for a short time windows, indicating most of the atomic motion is near the surface or along the grain boundary, in agreement with the findings from standard sintering theory.

STATUS OF SINTERING THEORY

Many individuals have worked to transform observations into a quantitative sintering theory. With any theory "the proof is in the pudding." Sintering theory must explain observations, and in that regard much progress has occurred.

Ashby made an important contribution via his computer models. Subsequent simulations added additional features, such as nonisothermal cycles [113–120]. The ability to accurately model nonisothermal sintering of nonspherical particles is evident in Figure 3.16. This is a plot of sintering shrinkage versus sintering time for flake tantalum powder, 1.9 μm in diameter and 0.4 μm thick, during 10°C/min heating to a 1350°C (1623 K). Good agreement between the model and experiment is oserved. Such findings illustrate how sintering theory has advanced.

A broad infrastructure was needed to create sintering theory. Because of the complexity and interlinked parameters, computer models are employed today. Much had to be determined to make those simulations effective.

To summarize, consider the sintered component shown in Figure 3.17. It is fabricated by injection molding a 16 μm stainless steel powder into a mold that is oversized by about 16% in each dimension. After vacuum sintering for two hours at 1325°C (1598 K),

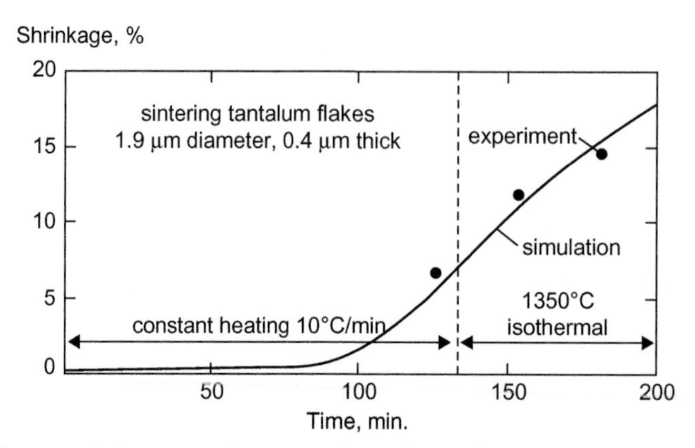

Figure 3.16 Real system behavior involves nonspherical particles and nonisothermal sintering. This study illustrates the maturation of sintering theory as applied to the behavior of tantalum flakes measured (symbols) and simulated (line) for a cycle of heating at 10°C/min to a hold temperature of 1350°C (1623 K) [120].

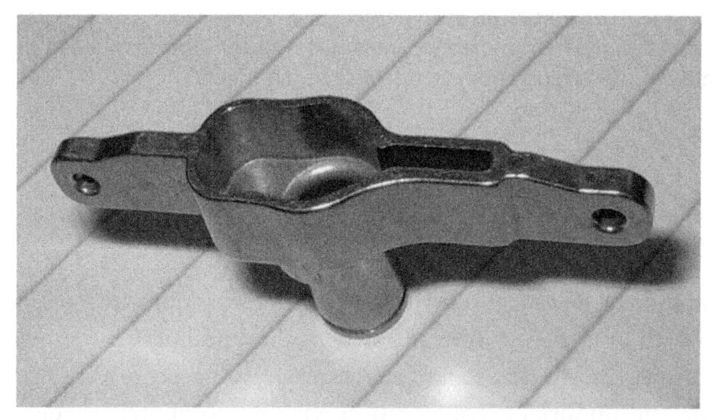

Figure 3.17 Photograph of an injection molded and sintered stainless steel cellular telephone swivel hinge. The ability to predict final size and properties is an important application of sintering computer simulations.

the device is dimensionally within specification. The ability to attain desired properties and dimensions on the first sintering cycle are major targets for sintering theory.

REFERENCES

[1] W.D. Jones, Principles of Powder Metallurgy with an Account of Industrial Practice, Edward Arnold, London, UK, 1937.

[2] W.D. Jones, Fundamental Principles of Powder Metallurgy, Edward Arnold Publishers, London, UK, 1960.

[3] F.N. Rhines, Seminar on the theory of sintering, Trans. TMS-AIME 166 (1946) 474–491.

[4] F.V. Lenel, Sintering in the presence of a liquid phase, Trans. TMS-AIME 175 (1948) 878–896.
[5] C.G. Goetzel, Treatise on Powder Metallurgy, vol. 1, Interscience Publishers, New York, NY, 1949, pp. 259–312.
[6] J. Frenkel, Viscous flow of crystalline bodies under the action of surface tension, J. Phys. 9 (1945) 385–391.
[7] M.M. Ristic, Sintering – past and present (On the 40th Anniversary of the Belgrade School of Sintering), Sci. Sintering 33 (2001) 143–147.
[8] V.V. Skorokhod, Development of the Ideas of Ya. I. Frenkel in the contemporary rheological theory of sintering, Powder Metall. Metal Ceram 34 (1995) 521–527.
[9] G.C. Kuczynski, Self-Diffusion in sintering of metallic particles, Trans. TMS-AIME 185 (1949) 169–178.
[10] W.D. Kingery, M. Berg, Study of the initial stages of sintering solids by viscous flow, evaporation-condensation, and self-diffusion, J. Appl. Phys. 26 (1955) 1205–1212.
[11] W.D. Kingery, Densification during sintering in the presence of a liquid Phase 1. Theory, J. Appl. Phys. 30 (1959) 301–306.
[12] W.D. Kingery, M.D. Narasimhan, Densification during sintering in the presence of a liquid Phase 2. Experimental, J. Appl. Phys. 30 (1959) 307–310.
[13] E. Pink, L. Bartha, The Metallurgy of Doped/Non-Sag Tungsten, Elsevier Applied Science, London, UK, 1989.
[14] R.L. Coble, Sintering crystalline solids. 1. Intermediate and final state diffusion models, J. Appl. Phys. 32 (1961) 787–792.
[15] R.L. Coble, Sintering crystalline solids. 2. Experimental test of diffusion models in powder compacts, J. Appl. Phys. 32 (1961) 793–799.
[16] R.L. Coble, A model for boundary diffusion controlled creep in polycrystalline materials, J. Appl. Phys. 34 (1963) 1679–1682.
[17] C.S. Smith, Further notes on the shape of metal grains: space-filling polyhedra with unlimited sharing of corners and faces, Acta Metall. 1 (1953) 295–300.
[18] A. Einstein, Investigations on the Theory of the Brownian Movement, R. Fuerth (Ed.), A.D. Cowper (translator), Dover Publications, New York, NY, 1956.
[19] P.E. Wretblad, J. Wulff, Sintering, in: J. Wulff (Ed.), Powder Metallurgy, American Society for Metals, Cleveland, OH, 1942, pp. 36–59.
[20] F.N. Rhines, H.S. Cannon, Rate of sintering of copper under a dead load, Trans. TMS-AIME 191 (1951) 529–530.
[21] R.A. Gregg, F.N. Rhines, Surface tension and the sintering force in copper, Metall. Trans. 4 (1973) 1365–1374.
[22] G. Herrmann, H. Gleiter, G. Baro, Investigation of low energy grain boundaries in metals by a sintering technique, Acta Metall. 24 (1976) 353–359.
[23] J.P. Hirth, A. Brief, History on dislocation theory, Metall. Trans. 16A (1985) 2085–2090.
[24] E. Friedrich, W. Schatt, Sintering of one-component model systems: nucleation and movement of dislocations in necks, Powder Metall. 23 (1980) 193–197.
[25] W. Schatt, E. Friedrich, Self-Activation of sintering processes in one-component systems, Powder Metall. Inter. 13 (1981) 15–20.
[26] W. Schatt, E. Friedrich, Sintering as a result of defect structure, Cryst. Res. Tech. 17 (1982) 1061–1070.
[27] E. Friedrich, W. Schatt, High temperature plasticity on solid phase sintering, Sci. Sintering 15 (1983) 63–71.
[28] F.V. Lenel, H.H. Hausner, E. Hayashi, G.S. Ansell, Some observations on the shrinkage behavior of copper compacts and of loose powder aggregates, Powder Metall. 4 (1961) 25–36.
[29] C.S. Morgan, Observation of dislocations in high temperature sintering, High Temp. – High Press 3 (1971) 317–324.
[30] D.L. Johnson, New method of obtaining volume, grain-boundary, and surface diffusion coefficients from sintering data, J. Appl. Phys. 40 (1969) 192–200.
[31] D.L. Johnson, T.M. Clarke, Grain boundary and volume diffusion in the sintering of silver, Acta Metall. 12 (1964) 1173–1179.

[32] L.L. Seigle, Atom movements during solid state sintering, Prog. Powder Metall. 20 (1964) 221–238.
[33] F.V. Lenel, G.S. Ansell, R.C. Morris, Theoretical considerations and experimental evidence for material transport by plastic flow during sintering, in: H.H. Hausner (Ed.), Modern Developments in Powder Metallurgy, vol. 4, Plenum Press, New York, NY, 1971, pp. 199–220.
[34] F.V. Lenel, G.S. Ansell, R.C. Morris, A bubble raft model to study sintering by plastic flow, in: J.S. Hirschhorn, K.H. Roll (Eds.), Advanced Experimental Techniques in Powder Metallurgy, Plenum Press, New York, NY, 1970, pp. 61–80.
[35] J.G. Early, F.V. Lenel, G.S. Ansell, The material transport mechanism during sintering of copper-powder compacts at high temperatures, Trans. TMS-AIME 230 (1964) 1641–1650.
[36] A.R. Hingorany, F.V. Lenel, G.S. Ansell, The role of plastic flow by dislocation motion in the sintering of calcium fluoride, in: T.J. Gray, V.D. Frechette (Eds.), Kinetics of Reactions in Ionic Systems, Plenum Press, New York, NY, 1969, pp. 375–390.
[37] H.B. Huntington, F. Seitz, Mechanism for self-diffusion in metallic copper, Phys. Rev. 61 (1942) 315–325.
[38] H.B. Huntington, Self-consistent treatment of the vacancy mechanism for metallic diffusion, Phys. Rev. 61 (1942) 325–338.
[39] A.D. Smigelskas, E.O. Kirkendall, Zinc diffusion in alpha brass, Trans. TMS-AIME, Met. Tech. XIII (1946), Technical Paper 2071.
[40] R.F. Mehl, The historical development of physical metallurgy, in: R.W. Cahn (Ed.), Physical Metallurgy, North Holland Publishing, Amsterdam Netherlands, 1965, pp. 1–31.
[41] B.Y. Pines, On sintering (In Solid Phase), Z. Tekh. Fiziki 16 (1946) 737–743.
[42] H. Meher, N.A. Stolwijk, Heroes and highlights in the history of diffusion, Diff. Fund. 11 (2009) 1–32.
[43] R.L. Coble, Initial sintering of alumina and hematite, J. Amer. Ceram. Soc. 41 (1958) 55–62.
[44] T.L. Wilson, P.G. Shewmon, The role of interfacial diffusion in the sintering of copper, Trans. TMS-AIME 236 (1966) 48–58.
[45] W. Schatt, H.E. Exner, E. Friedrich, G. Petzow, Versetzungsaktivierte Schwindungsvorgange Beim Einkomponenten-Sintern, Acta Metall. 30 (1982) 1367–1375.
[46] M.F. Ashby, A. First, Report on sintering diagrams, Acta Metall. 22 (1974) 275–289.
[47] F.B. Swinkels, M.F. Ashby, A second report on sintering diagrams, Acta Metall. 29 (1981) 259–281.
[48] D.S. Wilkinson, M.F. Ashby, The development of pressure sintering maps, in: G.C. Kuczynski (Ed.), Sintering and Catalysis, Plenum Press, New York, NY, 1975, pp. 473–492.
[49] A.S. Helle, K.E. Easterling, M.F. Ashby, Hot isostatic pressing diagrams new developments, Acta Metall. 33 (1985) 2163–2174.
[50] E. Arzt, M.F. Ashby, K.E. Easterling, Practical applications of hot-isostatic pressing diagrams: four case studies, Metall. Trans. 14A (1983) 211–221.
[51] C. Herring, Diffusional viscosity of a polycrystalline solid, J. Appl. Phys. 21 (1950) 437–445.
[52] J.K. Mackenzie, R. Shuttleworth, A phenomenological theory of sintering, Proc. Phys. Soc. 62 (1949) 833–852.
[53] R.M. German, Rheological model for viscous flow densification during supersolidus liquid phase sintering, Sci. Sintering 38 (2006) 27–40.
[54] E.B. Allison, P. Murray, A fundamental investigation of the mechanism of sintering, Acta Metall. 2 (1954) 487–512.
[55] R. Shuttleworth, The surface tension of solids, Proc. Phys. Soc. A63 (1950) 444–457.
[56] F.N. Rhines, C.E. Birchenall, L.A. Hughes, Behavior of pores during the sintering of copper compacts, Trans. TMS-AIME 188 (1950) 378–388.
[57] P.W. Voorhees, Ostwald ripening of two-phase mixtures, Ann. Rev. Mater. Sci. 22 (1992) 197–215.
[58] J.A. Ewing, W. Rosenhain, The crystalline structure of metals, Proc. Royal Soc. London 67 (1900) 112–117.
[59] C.S. Smith, Grains, phases, and interfaces: an interpretation of microstructure, Trans. TMS-AIME 175 (1948) 15–51.

[60] Lord Rayleigh: on the capillary phenomena of jets, Proc. Royal Soc. London 29 (1879) 71–97.
[61] F.N. Rhines, R.T. DeHoff, A Topological Approach to the Study of Sintering, in: H.H. Hausner (Ed.), Modern Developments in Powder Metallurgy, vol. 4, Plenum Press, New York, NY, 1971, pp. 173–188.
[62] F.N. Rhines, R.T. DeHoff, R.A. Rummel, Rate of densification in the sintering of uncompacted metal powders, in: W.A. Knepper (Ed.), Agglomeration, Interscience, New York, NY, 1962, pp. 351–369.
[63] D.A. Aboav, T.G. Langdon, The shape of grains in a polycrystal, Metallog 2 (1969) 171–178.
[64] D.A. Aboav, T.G. Langdon, The distribution of grain diameters in polycrystalline magnesium oxide, Metallog 1 (1969) 333–340.
[65] D.A. Aboav, T.G. Langdon, The planar distribution of grain size in a polycrystalline ceramic, Metallog 6 (1973) 9–15.
[66] E.N.D.C. Andrade, D.A. Aboav, Grain growth in metals of close-packed hexagonal structure, Proc. Royal Soc. London A291 (1966) 18–40.
[67] W. Beere, A unifying theory of the stability of penetrating liquid phases and sintering pores, Acta Metall. 23 (1975) 131–138.
[68] P.J. Wray, The geometry of two-phase aggregates in which the shape of the second phase is determined by its dihedral angle, Acta Metall. 24 (1976) 125–135.
[69] D.A. Aboav, The arrangement of cells in a net, Metallog 13 (1980) 43–58.
[70] A.J. Markworth, Growth of inert-gas bubbles in solids; behavior of non-uniform size distributions, J. Mater. Sci. 7 (1972) 1225–1228.
[71] A.J. Markworth, On the volume diffusion controlled final stage densification of a porous solid, Scripta Metall. 6 (1972) 957–960.
[72] A.J. Markworth, On the coarsening of gas-filled pores in solids, Metall. Trans. 4 (1973) 2651–2656.
[73] A.J. Markworth, Comments on foam stability, ostwald ripening, and grain growth, J. Colloid Interface Sci. 107 (1985) 569–571.
[74] R. Watanabe, Y. Masuda, Pinning effect of residual pores on the grain growth in porous sintered metals, J. Japan Soc. Powder Powder Metall. 29 (1982) 151–153.
[75] T. Kawashima, H. Okamoto, H. Yamamoto, A. Kitamura, Grain size dependence of the fracture toughness of silicon nitride ceramics, J. Ceram. Soc. Jpn. 99 (1991) 320–323.
[76] R. Haynes, The Mechanical Behavior of Sintered Metals, Freund Publishing House, London, UK, 1981.
[77] F.V. Lenel, Resistance sintering under pressure, Trans. TMS-AIME 203 (1955) 158–167.
[78] C.G. Goetzel, Structure and properties of copper powder compacts, J. Inst. Met. 66 (1940) 319–329.
[79] F.N. Rhines, R.A. Colton, Homogenization of copper-nickel powder alloys, in: J. Wulff (Ed.), Powder Metallurgy, American Society for Metals, Cleveland, OH, 1942, pp. 67–86.
[80] J.H. Dedrick, G.C. Kuczynski, Electrical conductivity method for measuring self-diffusion of metals, J. Appl. Phys. 21 (1950) 1224–1225.
[81] H.H. Hausner, J.H. Dedrick, Electrical properties as indicators of the degree of sintering, in: W.E. Kingston (Ed.), The Physics of Powder Metallurgy, McGraw-Hill, New York, NY, 1951, pp. 320–343.
[82] S. Brunauer, P.H. Emmett, E. Teller, Adsorption of gases in multimolecular layers, J. Amer. Chem. Soc. 60 (1938) 309–319.
[83] S. Prochazka, R.L. Coble, Surface diffusion in the initial sintering of alumina, Part 1. Model considerations, Phys. Sintering 2 (1970) 1–18.
[84] C.G. Goetzel, Hot pressed and sintered copper powder compacts, in: J. Wulff (Ed.), Powder Metallurgy, American Society for Metals, Cleveland, OH, 1942, pp. 340–351.
[85] C.G. Goetzel, Plastic deformation in powder metallurgy, in: J. Wulff (Ed.), Powder Metallurgy, American Society for Metals, Cleveland, OH, 1942, pp. 87–108.
[86] C.G. Goetzel, Some properties of sintered and hot pressed copper-tin powder compacts, Trans. TMS-AIME 161 (1944) 580–595.

[87] C.G. Goetzel, Principles and present status of hot pressing, in: W.E. Kingston (Ed.), The Physics of Powder Metallurgy, McGraw-Hill, New York, NY, 1951, pp. 256–277.
[88] G.E. Comstock, Molded Alumina, U. S. Patent 2,618,567, issued 18 November 1952.
[89] H.J. Hamjian, W.G. Lidman, Densification and kinetics of grain growth during the sintering of chromium carbide, Trans. TMS-AIME 197 (1953) 696–699.
[90] F.P. Bundy, H.T. Hall, H.M. Strong, R.H. Wentorf, Man made diamonds, Nature 176 (1955) 51–55. Errata 1993, vol. 365, p. 19.
[91] R.H. Wentorf, Cubic form of boron nitride, J. Chem. Phys. 26 (1957) 956.
[92] H. Katzman, W.F. Libby, Sintered diamond compacts with a cobalt binder, Science 172 (1971) 1132–1134.
[93] R.H. Wentorf, R.C. Devries, F.P. Bundy, Sintered superhard materials, Science 208 (1980) 873–880.
[94] P.W. Voorhees, R.J. Schaefer, In Situ observation of particle motion and diffusion interactions during coarsening, Acta Metall. 35 (1987) 327–339.
[95] M.P. Sweeney, D.L. Johnson, Microwave plasma sintering of alumina, Ceram. Trans. 21 (1991) 365–372.
[96] M. Nothe, K. Pischang, P. Ponizil, B. Kieback, J. Ohser, Study of particle rearrangement during sintering process by microfocus computer tomograph (micro-CT), Proceedings PM2004 Powder Metallurgy World Congress, vol. 2, European Powder Metallurgy Association, Shrewsbury, UK, 2004, pp. 221–226.
[97] P. Lu., J.L. Lannutti, P. Klobes, K. Meyer, X-ray computed tomograph and mercury porosimetry for evaluation of density evolution and porosity distribution, J. Amer. Ceram. Soc. 83 (2000) 518–522.
[98] S. Igarashi, M. Achikita, S. Matsuda: evolution of gases and sintering behavior in carbonyl iron powder for metal injection molding, P.H. Booker, J. Gaspervich, R.M. German (Eds.), Powder Injection Molding Symposium 1992, Metal Powder Industries Federation, Princeton, NJ, 1992, pp. 393–407.
[99] C.E.G. Bennett, N.A. McKinnon, L.S. Williams, Sintering in gas discharge, Nature 217 (1968) 1287.
[100] L. Froschauer, R.M. Fulrath, Direct observation of liquid-phase sintering in the system iron-copper, J. Mater. Sci. 10 (1975) 2146–2155.
[101] J.Y. Ying, J.B. Benzinger, Structural characterization of silica during sintering, Nanostr. Mater. 1 (1992) 149–154.
[102] H.N.G. Wadley, R.J. Schaefer, A.H. Kanh, M.F. Ashby, R.B. Clough, Y. Geffen, et al., Sensing and modeling of the hot isostatic pressing of copper powder, Acta Metall. Mater. 39 (1991) 979–986.
[103] G.A. Shoales, R.M. German, Combined effects of time and temperature on strength evolution using integral work-of-sintering concepts, Metall. Mater. Trans. 30A (1999) 465–470.
[104] D.C. Blaine, R. Bollina, S.J. Park, R.M. German, Critical use of video imaging to rationalize computer sintering simulations, Comput. Indust. 56 (2005) 867–875.
[105] H.C. Kim, I.J. Shon, Z.A. Munir, Rapid sintering of ultrafine WC-10 wt.% Co by high frequency induction heating, J. Mater. Sci. 40 (2005) 2849–2854.
[106] P.B. Linkson, Experimental sintering of iron pyrites, Ind. Eng. Chem. Proc. Design Dev. 9 (3) (1970) 379–385.
[107] M.A. Janney, H.D. Kimrey, Microwave sintering of alumina at 28 GHz, in: G.L. Messing, E.R. Fuller, H. Hausner (Eds.), Ceramic Transactions, vol. 1, American Ceramic Society, Westerville, OH, 1987, pp. 919–924.
[108] H. Zhu, R.S. Averback, Sintering processes of two nanoparticles, a study by molecular dynamics simulations, Phil. Mag. Lett. 73 (1996) 27–33.
[109] K.S. Hwang, H.S. Huang, Identification of the segregation layer and its effects on the activated sintering and ductility of Ni-Doped molybdenum, Acta Mater. 51 (2003) 3915–3926.
[110] J.L. Johnson, S.J. Park, Y.S. Kwon, Experimental and theoretical analysis of the factors affecting the thermal conductivity of W-Cu, in: P. Rodhammer (Ed.), Proceedings of the Seventeenth Plansee Seminar, vol. 1, Plansee Group, Reutte, Austria, 2009, p. RM02.

[111] G. Leitner, Application of Thermal Analysis to Material Science Case-study on hardmetals, J. Therm. Anal. Calorimetry. 56 (1999) 455–465.
[112] K.S. Hwang, R.M. German, Analysis of initial stage sintering by computer simulation, in: G.C. Kuczynski, A.E. Miller, G.A. Sargent (Eds.), Sintering and Heterogeneous Catalysis, Plenum Press, New York, NY, 1984, pp. 35–47.
[113] D.S. Wilkinson, M.F. Ashby, Mechanism mapping of sintering under an applied pressure, Sci. Sintering 10 (1978) 67–76.
[114] B.K. Lograsso, D.A. Koss, Densification of titanium powder during hot isostatic pressing, Metall. Trans. 19A (1988) 1767–1773.
[115] B.C. Mutsuddy, Manufacturing cost of injection molded Si_3N_4 prechamber combustion insert, Interceram 36 (5) (1987) 50–53.
[116] R.M. Bhatkal, T. Hannibal, The technical cost modeling of near net shape P/M manufacturing, J. Met. 51 (7) (1999) 26–27.
[117] R.M. German, Metal Injection Molding: A Comprehensive MIM Design Guide, Metal Powder Industries Federation, Princeton, NJ, 2011.
[118] S. Ahn, S.T. Chung, S.J. Park, R.M. German, Modeling and simulation of metal powder injection molding, in: D. Furrer, L. Semiatin (Eds.), Metals Process Simulation, ASM Handbook Volume 22B, ASM International, Materials Park, Oh, 2010, pp. 343–357.
[119] S.H. Chung, Y.S. Kwon, S.J. Park, R.M. German, Modeling and simulation of press and sinter powder metallurgy, in: D.U. Furrer, S.L. Semiatin (Eds.), Metals Process Simulation, ASM Handbook Volume 33B, ASM International, Materials Park, OH, 2010, pp. 323–334.
[120] S.G. Dubois, R. Ganesan, R.M. German, Sintering of high surface area tantalum powder, in: E. Chen, A. Crowson, E. Lavernia, W. Ebihara, P. Kumar (Eds.), Tantalum, The Minerals, Metals and Materials Society, Warrendale, PA, 1996, pp. 319–323.

[111] D. Lambert, Applications of Thermal Analysis for Material Science: Case study on [illegible] metals, J. Therm. Anal. Calorimetry 56 (1999) 455–465.

[112] K.S. Hwang, R.M. German, Analysis of initial stage sintering by computer simulation, in: G.C. Kuczynski, A.E. Miller, G.A. Sargent (Eds.), Sintering and Heterogeneous Catalysis, Plenum Press, New York, NY, 1984, pp. 35–47.

[113] D.S. Wilkinson, M.F. Ashby, Mechanism mapping of sintering under an applied pressure, Sci. Sintering 10 (1977) 67–76.

[114] H.K. Lorenzo, D.A. Koss, Densification of titanium powder during hot isostatic pressing, Metall. Trans. A 19A (1988) 1765–1773.

[115] D.C. Mosseddy, Manufacturing cost of injection molded Si_3N_4 production of combustion insert, Interceramic (5) (1987) 56–58.

[116] B. M. Zibabek, C. Hanmod, The technical cost modeling of near net shape PIM manufacturing, J. Met. 49 (1997) 26–33.

[117] R.M. German, Metal Injection Molding: A Comprehensive MIM Design Guide, Metal Powder Industries Federation, Princeton, NJ, 2011.

[118] S. Ahn, S.T. Chung, S.J. Park, R.M. German, Modeling and simulation of metal powder injection molding, in: D. Furrer, L. Semiatin (Eds.), Metals Process Simulation, ASM Handbook Volume 22B, ASM International, Materials Park, OH, 2010, pp. 343–354.

[119] S.H. Chung, Y.S. Kwon, S.J. Park, R.M. German, Modeling and simulation of press and sinter powder metallurgy, in: D.U. Furrer, S.L. Semiatin (Eds.), Metals Process Simulation, ASM Handbook Volume 22B, ASM International, Materials Park, OH, 2010, pp. 323–334.

[120] A.L. Dubois, R. Kahloun, R.M. German, Sintering of high surface area tungsten powder, in: L. Chen, V. Ginzburg, J. Lawrence, W. Bhat, B. Kumar (Eds.), Tungsten, Hard Materials, Metals and Materials Society, Warrendale, PA, 1996, pp. 317–323.

CHAPTER FOUR

Measurement Tools and Experimental Observations

CHANGES DURING SINTERING

Sintered products are durable and strong, but the green body prior to sintering is relatively weak. An increase in strength is one of the notable changes which occurs due to sintering. At the same time, the component often undergoes a dimensional change, usually shrinkage. These are bulk property changes, but the science that explains the changes requires focus at the particle level. Fundamentally, sinter necks between contacting particles provide the greatest insight in to the process. These bonds are evident in high magnification images, as illustrated in Figure 4.1, a scanning electron micrograph showing neck growth between sintered nickel particles. Frenkel [1] introduced neck size in his viscous flow sintering model, and since then the concept of sintering is analogous with neck growth.

The strength improvement due to sintering is evident in Figure 4.2. This plot shows data for bronze powder during heating to progressively higher temperatures, for

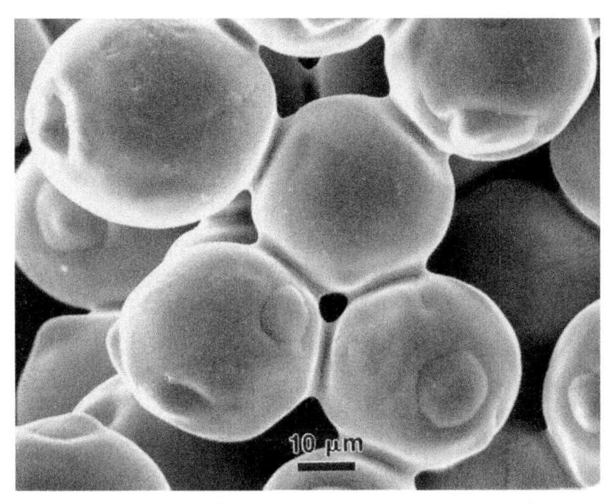

Figure 4.1 Scanning electron micrograph of neck growth between 32 μm nickel spheres during sintering at 1050°C (1323 K) for 30 min in a vacuum. Prior to sintering the particles were loosely packed into a crucible.

Sintering: From Empirical Observations to Scientific Principles
DOI: http://dx.doi.org/10.1016/B978-0-12-401682-8.00004-5

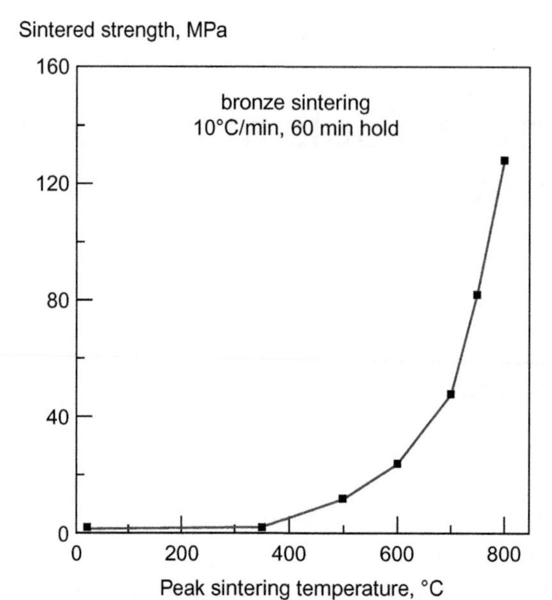

Figure 4.2 Sinter bonding greatly increases the strength, as evident here by strength data collected after heating 26 μm bronze (Cu-10Sn) spheres at 10°C/min to various peak temperatures with a 60 min hold. This bronze starts to sinter near 400°C (673 K).

60 min hold at the peak temperature. There are many means to quantify sintering; both *in situ* and *post mortem* data give insights. For example, besides the degree of sintering, hardness is sensitive to chemical reactions between powders, and to phase transformations during cooling. An *in situ* test provides information on bonding, but a *post mortem* test is probably more relevant to an application. A large variety of approaches now exist, giving an extensive toolbox of techniques. However, more than anything else, mechanical properties and density are focal points for sintering studies.

The sintering response parameters, such as density or hardness, are measured with respect to independently adjustable parameters, such as time or peak temperature. As an example, Figure 4.3 plots the change in sintered density for titanium powder as a function of sintering time for three different sintering temperatures [2]. A higher temperature increases the density, while a longer time adds densification, but with less dramatic impact.

The starting condition is determined by the component's shaping process—that is how the particles are formed into a green body [3]. Pressure compresses particles to prescribed dimensions with a higher green density than the loose powder [4]. For a ductile material, deformation during compaction occurs at the particle contact points. Higher pressures increase the number of contact points per particle, known as the coordination number, because particles deform and rearrange to increase density.

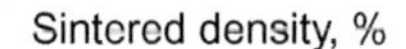

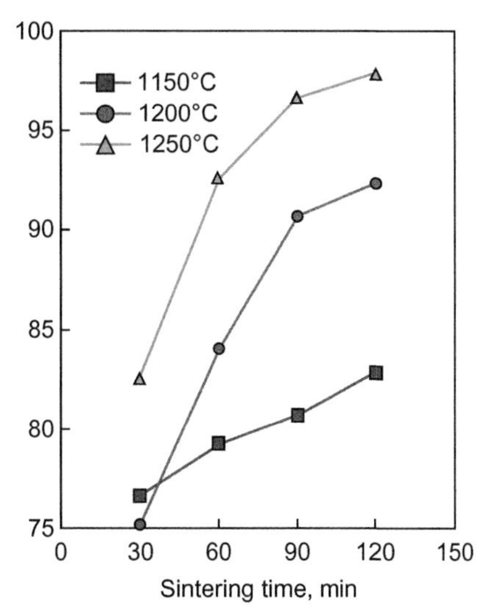

Figure 4.3 Sintered density for injection molded 42 μm titanium powder showing the density increase versus sintering hold time in vacuum at three temperatures [2].

If the powder is ductile, then each contact enlarges due to the concentrated stress at the contact points, while there is a concomitant reduction in pore size between particles. Compacted powders exhibit higher sintered densities when compared to loose powders. This is illustrated in Figure 4.4, a plot of the sintered density for 4.3 μm nickel powder sintered in hydrogen at 900°C (1173 K) for either 5 or 20 minutes using three different green densities [5]; the sintered density increases with the green density. Since attributes such as the neck size, porosity, and dimensions are common monitors of sintering, it is important to know their condition prior to sintering. Once a powder is shaped, sintering is tracked by the changes in the compact from the initial condition.

Sintering models often assume monosized spherical particles. The term "monosize" implies that all of the particles are the same size, yet most powders have a particle size distribution, and the particles shapes are often not spherical. In spite of this gap between model and reality, the general applicability of the sintering models is well recognized.

When surface forces are weak, which is true for larger particle sizes that pack to about 64% density, the monosized spherical structure will take on an initial packing coordination number (number of particle contacts on each particle) of near seven [6]. On the other hand, for nanoscale particles with strong surface forces, the packing density might be just 4% of the theoretical packing area and the coordination will average two contacts per particle [7]. At 100% density, the coordination number is near 14.

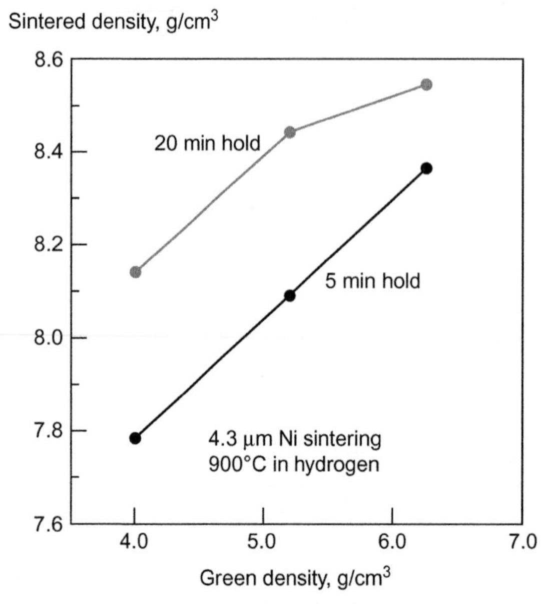

Figure 4.4 Sintered density for 4.3 μm nickel powder held for 5 min or 20 min at 900°C (1173 K) in hydrogen. The sintered density increases with a higher compacted density prior to sintering.

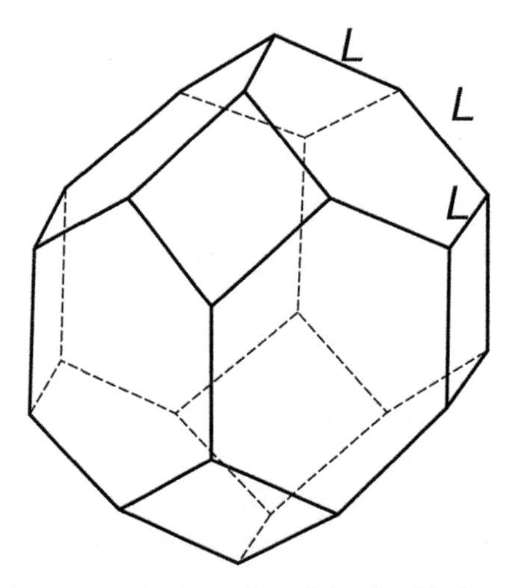

Figure 4.5 The tetrakaidecahedron grain shape for a fully densified crystalline material is the basis for many sintering models.

With densification, by either compaction or sintering, the particles become polygonal and approach a grain shape that is tetrakaidecahedronal. A tetrakaidecahedron is a 14-sided polyhedron composed of a mixture of faces having four and six edges (squares and hexagons) as illustrated in Figure 4.5 (it is also known as a truncated octahedron).

It consists of 36 edges constructed from eight hexagons and six squares. For this geometry, the surface area is 26.78 L^2 and the volume is 11.31 L^3, where L is the edge length. The grain size G based on the equivalent spherical diameter is 2.78 L.

Besides density, valuable information is gained from examination of the sintered microstructure [8,9]. The examinations include a determination of the amount and placement of each phase, including the pore structure. Mechanical properties are difficult to interpret without parallel microstructural measurements.

PARTICLE BONDING

Sintering quantification relies on measurement tools. Microstructural parameters such as neck size, grain size, and pore size are distributed parameters. This implies a characteristic median or mean size with a distribution around that value. Bulk parameters are the density, surface area, shrinkage, electrical conductivity, magnetic permeability, hardness, strength, or elastic modulus. In combination, the bulk and microstructural parameters provide a high fidelity picture of the sintering process. Inherently, the neck size is one of the most important measures.

Neck Size

Necks grow between particles during sintering. The neck diameter X is measured at the saddle point, as sketched in Figure 4.6, corresponding to the contact at the particle junction. This neck measurement is a dimensionless parameter widely used in sintering, termed the neck size ratio X/D, reflecting the neck diameter divided by the particle or grain diameter. As sintering progresses, the grain size enlarges and the initial particle size is not evident in the microstructure, so grain size is used. Other monitors of sintering are related to neck size.

The drawing in Figure 4.6 assumes that the two particles are amorphous, so there is no grain boundary associated with the contact. This is the idea introduced by Frenkel [1] for viscous flow sintering. As the neck size enlarges, the surface energy

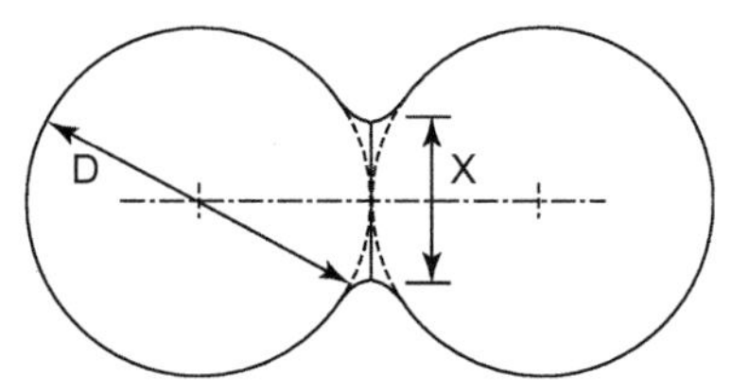

Figure 4.6 The interparticle neck size *X* is a common feature used to monitor sintering, shown here for a two sphere contact. The grain or particle size is denoted as *D*, so the neck size ratio is *X/D*.

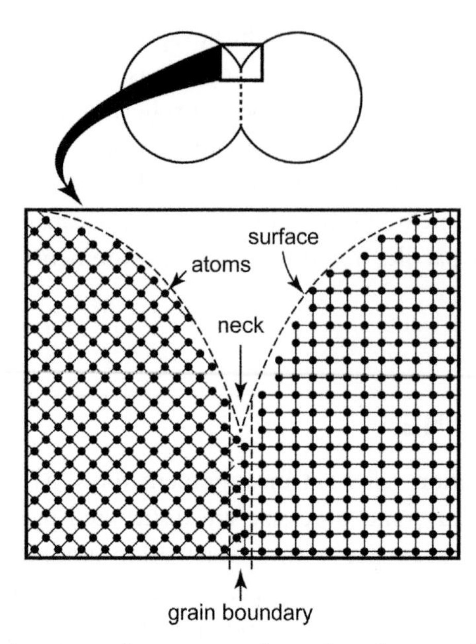

Figure 4.7 Crystalline solids join at the interparticle neck with a misalignment of crystal planes, resulting in a grain boundary where defective atomic bonding enables rapid diffusion that is often critical to determining the rate of sintering.

declines. Frenkel assumed that the surface energy dissipated by neck growth then provided energy for viscous deformation in the form of shrinkage.

However, most particles are crystalline. Thus, particles contacts form with random crystal orientations, resulting in a grain boundary as sketched in Figure 4.7. Since there are many possible grain-grain orientations, the grain boundary character varies from neck to neck within the sintering body. Likewise, the energy associated with the grain boundary varies from neck to neck. Sintering theory assumes an average taken over all of the possible orientations.

Because of grain boundary energy from the disrupted atomic bonding, the neck between crystalline particles is not smooth. Instead it is characterized by a dihedral angle ϕ that emerges at the saddle point. Figure 4.8 shows the refined neck shape model with the dihedral angle where the grain boundary intersects the surface. The scanning electron micrograph in Figure 4.9 illustrates this feature of the angular neck typical to sintering.

The dihedral angle ϕ is a reflection of the balance of solid-vapor surface energy and grain boundary energy is determined by the vertical energy resolution, given as:

$$\gamma_{SS} = 2\gamma_{SV}\cos(\phi/2) \tag{4.1}$$

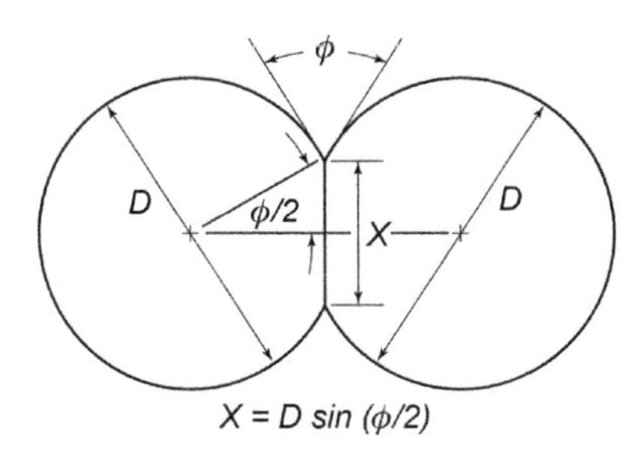

Figure 4.8 The refined neck geometry includes a dihedral angle to reflect the grain boundary energy. During sintering, neck growth progresses until the point where the dihedral angle limits the neck size ratio at sin(ϕ/2). This then becomes a terminal point for neck growth during sintering.

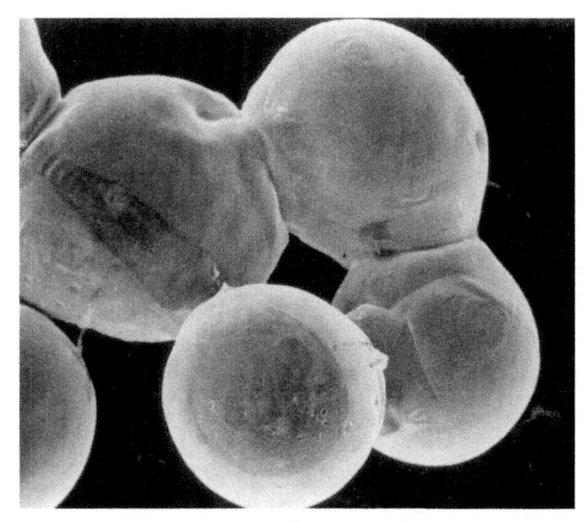

Figure 4.9 This scanning electron micrograph of a sintered neck between 20 μm copper spheres shows the terminal condition for a dihedral angle of about 140° for one grain pair and 90° for another pair, reflecting variations in the grain boundary energy.

where γ_{SS} is the solid-solid grain boundary energy and γ_{SV} is the solid-vapor surface energy. The surface energy reflects the unsaturated surface bonds, and the grain boundary energy depends on the number of dangling bonds across the interface. In liquid phase sintering, a dihedral angle forms at the necks, but reflects the solid-liquid interfacial energy γ_{SL} with a similar relation:

$$\gamma_{SS} = 2\,\gamma_{SL} cos(\phi/2) \tag{4.2}$$

The liquid phase sintering dihedral energy balance is diagramed in Figure 4.10, in which it can be seen that the sum of the two resolved solid-liquid surface energies balances the opposing solid-solid grain boundary energy. The neck size ratio X/D

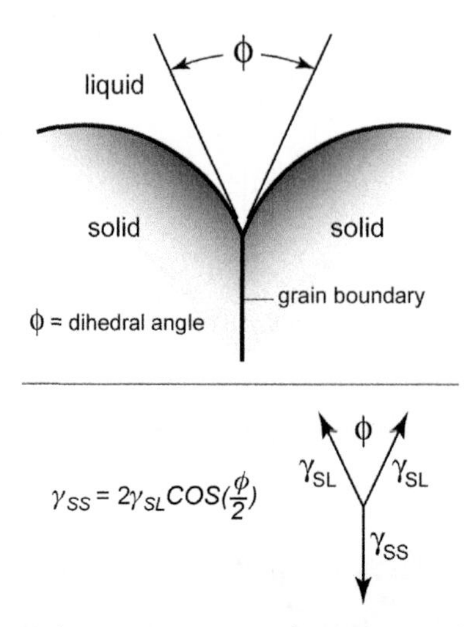

Figure 4.10 The interparticle neck contains a grain boundary and the energy of the solid-solid grain boundary γ_{SS} balanced against the solid-vapor γ_{SV} or solid-liquid γ_{SL} surface energy determines the dihedral angle ϕ. This drawing shows the solid-liquid case.

reflects the degree of sintering, but it took considerable time for theory to include the dihedral angle in the models [10].

For loose powders, neck growth starts at the particle contact points, often approximated as $X/D = 0.01$ to allow for weak attractive forces. In the case where all neck growth is associated with sintering densification, the relation between the fractional sintered density f and neck size ratio X/D is:

$$\frac{X}{D} = b\sqrt{1 - (f_O/f)^{1/3}} \tag{4.3}$$

where f_O is the initial loose powder fractional packing density (at $X = 0$) and b is a constant estimated at 3.6. This relation is valid for a neck size ratio up to about 0.33. It assumes all neck growth occurs due to mass transport associated with the grain boundary.

In many cases, neck growth also arises from atoms moving along the pore surface by surface diffusion or evaporation-condensation routes. As more relative surface transport occurs, Equation 4.3 is inaccurate, so it is a limiting case. Some substances sinter totally by surface diffusion, giving neck growth with no change in sintered density.

Much effort is given to measuring density, a relatively easy sintering monitor. Density is mass divided by volume, and fractional density is the ratio of sintered density with respect to the theoretical density. If the test geometry is simple, then the density is easily calculated by weighing and dimensioning the sample. For some shapes, the Archimedes approach based on mass determinations in and out of water is used.

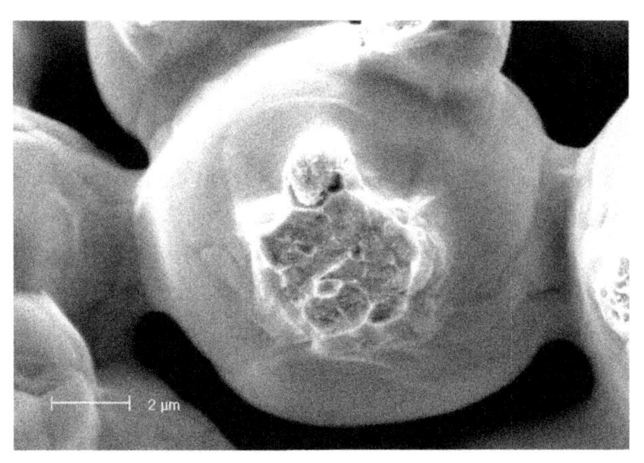

Figure 4.11 Sintered materials that are less than full density show a propensity to fracture along the weak path and predominantly fail at the interparticle neck as evident in this scanning electron microscope image.

The initial density depends on several factors, including the particle size and powder consolidation. If the particles are deformable, compaction produces a higher green density and an enlarged initial contact. During sintering, each particle bonds to its neighbors, and as densification occurs new bonds are formed. Thus, the mean coordination number N_c increases approximately with the fractional density f as follows:

$$N_C = 14 - 10.3(1-f)^{1/3} \tag{4.4}$$

up to full density ($f = 1$ or 100%).

Neck size is measured by electron or optical microscopy. Fractured material usually fails along the saddle points to reveal the neck size, as illustrated in Figure 4.11. An alternative is to rely on hot stage microscopy to track neck size during sintering. The advantage of hot stage microscopy is that is can continuously track the behavior during the heating cycle. Besides neck size, it is common to observe pore structure changes and particle rearrangement. For small or nanoscale powders, hot stage transmission electron microscopy is most useful to track neck size versus time or temperature [11,12]. This way each particle, its size, location, and relative motion is accessible, including its crystal structure and local chemistry. A new approach relies on synchrotron light sources to monitor neck growth for larger particles [13].

Neck Shape

Prior to sintering the principal geometric feature is the particle size. As necks grow, the particles are still evident, but they become lost with neck growth. Sintering connects the solid grains into a polycrystalline solid. Grain growth occurs in parallel with sintering, destroying the microstructure scale associated with the initial particle size. Thus, the protocol is to speak of the grain size, not the particle size, once sintering initiates.

Early sintering observations identified the neck shape as a saddle surface. Kuczynski [14] used a circle approximation for his early model. However, the abrupt change in curvature where the circular neck and spherical grain intersect produces a mathematical instability. Early efforts tried to calculate the idealized neck shape using computer simulations [15]. Unfortunately, these simulations were unstable and created false undercutting profiles that would cause desintering. Later models corrected these problems, and eventually added a dihedral angle [16–18]. Thus, the neck shape is defined by a constant, smooth curvature gradient, without undercutting. This is illustrated in Figure 4.12 for surface diffusion controlled sintering [19]. The mass forming the neck is redistributed from the neighboring grain surface.

Initial sintering mass flow into the neck is rapid compared to grain boundary grooving, so it is acceptable to assume a smooth neck. Later in sintering, as neck growth slows, the pores take on a lenticular shape, reflecting the dihedral angle at the grain boundary. When two contacting grains are different in size, the larger grows at the expense of the smaller, giving grain coalescence. A schematic of the process is given in Figure 4.13. The dihedral angle initially prevents coalescence, but grain rotation aligns the grains to enable coalescence. This fusing of two grains into one larger

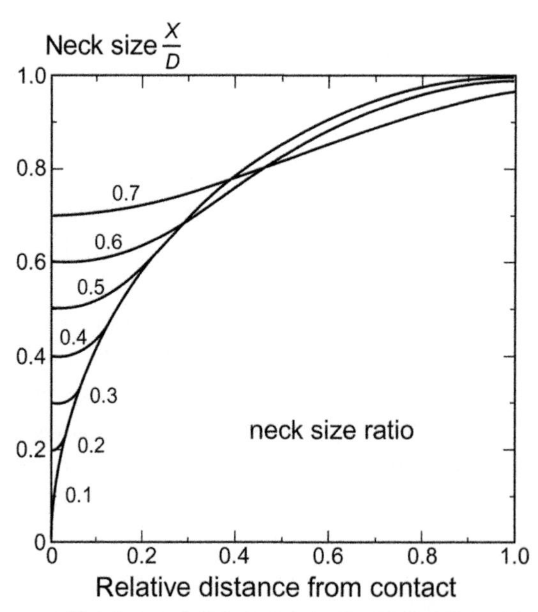

Figure 4.12 Neck shape profiles for neck size ratios (*X/D*) ranging from 0.1 to 0.7 [19]. These profiles correspond to quarter sections of a single sphere to show the neck growth by progressive surface diffusion of mass from the particle surface far removed from the neck.

grain is accompanied by elimination of the interparticle grain boundary and occurs throughout sintering,

Neck growth occurs at the contact points on the grain surface. If densification occurs, then the coordination number increases and new necks are initiated. Once the necks reach a size near $X/D = 1/3$, then the curvatures fields from neighboring grains overlap. Now the pore structure becomes rounded as illustrated in Figure 4.14, and the pores shrink to a point where the grains polygonize with a few spherical pores at the grain corners. This is the final stage of sintering, and an example of the microstructure of this stage is shown in Figure 4.15. The fracture surface in this image shows spherical pores located on grain boundaries of the polygonal grains. In this case approximately 64,000 particles coalesced to form each grain, so the initial particle character is lost due to microstructure coarsening.

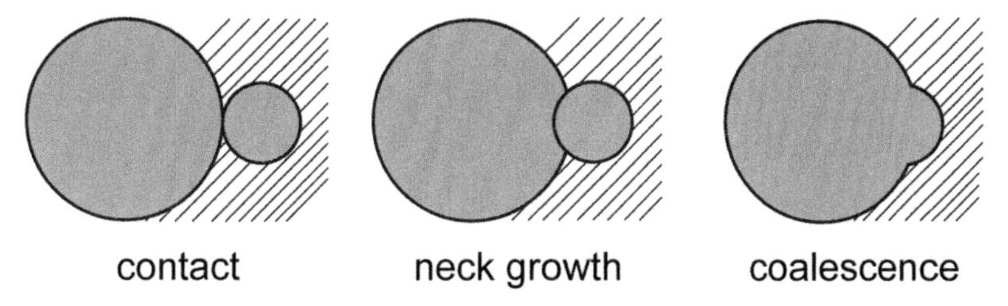

Figure 4.13 Grain contact of two dissimilar size crystal regions (two differently sized particles) results in growth of the larger grain at the expense of the smaller grain, and if the grains rotate or migrate, the interparticle grain boundary between the two grains coalesces, giving a larger average grain size and a decrease in the number of grains as part of the sintering process.

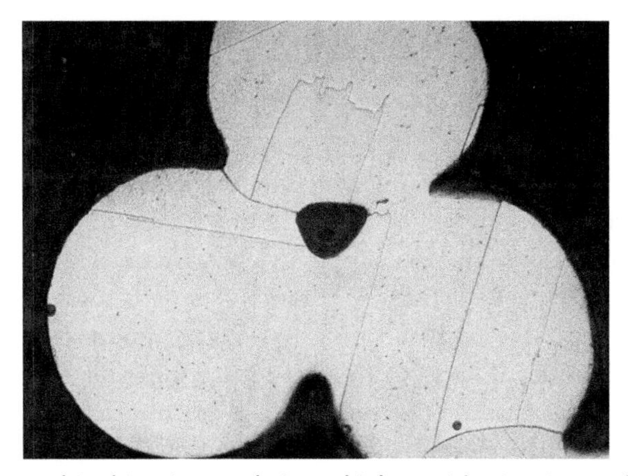

Figure 4.14 As illustrated in this micrograph, in multiple particle sintering neck growth, neighboring grains impinge to round the pore, leading to progressive smoothing of the surfaces and spheroidization of the pores.

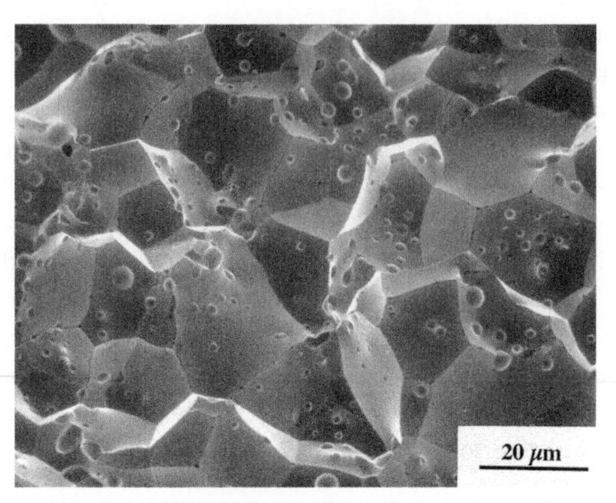

Figure 4.15 Final stage sintering corresponds to spherical pores located on grain boundaries, and a practical example of the final stage microstructure is shown by this tungsten fracture surface. Coarsening has enlarged the initial 0.5 μm particles to 20 μm grains, giving one grain in the sintered microstructure for every 64,000 initial particles.

MECHANICAL PROPERTIES

Sintered materials are frequently used in structural applications, so sintering cycles are designed around combinations of strength and cost. Porosity is detrimental since pores reduce the load bearing cross-sectional area and act as incipient cracks. The elimination of pores is an obvious requirement for advatangeous mechanical properties. Two mechanical properties are commonly used to track sintering—strength (fracture resistance) and hardness (penetration resistance).

Strength

The powder starts with little strength, which grows with sintering. Tests for strength are well developed, but sintered materials testing tends to rely on a few simple measurements. One test is performed using a right circular cylinder which is crushed by loading the flat faces until failure. A common strength measure used for sintered brittle materials is the three-point bending transverse rupture test. Prior to fracture, the compact is dimensioned, then it is loaded to fracture as illustrated in Figure 4.16. The calculated strength σ is proportional to the fracture load F, sample width W, thickness T, and span length between the lower two supports L:

$$\sigma = \frac{3FL}{2WT^2} \tag{4.5}$$

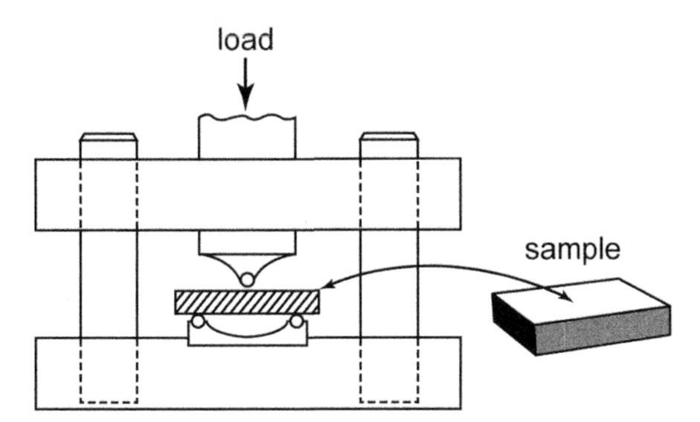

Figure 4.16 One of the common strength tests for sintered materials is the transverse rupture strength test (TRS or modulus of rupture). As illustrated here the rectangular sample is subjected to three-point loading and the dimensions and fracture load are used to calculate the strength, assuming brittle failure.

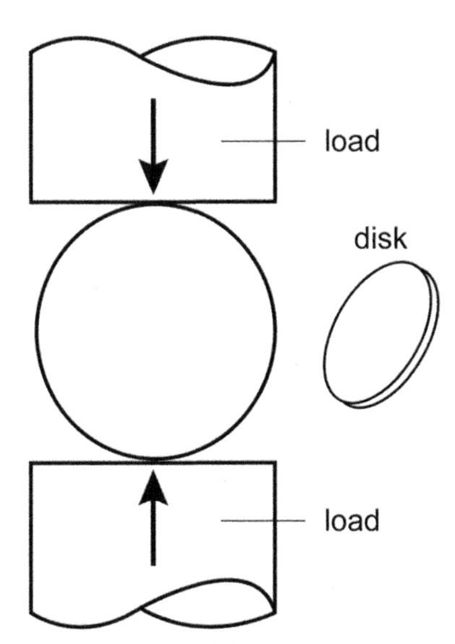

Figure 4.17 A disk shaped sample is crushed between two platens to determine the fracture strength in a test termed the Brazilian test. The samples are easy to fabricate from powder, so this and the transverse rupture tests are popular for measuring sintered strength.

Another strength tests is shown in Figure 4.17, where a thin sintered cylinder is compressed on its side. For this test the sintered strength σ is calculated from the facture load F, disk diameter D, and thickness T as follows:

$$\sigma = \frac{2F}{\pi DT} \tag{4.6}$$

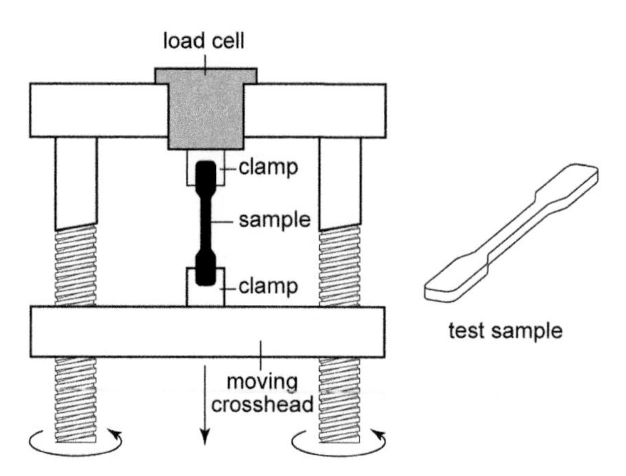

Figure 4.18 Tensile testing of sintered ductile materials relies on a flat sample that is die compacted from the powder prior to sintering, usually with about a 6 mm by 6 mm cross-section in the center.

A flat tensile geometry is used to measure strength in tension. For ease of fabrication, the powder is pressed into a sample as diagrammed in Figure 4.18. The sintered tensile bar is pulled on the ends and strength is calculated from the fracture load divided by cross-sectional area at the center of the bar. Tensile testing is reserved for ductile materials, since gripping the bars induces damage, and the transverse rupture strength is reserved for brittle materials, since the strength calculation assumes trivial deflection prior to fracture.

Strength varies significantly with the degree of sintering, and correlates with the sintered density. As an example, in sintered steels the tensile strength doubles as the sintered density increases from 90% to 100%. For alumina, the sintered strength versus the degree of sintering is illustrated in Figure 4.19 after 100 min hold at various temperatures [20].

Strength increases at higher temperatures where atomic motion is more active. It correlates with neck size, since fracture favors a path through the sinter contacts. Even though the density is high, strength can still be low.

Although strength is often a measure of success in sintering, it also depends on many other factors, such as the grain size and pore size. Composition is another factor and processing defects, such as forming cracks, are dominant.

Most often, simple models of strength σ versus fractional density f are successful:

$$\sigma = \sigma_O f^M \tag{4.7}$$

The exponent M depends on the sintering conditions and σ_O is the full density strength. Early data supporting this relation were provided by Coble and Kingery [21], as plotted in Figure 4.20. Rearranging the above equation gives a linear fit

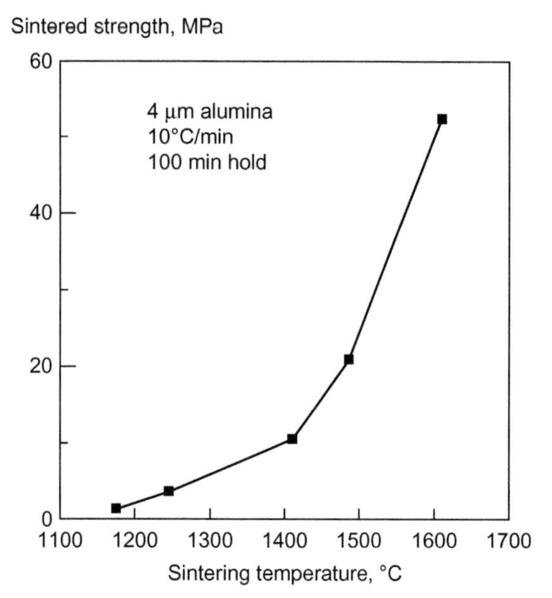

Figure 4.19 Sintered strength improves with higher sintering temperatures as illustrated by the data for 4 μm alumina powder heated at 10°C/min to various hold temperatures for 100 min [20].

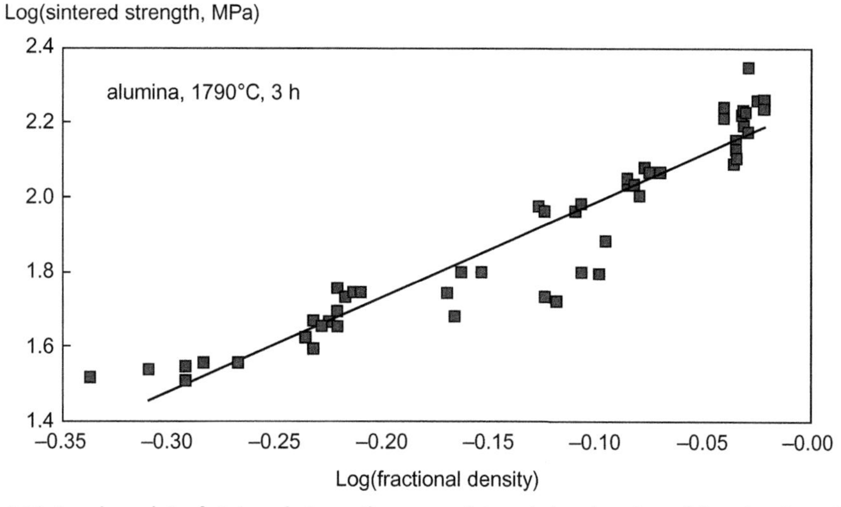

Figure 4.20 Log-log plot of sintered strength versus sintered density plotted for alumina with variations in initial density after sintering for 360 min at 1790°C (2063 K). Although the data are scattered, the role of sintered density is evident.

between the logarithm of the strength and logarithm of the fractional density, with a slope of M. For the case of alumina, the exponent M is 2.5; various studies find values up to about $M = 6$.

The scatter in Figure 4.20 indicates that other factors besides fractional density are at work in determining sintered strength. Most typically pore size and grain size are the required secondary factors. In the absence of microstructural abnormalities, sintered strength σ best correlates with fractional sintered density f if shrinkage or neck size is added [22,23]:

$$\sigma = \sigma_O f^M \left(\frac{\Delta L}{L_O}\right) \tag{4.8}$$

or:

$$\sigma = \sigma_O f^M \frac{2N_C}{3\pi} \left(\frac{X}{D}\right)^2 \tag{4.9}$$

where σ_O is the full density strength for the material, $\Delta L/L_O$ is the sintering shrinkage, N_C is the grain coordination number (depends on density), X/D is the neck size ratio, and the factor 2/3 reflects stress concentration and the preference for facture to occur at interparticle sinter bonds. The terms in these relations relate to fractional density, giving a power law relation between sintered strength and fractional density, except for cases where sintering occurs without densification.

Fracture tests are also performed *in situ* during sintering [24]. Early neck growth gives strengthening without a change in density. At high temperatures the compact weakens due to thermal softening. So, *in situ* strength starts low prior to significant neck growth, increases with neck growth, subsequently further increases due to density gains, but falls at high temperatures due to thermal softening. The full density strength is essentially zero at the melting point.

The resulting *in situ* strength variation during heating is plotted in Figure 4.21. The strength increases and then decreases during heating, with thermal softening causing a strength decrease at the higher temperatures. Note that the sintered strength increases with higher peak temperatures as shown by the room temperature values in Figure 4.22. The peak strength *in situ* is just over 100 MPa near 600°C (873 K), but the peak strength after sintering at 800°C (1073 K) is 677 MPa; almost sevenfold higher. The slight strength loss at the very highest temperatures comes from microstructure coarsening.

Those conditions which improve the degree of sinter bonding improve the mechanical properties. For example, Fe-0.5Sn with a green density of 85% sintered at 1100°C (1373 K) for 1 h in hydrogen gives 145 MPa strength. If ammonium chloride

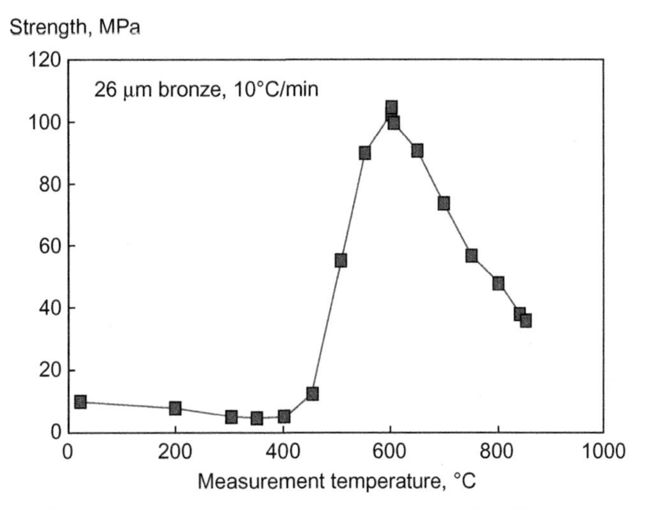

Figure 4.21 A plot of the *in situ* transverse rupture strength for 26 μm bronze powder versus temperature while heating at 10°C/min. Sinter bonding leads to a dramatic strength increase, but at higher temperatures the material softens so *in situ* strength subsequently decreases.

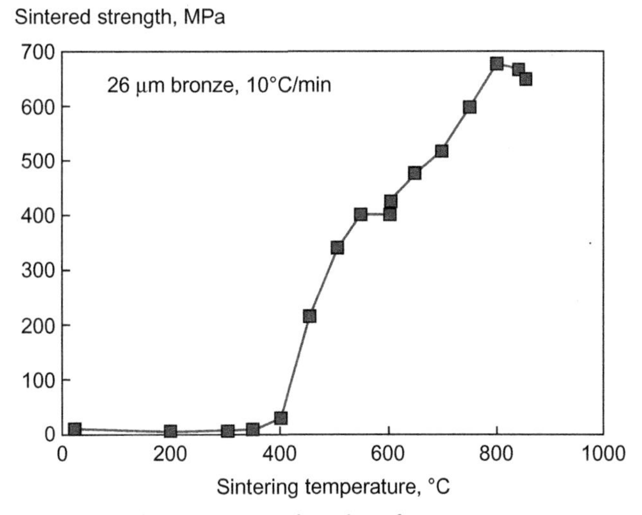

Figure 4.22 In comparison with Figure 4.21, this plot of transverse rupture strength for a 26 μm spherical bronze powder gives the room temperature sintered strength after heating to various temperatures.

is added to the atmosphere, the strength increases to 222 MPa due to enhanced neck growth. Strength models which only include density are unable to explain such variations, so surface area or pore structure data are needed to explain the behavior.

Hardness

Like strength, sintered hardness is dominated by the sintered density [25]. Since hardness is a straightforward measurement, hardness testing is routinely used as a screening tool for proper sintering. It is often cited in property standards for sintered materials to provide assurance on proper sintering.

Superficial hardness testing, with low applied loads, enables mapping of density gradients in sintered bodies. A correlation is created between fractional density and hardness, as illustrated for WC-10Co in Figure 4.23 [26]. Once this correlation is established, then a test body is mapped for density gradients using point-to-point hardness tests. The approach is nominally accurate in measuring density gradients arising from forming tool friction. Since sintering shrinkage varies with initial density, hardness maps combined with finite element analysis provide a means to predict distortion during sintering.

Elastic Properties

The elastic properties vary with both the sintered density and microstructure [27]. As a simple relation, independent of microstructure, elastic modulus E varies with the sintered fractional density f as follows [21,28,29]:

$$E = E_O f^Z \tag{4.10}$$

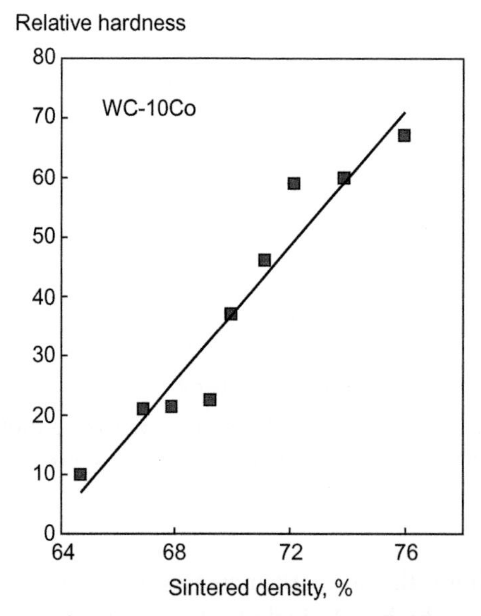

Figure 4.23 Relative hardness of WC-10Co versus fractional density, and like strength generally a linear relation is observed.

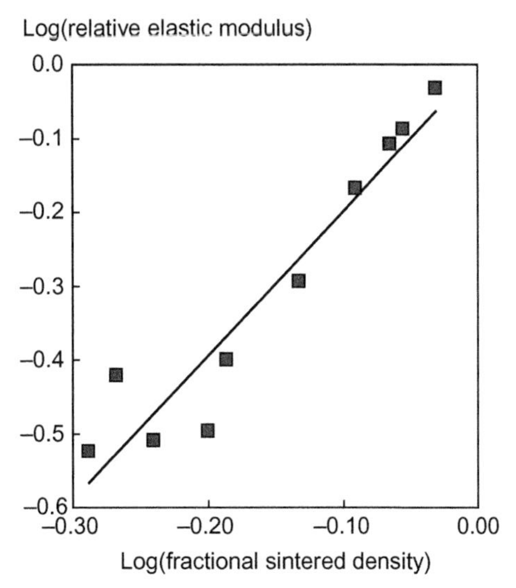

Figure 4.24 Log-log plot of sintered elastic modulus for copper versus the fractional density [30], corresponding to a power law relation.

where E_O is the elastic modulus at full density and Z depends on the pore structure; it is usually less than 4. A log-log plot of elastic modulus versus fractional density produces a straight line of slope Z. Figure 4.24 is an example plot taken from sintered copper [30]. The straight line fit has a slope of 2. Similar behavior is evident in a wide variety of sintered materials.

In the case of sintered steel the elastic modulus at full density is 207 GPa, but at 90% density the elastic modulus is 30% lower at 147 GPa, and at 80% density it is 89 GPa, less than half of the full density value. The corresponding exponent Z is 3.4. Parallel determinations of the shear modulus show essentially the same pattern.

In situ elastic modulus determinations are another means to track sintering [31]; however, most determinations are made outside of the sintering environment. The measurements rely on nondestructive sonic or ultrasonic waves. An example test arrangement is shown in Figure 4.25, where the sample is held in a furnace and resonated to measure the elastic modulus. Resonant frequency testing relies on a dynamic elastic modulus determination to link the sound velocity to the material density. The governing relation is:

$$\omega = \frac{K}{2L}\left(\frac{E}{f}\right)^{1/2} \tag{4.11}$$

where ω is the longitudinal resonant frequency, L is the sample length, E is the elastic modulus, f is the fractional density, and K is a constant that depends on the material and test conditions.

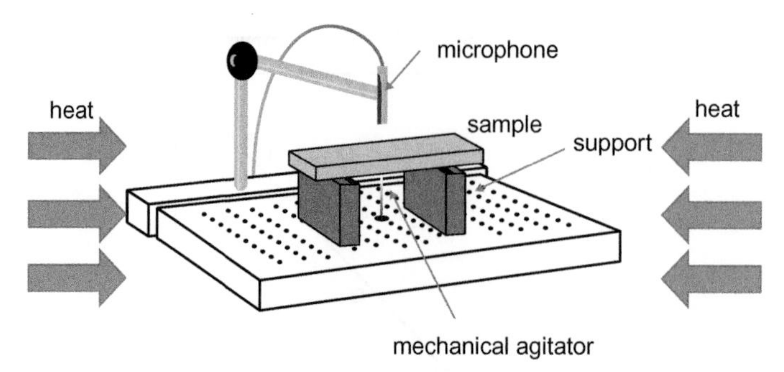

Figure 4.25 *In situ* testing for elastic modulus is possible using an assembly in a furnace where a microphone measures the sound wave characteristics from a periodic mechanical impact on the sintering sample.

Other Mechanical Properties

Other mechanical properties are occasionally used to monitor sintering, but mostly because of specific performance concerns. Parameters such as ductility, impact toughness, or fracture toughness might be included in technical specifications, but little effort is made to predict these attributes with respect to sintering parameters.

Dynamic mechanical properties associated with fracture, such as impact toughness, fracture toughness, and even fatigue strength, vary with the sintered density and microstructure. At lower sintered densities, the fatigue behavior is dominated by fractional density, but at high densities it depends more on the microstructure, notably the pore size and pore spacing, since crack propagation follows the easy path between pores.

DIMENSIONAL CHANGE

Neck growth is fundamental to sintering. Mass transfer to the neck ultimately comes from two sources—the pore surfaces or the grain boundaries. Mass transfer by surface transport moves atoms or molecules on the solid-vapor surface from convex to concave positions. The neck is concave while regions away from the neck are convex. Bulk transport corresponds to mass transport involving the grain boundaries connecting the pores, providing a pipeline for mass flow to the neck. This results in a progressive approach of the grain centers or shrinkage.

Surface transport controlled sintering gives no shrinkage, but bulk transport controlled sintering induces shrinkage. When captured over many grain contacts, a net dimensional change is evident at the macroscale. As illustrated in Figure 4.26, the component shrinks. In this photograph, the upper component is shown after sintering and although the same shape, it is significantly smaller due to shrinkage.

Figure 4.26 An example of sintering shrinkage in a stainless steel component, where the lower piece is prior to sintering (50% dense) and the upper pieces is after sintering (100% dense).

Shrinkage

Dimensional change and concomitant density changes are widely used sintering measures. In some situations, sintering is intentionally performed to bond particles without significant dimensional change, such as in the production of automotive gears, filters, and capacitors, but in other situations densification is desired to improve properties.

Sintering shrinkage is the change in component size divided by the initial size prior to sintering. Mathematically, this is expressed by the non-dimensional linear dimensional change $\Delta L/L_O$, reflecting the change in an initial length L_O to a final sintered length L_S given as ΔL. Usually the sintered dimension is smaller so $\Delta L/L_o$ is negative. Often shrinkage is expressed as a positive attribute, even though the size change is negative. For example, a powder compact with an initial length of 10 mm which shrinks to 8.5 mm length after sintering has 15% shrinkage. In some cases sintering shrinkage can be as much as 30%.

In a homogeneous green body, each dimension should exhibit the same shrinkage. The final volume is much smaller than the initial volume, while the mass remains the same, hence a density increase is associated with shrinkage. Any convenient dimension is measured—length, thickness, height, or diameter. When compaction tooling is designed, shrinkage is taken from the cavity dimensions to enable proper tool design to deliver final sintered size, so it includes a small elastic expansion of the powder after compaction that makes the green compact larger than the die size.

Early in sintering densification, the correlation between neck size ratio and shrinkage as plotted in Figure 4.27 occurs. This is only applicable to bulk transport controlled

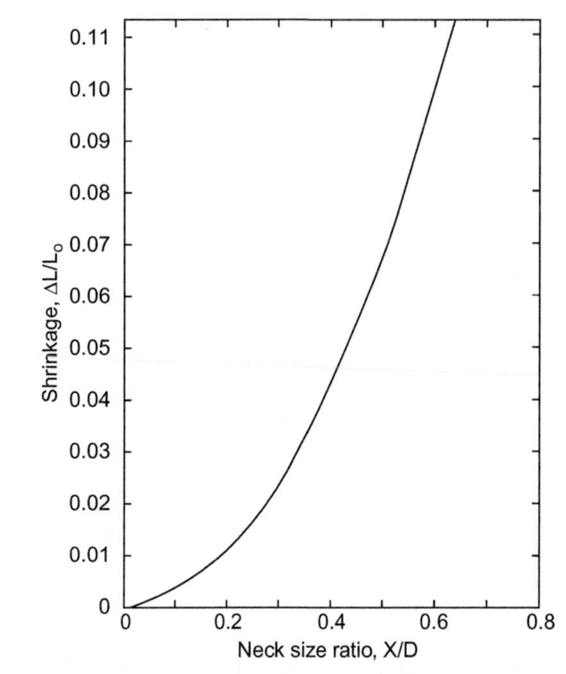

Figure 4.27 Model relation between sintering shrinkage and sintered neck size for spherical particles sintering by bulk transport processes such as grain boundary diffusion.

sintering. Based on the sintering geometry the relation between shrinkage $\Delta L/L_O$ and neck size ratio X/D is as follows:

$$\frac{\Delta L}{L_O} = \frac{1}{b}\left(\frac{X}{D}\right)^2 \tag{4.12}$$

The negative sign for sintering shrinkage is ignored. Depending on the dihedral angle, the parameter b is 3.6 based on experimental data. The neck size ratio remains below 0.5, so this shrinkage relation is only valid to about 6% shrinkage. Larger shrinkages occur without neck growth as grain coordination, grain size, and pore annihilation effects dominate.

The usual assumption is that the volume change during sintering $\Delta V/V_O$ is isotropic, thus if the mass remains constant during sintering, the sintered fractional density f depends on the initial unsintered fractional density f_O and shrinkage as:

$$f = \frac{f_O}{\left(1 - \frac{\Delta L}{L_O}\right)^3} \tag{4.13}$$

This relation is generally valid up to full density.

Dimensional change in sintering is measured using micrometers, calipers, or coordinate measuring machines. Since shrinkage is based on the difference in measurements, an accuracy of 0.01% is about the best possible.

For research and development activities, a recording dilatometer is employed to make dimensional measurements *in situ* during heating. Figure 4.28 diagrams a vertical arrangement. The powder compact is situated in a furnace with a counter-balanced measuring probe contacting the compact. An alternative is to rely on a noncontact laser micrometer. As the compact expands or shrinks, the dimension is constantly measured. Constant heating rate profiles are useful in scanning dimensional change versus temperature, as illustrated in Figure 4.29 for a 0.4 μm alumina powder. The plot shows no measurable shrinkage below approximately 900°C (1173 K) and near complete densification by 1600°C (1873 K). The measuring device thermal expansion is subtracted from the signal using calibration runs with known standards.

The dilatometer signal correlates to dimensional change, and three approaches are in use—a variable displacement transformer, optical grating, or laser interferometers. A dual beam system is the most recent addition [32]. Dimensional changes down to 0.1 μm can be measured. Typically measurements are performed using linear heating, for example at 10°C/min. A plot of the first derivative versus time gives the shrinkage rate, and when plotted versus temperature, gives an indication of the temperature of

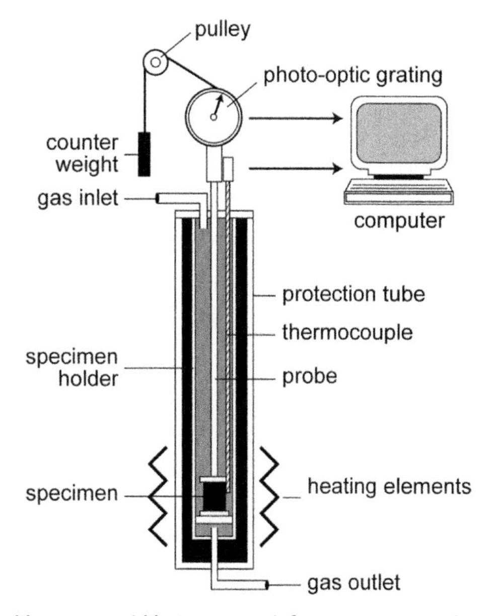

Figure 4.28 A recording dilatometer (dilation meter) for sintering studies relies on a test specimen located inside the furnace with dimensional measurements made by a probe or push rod that feeds displacement data to an external recording device. Displacements using photo-optic gratings, eddy current, or linear variable differential transducers are in common use.

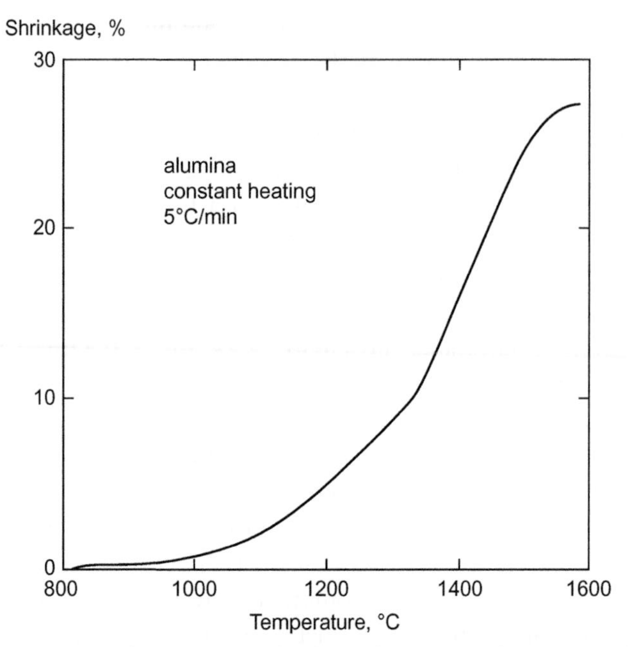

Figure 4.29 Dilatometer recorded sintering shrinkage for 0.4 μm alumina powder, as measured during a constant heating rate experiment.

most intense sintering. Figure 4.30 is an example for a mixture of powders (88W-8Ni-4Cu); the peak in the shrinkage rate occurs at 1462°C (1735 K) for a 1°C/min heating rate and 1479°C (1752 K) for a 15°C/min heating rate.

Dilatometry enables scanning over a variety of conditions to map sintering response. Time-based experiments show that the rate of sintering declines. Generally, the cumulative shrinkage $\Delta L/L_O$ follows a power law relationship with the isothermal sintering time t:

$$\frac{\Delta L}{L_O} \approx t^N \tag{4.14}$$

where N is frequently near 0.33. Shrinkage rate declines with extended hold times. This is evident in the 45 nm urania (UO_2) sintering results shown in Figure 4.31 at 838°C (1111 K). At shorter times, the rate of change is large, but as the system consumes surface energy the rate slows, similar to the way that fires burn. The line in this plot corresponds to $N = 0.33$.

Temperature sensitivity is extracted using constant heating rate experiments, assuming an Arrhenius response:

$$\frac{1}{L_O}\frac{dL}{dt} \approx exp\left[-\frac{Q}{RT}\right] \tag{4.15}$$

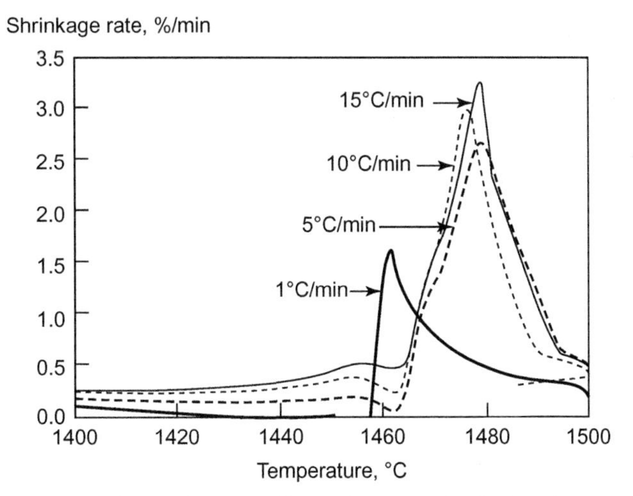

Figure 4.30 Dilatometer shrinkage data are differentiated to obtain shrinkage rate plots, showing temperatures of intense sintering densification, as illustrated in this figure for heating rates from 1 to 15°C/min for a mixture corresponding to 88W-8Ni-4Cu heated in hydrogen.

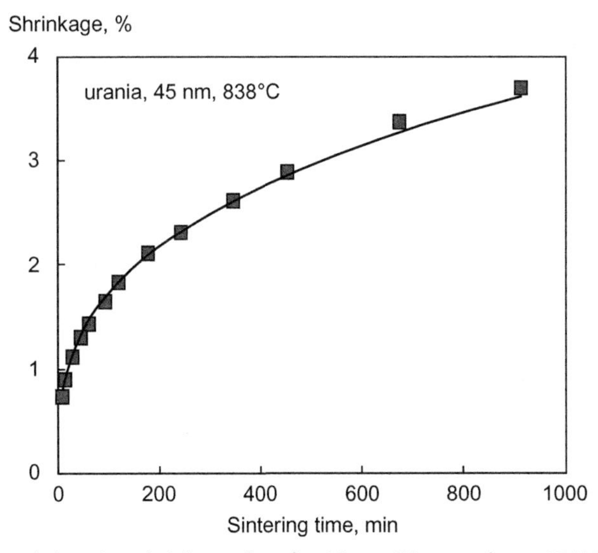

Figure 4.31 Isothermal sintering shrinkage data for 45 nm UO_2 powder at 838°C (1111 K) [35]. The symbols represent shrinkage measurements and the solid line corresponds to a cube-root time dependence as expected for grain boundary diffusion.

where $1/L_O \, dL/dt$ is the shrinkage rate, and Q is the activation energy for the process, R is the gas constant, and T is the absolute temperature. Figure 4.32 demonstrates the use of an Arrhenius analysis using shrinkage data for a 15 to 25 μm glass powder heated at 0.8°C/min [33]. In this plot the y-axis is the shrinkage divided by the square

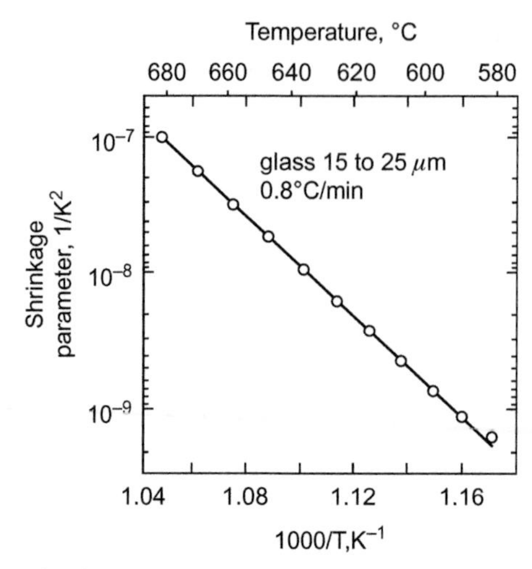

Figure 4.32 An inverse absolute temperature plot used to extract process activation energy for sintering 15 to 25 μm glass powder scanned at 0.8°C/min over a range of temperatures [33].

of the absolute temperature and the x-axis is the inverse of the absolute temperature. The basis for this plotting conforms to the anticipated Arrhenius temperature dependence for viscous flow. The shrinkage response is more sensitive to temperature changes than time changes.

Dilatometers are constructed from inert, high temperature materials. Alumina is common, although refractory metals or graphite are also used. The contact probe must be inert with respect to the sintering material. Since the thermal shock resistance of alumina is relatively poor, heating rates need to be kept below 25°C/min. For high temperatures and high-pressures systems fabricated from graphite are used, and the dilatometer is enclosed in a pressure vessel. Dilatometers reach maximum temperatures of up to 2300°C (2573 K) and gas pressures up to 500 MPa.

In dilatometry, the process atmosphere can be made up of air, vacuum, nitrogen, argon, and hydrogen. Contact probes exert a small force on the sintering compact, adding a creep component to the sintering data. By varying the probe force the effect is extracted by extrapolation to zero creep condition or the enhanced sintering is extracted by applying progressively higher loads. The higher the applied load, the greater the shrinkage during heating. Cyclic stress is another option [34].

Instantaneous changes during dilatometry allow process sensitivities to be extracted. For example, heating rate, atmosphere, pressure, or temperature can be adjusted fairly quickly. Consequently, the density and state of the compact are the same before and after the change. The Dorn technique uses this idea to determine the activation energy [35]. In this way the activation energy is calculated from the ratio of shrinkage rates:

$$Q = \frac{RT_1T_2}{T_1 - T_2} \ln\left(\frac{V_1}{V_2}\right) \tag{4.16}$$

Again, R is the universal gas constant, T_1 and T_2 are the temperatures (absolute), and V_1 and V_2 are the sintering shrinkage rates. This technique provides insight into the fundamental sintering mechanism by identifying the activation energy for shrinkage using *in situ* measurements.

Dimensional control is important to sintering, not only as a means of monitoring the process but also because final dimensions are a main acceptance criterion for sintered products. For precision components, that means the forming tools are sized to deliver the specified size and shape after sintering. Determination of the way to sinter to the prescribed size is a necessary sintering skill. Complications arise since dimensional change is anisotropic due to a variety of factors:

- anisotropic pore shapes (due to flattening of the pores in the compaction direction)
- thermal gradients in the furnace leading to non-uniform heating
- gravity (especially for larger components)
- substrate friction
- particle segregation in the forming process.

In some cases the anisotropy of the shrinkage is very noticeable.

Swelling

Materials that undergo reactions or phase transformations swell instead of shrinking. An example is when one constituent melts and dissolves into another, creating pores in the microstructure, as seen in sintering mixtures of iron and aluminum. The aluminum forms a pore when it melts, as shown in Figure 4.33. Other causes of swelling come from trapped gases or reactions that produce gases—dissolved carbon and oxygen react to form gaseous carbon monoxide or carbon dioxide (depending on concentration and temperature) that cause swelling.

Like shrinkage, swelling involves a dimensional change. Swelling causes the dimensions to increase, so the parameter $\Delta L/L_o$ is positive, while if the dimension is smaller after sintering the process is termed shrinking and $\Delta L/L_o$ is negative. With swelling, the volume expands and the density decreases. Usually, the mass remains constant (except for volatile lubricants and binders), so the sintered fractional density is likewise a function of the degree of dimensional change on swelling. It is possible that the structure then densifies after swelling, as is the case in sintering mixtures of nickel and aluminum (Ni_3Al) or nickel and titanium (NiTi).

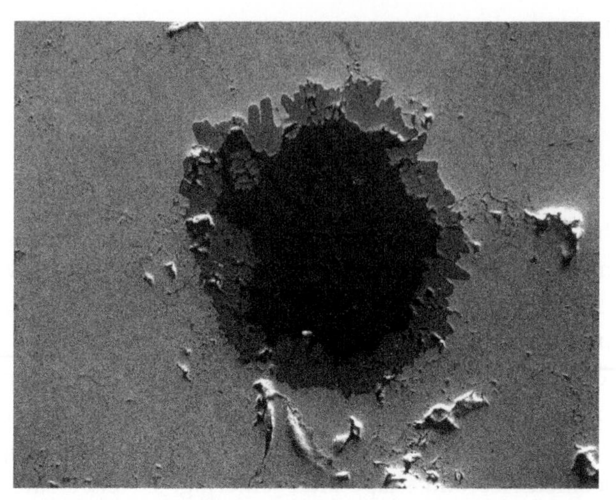

Figure 4.33 Micrograph showing the diffusion zone around an aluminum particle in an iron matrix, where the outward diffusion of the aluminum caused swelling and produced a pore due to unbalanced diffusion.

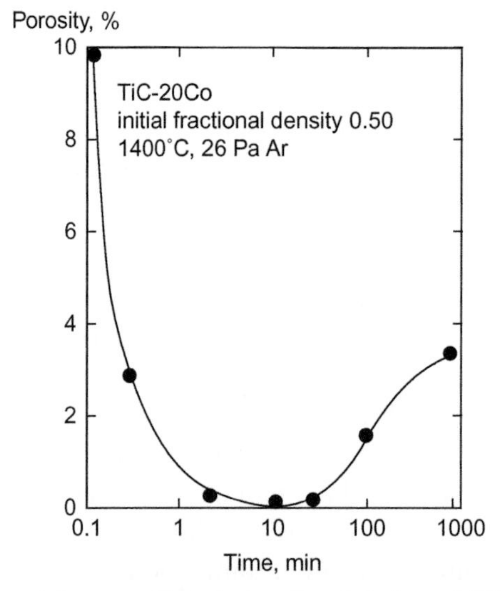

Figure 4.34 Densification and then swelling during the sintering of TiC-20Co at 1400°C (1673 K) for a compact that was initially 50% dense [36]. The time axis is on a logarithmic scale.

An example of swelling during sintering is seen in sintering TiC-Co, shown in Figure 4.34. This is a plot of sintered density versus time for a compact starting with a fractional density of 0.5 [36]. After about 30 min the porosity increases due to internal swelling reactions associated with gas production in the pores.

Direct Imaging

Dilatometry provides one-dimensional data during sintering. In many situations the sintering changes are not uniform over the body being processed, so the dilatometry results tell only part of the story. An alternative is *in situ* video imaging, for obtaining two-dimensional data from sequential photographs [37]. A strobe light and video camera are aligned with the test material inside a furnace. At high temperatures the optical emissions inside a furnace obscure imaging. This problem can be compensated for by the use of high intensity strobe lighting in combination with the shuttering of the imaging system, providing a means to generate high definition images at high temperatures. The camera and strobe are triggered in unison and a frequency at 1 Hz provides high fidelity data.

Data taken from the sintering profile using video imaging is given in Figure 4.35 where shape distortion is evident as sintering time progresses. From these data, models provide a means to extract the effective system viscosity for modeling sintered dimensions [38]. As an example of the possible imaging, Figure 4.36 is a photograph taken during liquid phase sintering at 1500°C (1773 K), showing a progressive liquid formation. A difference in reflectivity demarcates between the solid top and the liquid at the

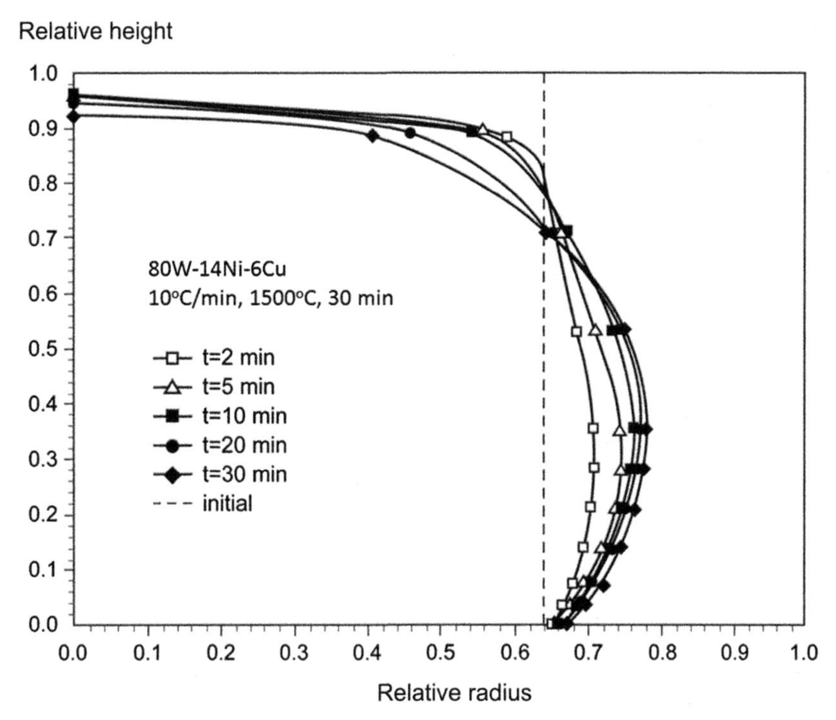

Figure 4.35 Axial x-y dimensional profiles of 80W-14Ni-6Cu compacts sintered for times up to 30 min at 1500°C (1773 K). These profiles were generated to show the progressive distortion of the compacts from the initial right circular cylinder geometry indicated by the dashed line.

bottom. The liquid nucleated at a hot spot and propagated as heat formed liquid, a process that propagated at about 1 mm/s.

One use for video imaging is in capturing the deflection in sintering experiment involving a bending beam. The experiment assumes that the sintering body is viscous, and that gravity induces a deflection. As sketched in Figure 4.37, a rectangular bar of

Figure 4.36 An *in situ* video image taken at approximately 1500°C (1773 K) using strobe flashing to generate contrast in the furnace. Liquid formation is spreading from the bottom to the top of the component, as shown by the difference in reflectivity.

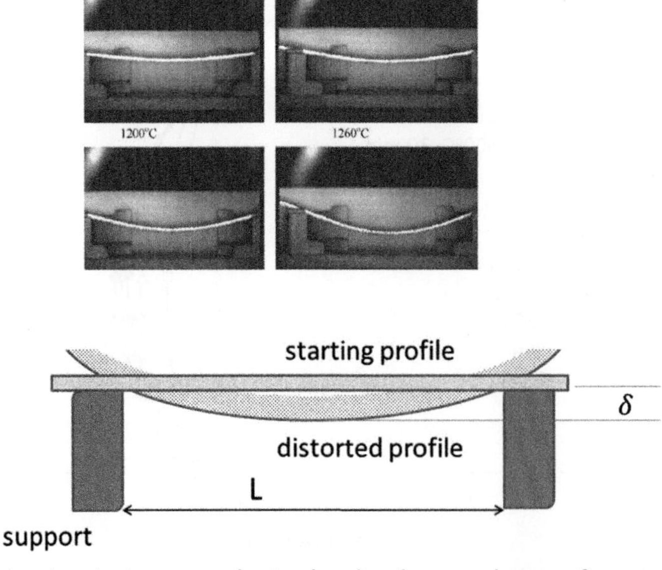

Figure 4.37 *In situ* viscosity is measured using bending beam techniques for rectangular, flat, thin samples spanning between two end supports. The viscosity is calculated from the time dependent sagging of the samples. At the top is a sequence of images taken during constant rate heating.

thickness h is suspended over a gap of span length L. The midpoint deflection δ from the horizontal is recorded versus time, temperature, and density (calculated from the sample size). The *in situ* viscosity η as a function of sintered fractional density f is given as follows:

$$\eta = \frac{5f\rho g L^4}{32(d\delta/dt)h^2} \tag{4.17}$$

where ρ is the theoretical density of the material (kg/m^3), g is gravitational acceleration (9.8 m^2/s), and $d\delta/dt$ is the deflection rate (m/s). Both dimensional change and deflection rate are recorded via video imaging [37]. The fractional density is calculated from the dimensional change assuming that the starting density is known, and the use of frequent images to capture deflection provides data on viscosity. As a consequence, plots of sintering viscosity versus fractional density are generated. Data for a 49 μm stainless steel powder sintering in the 1200 to 1400°C (1473 to 1673 K) range using a 5°C/min heating rate in hydrogen are shown in Figure 4.38. Densification accelerated as the viscosity decreased.

Such tests generally find sintering densification occurs when the material viscosity falls to about 10^{10} Pa · s. This is about the same viscosity as modeling clay. It seems that densification depends on heating the material to a temperature at which atomic flow softens the material to this effective viscosity range. Based on the Stokes-Einstein relation, diffusivity is inversely related to viscosity, so it is possible to estimate a temperature range for sintering densification if the material diffusivity is known.

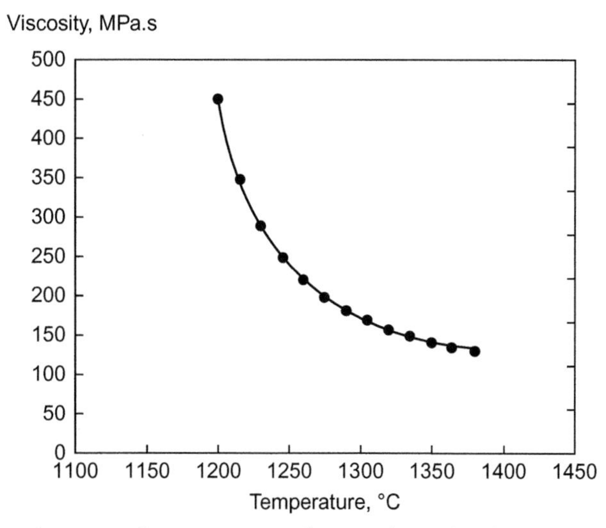

Figure 4.38 Calculated viscosity for a 49 μm stainless steel powder during heating over a range of temperatures from 1200 to almost 1400°C (1473 to 1673 K) at 5°C/min. Significant sintering took place as the viscosity dropped.

Synchrotron tomography is an option for *in situ* tracking of sintering events [13,39,40]. To fit into the technical window, the experiments are limited to lower temperatures, smaller samples, and larger particles. Measurements over time have confirmed neck growth, grain growth, grain rotation, and even neck rupture during sintering.

DENSITY, DENSIFICATION, POROSITY

Density is the most widely reported property in sintering. It is defined as the mass per unit volume. Unlike the absolute density (Mg/m^3 or g/cm^3), which depends on composition, fractional density gives evidence of the fundamental events occurring during sintering, independently of the material. Fractional density f is the measured density ρ divided by the theoretical density ρ_T for the same material.

Density is best measured by direct determination of the mass and volume. For simple geometries, the volume is calculated from the dimensions. However, if the shape is complex, then the immersion density is obtained by using the Archimedes technique. Before immersion, the open pores are filled with a fluid, such as water, mineral oil, silicone oil, or paraffin oil under vacuum. The Archimedes technique requires a series of weight determinations. In the test, the sample is first weighed dry (W_1), then weighed after fluid impregnation (W_2), and finally weighed while immersed in water (W_3). A wire is used to suspend the sample in the water, and its weight W_w is also measured in water. The density ρ is calculated from the weight determinations as:

$$\rho = \frac{W_1 \rho_W}{W_2 - (W_3 - W_W)} \tag{4.18}$$

where ρ_w is the density of water (which is slightly temperature dependent), for example:

$$\rho_W = 1.0017 - 0.0002315\,(T - 273) \tag{4.19}$$

with T being the water temperature in K. This series of weight measurements determines the density of an irregular shape with an accuracy ranging from three to six significant digits. The fractional density is defined as $f = \rho/\rho_T$, where ρ_T is the theoretical density.

Other density measurement techniques include X-ray absorption, magnetic resonance imaging, small angle neutron scattering, ultrasonic attenuation, and gamma ray absorption. They are used less frequently, since their resolution is lower and the equipment is expensive; however, such approaches prove useful for assessing density gradients in green and sintered structures. For example, intensity variations from X-ray examinations show point-to-point density variations are usually about 2% in green bodies.

Gas pycnometry relies on helium to measure sample volume, as long as the pores are closed. Using two chambers of known volume, a quantity of pressurized helium is introduced from the first to the second chamber holding the sample. The gas pressure change depends on the volume displaced by the sintered compact. The larger the volume, the higher the new pressure. Independent determination of sample mass then gives the density.

Eddy currents are also used to measure density *in situ*. The sample is surrounded by an inducing electrical coil, and the eddy currents are measured through a secondary coil, both of which are related to the compact density [41]. The measurements are not accurate, but have value for measuring density in pressurized furnaces, such as hot isostatic presses.

Densification Ψ indicates the relative change in density normalized to the density change needed to reach full density:

$$\Psi = \frac{f - f_O}{1 - f_O} \tag{4.20}$$

where f is the fractional sintered density and f_O is the initial density. At the start of sintering the densification is zero. If the material is sintered to full density ($f = 1$), then densification is unity. Like an exhaustion parameter, densification measures how much of the potential change has been achieved. Figure 4.39 is an example of sintering densification for a mixture of tungsten and copper sintered at 1150 and 1250°C in hydrogen for 120 min [42]. Variation in the green density, that give scatter to the sintered density, is minimized by resort to densification.

Porosity is closely related to density, where the fractional porosity is simply $1 - f$, f being the fractional density. Porosity is a measure of the pore volume. It is either expressed as a percentage or fraction. For simple geometries, porosity is measured by determining the weight and dimensions and comparing the density to the theoretical value. For more complex shapes or inhomogeneous structures, it is necessary to rely on microscopy [43]. It is important to preserve the pore structure during specimen preparation. Artifacts introduced during preparation easily change the apparent structure:

- smeared material covers over pores giving a low porosity
- grains pulled out during preparation appear as pores
- etching enlarges pores, leading to an overestimation of porosity.

The best microstructure preservation occurs by first polishing and etching the sample. Then liquid epoxy is impregnated into the polished surface to stabilize the pores. After hardening the epoxy, the structure is repolished. These steps are repeated until the image is stabilized. The microscopic contrast between the solid and pores allows quantitative analysis of the image for porosity. Porosity maps are created by performing the analysis over a section plane, leading to plots of porosity profiles versus location in the component.

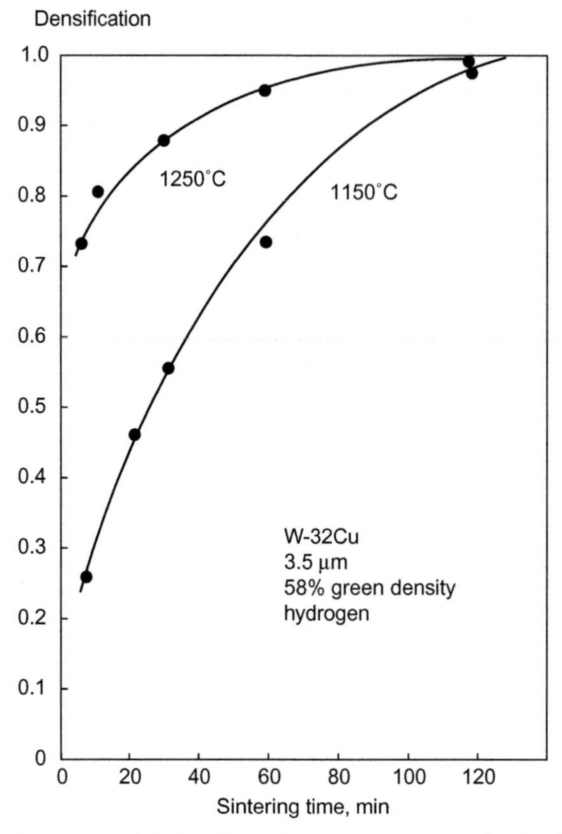

Figure 4.39 Densification versus sintering time at two temperatures for the liquid phase sintering of W-32Cu compacts formed from 3.5 μm tungsten powder starting with 58% green density [42].

CONDUCTIVITY

Thermal and electrical properties are important for applications in computers, electronics, and electric power. Pores are detrimental to conductivity. This is important during sintering, since thermal stresses from non-uniform heating can damage a component. To avoid distortion, heating rates are often restricted to 5 to 25°C/min.

Conductivity decreases as the porosity increases. Early in sintering, during neck growth prior to significant densification, the conductivity K varies with the square of the neck size ratio X/D as follows [44]:

$$K = K_{Og}\left(\frac{X}{D}\right)^2 \tag{4.21}$$

where K_O is the bulk conductivity at the measurement temperature. The parameter g depends on the grain size and initial packing density, since grain boundaries and pores both lower conductivity. As sintering progresses, the necks enlarge to a limit near $X/D = 0.5$. Further conductivity arise with densification, giving [45]:

$$K = K_O \frac{f}{1 + \chi(1-f)^2} \tag{4.22}$$

The empirical coefficient χ expresses the pore sensitivity. Unfortunately, this model lacks internal structure-dependent parameters, so it is only nominally accurate. Analysis of several materials, representing a variety of pore sizes and shapes, gives a best fit value of 11 for χ. As full density is attained the pores are spherical and the conductivity depends on fractional density:

$$K = K_O[1 - \omega(1 - f)] \tag{4.23}$$

where ω is between 1 and 2 [46]. Thus, prior to significant densification the conductivity increases with neck size, and as densification progresses it depends on density, and as full density is approached conductivity is a linear function of density. The behavior of sintered copper, alumina, and stainless steel based on relative conductivity (K/K_O) is illustrated in Figure 4.40. The solid line is Equation 4.22. Strength evolution during sintering is similar, so conductivity is also means for assessing strength.

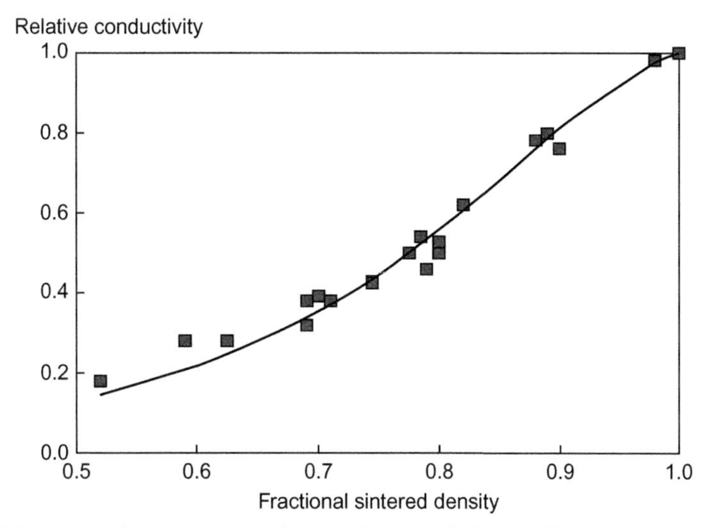

Figure 4.40 Relative conductivity versus fractional sintered density for several materials, showing the nearly linear behavior at high densities.

MAGNETIC PROPERTIES

The magnetic properties of sintered materials include both the soft and hard magnetic response. Because magnetic properties are lost at elevated temperatures, they are measured after sintering. Although the degree of sintering is important, impurity effects are condsiderable and require both densification and contamination control. Secondary factors include the pore shape and grain size. Magnetic properties are not good measures for sintering except as related to performance.

SURFACE AREA AND GAS PERMEABILITY

Surface area is lost during sintering. It is a good measure of sintering since it includes a large number of sintering necks. As plotted in Figure 4.41, early neck growth contributes to a rapid surface area loss, and as surface area is consumed the rate of reduction slows [47]. The plot in this figure shows the surface area during sintering 20 nm ZnO in oxygen versus sintering time at four temperatures. Surface

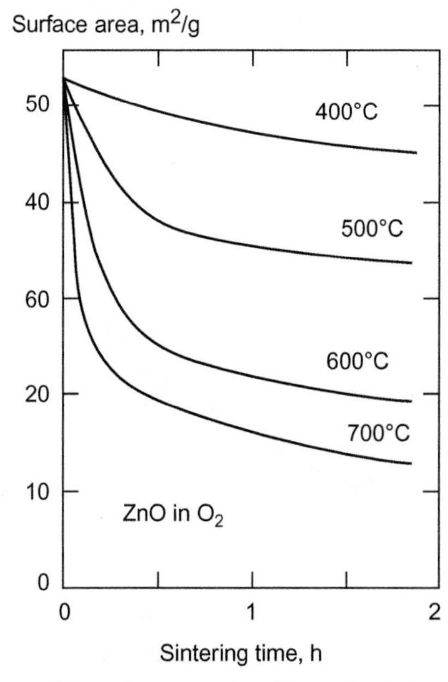

Figure 4.41 Gas absorption specific surface area for 20 nm ZnO sintered in a 50 kPa partial pressure of oxygen (half atmosphere) for various times up to 120 min and four temperatures [47]. The surface area loss is faster early in sintering and is accelerated by higher temperatures.

area is an effective way of tracking sintering, since all mechanisms can be monitored, even when there is no shrinkage. Even if a powder fails to shrink during sintering, the interparticle bonds reduce the surface area.

Like shrinkage, the surface area S is normalized by the initial surface area S_O. The surface area reduction $\Delta S/S_O$ is a dimensionless parameter which is useful for monitoring sintering. Initially there is a strong correlation between the neck size ratio X/D and the surface area reduction $\Delta S/S_O$, where ΔS is the change in surface area from the initial surface area S_O, treated as a positive change [48]:

$$\frac{\Delta S}{S_O} = k\left(\frac{X}{D}\right)^M \tag{4.24}$$

The exponent M ranges from 1.8 to 2.0 and depends slightly on the transport mechanism and initial compact density. It tends toward 2 for sintering that is controlled by surface diffusion or evaporation-condensation. Figure 4.42 is a plot of the relation between surface area reduction $\Delta S/S_O$ and neck size ratio X/D for spheres starting with an initial packing density of 60%. A high initial packing density correspond to more necks per particle and a faster loss of surface area, and a lower initial packing density give a correspondingly lower rate of surface area loss.

Three approaches are used to measure surface area—gas absorption, gas permeability, and quantitative microscopy. Both gas adsorption and gas permeability only measure the open pore structure. Quantitative microscopy is appropriate for cases where there is a low surface area or closed pores. The latter happens as full density is approached.

The gas adsorption approach starts with vacuum or inert gas bakeout of the sample. The clean surface is exposed to partial pressures of adsorbing vapors in an arrangement such as that sketched in Figure 4.43. A measurement is made of the amount of gas adsorbed on the surface versus the partial pressure, in this case using

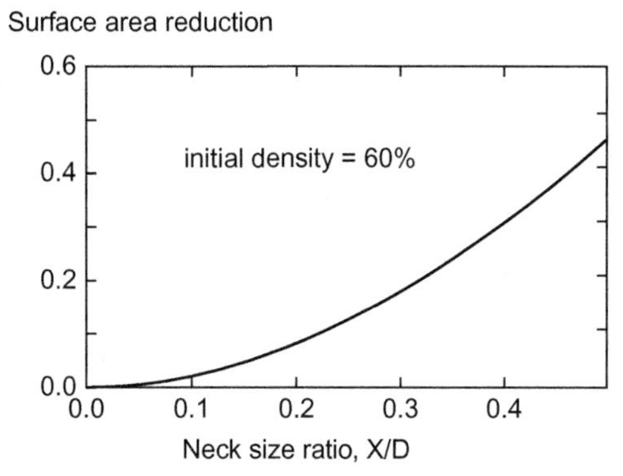

Figure 4.42 The relation between surface area reduction ($\Delta S/S_O$) and neck size ratio (X/D) for sintering a compact of spherical particles starting at a 0.6 fractional density.

the change in gas thermal conductivity (helium is more thermally conductive versus nitrogen, so as nitrogen is absorbed the gas thermal conductivity changes). From the amount of gas absorbed, the approach calculates what Brunauer, Emmett and Teller referred to as specific surface area [49]. An example of the surface area reduction during sintering was shown earlier in Figure 4.41 for zinc oxide.

An alternative is to use gas permeation through the open pores, since the gas permeability depends on the surface area. A gas flows through the sintered structure from an upstream pressure P_1 to a downstream pressure P_2, as sketched in Figure 4.44.

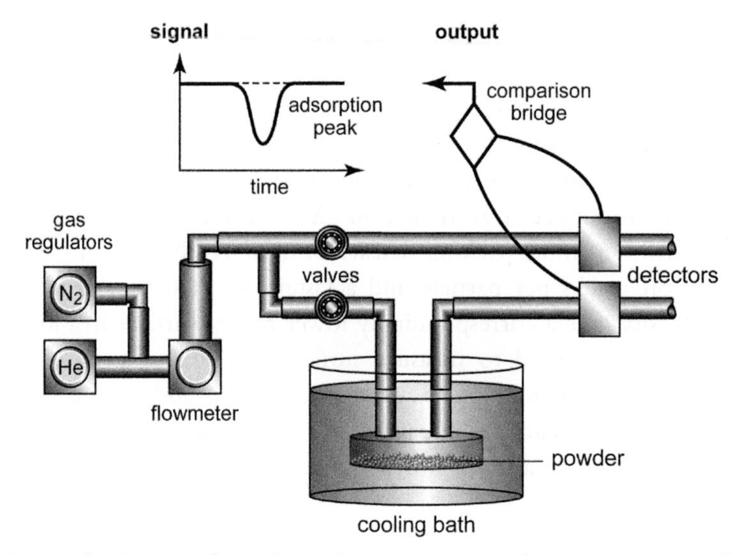

Figure 4.43 Layout for the use of gas absorption to measure surface area on sintered powder. The powder compact is immersed in a cooling bath (liquid nitrogen is common) and that causes absorption of nitrogen from the gas flow. The difference in thermal conductivity between helium and nitrogen results in a measurable absorption peak that is used to quantify the amount of nitrogen absorbed and the surface area.

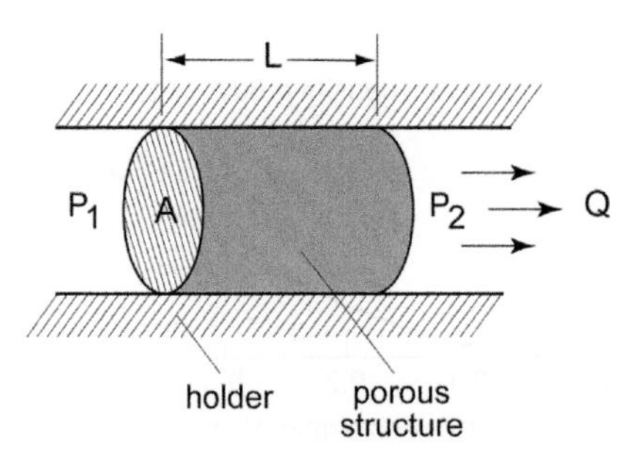

Figure 4.44 Gas permeability is a means for measuring surface area on a porous compact based on Darcy's law where the pressure drop and flow rate are related to measure the permeability and surface area is calculated from the permeability.

Darcy's equation states that the gas flow velocity v based on the pressure drop $\Delta P = P_1 - P_2$ and the measured gas flow velocity v (Q/A where Q is the volumetric gas flow rate measured at standard pressure and A is the cross-sectional area) is as follows:

$$v = \frac{\Delta P \alpha}{L \mu} \tag{4.25}$$

The sample length is L and the gas viscosity is μ. The permeability coefficient α is used to calculate the surface area. The gas flow rate depends on the surface area and permeability based on the Kozeny-Carman relation:

$$S = \frac{1}{\rho_T} \sqrt{\left[\frac{(1-f)^3}{5\alpha f^2}\right]} \tag{4.26}$$

with ρ_T being equal to the theoretical density of the material and f equal to the fractional density. The behavior is shown in Figure 4.45 as a log-log plot of permeability measured surface area reduction for 0.2 μm alumina sintered for at 750°C (1023 K) for various times [50]. Permeability provides a gauge of surface area as long as the pores are open, meaning that the technique does not assess dead-ended pores.

Microscopy surface area measurements are made on cross-sectioned and polished samples. It is most useful for low surface area samples. Computer based image analysis programs extract the surface area using the pore perimeter per unit cross-sectional

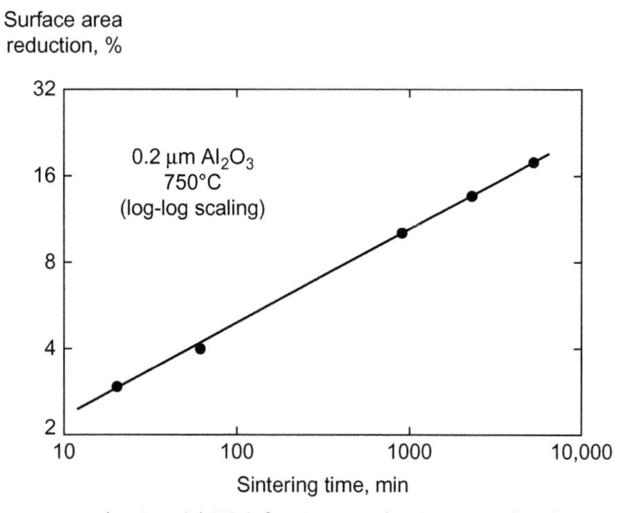

Figure 4.45 Surface area reduction ($\Delta S/S_O$) for 0.2 μm alumina powder during sintering at 750°C (1023 K) for various times [50]. Here a log-log plot is used to extract the sintering time dependence.

area. The number of pore-solid intersections per unit test line length N_L equals the surface area S_V per unit volume:

$$S_V = \frac{2N_L}{f} \tag{4.27}$$

Gas absorption surface area is based on area per unit mass, while quantitative microscopy is based on area per unit volume, so the two values differ but can be compared using the sample density.

PORE STRUCTURE

The pores between grains are an inherent part of sintering. Pores are involved in mass transport, densification, grain coarsening, and gas reactions during sintering. When a liquid forms, the pores distribute the liquid throughout the grain structure.

Pores are characterized by the amount, size, shape, and distribution of voids in the compact. Up to approximately 90 to 95% density, the pore space consists of interconnected tubes as illustrated in Figure 4.46. Ideally the pores occupy the grain edges, so it is incorrect to think of the pores as isolated objects. Usually there is a characteristic

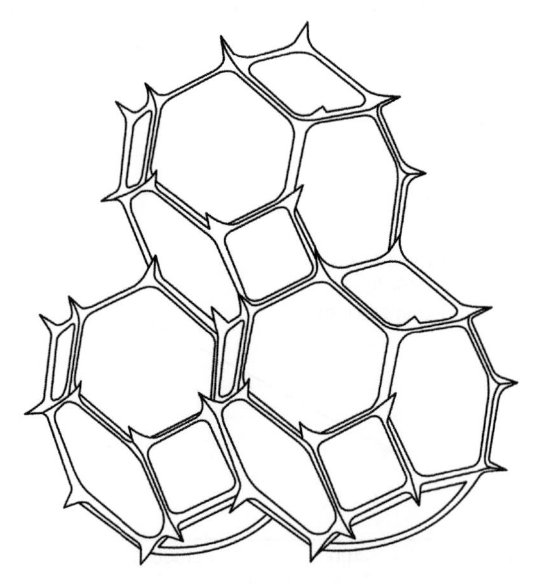

Figure 4.46 During sintering the necks grow to form a network of pores on the grain edges. An example geometry is shown here, where the grain shape is a contacting tetrakaidecahedron and the pores form a network of tubes on the grain edges.

size for the pores, but the tubes form a complex three-dimensional network. The pore structure is modified by sintering. Often, the average pore size remains relatively constant but the number of pores declines. For example, the data in Figure 4.47 for 7 μm iron shows little change in pore size while the porosity declines and the grain size increases [51].

There are two types of pores: Open pores imply the pore is connected to the outside and allows fluids to move in or out of the body, while closed pores are sealed and not in communication with the external surface. Open pores close as the porosity declines during sintering, usually starting at about 85% density and reaching total closure by about 95% density. Such data are given in Figure 4.48 for copper during sintering using a range of compaction pressures [52].

The measurement of pore size, pore shape, and pore connectivity is performed by microscopy—optical microscopy for larger features and electron microscopy for smaller features (generally below a few micrometers). Pore size and pore shape measurements by quantitative microscopy are accurate, but tedious to perform. The pores are sliced randomly. Serial sectioning allows for three-dimensional reconstructions using closely spaced slices [53]. Random cross-sections do not necessarily capture the largest pore dimension, so the pores usually appear smaller than their actual size. Even so, random two-dimensional cross-sections allow accurate deduction of the mean pore size and spacing.

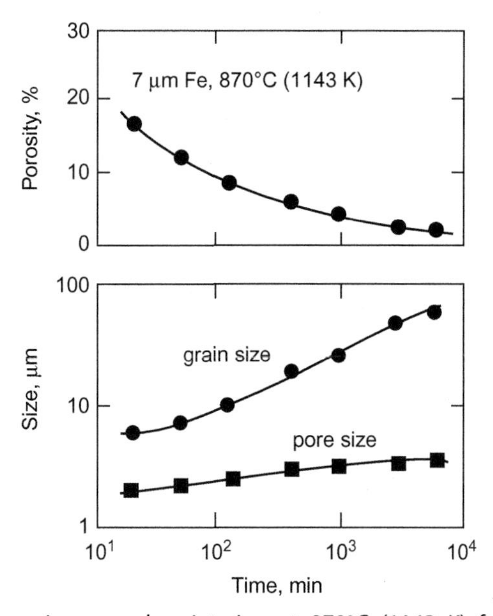

Figure 4.47 Data for 7 μm iron powder sintering at 870°C (1143 K) for prolonged times [51]. Plotted here is porosity, grain size, and pore size—where the porosity decreases but both the grain and pore size increase.

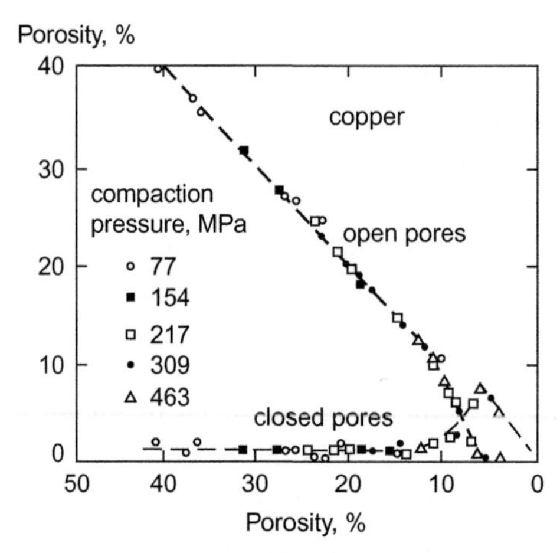

Figure 4.48 Copper sintering data showing the decline in porosity and the fraction of open and closed pores during sintering for five different initial compaction pressures [52]. Pore closure becomes evident below 15% porosity and by 5% porosity all of the remaining pores are closed.

Mercury porosimetry is one means for measuring a pore size distribution for open pore structures. It is applicable to pores as small as 1 nm. A mercury porosimeter measures the amount of mercury intruded into the porous material versus the applied pressure. It assumes the pores are cylindrical in shape. For most materials the contact angle with mercury is high, near 130° and the surface energy ranges near 0.48 J/m^2. Because of the high contact angle, mercury resists pressurized intrusion into pores in an inverse relation to the pore size. Assuming a cylindrical pore, the pressure P to cause mercury intrusion depends on the contact angle θ, pore diameter d and surface energy γ according to the Washburn equation, as follows:

$$P = -\frac{4\gamma \cos(\theta)}{d} \tag{4.28}$$

The negative sign indicates a compressive pressure is required to intrude mercury into the pore. A mercury intrusion experiment is carried out by surrounding the material with mercury. The sample is evacuated to remove gases and oil pressure is applied to the mercury to force it into the pores; the large pores fill at lower pressures. A record of the mercury volume intruded versus the pressure enables a calculation of the pore size distribution. The measurement is conducted in a high pressure chamber. The sample is placed in a calibrated tube and the volume of mercury intruded versus pressure is sensed similar to how a mercury thermometer operates. The pore size distribution data can be expressed on either a cumulative or differential basis.

Plots of cumulative pore size distribution are given in Figure 4.49 for titania sintered at temperatures from 900 and 1200°C (1173 and 1473 K) [54]. Sintering at the lower temperature results in a 55% density and pores in the 0.1 μm size range.

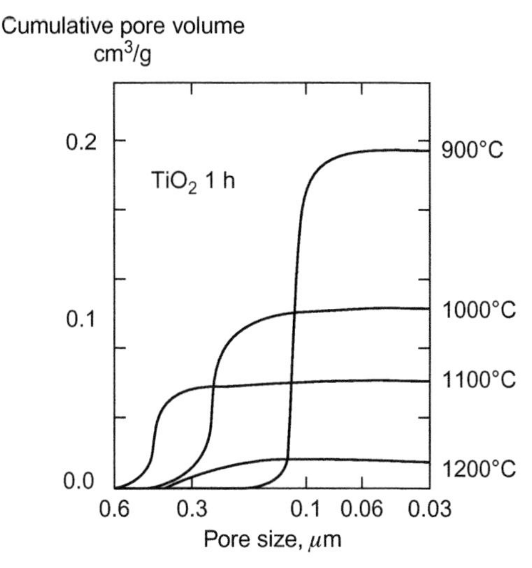

Figure 4.49 Pore size distributions showing the cumulative fraction of pores versus the pore size for titania sintered 60 min at four temperatures [54]. Note the total pore volume decreases with higher sintering temperatures, but the pore size increases from 900°C (1173 K) to 1000°C (1273 K) to 1100°C (1373 K).

As the sintering temperature increases there is less porosity, but the pore size distribution moves to a larger average size. By 1100°C (1373 K) the density is 79% but the pores are 0.4 μm in size. Only at 1200°C (1473 K) when the density is 93% does shrinkage of the largest pores take place. Mercury porosimetry data is difficult to interpret, because assumptions of a simple pore shape ignores their undulating and multiply connected character.

Pore shape and open pore surface area are also calculated in mercury porosimetry. Pore shape comes from the hysteresis on depressurization [55,56]. The greater the hysteresis, the greater the departure from a cylindrical pore shape. A non-uniform pore limits the flow of mercury in and out of the pore based on any constrictions. Thus, the volume of an irregular pore is recorded at the pressure necessary to break through the minimum size. On depressurization, mercury retention also depends on the pore shape.

Generally the sintered pore size distribution conforms to a Weibull model. The cumulative fraction of pores F larger than size d depends on the median pore size d_M and an exponent M. The corresponding distribution is expressed as:

$$F(d) = 1 - exp\left(-g\left(\frac{d}{d_M}\right)^M\right) \tag{4.29}$$

where g equals 0.6931 (ln 0.5). Figure 4.50 plots data on the cumulative pore size distribution for sintered zirconia [57]. The solid line is the fit to the pore size data based on the above distribution using $M = 2$.

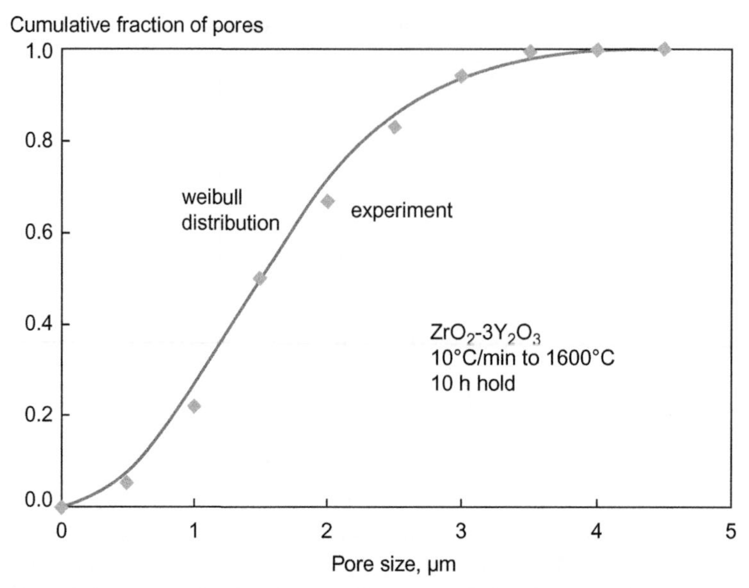

Figure 4.50 Cumulative pore size distribution for yttria stabilized zirconia heated at 10°C/min to 1600°C (1873 K) for 600 min [57]. The solid line shows the fit based on an assumed Weibull distribution with a modulus $M = 2$.

The bubble point test is a related pore size measure based on the displacement of a wetting fluid from open pores. It measures the maximum interconnected pore size and is best applied to thin structures. A gas pressurized to P is used to displace a wetting fluid, such as ethanol, from the pores (contact angle of zero or $cos(\theta) = 1$ in the Washburn equation). The pore diameter d is estimated as follows:

$$d = \frac{4\gamma}{P} \tag{4.30}$$

where γ is the liquid surface energy. The pressure corresponds to the point where the gas first bubbles through the pores; this first bubbling corresponds to the largest pores.

Analysis techniques for pore size are sensitive to pore closure. Only open pores contribute to the bubble point, permeability, or gas absorption tests. Likewise, mercury porosimetry only accesses the open pore structure. Optical microscopy is unable to distinguish between open and closed porosity. However, the smaller pores are typically closed pores. The total porosity does not distinguish between the open and closed pores, but is the most important parameter.

Pore size and shape vary during sintering. Initially, the pores are irregular, but as sintering progresses the pores form a smooth, near-cylindrical network. At roughly 85% density, the tubular open pores start to collapse into near spherical closed pores. Because of the changes in pore size, pore shape, and pore connectivity, porosity is an incomplete measure of sintering response. Indeed, inhomogeneity in the starting pore structure leads to pore growth and inhibited densification. High green densities with small and homogeneous pores are preferred for sintering. Late in sintering the pore size distribution has a dominant role in sintering densification.

MICROSTRUCTURE

The sintered microstructure is dominated by its pore and grain structure. The adage of a picture being worth a thousand words is evident in sintering. For example, Figure 4.51 contrasts two sintered stainless steel microstructures with differences in porosity (black region), pore size, pore shape, grain boundary attachment to the pores, and overall microstructure. Sintering produces many changes measurable by micro-structure parameters [9,43,53,58–61].

Composition influences microstructure by changing the ratio of phases. For example, in liquid phase sintering at relatively low solid levels, the solid grains are small, separated by the matrix, and are nearly spherical. As the solid content increases, the grain size increases, grain contacts increase, and the grains become polygonal. A quantitative description of these several microstructural changes is important for monitoring sintering.

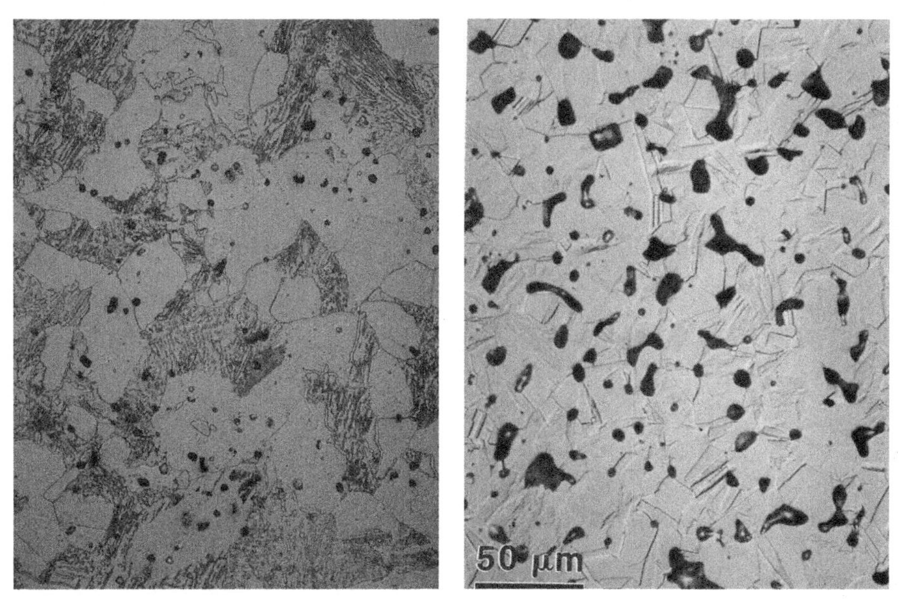

Figure 4.51 Two sintered stainless steel microstructures, where the pore size (black), pore shape, and porosity are dramatically different. There is also a difference in pore attachment to the grain boundaries.

Interfacial energies determine microstructure; in turn microstructural measurements provide a means to extract interfacial energy data. For example, the dihedral angle is extracted by using two-dimensional microscopy [60]. Furthermore, the phase distribution, including the location of residual pores, depends on the interfacial energies.

Porosity and pore size measurements are performed by quantitative microscopy. An important parameter with regard to fracture behavior is the distance between pores, which is measured by the mean free separation λ as follows:

$$\lambda = \frac{1-f}{N_L} \tag{4.31}$$

where f is the fractional density (or volume fraction of solid phase in a liquid phase sintered material) and N_L is the number of solid grain intercepts per unit of test line length.

Microscopy measurements are performed on a random cross-section of the materal. Accordingly, the feature size is smaller than the true sizes. A sectioning plane might miss the neck bonding to grains together, as sketched in Figure 4.52. Thus, neck size measurements from a two-dimensional section are in error. Derivation of three-dimensional information is possible via serial sectioning, where small incremental planes are removed, polished and photographed to reconstruct the true object's shape and size [61].

Many sintered materials are formed from mixed powders, including bronze, steel, mullite, silicon nitride, and cemented carbides. Homogenization during sintering is usually slower than densification. Microstructural analysis provides a means to monitor homogenization by determining the composition gradients.

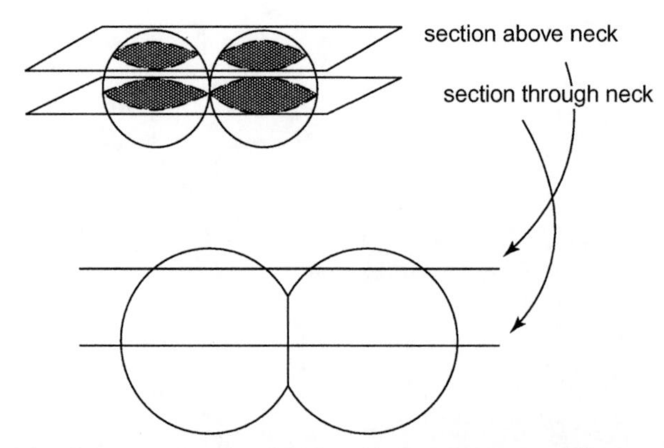

Figure 4.52 A sketch of two sintered spheres and the apparent difference seen in neck size depending on the position of the random section plane. It is for this reason two-dimensional neck size data are unreliable.

Grain Size Distribution

Changes in grain size and grain shape occur during sintering. Usually grain growth follows a law by which the average grain volume enlarges linearly with time t. Accordingly, the cube of the linear grain size G is proportional to the grain volume, giving:

$$G^3 \approx t \tag{4.32}$$

Figure 4.53 is a logarithmic plot of grain size and time for 25 μm stainless steel sintering in vacuum at 1320°C (1593 K). The straight line has a slope of 3. During these experiments the density increased from 91 to 98%.

Grain size is estimated from the microstructure cross-section using random intercept, equivalent circle area, grain perimeter, or embracing circle techniques [62–64]. The random intercept grain size G_R is based on the length of test line falling on the grain image. A more accurate grain size measure on two dimension sections is by calculating the diameter of a circle with the same area as the grain, giving the equivalent circular grain size G_C:

$$G_C = \sqrt{\frac{4A}{\pi}} \tag{4.33}$$

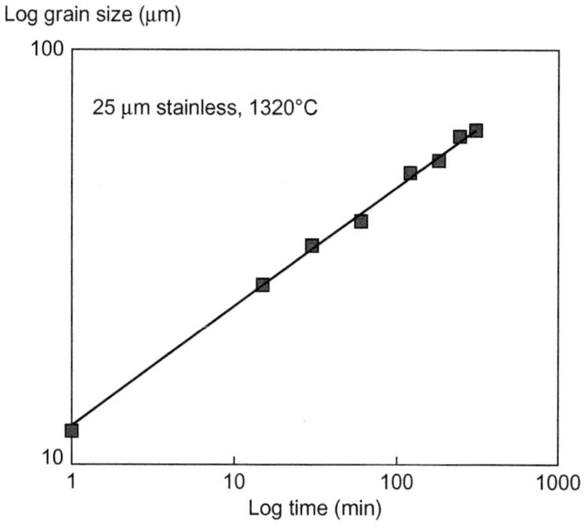

Figure 4.53 Log-log plot of grain size versus vacuum sintering time at 1320°C (1593 K) for a 25 μm stainless steel powder. The symbols are measured sizes and the solid line corresponds to a cubic relation between grain size and sintering time. During this time the sintered fractional density increased from 0.91 to 0.98.

The true three-dimensional grain size G is used in theories, but grain size measurements are taken from random two-dimensional cross-sections. A transformation is needed to convert to the true three-dimensional values [65]. It is improper to compare intercept size distributions with theoretical results based on three-dimensional distributions.

The relations between the true average 3-D grain size and the 2-D equivalent circle average or intercept size is:

$$G = 1.27 G_C = 1.65 G_R \tag{4.34}$$

Transforming two-dimensional data for comparison with three-dimensional models encounters difficulties because of the random section plane and nonspherical grain shape.

Grain size is a distributed parameter, meaning there is a natural distribution in the microstructure. Sintering progresses toward a self-similar grain size distribution, meaning that the distribution is the same shape at long sintering times. The location of the median grain size is the only adjustable parameter. Such behavior is evident in the magnesia data plotted in Figure 4.54 [66]. The symbols are experimental two-dimensional grain size measurements, whereas the solid line corresponds to the Weibull exponential fit. Let $F(G)$ be the cumulative fraction of grains with size of G, where G_M is the median size, then:

$$F(G) = 1 - exp\left[\beta\left(\frac{G}{G_M}\right)^M\right] \tag{4.35}$$

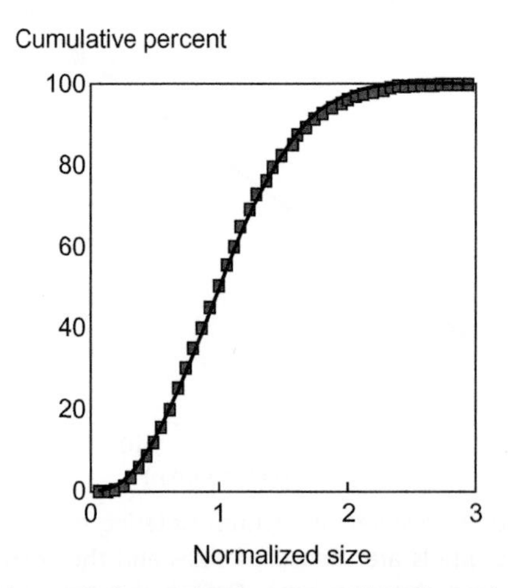

Figure 4.54 Two-dimensional cumulative grain size data taken from sintered magnesia (MgO) [66]. The symbols are the measured values and the solid line is a Weibull distribution fit.

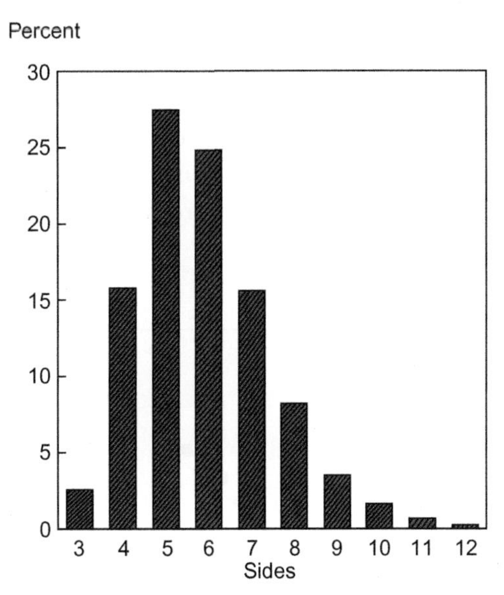

Figure 4.55 Grain shape data taken from sintered magnesia using two-dimensional sections [67], showing pentagons are the most common grain shape.

where the factor $\beta = -ln2$ (or -0.6931) to ensure that $F(G) = 0.5$ at the median or 50% size when $G = G_M$. The exponent M reflects the dispersion of the distribution. For 2-D data M tends toward 2 and for 3-D data M tends to be near 3.

Grain shape also converges to a self-similar distribution. Grain shape is measured in two dimensions by the number of grain sides. Figure 4.55 is a histogram of the number of grain sides in sintered magnesia [67].

Grain growth is influenced by pores. An important relation between grain size G and fractional density f during sintering is as follows [68]:

$$G = \frac{G_O \theta}{\sqrt{1 - f}} \tag{4.36}$$

where G_O is the initial grain size and θ is near 0.6. This relation applies up to near full densification, at which point grain growth continues without a change in density. Figure 4.56 plots an example fit for nickel during sintering [5]. These data were collected for a 4.3 μm powder sintered at 900°C (1173 K) for up to 40 h, giving a final density of 93%. The relation between grain size and fractional density works since the average pore size changes slowly when compared to the changes in porosity and grain size during sintering.

Pore Interaction with Grain Boundary

Late in sintering, an important microstructure divergence occurs in some sintering cycles. Commonly, grain growth occurs with pore drag, but it can also strand pores [69]. This becomes evident at 90 to 95% density. The two variants are sketched in Figure 4.57. Full density sintering is not possible (without external pressure) if the pores separate from the grain boundaries, since pore annihilation requires grain boundaries [70].

As grain growth occurs, the slowly moving pores retard the moving grain boundaries, leading to distorted pore shapes. A schematic of the interaction is given in Figure 4.58 for a dihedral angle of 140° and pore size to grain size ratio of 0.4. The

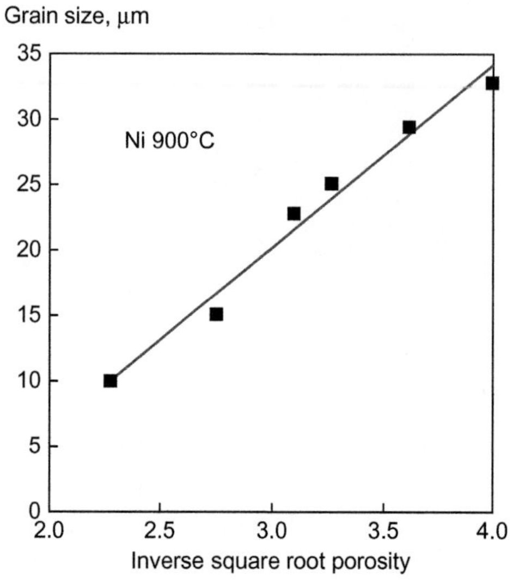

Figure 4.56 Nickel sintering data relating the grain size to the inverse square-root of the fractional porosity for samples prepared at 900°C (1173 K) [5]. The initial powder was 4.3 μm prior to sintering, so the grain coarsening is extensive, reaching a grain size of 33 μm at 6.3% porosity.

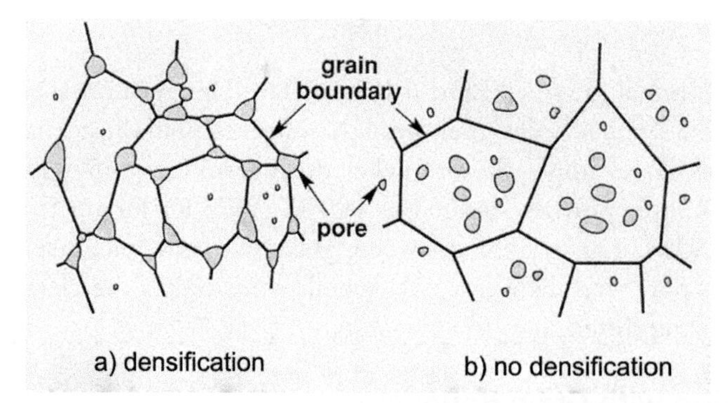

Figure 4.57 Pore attachment to grain boundaries is critical to sustained sintering densification and this figure schematically shows two distinct possibilities. The configuration on the left is most desired for densification.

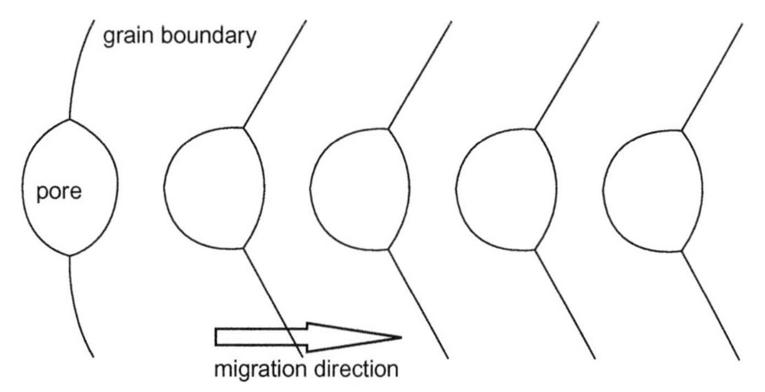

Figure 4.58 Moving grain boundaries can drag a pore. The resulting distorted pore shape enables pore migration by transport across the pore due to differing curvature. But as grain growth progresses the grain boundary separates from the pore, leading to hindered sintering densification.

grain boundary is moving, and since pore migration is slow the pore is distorted. Eventually the two will separate.

Microstructural measurement of the pore-boundary attachment is performed using quantitative microscopy [71]. The dihedral angle determines the attachment, where 180° corresponds to no grain boundary energy, spherical pores, and no attachment. More typical in sintering are pores attached to the grain boundaries with a lenticular pore shape. In this case, microstructural differences arise depending on the relative rates of pore motion (via transport across the pore) and grain boundary motion (via transport across the boundary), and pore attachment [72]. Quantitative microscopy is used to calculate the relative pore-boundary interaction in sintering, generally showing preferential attachment of pores on boundaries over random placement [73].

THERMAL PROPERTIES

Thermal characteristics prove important to understanding the reactions which occur during sintering. A variety of approaches are used, including the following:

- Differential thermal analysis or differential scanning calorimetry, where exothermic (reaction) or endothermic (melting) events are detected during a heating cycle—used to identify reaction points or phase transformations [74].
- Mass spectroscopy, infrared spectroscopy, or gas chromatography, where the effluent gas from a sintering experiment is analyzed to measure the atomic mass of the reaction species, such as CO or CO_2 formation [75].
- Thermogravimetric analysis, showing mass changes during heating that might include the formation of volatile reaction products (H and O to form steam) or atmospherical reactions with the sintering material (Al heated in N_2 to form AlN) [76].

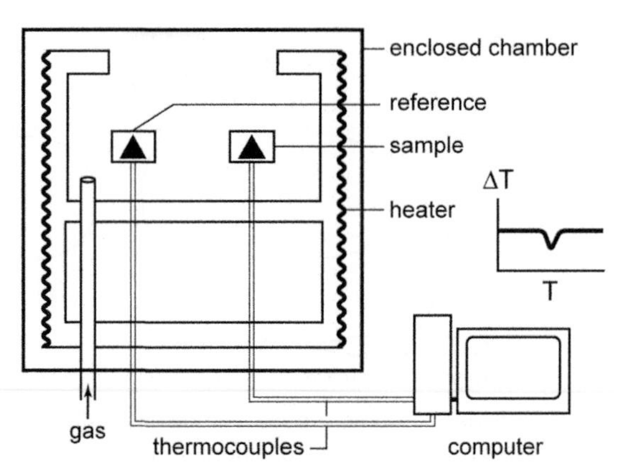

Figure 4.59 Thermal analysis relies on differential thermal analysis (DTA), thermogravimetric analysis (TGA), or differential scanning calorimetry (DSC). Shown here is a schematic for how DTA or DSC data are collected using a sintering sample in the same furnace environment as a reference. If the mass is recorded simultaneously, then TGA data are also generated.

These tools are potent when coupled with dilatometry or other measurement tools. For example, thermal analysis is helpful in determining the temperature of first melt formation during liquid phase sintering. As illustrated in Figure 4.59, thermal analysis relies on two samples in the same thermal environment. The samples are surrounded by protective gas. One of the crucibles holds the sintering sample while the other contains an inert standard, such as platinum or alumina. As the samples are heated, energy flows into both crucibles at the same rate to give similar temperatures. By measuring the temperature difference between the samples, any perturbations due to exothermic or endothermic reactions are detected if they heat up faster or slower. A plot of temperature change versus temperature identifies where reactions occur.

Another helpful measure is the vapor pressure during vacuum sintering, with the possible addition of mass spectroscopy or gas chromatography [77]. Figure 4.60 shows experiments conducted on 2 μm hydroxyapatite powder heated to 1000°C (1273 K) and compares the moisture evolution with the same powder mixed with 40 vol.% of 70 nm zirconia [78]. The combination of gas pressure and gas composition measurements provides information on pore closure and chemical reactions during sintering.

For metals, a vapor pressure decrease is associated with the point of pore closure. In reactive situations, mass gains are possible but is a mass loss more typical, as volatile reaction species evolve. As an example, decarburization in sintered steels is traceable to the initial oxygen level in the powder. Oxygen and carbon react, leading to low hardness surfaces after sintering. Figure 4.61 plots data for Fe-1C, where the thermogravimetric analysis traces the reaction of added carbon with residual oxygen during heating [79]. The decarburization reduced the final carbon level by about 0.1 wt.%.

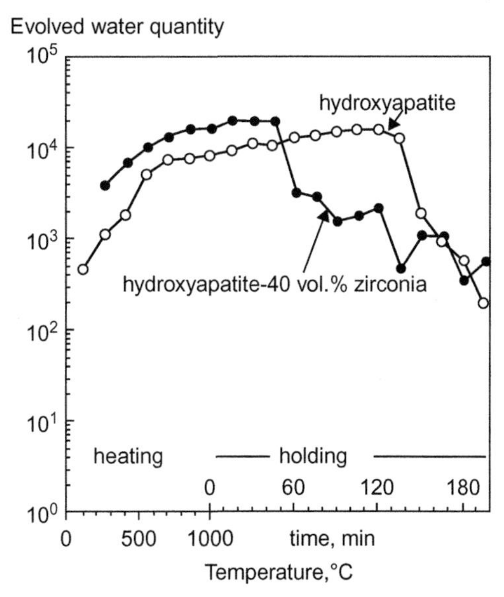

Figure 4.60 Data on gas evolution during heating and hold for two powder systems, hydroxyapatite (calcium phosphate) and hydroxyapatite with zirconia. These data are recorded using gas chromatography to determine water evolution in heating and holding at 1000°C (1273 K) [78].

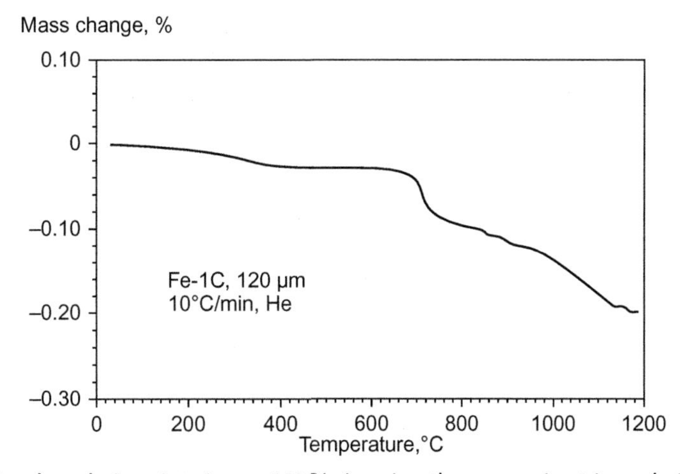

Figure 4.61 Mass loss during sintering at 10°C/min using thermogravimetric analysis of Fe-1C mixture heated to 1200°C (1473 K) [79].

SUMMARY

Sintering practice remains far ahead of a predictive sintering science, but theory has emerged to explain the observations outlined in this chapter. This chapter has given attention to the tools and data available to establish a quantitative conceptualization of the sintering process. The complexity of the story is captured by the micrographs in Figure 4.62. These were taken using 16 μm average particle size stainless steel powder. For these micrographs, samples were heated at 10°C/min in hydrogen to

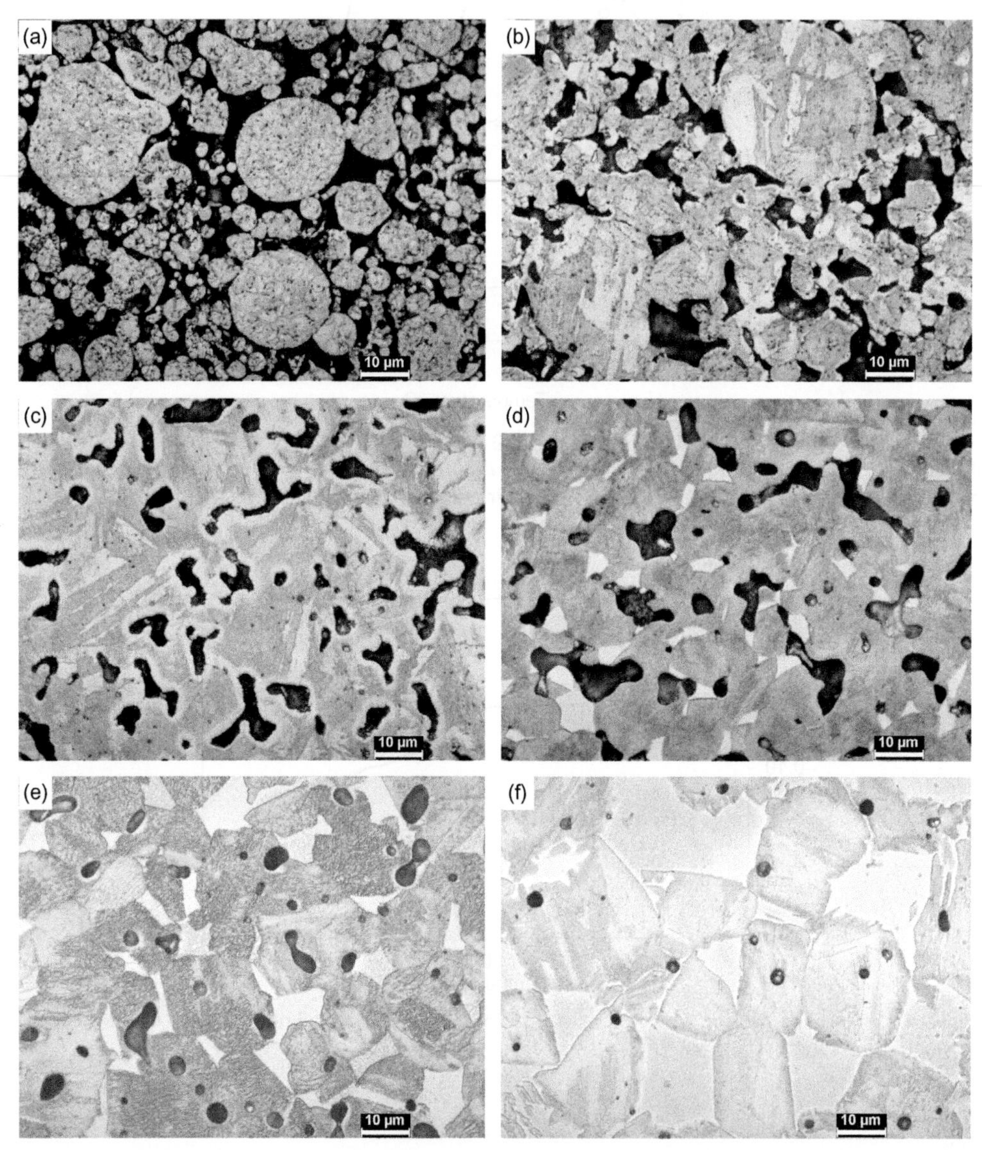

Figure 4.62 Polished and etched micrographs capturing sintering for a stainless steel powder heated to various temperatures in hydrogen; a) heated to 1000°C, b) 1100°C, c) 1200°C, d) 1260°C, e) 1300°C, and f) 1365°C (1273 to 1538 K). The black phase is porosity and a metallurgical transformation to delta-ferrite gives the two phase final microstructure.

Table 4.1 Some Measurement Techniques and Observations Used in Sintering Studies

Parameters	Measurement Techniques
density	dimensions and weight
	Archimedes water immersion
	attenuation: X-ray, gamma ray, neutron absorption
	quantitative microscopy
	helium pycnometry
	ultrasonic attenuation
pore size	quantitative microscopy
	bubble point
	mercury porosimetry
	small angle neutron scattering
grain size	quantitative microscopy
neck size	quantitative microscopy
	microfocus computer tomography
	synchrotron imaging
shrinkage/swelling	bulk linear dimensions
	dilatometry
surface area	gas adsorption
	mercury porosimetry
	gas permeability
	quantitative microscopy
thermal and electrical conductivity	eddy current
	differential thermal analysis
	thermogravimetric analysis
	magnetic permeability
mechanical	rupture, tensile, or fracture strength
	hardness
	elastic modulus, resonant frequency, or sonic velocity
	fracture toughness, impact energy
reactions, homogenization	X-ray diffraction
	thermal analysis, thermogravimetric or differential analysis
	chromatography
	microscopy, microprobe, or electron microscopy
	positron annihilation
	mass spectroscopy

various temperatures, and then water quenched to preserve the evolving microstructure. Polished cross-sections then show the microstructural changes in reaching the peak sintering temperature of 1365°C (1538 K). Small sinter bonds form and grow as temperature increases. Eventually the sinter bonds overlap, resulting in rounded pores. The porosity decreases as the temperature increases. A second phase forms corresponding

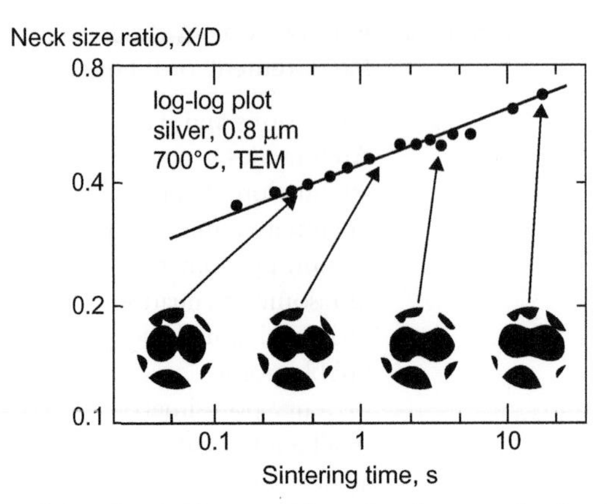

Figure 4.63 The essence of sintering is illustrated here for the growth of a sinter neck during heating for 0.8 μm silver particles in a hot stage transmission electron microscope at 700°C (973 K) [80].

to delta-ferrite and the final sintered microstructure consists of closed pores in a two phase solid.

A key objective of sintering theory is to explain this progression and to predict the resulting density, properties, component size, and microstructure as related to the process parameters such as particle size, green density, heating rate, peak temperature, hold time, and atmosphere.

Several tools are available for such sintering studies, as summarized in Table 4.1. They often provide complimentary insight. Selection of the appropriate tool and proper experimental protocol depends on the application and sintering behavior. It would be inappropriate to anticipate which observation route is useful, so tools need to be selected to ensure critical insight.

Besides sintering optimization, the observations and tools outlined in this chapter provide a sense of what needs to be explained by sintering theory. Early observations demonstrated neck growth as the fundamental event in sintering [80]; Figure 4.63 shows early neck growth measurements for silver. Explaining these data is the core of modeling efforts and is a focus for much of this book. Sintering observations were advanced prior to the emergence of a robust theory. Observations covered in this chapter are litmus tests for theoretical efforts covered in subsequent chapters.

REFERENCES

[1] J. Frenkel, Viscous flow of crystalline bodies under the action of surface tension, J. Phys. 9 (1945) 385–391.

[2] R.F. Wang, Y.X. Wu, X. Zhou, C.A. Tang, Debinding and sintering processes for injection molded pure titanium, Powder Metall. Tech. 24 (2) (2006) 83–93.

[3] R.M. German, Powder Metallurgy and Particulate Materials Processing, Metal Powder Industries Federation, Princeton, NJ, 2005.

[4] C.D. Turner, M.F. Ashby, The cold isostatic pressing of composite powders – I. Experimental investigations using model powders, Acta Mater. 44 (1996) 4521–4530.
[5] P.E. Evans, D.W. Ashall, Grain growth in sintered nickel powder, Inter. J. Powder Metall. 1 (1) (1965) 32–41.
[6] R.M. German, Particle Packing Characteristics, Metal Powder Industries Federation, Princeton, NJ, 1989.
[7] M.J. Mayo, Processing of nanocrystalline ceramics from ultrafine particles, Inter. Mater. Rev. 41 (1996) 85–115.
[8] Y. Liu, B.R. Patterson, A. Stereological, Model of the degree of grain boundary-pore contact during sintering, Metall. Trans. 24A (1993) 1497–1505.
[9] F.N. Rhines, R.T. DeHoff, A. Topological, Approach to the study of sintering, in: H.H. Hausner (Ed.), Modern Developments in Powder Metallurgy, vol. 4, Plenum Press, New York, NY, 1971, pp. 173–188.
[10] D.L. Johnson, Sintering kinetics for combined volume, grain boundary and surface diffusion, Phys. Sintering 1 (1969) B1–B22.
[11] S.B. Boskovic, M.M. Ristic, Sintering of nonstoichiometric nickel oxide, Powder Metall. Metal Ceram 11 (1972) 755–759.
[12] M. Yeadon, J.C. Yang, R.S. Averback, J.W. Bullard, J.M. Gibson, Sintering of silver and copper nanoparticles on (001) copper observed by in situ ultrahigh vacuum transmission electron microscopy, Nano. Mater. 10 (1998) 731–739.
[13] O. Lame, D. Bellet, M. Di Michiel, D. Bouvard, Bulk observation of metal powder sintering by X-ray synchrotron microtomography, Acta Mater. 52 (2004) 977–984.
[14] G.C. Kuczynski, Self-diffusion in sintering of metallic particles, Trans. TMS-AIME 185 (1949) 169–178.
[15] F.A. Nichols, W.W. Mullins, Morphological changes of a surface of revolution due to capillarity-induced surface diffusion, J. Appl. Phys. 36 (1965) 1826–1835.
[16] R.M. German, Z.A. Munir, Morphology relations during surface-transport controlled sintering, Metall. Trans. 6B (1975) 289–294.
[17] R.M. German, Z.A. Munir, Morphology relations during bulk-transport sintering, Metall. Trans. 6A (1975) 2229–2234.
[18] J. Bernholc, P. Salamon, R.S. Berry, Annealing of fine powders: initial shapes and grain boundary motion, in: P. Jena, B.K. Rao, S.N. Kahanna (Eds.), Physics and Chemistry of Small Clusters, Plenum Press, New York, NY, 1987, pp. 43–48.
[19] R.M. German, J.F. Lathrop, Simulation of spherical powder sintering by surface diffusion, J. Mater. Sci. 13 (1978) 921–929.
[20] P.D. Wilcox, I.B. Cutler, Strength of partly sintered alumina compacts, J. Amer. Ceram. Soc. 49 (1966) 249–252.
[21] R.L. Coble, W.D. Kingery, Effect of porosity on physical properties of sintered alumina, J. Amer. Ceram. Soc. 39 (1956) 377–385.
[22] J.V. Kumar, The hypothesis of constant relative responses and its application to the sintering process of spherical powders, Solid State Phen 8 (1989) 125–134.
[23] X. Xu, P. Lu, R.M. German, Densification and strength evolution in solid-state sintering: Part II, strength model, J. Mater. Sci. 37 (2002) 117–126.
[24] G.A. Shoales, R.M. German, In situ strength evolution during the sintering of bronze powders, Metall. Mater. Trans. 29A (1998) 1257–1263.
[25] Z.Z. Fang, Correlation of transverse rupture strength of WC-Co with hardness, Inter. J. Refract. Met. Hard Mater 23 (2005) 119–127.
[26] S.J. Park, S.T. Chung, Y.S. Kwon, R.M. German, Press – sinter simulation tool and its applications, *Proceedings PM 2010 World Congress on Powder Metallurgy*, Florence, Italy, October, European Powder Metallurgy Association, Shrewsbury, UK, 2010.
[27] R. Haynes, The Mechanical Behavior of Sintered Metals, Freund Publishing House, London, UK, 1981.
[28] R. Haynes, J.T. Egediege, Effect of porosity and sintering conditions on elastic constants of sintered irons, Powder Metall. 32 (1989) 47–52.

[29] L.P. Martin, D. Dadon, M. Rosen, Evaluation of ultrasonically determined elasticity-porosity relations in zinc oxide, J. Amer. Ceram. Soc. 79 (1996) 1281–1289.
[30] G.F. Huettig, K. Adlassnig, O. Foglar, Relationships between properties and structure of sintered material made up of single atom type, in: W.E. Kingston (Ed.), The Physics of Powder Metallurgy, McGraw-Hill, New York, NY, 1951, pp. 180–188.
[31] G. Roebben, B. Bollen, A. Brebels, J. Van Humbeeck, O. Van Der Biest, Impulse excitation apparatus to measure resonant frequencies, elastic moduli, and internal friction at room and high temperature, Rev. Sci. Inst. 68 (1997) 4511–4515.
[32] M. Paganelli, Using the optical dilatometer to determine sintering behavior, Ceram. Bull. 81 (11) (2002) 25–30.
[33] I.B. Cutler, Sintering of glass powders during constant rates of heating, J. Amer. Ceram. Soc. 52 (1969) 14–17.
[34] P.Z. Cai, G.L. Messing, D.L. Green, Determination of the mechanical response of sintering compacts by cyclic loading dilatometry, J. Amer. Ceram. Soc. 80 (1997) 445–452.
[35] J.J. Bacmann, G. Cizeron, Dorn method in the study of initial phase of uranium dioxide sintering, J. Amer. Ceram. Soc. 51 (1968) 209–212.
[36] B. Merideth, D.R. Milner, The liquid-phase sintering of titanium carbide, Powder Metall. 19 (1976) 162–170.
[37] D.C. Blaine, R. Bollina, S.J. Park, R.M. German, Critical use of video imaging to rationalize computer sintering simulations, Compt. Ind. 56 (2005) 867–875.
[38] C. Binet, K.L. Lencoski, D.F. Heaney, R.M. German, Modeling of distortion after densification during liquid phase sintering, Metall. Mater. Trans. 35A (2004) 3833–3841.
[39] L. He., Y. Zhou, Y. Bao, Z. Lin, J. Wang, Synthesis, physical, and mechanical properties of bulk $Zr_3Al_3C_5$ ceramic, J. Amer. Ceram. Soc. 90 (2007) 1164–1170.
[40] F. Bernard, E. Gaffet, M. Gramond, M. Gailhanou, J.C. Gachon, Simultaneous IR and time-resolved X-ray diffraction measurements for studying self-sustained reaction, J. Synch. Rad. 7 (2000) 27–33.
[41] N.M. Wereley, T.F. Zahrah, F.H. Charron, Intelligent control of consolidation and solidification processes, J. Mater. Eng. Perform. 2 (1993) 671–682.
[42] K.V. Sebastian, G.S. Tendolkar, Densification in W-Cu sintered alloys produced from coreduced powders, Plansee. Pulver 25 (1977) 84–100.
[43] R.T. DeHoff, E.H. Aigeltinger, Experimental quantitative microscopy with special applications to sintering, in: J.S. Hirschhorn, K.H. Roll (Eds.), Advanced Experimental Techniques in Powder Metallurgy, Plenum Press, New York, NY, 1970, pp. 81–137.
[44] C.E. Schlaefer, R.M. German, Thermal conductivity evolution during initial stage sintering, *Advances in Powder Metallurgy and Particulate Materials – 2003*, Part 5, Metal Powder Industries Federation, Princeton, NJ, 2003, pp. 32–40.
[45] J. Gurland, Application of dihedral angle measurements to the microstructure of cemented carbides WC-Co, Metallog. 10 (1977) 461–468.
[46] J.H. Enloe, R.W. Rice, J.W. Lau, R. Kumar, S.Y. Lee, Microstructural effects on the thermal conductivity of polycrystalline aluminum nitride, J. Amer. Ceram. Soc. 74 (1991) 2214–2219.
[47] T.J. Gray, Sintering of zinc oxide, J. Amer. Ceram. Soc. 37 (1954) 534–539.
[48] R.M. German, Z.A. Munir, Surface area reduction during isothermal sintering, J. Amer. Ceram. Soc. 59 (1976) 379–383.
[49] S. Brunauer, P.H. Emmett, E. Teller, Adsorption of gases in multimolecular layers, J. Amer. Chem. Soc. 60 (1938) 309–319.
[50] S. Prochazka, R.L. Coble, Surface diffusion in the initial sintering of alumina, Part 3. Kinetic study, Phys. Sintering 2 (1970) 15–34.
[51] R. Watanabe, Y. Masuda, Quantitative estimation of structural change in carbonyl iron powder compacts during sintering, Trans. Japan Inst. Met. 13 (1972) 134–139.
[52] F. Thummler, W. Thomma, The sintering process, Metall. Rev. 12 (1967) 69–108.
[53] A.S. Watwe, R.T. DeHoff, Metric and topological characterization of the advanced stages of loose stack sintering, Metall. Trans. 21A (1990) 2935–2941.

[54] O.J. Whittemore, Pore morphography in ceramic processing, in: H. Palmour, R.F. Davis, T.M. Hare (Eds.), Processing of Crystalline Ceramics, Plenum Press, New York, NY, 1978, pp. 125–133.
[55] S.M. Sweeney, C.L. Martin, Pore size distributions calculated from 3-D images of DEM-simulated powder compacts, Acta Mater. 51 (2003) 3635–3649.
[56] P. Lu., J.L. Lannutti, P. Klobes, K. Meyer, X-ray computed tomograph and mercury porosimetry for evaluation of density evolution and porosity distribution, J. Amer. Ceram. Soc. 83 (2000) 518–522.
[57] G.J. Wright, J.A. Yeomans, Constrained Sintering of yttria stabilized zirconia electrolytes: the influence of two step sintering profiles on microstructure and gas permeance, Inter. J. App. Ceram. Tech. 4 (2008) 589–596.
[58] J. Soares, L.F. Malheiros, J. Sacramento, M.A. Valente, F.J. Oliveira, Microstructure and properties of submicrometer carbides obtained by conventional sintering, J. Amer. Ceram. Soc. 94 (2011) 84–91.
[59] H.F. Fischmeister, Characterization of porous structures by stereological measurements, Powder Metall. Inter. 7 (1975) 178–188.
[60] J. Liu, R.M. German, Three-dimensional coordination number in liquid phase sintered microstructures from two-dimensional connectivity, P/M Sci. Tech. Briefs 6 (2) (2004) 19–21.
[61] A. Tewari, A.M. Gokhale, R.M. German, Effect of gravity on three-dimensional coordination number distribution in liquid phase sintered microstructures, Acta Mater. vol. 47 (1999) 3721–3734.
[62] S.J. Park, K. Cowan, J.L. Johnson, R.M. German, Grain Size Measurement Methods and Models for Nanograined WC-Co, Inter. J. Refract. Met. Hard Mater. vol. 26 (2008) 152–163.
[63] M. Brieseck, B. Gneis, K. Wagner, S. Wagner, W. Lengauer, A straightforward method for analyzing the grain size distribution in WC-Co hardmetals, in: P. Rodhammer (Ed.), Proceedings of the Seventeenth Plansee Seminar, vol. 1, Plansee Group, Reutte, Austria, 2009, pp. AT13.1-AT13.10.
[64] P. Louis, A.M. Gokhale, Application of image analysis for characterization of spatial arrangements of features in microstructure, Metall. Mater. Trans. 26A (1995) 1449–1456.
[65] Y. Liu, R.M. German, R.G. Iacocca, Microstructure quantification procedures in liquid-phase sintered materials, Acta Mater. 47 (1999) 915–926.
[66] D.A. Aboav, T.G. Langdon, The distribution of grain diameters in polycrystalline magnesium oxide, Metallog. 1 (1969) 333–340.
[67] D.A. Aboav, T.G. Langdon, The shape of grains in a polycrystal, Metallog. 2 (1969) 171–178.
[68] R.M. German, Coarsening in sintering: grain shape distribution, grain size distribution, and grain growth kinetics in solid-pore systems, Crit. Rev Solid State Mater. Sci 35 (2010) 263–305.
[69] W.D. Kingery, B. Francois, The sintering of crystalline oxides, 1. Interactions between grain boundaries and pores, in: G.C. Kuczynski, N.A. Hooton, C.F. Gibbon (Eds.), Sintering and Related Phenomena, Gordon and Breach, New York, NY, 1967, pp. 471–496.
[70] J. Svoboda, H. Riedel, Pore-boundary interactions and evolution equations for the porosity and grain size during sintering, Acta Metall. Mater. 40 (1992) 2829–2840.
[71] B.R. Patterson, Y. Liu, Quantification of grain boundary-pore contact during sintering, J. Amer. Ceram. Soc. 73 (1990) 3703–3705.
[72] T. Sone, H. Akagi, H. Watarai, Effect of pore-grain boundary interactions on discontinuous grain growth, J. Amer. Ceram. Soc. 74 (1991) 3151–3154.
[73] B.R. Patterson, Y. Liu, J.A. Griffin, Degree of pore-grain boundary contact during sintering, Metall. Trans. 21A (1990) 2137–2139.
[74] M. Whitney, S.F. Corbin, R.B. Gorbet, Investigation of the mechanisms of reactive sintering and combustion synthesis of NiTi using differential scanning calorimetry and microstructural analysis, Acta Mater. 56 (2008) 559–570.
[75] D.S. Janisch, W. Lengauer, A. Eder, K. Dreyere, K.l. Doedinger, H.W. Daub, et al., Nitridation of sintering of WC-Ti(C,N)-(Ta,Nb)C – Co hardmetals, Inter. J. Refract. Met. Hard Mater. 36 (2013) 22–30.
[76] R.K. Enneti, S.J. Park, R.M. German, S.V. Atre, Review – thermal debinding process in particulate materials processing, Mater. Manuf. Proc. 27 (2012) 103–118.

[77] H. Atsushi, S. Yabe, Measurement of sintering gas release behavior of Fe powder by gas chromatograph method, J. Japan Soc. Powder Powder Metall. 56 (2009) 18–25.
[78] Y. Yamada, R. Watanabe, Gas chromatographic study of decomposition of hydroxyapatite in the presence of dispersed zirconia at elevated temperatures, J. Ceram. Soc. Japan 103 (1995) 1264–1269.
[79] H. Danninger, C. Gierl, Processes in PM steel compacts during the initial stages of sintering, Mater. Chem. Phys. 67 (2001) 49–55.
[80] S.M. Kaufman, T.J. Whalen, L.R. Sefton, E. Eichen, The utilization of electron microscopy in the study of powder metallurgical phenomena I. Neck growth measurements for submicron copper and silver spheres, in: J.S. Hirschhorn, K.H. Roll (Eds.), Advanced Experimental Techniques in Powder Metallurgy, Plenum Press, New York, NY, 1970, pp. 25–39.

CHAPTER FIVE

Early Quantitative Treatments

INTRODUCTION

Sintering science is the product of increasingly quantitative waves of development. The first wave was totally empirical and lasted thousands of years, the efforts of which are only documented by artifacts in museums. The second development wave recognized sintering as a fabrication process important in special applications. Even with limited measurement tools, useful products emerged, such as platinum crucibles for melting glass. The third development wave arose with the advent of quantitative measurement tools. Many of these tools are still in use. The resulting burst of quantitative reports proved crucial to developing phenomenological concepts and the predictive atomistic models that underpin our modern understanding of sintering.

Sintering science followed within a decade of the quantitative observations. The measurements aroused conjectures, and in turn the conjectures generated insights that formed the platform for sintering science. This was a simple example of the old rule of "getting what you measure;" for example, dilatometry data came before shrinkage models.

Quantitative experimental efforts looked to improve performance. The drive was for commercial products. Thus, data on fatigue life versus sintering cycles came prior to neck growth models. Qualification for an application relied on data such as hardness, strength, electrical or thermal conductivity, ductility, wear resistance, or radiation absorption [1–10]. Microstructural examinations arose when performance attributes seemed confused—something was missing.

This chapter highlights the transition period that was critical to building theories after key process parameters had been identified. During this time, the default test bed for sintering quantification was copper, a material of great importance during early industrial electrification efforts. It was a powder readily available from a number of sources, with good purity, and it was widely used for quantitative experiments. Curiously, copper appears in many of the early critical studies.

Sintering: From Empirical Observations to Scientific Principles
DOI: http://dx.doi.org/10.1016/B978-0-12-401682-8.00005-7

ONSET OF SINTERING SCIENCE

Sintering advances fed on one another, as successive layers of knowledge developed. Empirical advances in sintered platinum subsequently seeded the way that tantalum, tungsten, and osmium were sintered to form lamp filaments. Wire drawing dies for lamp filament production required new hard materials, leading to liquid phase sintered WC-Co. Sintered diamond composites were needed to finish the carbides drawing dies. When high pressure processing was realized, artificial diamonds became available through thc use of sintered cemented carbide anvils. In retrospect it can be seen that a cascade of interdependent developments occurred in sintered materials.

Before the 1940s, most of the sintering literature was qualitative. By the late 1940s, quantitative models started to emerge as a diversity of applications arose. The electrical contacts used in power switching relied on composites of copper and an arc erosion resistant powder (tungsten, molybdenum, tungsten carbide) [11,12]. Empirical gains spread to include filters, nuclear fuels, insulators, transducers, and sintered polymers [13–16].

Considerable development was required to create this measurement framework. With this framework quantitative measures of sintering arose. In preparing this book it was evident that early work on copper sintering provided insight to the changes.

COPPER SINTERING

Copper powders were available via several fabrication routes. Their high purity and low oxide stability allowed sintering without complications from surface oxides.

In 1940, Goetzel [17] assessed copper sintering to reflect on the maturation of sintering. He relied on reduced and electrolytic copper powders subjected to an experimental matrix that included particle type, green density (compaction pressure), peak temperature, hold time, and atmosphere. His experiments included compaction pressures from 70 to 620 MPa, sintering temperature variations from 600 to 900°C (873 to 1173 K), hold times from 1 to 16 h, and vacuum or hydrogen atmospheres. Response variables included density, hardness, strength, ductility, electrical conductivity, fatigue strength, impact toughness, and deformation response. Goetzel delineated the combination of temperature, time, and pressure associated with 100% density. Compaction pressure had the largest influence, as illustrated by the data in Figure 5.1 [17]. Curiously, in experiments performed at 750°C (1023 K) using 350 MPa compaction pressure, sintered density increased but sintered strength fell from 97 MPa at 6 hours hold to 89 MP at 15 hours hold. Mechanical and electrical property optimization dominated his research, as opposed to any search for the sintering mechanisms.

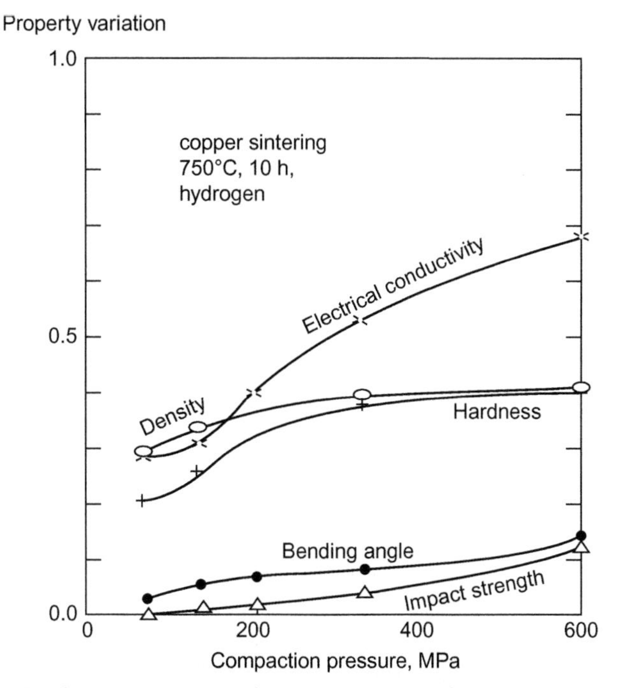

Figure 5.1 Property development in sintered copper as reported by Goetzel in 1940 [17]. This plot gives the electrical conductivity, density, hardness, bending angle (ductility), and impact strength versus compaction pressure following a 750°C (1023 K), 10 h, hydrogen sintering.

This early quantitative sintering treatment was remarkable, considering the context. Only recently was it recognized that sintering related to the homologous temperature [18,19]. Yet, atomic diffusion was not accepted, so copper sintering was seen as a viscous flow process [20–25].

By 1942, anomalous sintering results had been identified; copper compacts swelled under certain firing conditions [25,26]. This is evident in Figure 5.2. The processing details are not given, but swelling varies with the green density. Currently, swelling is attributed to delayed gas reactions, but this behavior perplexed the sintering community, as is evident from the proliferation of powder specifications attempting to avoid the problem [7].

As performance data on strength, hardness, and conductivity emerged, questions arose about what was happening to cause the changes. Kuczynski [27] relied on copper for his neck growth model and Duwez and Martens [28] relied on copper for early dilatometry. Quickly copper became the *de facto* basis for sintering research. Basic parameters such as neck size then became the focus. Figure 5.3 shows an early example of neck growth during copper sphere sintering [29]. In this case 640 to

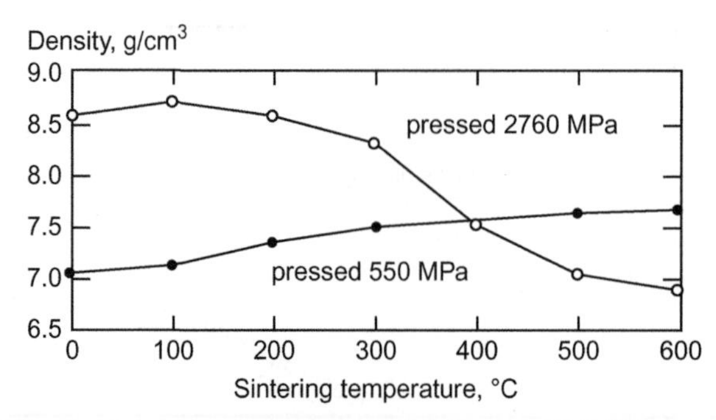

Figure 5.2 Sintered density for copper pressed at 550 MPa and 2760 MPa, sintered at various temperatures up to 600°C (873 K) [25]. The higher compaction pressure causes swelling at the higher temperatures, resulting in a lower final density. Early efforts did not rationalize swelling to gas reactions and suspected some novel mechanism.

Figure 5.3 Necks observed in optical microscopy after sintering [29]. For these experiments 640 to 560 μm copper spheres were heated to 900°C (1173 K) for 24 h.

560 μm powder was heated to 900°C (1173 K) for 24 h to induce necks. Repeated runs at different temperatures and times allowed comparison with models for volume diffusion as expected with the low surface and grain boundary area associated with large particles. Checks against radioisotope diffusion rates showed agreement. For example, Figure 5.4 plots the diffusion coefficient extracted from neck growth data in 1949 [29]. The 230 kJ/mol activation energy for sintering is similar to reports for copper volume diffusion at 226 kJ/mol. This was a significant victory for diffusion over viscous flow.

Quantitative data enabled testing of the new models, but bulk attributes measured after sintering did not give the required insight. For example, in a study of copper sintering by Okamura et al. [30], strength and shrinkage failed to correlate;

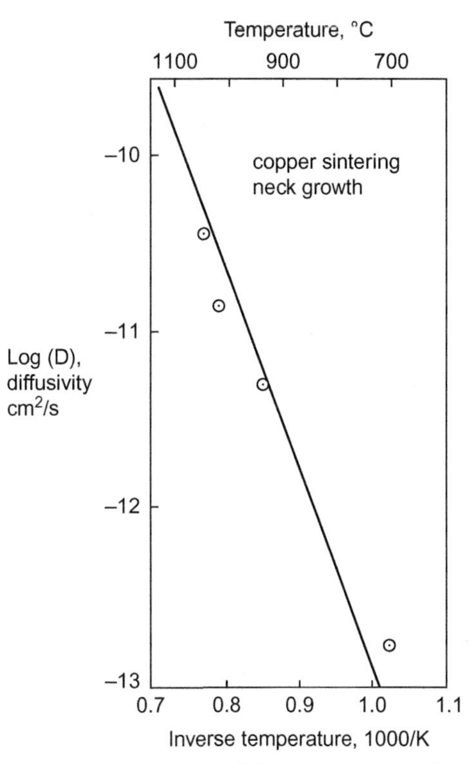

Figure 5.4 Diffusivity data for copper as extracted from sintering neck growth, plotted in terms of the logarithm versus inverse absolute temperature [29]. The extracted activation energy is close to that found for volume diffusion.

but shrinkage did provide an activation energy in line with volume diffusion. Parameters such as shrinkage and neck size correlated with diffusion, but strength and bulk properties did not. This gave focus to new tools that properly tracked sintering, such as dilatometry.

Dilatometry is a means to assess several influences—particle size, green density, heating rate, peak temperature, hold time, oxide content, and alloying. It helped isolate the way that these parameters influenced copper sintering [31–42]. Figure 5.5 plots the dimensional change for copper powder heated at 4°C/min to 980°C (1253 K) [28]. It shows thermal expansion on initial heating and the onset of shrinkage near 500°C (773 K). Shrinkage accelerates and continues through the heating period, followed by contraction on cooling. Parallel experiments identified compaction pressure effects. For example, sintering shrinkage is plotted in Figure 5.6 for 74 to 43 μm copper powder formed at 140 MPa sintered at six temperatures [43]. The acceleration of sintering at higher temperatures is very evident.

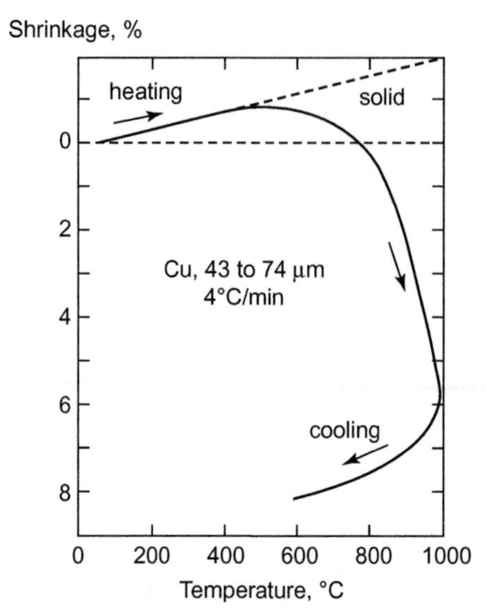

Figure 5.5 Dilatometry data for sintering 74 to 43 μm copper powder at a heating rate of 4°C/min to a peak temperature of 980°C (1253 K) [28]. First sintering shrinkage is evident at about 500°C (773 K).

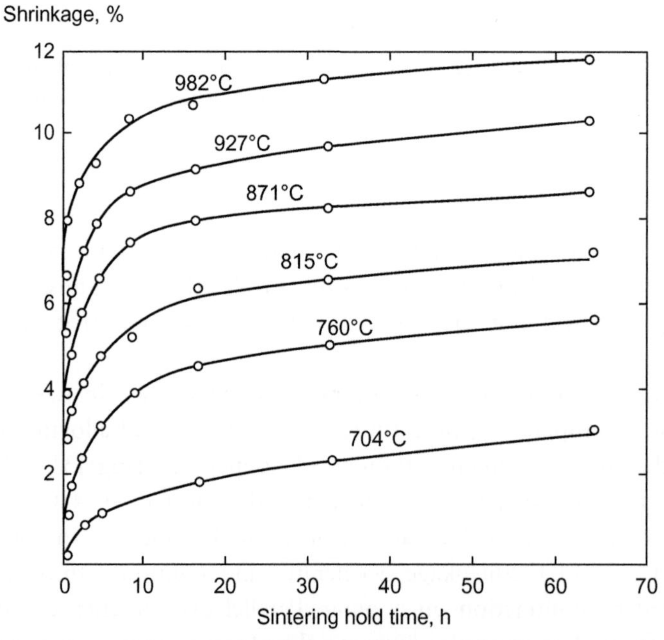

Figure 5.6 Sintering shrinkage for copper powder compacts formed by pressing at 140 MPa using 74 and 43 μm powder and heated for various times and six temperatures [43].

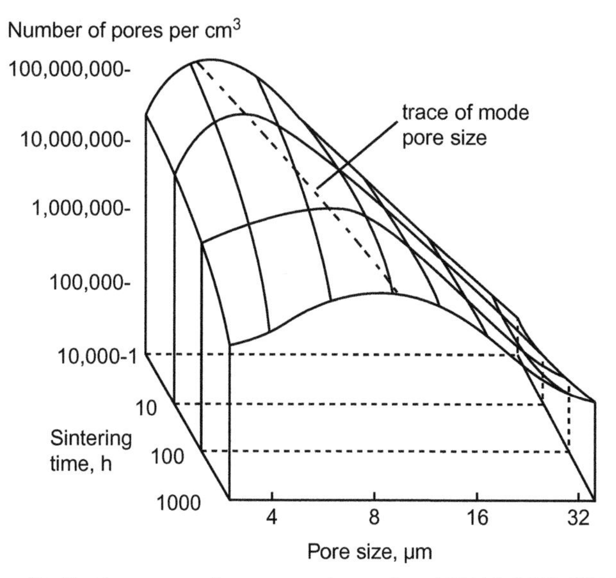

Figure 5.7 Pore size distribution curves for copper sintered at 1000°C (1273 K)[44]. The distribution shows declining porosity (area under the curve), declining pore population, and pore coarsening with extended sintering times. Microstructure insights led to advanced sintering models.

Other parameters added quantitative insight. For example, Shaler [24] tracked the pore size for copper, noting which sintering conditions gave shrinkage versus swelling. Several investigators recognized the way that the pore number and pore size varied with sintering time for copper [43–45]. Sintering has a significant pore coarsening character, as evident by the 1000°C (1273 K) data from Duwez [44] in Figure 5.7. As the number of pores and the porosity decline, the mode pore size increases. Such coarsening behavior illustrated the need to move beyond bulk properties to understand the changes associated with sintering.

By 1949, Alexander and Balluffi [46] recognized the critical role of grain boundaries in pore annihilation. This set the stage for adding grain boundary diffusion and inclusion of grain growth in sintering models. This quantitative footing was in place by the 1950s. These advances depended on quantitative data, and copper provided the most widely employed basis for the experiments and models. The agreement between models based on atomic diffusion and measured sintering rates solidified theory. In turn, the explanations arose for how properties such as strength changed with sintering parameters [43]. Much of the quantitative basis for sintering theory arose from sintering experiments on copper. From that base, a solid framework arose; as treated in subsequent chapters of this book.

REFERENCES

[1] W.D. Coolidge, Ductile tungsten, Trans. Amer. Inst. Elect. Eng. 29 (1910) 961–965.
[2] B.G. Klugh, The microstructure of sintered iron bearing materials, Trans. TMS-AIME 45 (1913) 330–345.
[3] J.B. Ferguson, Note on the sintering of magnesia, J. Amer. Ceram. Soc. 1 (1918) 439–440.
[4] R.C. Smith, Sintering: its nature and causes, J. Chem. Soc. Trans. 123 (1923) 2088–2094.
[5] S.L. Hoyt, Metal carbides and cemented tungsten carbide, Trans. AIME 89 (1930) 9–58.
[6] C.J. Smithells, A new alloy of high density, Nature 139 (1937) 490–491.
[7] J.E. Drapeau, Sintering characteristics of various copper powders, in: J. Wulff (Ed.), Powder Metallurgy, American Society for Metals, Cleveland, OH, 1942, pp. 323–331.
[8] G.H.S. Price, C.J. Smithells, S.V. Williams, Sintered alloys. part I - copper-nickel-tungsten alloys sintered with a liquid phase present, J. Inst. Met 62 (1938) 239–264.
[9] H.E. Hall, Sintering of copper and tin powders, Met. Alloys 14 (1939) 297–299.
[10] H.E. White, R.M. Shremp, Beryllium oxide: I, J. Amer. Ceram. Soc. 22 (1939) 185–189.
[11] C.G. Goetzel, Treatise on Powder Metallurgy, vol. I, Interscience Publishers, New York, NY, 1949, pp. 259–312.
[12] C.G. Goetzel, Treatise on Powder Metallurgy, vol. III, Interscience Publishers, New York, NY, 1952.
[13] G.A. Geach, A.A. Woolf, The sintering behavior of organic materials, in: W. Leszynski (Ed.), Powder Metallurgy, Interscience, New York, NY, 1961, pp. 201–206.
[14] P.D.S. St Pierre, Constitution of bone china: I, high temperature phase equilibrium studies in the system tricalcium phosphate-alumina-silica, J. Amer. Ceram. Soc. 37 (1954) 243–258.
[15] P. Duwez, H.E. Martin, The powder metallurgy of porous metals and alloys having a controlled porosity, Trans. TMS-AIME 175 (1948) 848–877.
[16] T.J. Gray, Sintering of zinc oxide, J. Amer. Ceram. Soc. 37 (1954) 534–539.
[17] C.G. Goetzel, Structure and properties of copper powder compacts, J. Inst. Met. 66 (1940) 319–329.
[18] G.F. Huttig, Die Frittungsvorange innerhalb von Pulvern, weiche aus einer einzigen Komponente bestehen – Ein Beitrag zur Aufklarung der Prozesse der Metall-Kermik und Oxyd-Keramik, Kolloid Z. 98 (1942) 6–33.
[19] F. Sauerwald, Uber de Elementarvorgange beim Fritten and Sintern von Metallpulvern mit besonderer Berusksichtigung der Realstruktur ihere Oberflachen, Kolloid Z 104 (1943) 144–160.
[20] A.J. Shaler, J. Wulff, Rate of sintering of copper powder, Phys. Rev. 72 (1947) 79–80.
[21] A.J. Shaler, J. Wulff, On the rate of sintering of metal powders, Phys. Rev. 73 (1948) 926.
[22] C. Herring, Diffusional viscosity of a polycrystalline solid, J. Appl. Phys. 21 (1950) 437–445.
[23] J.K. Mackenzie, R. Shuttleworth, A phenomenological theory of sintering, Proc. Phys. Soc 62 (1949) 833–852.
[24] A.J. Shaler, Seminar on the kinetics of sintering, Trans. TMS-AIME 185 (1949) 796–804.
[25] P.E. Wretblad, J. Wulff, sintering, in: J. Wulff (Ed.), Powder Metallurgy, American Society for Metals, Cleveland, OH, 1942, pp. 36–59.
[26] C.G. Goetzel, Hot pressed and sintered copper powder compacts, in: J. Wulff (Ed.), Powder Metallurgy, American Society for Metals, Cleveland, OH, 1942, pp. 340–351.
[27] G.C. Kuczynski, Self-diffusion in sintering of metallic particles, Trans. TMS-AIME 185 (1949) 169–178.
[28] P. Duwez, H. Martens, A dilatometric study of the sintering of metal powder compacts, Trans. TMS-AIME 185 (1949) 572–576.
[29] J.H. Dedrick, A. Gerds, A study of the mechanism of sintering of metallic particles, J. Appl. Phys. 20 (1949) 1042–1044.
[30] T. Okamura, Y. Masuda, Experimental study on the kinetics of sintering metal powder at constant temperature, I, Trans. Japan Inst. Met. 1 (1949) 357–363.
[31] T.P. Hoar, J.M. Butler, Influence of oxide on the pressing and sintering of copper compacts, J. Inst. Met. 78 (1950) 351–393.

[32] A. Duffield, P. Grootenhuis, The effect of particle size on the sintering of copper powder, J. Inst. Met. 87 (1958) 33–41.
[33] H. Mitani, Abnormal expansion of cu-sn powder compacts during sintering, Trans. Japan Inst. Met. 3 (1962) 244–251.
[34] P. Duwez, C.B. Jordan, Sintering of Copper Gold Alloys, in: W.E. Kingston (Ed.), The Physics of Powder Metallurgy, McGraw-Hill, New York, NY, 1951, pp. 230–237.
[35] J. Gurland, J.T. Norton, Role of the binder phase in cemented tungsten carbide-cobalt alloys, Trans. TMS-AIME 194 (1952) 1051–1056.
[36] E.B. Allison, P. Murray, A fundamental investigation of the mechanism of sintering, Acta Metall. 2 (1954) 487–512.
[37] G. Cizeron, P. Lacombe, Influence des phenomenes d'autodiffusion dans le frittage du fer pur d'origine carbonyle en dessous et au-dessus du poin, Rev. Metall. 52 (1955) 771–783.
[38] G. Cizeron, P. Lacombe, Influence des chauffages de part et d'autre du point de transformation alpha-gamma du fer sur les processus d'autodiffus, Rev. Metall. 53 (1956) 819–829.
[39] G. Cizeron, Role de la structure et de l'atmosphere sur la cinetique du frittage du fer carbonyle, Compt. Rendus 245 (1957) 2047–2050.
[40] G. Cizeron, Influence de la grosseur du grain sur la cinetique du retrait d'agglomeres de fer ex-carbonyle au cours du frittage ene, Compt. Rendus 245 (1957) 2051–2054.
[41] G. Cizeron, Influence d'une double compression sur la cinetique du frittage d'agglomeres de fer ex-carbonyle, Compt. Rendus 244 (1958) 3060–3063.
[42] J. Gurland, Observations on the structure and sintering mechanism of cemented carbides, Trans. TMS-AIME 215 (1959) 601–608.
[43] R.T. DeHoff, J.P. Gillard, Relationship between microstructure and mechanical properties in sintered copper, in: H.H. Hausner (Ed.), Modern Developments in Powder Metallurgy, 4, Plenum Press, New York, NY, 1971, pp. 281–290.
[44] P. Duwez, Diffusion in sintering, Atom Movements, American Society for Metals, Metals Park, OH, 1950, pp. 192–208.
[45] A.W. Postlethwaite, A.J. Shaler, Shrinkage of synthetic pores in copper, in: W.E. Kingston (Ed.), The Physics of Powder Metallurgy, McGraw-Hill, New York, NY, 1951, pp. 189–201.
[46] B.H. Alexander, R. Balluffi, Experiments on the mechanism of sintering, J. Met. 2 (1950) 1219.
[47] C.B. Jordan, P. Duwez, The densification of copper compacts in hydrogen and in vacuum, Trans. TMS-AIME 185 (1949) 96–99.

CHAPTER SIX

Geometric Trajectories during Sintering

OVERVIEW

Sintering bonds particles together with simultaneous changes in the pore and grain structures. These geometric changes reflect migration toward a low interfacial energy. Surface energy loss is evident by the reduction in surface area. Particle bonding associated with surface area reduction also strengthens the compact. For crystalline materials, sinter bonding increases grain boundary area; surface area transforms into grain boundary area. Later, grain boundary area decreases via grain growth while the number of pores and grains decrease. Thus, the geometric trajectory associated with sintering is manifested by changes in several features.

Many materials are sintered to near full density. Rationalization of the sintering cycles over this wide range of materials is possible using normalized temperature. The homologous temperature is the absolute temperature normalized to the absolute melting temperature for a given material. For copper, melting occurs at 1356 K and full density sintering at 1318 K; this corresponds to a homologous temperature of 0.972 (obtained from 1318/1356). Likewise, the geometric progression in sintering is similar across a wide range of materials. This is true even at the nanoscale size range [1]. This leads to the concept of self-similar geometric behavior in sintering: Microstructures are similar except for scaling parameters. This is because sintering involves slow energy release, allowing similar evolution trajectories. A 1 μm particle with a 1 J/m^2 surface energy releases 6 J/cm^3 during sintering over maybe an hour. The energy release progression is slow, and consequently the geometric progression can be generalized in terms of similar features. In this chapter, the discussion is restricted to the geometric progression.

Not all sintering is geared to achieving high densities; a variety of structures rely on intentional porosity for their proper function. Applications for porous materials include capacitors, bearings, filters, sound absorbers, ionizers, and nuclear fuels. In these cases sinter bonding is required for strength, but densification is minimized. Even without densification, significant changes take place as is evident

Sintering: From Empirical Observations to Scientific Principles
DOI: http://dx.doi.org/10.1016/B978-0-12-401682-8.00006-9

from the improved strength. Sintering involves significant shifts in the microstructure:

- neck size enlarges
- number of grains declines
- grain size enlarges; in some cases thousands of grains meld into a single large grain
- grain shape becomes polygonal
- surface area decreases
- pore population declines.

STAGES OF SINTERING

The sintering stages represent the geometric progression of transforming a weak powder compact into a strong object. Depending on the conditions at which sintering is initiated; the sintering stages might begin with loose or deformed particles. The former is the case for bodies formed using slip casting, tape casting, slurry casting, extrusion, injection molding, and low pressure die compaction—all cases where particle shaping occurs without significant pressurization. Particle deformation to give a high starting density is associated with high pressure forming, such as die compaction, cold isostatic pressing, and cycles where pressure is applied prior to sintering. For sintering models the particles are assumed to be spherical, starting with point contacts between neighboring particles. Under these conditions all stages of sintering are encountered. Starting at a high green density, the early sintering stages are absent [2].

Sintering is divided into four stages:

- Contact formation—weak atomic forces at particle contacts hold the particles together prior to sintering.
- Neck growth—the initial stage where each contact enlarges without interaction with neighboring contacts.
- Pore rounding—the intermediate stage where neighboring necks grow and interact to give a network of tubular pores.
- Pore closure—the final stage where tubular pores pinch closed to form discrete spherical or lenticular pores.

Figure 6.1 is a representation of these sintering stages. Initial stage sintering starts with contacting particles where a neck grows via short range atomic motion. At the same time a grain boundary forms in the neck, since the grains have random crystal orientations with respect to one another. The sinter necks are saddle surfaces due to the mixture of concave and convex surface curvatures present. Figure 6.2 is a scanning electron micrograph of the necks growing during sintering. Up to a neck size of about one-third of the particle size, each neck grows in isolation from its neighbors. The neck diameter is designated X and the particle diameter is designated D. Initial

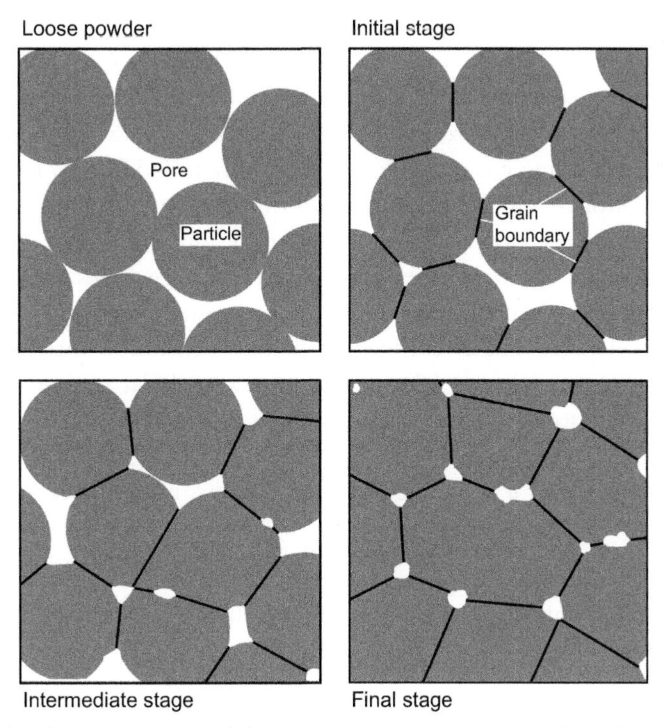

Figure 6.1 Graphical representation of the sintering stages, assuming the spherical particles start as a loose structure. Neck growth produces densification, round pores, and a strong body. Grain boundaries form at the particle contacts and grow as the necks merge.

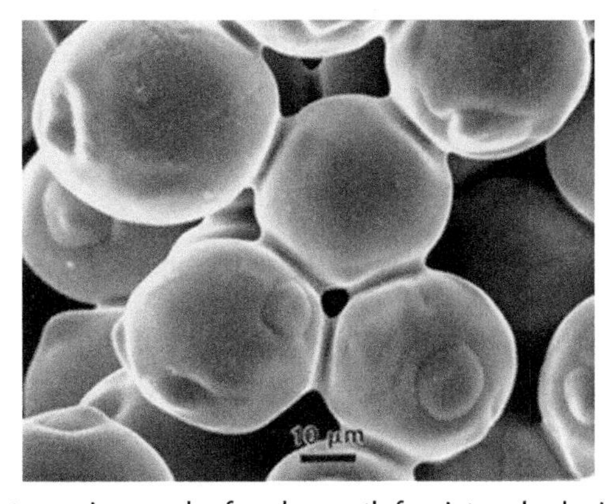

Figure 6.2 Scanning electron micrograph of neck growth for sintered spherical 32 μm nickel particles.

stage sintering corresponds to $X/D \leq 0.33$. The pore structure is open, meaning that gas can permeate through the body.

In the intermediate stage of sintering, the necks continue to enlarge and overlap to generate a smooth pore. The grain boundary area in the neck enlarges while neighboring necks impinge. The pores are still open, so the sintering body is not air-tight (see Figure 4.46). The grains are polygons with flat grain boundaries on the contact faces, while the pores form an interconnected three-dimensional array on the grain edges. Usually the neck size ratio X/D reaches upwards to about 0.5; that is, intermediate stage sintering corresponds approximately to $0.33 \leq X/D \leq 0.50$.

The transition from a tubular pore network on the grain edges to a spherical pore network on the grain corners occurs as densification reaches a critical point. Densification shrinks the pore diameter while grain growth extends the pore length. As a model for this transition, consider a dodecahedron grain shape. It is a polygon with 12 sides or faces, 30 edges, and 20 corners. Each edge is shared by three grains and each corner is shared by four grains, so effectively every two tubular pores (on the edges) in the intermediate stage forms one spherical pore (on the corners). Long thin pores associated with the intermediate stage pinch into spheres according to the Rayleigh instability when the pore diameter d and the grain edge length L reach a condition when:

$$L \geq \pi d \tag{6.1}$$

The pore diameter after spheroidization is 3/2 that of the tubular pore diameter. In the intermediate stage, the fractional porosity ε is given as:

$$\varepsilon = \left(\frac{30}{3}\right) \frac{\pi d^2 L}{4(7.66 L^3)} = 1.025 \left(\frac{d}{L}\right)^2 \tag{6.2}$$

where 30/3 reflects the 30 grain edges shared by three grains each, $\pi d^2 L/4$ is the volume of each tube, and $7.66\, L^3$ is the dodecahedron grain volume. Substitution of $L = \pi d$ gives the porosity for the transition to the final stage at 0.1 or 90% density. Other assumptions lead to predictions ranging from 79 to 92% theoretical density for pore closure [3]. In final stage sintering, the grain shape converges toward a 14-sided tetrakaidecahedron with 36 edges shared by three grains each and 24 corners shared by four grains each. The critical transition to final stage sintering is at 92% density.

The idealized final stage situation is illustrated in Figure 6.3, in which a tetrakaidecahedronal grain is decorated with round pores on each of the 24 corners; each pore is shared by four grains so there is a net of six pores per grain. Necks are no longer discernible at this point. A generally accepted transition from intermediate to final stage sintering is 90 to 92% density. In experimental measurements, closed pores are detected by 85% density, about half are closed by 92% density, and essentially all remaining pores are closed by 95% density. This range of pore closure conditions reflects the distribution of particle size and packing density inherent to sintered structures. Closed pores do not allow gas permeation into the body.

To summarize the sintering stages, Table 6.1 gives the approximate geometric changes associated with each stage. Actual sintering involves a range of particle sizes,

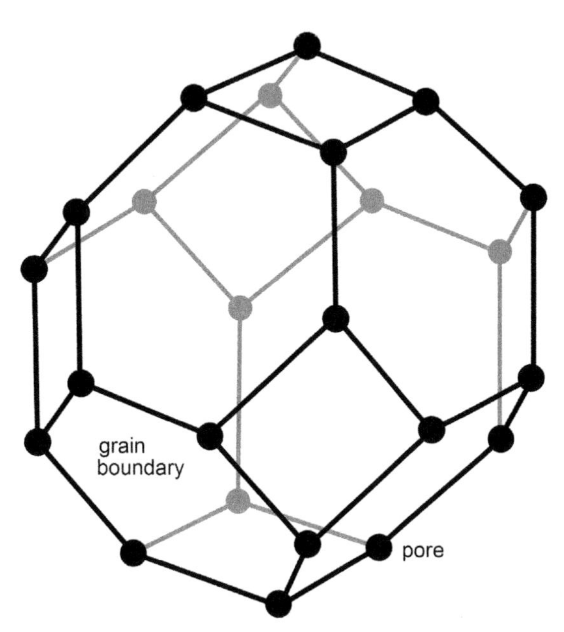

Figure 6.3 Final stage sintering is associated with a polygonal grain and spherical pores located on the grain corners. The pores are not connected, so it is not possible to exchange gas through the pore network. This drawing shows a 14-sided tetrakaidecahedron with a mixture of hexagon and square faces.

Table 6.1 Geometric Changes in the Three Stages of Sintering for Monosize Spheres

	Initial Stage	Intermediate Stage	Final Stage
neck size ratio, X/D	<0.33	0.33 to 0.5	>0.5
coordination, N_C	<7	8 to 12	12 to 14
density, %	60 to 66	66 to 92	>92
shrinkage, $\Delta L/L_O$ %	<3	3 to 13	>13
surface area, S/S_O %	100 to 50	50 to 10	<10
grain size ratio, G/D	≈1	>1	>>1
pore size ratio, d/G	<0.2	near constant	shrinking

X = neck diameter, D = particle diameter, N_C = coordination number.
$\Delta L/L_O$ = sintering dimensional change over initial size, commonly termed shrinkage.
S = specific surface area, S_O = initial specific surface area prior to sintering.
G = grain size, G_O = starting grain size, d = pore diameter.

grain contacts, and pore shapes, so the transitions are not crisp. Such a conceptualization helps understand the changes that impact mathematical modeling efforts.

Contact Formation

Early sintering initiates at the points where particles make contact. Weak cohesive forces attract the particles to each other, such as van der Waals forces, producing a small bond. The contact stress is often estimated at 1 MPa. Initial sintering enlarges the contact. As the neck becomes larger it progressively requires more mass to enlarge. For example, approximating the neck shape to circular gives a link between the neck volume V and the neck diameter X as:

$$V = \frac{\pi}{8}\frac{X^4}{D} \tag{6.3}$$

This is valid for small neck sizes. When X is small, the incremental volume dV required to give a neck size gain dX is small, but as neck growth continues the added volume increases rapidly, since an incremental change in neck volume scales with the cube of the neck size:

$$\frac{dV}{dX} \approx \frac{X^3}{D} \tag{6.4}$$

In other words, a small volume dV is required to grow a small neck but a large volume is required to give enlargement to a large neck. The neck growth rate naturally slows as more volume is required to enlarge the neck.

In practice, much sintering is performed using non-spherical powders, with wide particle size distributions and compaction prior to sintering. External pressure repacks the particles, collapses large pores, enlarges the contacts, and most importantly increases the number of contacts per particle. This latter parameter is the coordination number. Each contact is an independent site for sintering, so higher compact densities imply more contacts and faster sintering. The contacts enlarge during heating to increase component strength. Curiously, this increase in strength tends to resist further densification until a relatively high temperature is achieved and thermal softening occurs.

Prior to sintering the particles form random contacts with each other. When the necks grow, a grain boundary emerges at the contact plane. The resulting grain boundary structure is an important feature in determining final density.

Neck Growth–Initial Stage

Neck growth between contacting particles is an obvious aspect of sintering, especially in its initial stage. The neck size is sufficiently small that neighboring necks grow independently of one another. This stage ends when the necks impinge at approximately a neck size ratio X/D of 0.33. In a typical powder compact, the contact growth occurs at several sites simultaneously. Each neck grows to eventually merge with its neighbor to obscure individual necks.

Early models approximated the neck saddle surface as a circle of diameter p, as illustrated in Figure 6.4 [4]. A right triangle is formed with $(D+p)^2 = (D)^2 + (X+p)^2$, enabling an approximate solution for the neck volume, neck area, and curvature. The

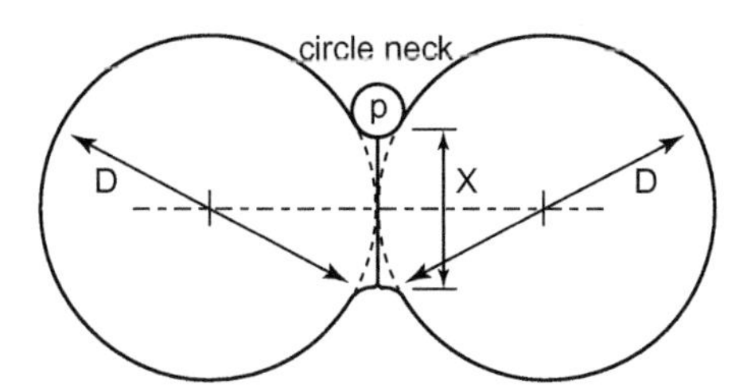

Figure 6.4 The neck geometry applied to initial stage sintering of two spheres. The circle approximation for the neck saddle surface (convex curvature is described by the diameter *X* while concave curvature is described by the diameter *p*) enables linkage between neck size and neck volume. A modified geometry with center-center attraction links neck size to sintering shrinkage.

small circle fit is has been modified to minimizes neck surface area, conserves mass, and ensures a smooth curvature gradient [5–7].

For a crystalline material, a grain boundary forms at the contact plane and grows as the neck enlarges. This means that the solid-vapor surface area is annihilated, but the solid-solid (grain boundary) interface area increases. Both depend on crystal orientation. The grain boundary forms randomly. If sintering is slow, then it forms a groove at the saddle point. Usually grooving is ignored, since the rate of mass arrival to the neck is faster than the time required for grooving. The solid-solid grain boundary energy γ_{SS} and solid-vapor surface energy γ_{SV} strike a balance, formulated as:

$$\gamma_{SS} = 2\gamma_{SV} \cos\left(\frac{\phi}{2}\right) \tag{6.5}$$

where ϕ is the dihedral angle. A limiting neck size X is set by the dihedral angle and the particle size D as follows:

$$X = D \sin\left(\frac{\phi}{2}\right) \tag{6.6}$$

With grain growth, grain size G determines the neck size, not the particle size D. Grain coarsening allows for neck growth, letting X further enlarge. Necks with grooves indicate terminated neck growth. Most materials have low grain boundary energies and sinter to full density.

For a crystalline solid, the grain boundary energy depends on the crystal misorientation across the boundary. Certain misorientations have lower energies, as illustrated in Figure 6.5 [8]. Consequently, particles rotate during initial stage sintering to obtain a higher packing density and lower energy [9–11]. Initial stage particle rearrangement is depicted in Figure 6.6 [12]. The three copper particles bond and rotate as the necks grow, producing a new contact. Rotation rates are slow; around 0.03 degrees per

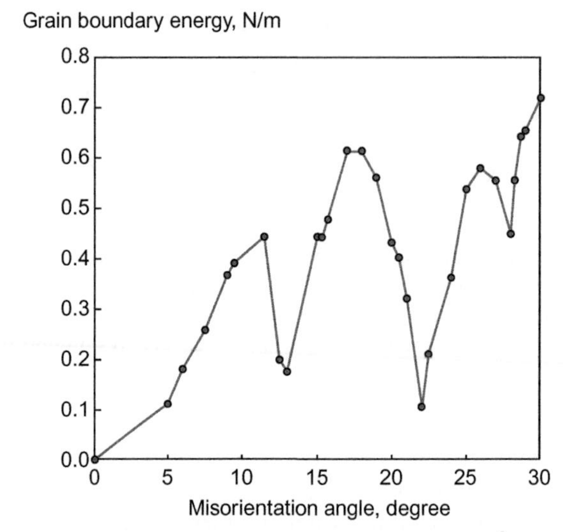

Figure 6.5 An example of how grain boundary energy (N/m or J/m^2) varies with the crystallographic misorientation angle, showing preferred configurations [8]. Grain rotation relies on vacancy annihilation to move into preferred orientations. This leads to a distribution in dihedral angles since not all sinter contacts have the same grain boundary energy.

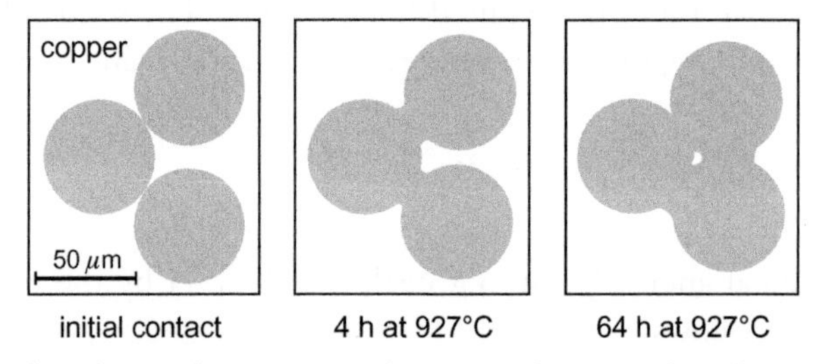

Figure 6.6 Images taken during the sintering of copper spheres to show how grain rotation occurs to lower grain boundary energy while forming new contacts [10].

second. In some cases the orientation after rearrangement is coincident, meaning there is no grain boundary between contacting grains and the crystal structures align. Otherwise, the grains continue to rotate slowly during sintering to minimize energy.

Monosized spheres at a green density of 64% of theoretical average seven contacts per particle. As sintering shrinkage occurs, new contacts form, reaching a coordination average of 14 at full density. These newly formed contacts start sintering at the initial stage even though the primary contacts are much further along in the sintering process. Consequently, the stages of sintering overlap.

Most sintering measures relate to the neck size ratio X/D. However, as the necks grow and overlap, attention shifts to the pore structure in the intermediate stage of sintering.

Pore Rounding–Intermediate Stage

In the intermediate stage of sintering the open pore network progressively becomes smoother. Although the total porosity declines, concomitant grain growth enlarges the pores. This combination of densification, pore rounding, and grain growth continues throughout sintering.

The structure consists of rounded pores occupying the grain edges. The grain shape progresses toward a tetrakaidecahedron at full density. As the solid-solid grain boundary area increases, the increased area enables grain growth. The net result is a larger average grain size with fewer grains as porosity declines. The driving force for continued sintering is the elimination of interfacial area, first surface area and eventually grain boundary area.

For sintering systems that densify, the intermediate stage continues up to about 92% density. Grain growth accelerates as pores are eliminated, so the grain size progressively increases and the grain size exceeds the initial particle size. During this stage most of the property development occurs. By the end of the intermediate stage the pores are spherical and closed.

Sintering geometry in intermediate stage sintering consists of pores located on the grain edges. The pores are not perfect cylinders and vary in size, while the grains are distributed in size. It is common to assume that the grains are tetrakaidecahedrons with connected tubular pores on the edges. For such geometry the pore diameter d, grain size G, and fractional density f are related as follows:

$$f = 1 - 2\pi\left(\frac{d}{G}\right)^2 \tag{6.7}$$

This gives a relation between porosity ($\varepsilon = 1 - f$), pore size d, and grain size G, as long as the pores remain attached to the grain boundaries. A demonstration of such behavior is given in Figure 6.7 for iron [13]. The square of the pore size divided by the grain size is a linear function of porosity. Later, in the final stage of sintering, this relation breaks down due to pore separation from the grain boundaries. The smallest pores are eliminated preferentially while at the same time pore coarsening occurs to give both pore growth and pore shrinkage.

Pore Closure–Final State

The open pore network of the intermediate stage becomes geometrically unstable as the pore diameter shrinks and the pore length grows due to grain growth. This instability causes the pores to pinch into closed, spherical voids at 90 to 92% density. It is analogous to the breakup event that changes a long, thin liquid stream into discrete droplets.

These closed pores are easily identified in cross-section microscopy images, since they are circular. Spherical pores on grain corners are less effective in slowing grain growth. Since the pores are closed, any trapped gas limits the end-point density;

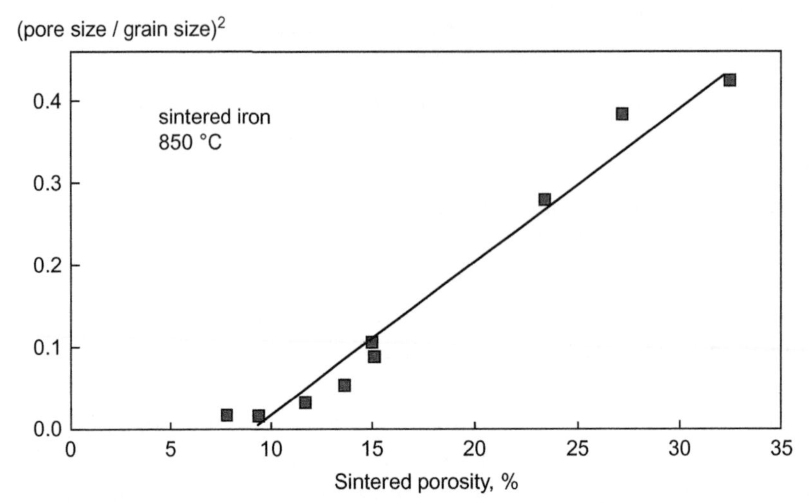

Figure 6.7 A plot of the pore size to grain size ratio squared $(d/G)^2$ versus the porosity for iron powder at 850°C (1123 K) [13]. The behavior corresponds to rounded tubular pores connected on the grain edges.

accordingly, vacuum sintering is generally most successful in attaining high sintered densities, as long as there is no vapor filling the pores.

There is no sharp change in the pore structure to form closed pores. Because of gradients in the microstructure, pore closure occurs over a range of porosities. Generally, they first start to form at 15% porosity and all pores are closed by 5% porosity. As noted via quantitative microscopy, the typical transition occurs between 90 and 92% density. If the grain boundary energy is ignored, then the pores are spherical. This gives an apparent increase in pore size when using cross-section microscopy.

In an early model for final stage sintering, Coble [14] assumed spherical pores and noted fractional density f, pore diameter d (now assumed to be a sphere), and grain edge length L were related when the pores were in ideal positions:

$$f = 1 - \frac{\pi}{\sqrt{2}}\left(\frac{d}{2L}\right)^3 \tag{6.8}$$

The grain edge length L is a proportional to the grain size G. Assuming a spherical equivalent grain size gives $G = 8^{1/2}\,L$ which corresponds to $G = 2.83\,L$. Thus, the rearranged relation between pore size, grain size, and fractional density gives:

$$d = G\left(\frac{1-f}{2\pi}\right)^{1/3} \tag{6.9}$$

Thus, porosity ($\varepsilon = 1 - f$) is proportional to the cube of the ratio of the pore size to grain size.

INTERFACE CURVATURE AND ENERGY

The energy release involved in sintering is small; for solid-state sintering it is of the order of 6 MJ/m^3. For comparison, gasoline has an energy density 6000 times higher. Thus, the release rate in terms of surface area and grain boundary area annihilation is slow. Even so, the sintering trajectory tracks to the lowest energy geometry.

Early sintering is associated with the elimination of curvature gradients, especially those near the neck. By the intermediate stage of sintering the curvature gradients are diminished and the pores are smooth. The driving force is then the elimination of interfacial area by pore elimination and grain growth.

Curvature Gradients

Initial sintering is associated with neck growth. Early neck growth is rapid [15]. This can be shown using the neck geometry shown in Figure 6.8 without a grain boundary groove (when neck growth is rapid the dihedral angle does not form [16]). Let the center-center axis be z, and the symmetric axes be w, y, then the minimum energy neck profile is given as:

$$1 + \sqrt{(w^2 + y^2)} = \cosh(\beta z) + X \tag{6.10}$$

where w, y, and z are the coordinates and β describes the neck shape while being constrained by volume conservation and continuity. This neck profile has a constant curvature gradient as anticipated for continued mass flow during sintering.

Curvature is described using two perpendicular radii K_1 and K_2 fitted to any point on a surface, as described in Figure 6.9. The mean curvature M is the average of these two:

$$M = \frac{1}{2}(K_1 + K_2) \tag{6.11}$$

There is a change in sign for the curvature if it is located inside the mass versus outside the mass, so the initial stage sintering saddle surface is a mixture of

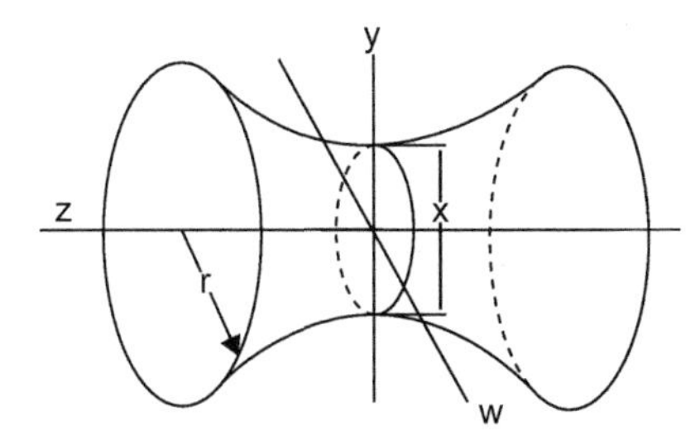

Figure 6.8 Neck geometry for calculation of the initial stage curvature gradient, in this case without a grain boundary dihedral angle groove.

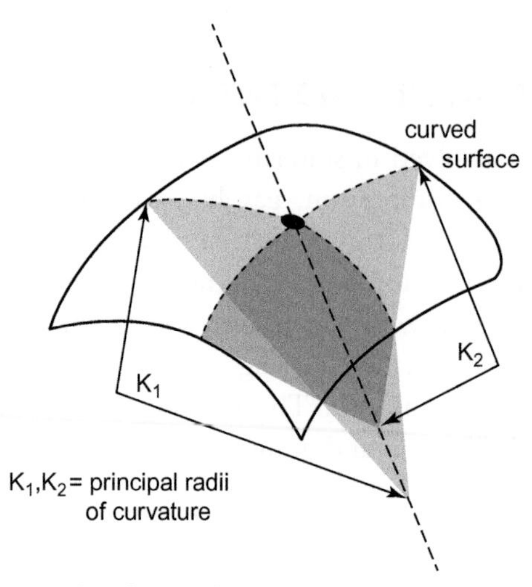

Figure 6.9 Illustration of a curved surface and two perpendicular curvature radii used to determine the mean curvature.

both positive and negative curvatures. For the neck geometry, the curvature is expressed as:

$$M = \frac{f' + (f')^3 + rf''}{2r(1+(f')^2)^{3/2}} \tag{6.12}$$

where $z = f(r)$ relates the z-axis to the radial distance r. On the concave surface, the mean curvature is positive, and applying a cylindrical coordinate system relates the radius r to the w and y coordinates through the following:

$$r = \sqrt{(w^2 + y^2)} \tag{6.13}$$

and:

$$f(r) = \frac{1}{\beta}\cosh^{-1}\left(r - \frac{X}{2} + 1\right) \tag{6.14}$$

or:

$$f(r) = \frac{1}{\beta}\cosh^{-1} V \tag{6.15}$$

where $V = (r - X/2 + 1)$. If sintering is by surface diffusion there is no change in density during neck growth so:

$$\beta \cong \frac{3}{4X^{3/2}} \tag{6.16}$$

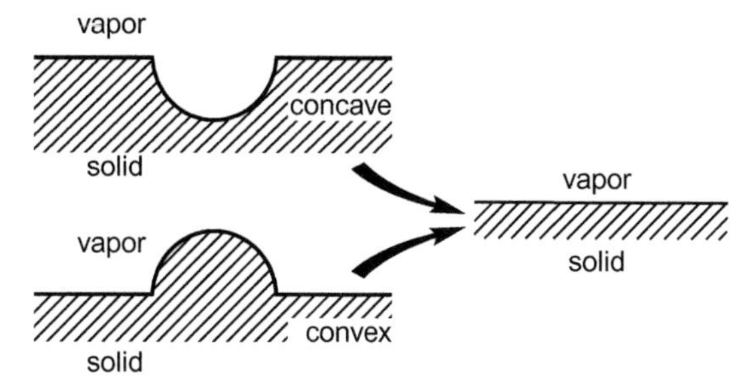

Figure 6.10 The combination of convexity and concavity leads to mass flow to form a flat surface.

Curvature in sintering is such that concave or convex surfaces move to become flat, as illustrated in Figure 6.10. In other words, the neck enlarges to fill and the convex particle surface shrinks to provide mass for the neck. The normal surface motion N at any point depends on the curvature gradient along the surface profile S:

$$\frac{dN}{dt} = B\frac{d^2M}{dS^2} \tag{6.17}$$

The parameter B is related to mass transport, for example by surface diffusion, while S is measured by the distance along the surface. Note that the second derivative of the mean curvature along the surface is the primary motivation for mass flow in early sintering, geared toward the elimination of curvature gradients. Analysis of early sintering shows:

$$M = a + bS + cS^2 \tag{6.18}$$

which gives $d^2M/dS^2 = 2c$, a constant for any given neck size. The curvature gradient declines by a factor 10^5 from a neck size ratio $X/D = 0.05$ to $X/D = 0.3$. At small necks the curvature gradient is large and dominates sintering. As neck growth proceeds, the gradient relaxes. By a neck size ratio X/D of 0.3 (about half the initial surface area is lost), the gradient is relaxed to the point where the key motivation for mass flow is surface area elimination. An example of curvature change during sintering is plotted in Figure 6.11 for the sintering of nickel powder [17]; curvature declines with densification.

Another aspect of surface curvature is stress. Flat surfaces are assumed to be stress-free, while convex surfaces are under tension and concave surfaces are under compression. The saddle surface of the neck is then a mixture of both concave and convex curvatures. The generalized Laplace relation links stress σ and curvature:

$$\sigma = \gamma\left(\frac{1}{K_1} + \frac{1}{K_2}\right) \tag{6.19}$$

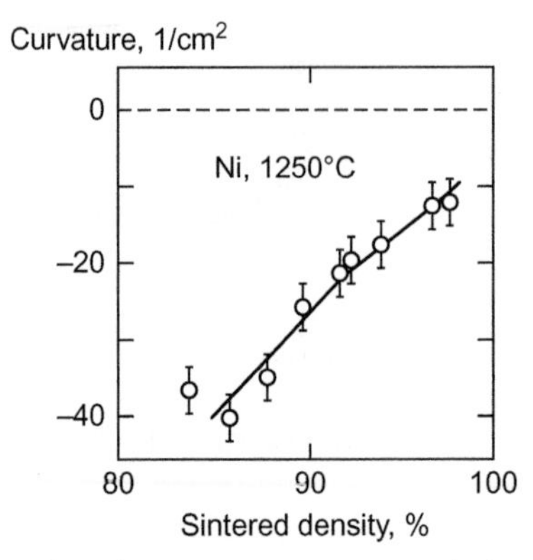

Figure 6.11 Nickel sintering data for microstructure curvature during densification [17].

where K_1 and K_2 are the radii of arcs on a surface and γ is the surface energy. For a sphere the two radii are the same. This is the case for spherical pores in final stage sintering. In the initial stage the neck is a saddle surface. so the stress is a mixture of concave (radius outside solid) and convex (radius inside solid) contributions, giving:

$$\sigma = \gamma\left(\frac{2}{X} - \frac{4D}{X^2}\right) \tag{6.20}$$

This is the sintering stress. It is large for small necks. Sometimes an external pressure is applied to increase sintering densification. Compared to the sintering stress, low applied pressures have little influence. For example, a 1 μm powder with a neck size ratio $X/D = 0.1$ and surface energy of 1 J/m^2 has an inherent sintering stress of 380 MPa, while many pressure-assisted sintering devices supply only 50 MPa. The self-induced sintering stress from nanoscale powders provides a means to drive rapid densification. On the other hand, larger powders have small sintering stresses. An example of the sintering stress for 30 μm copper powder versus density is plotted in Figure 6.12 [18]. In this case the sintering stress is in the 0.1 MPa range. Further, it declines as surface curvature is eliminated.

Interfacial Energy Changes

As sintering progresses, the driving force shifts from elimination of curvature gradients to elimination of interfacial energy. Two factors are evident—a change in surface area and a change in grain boundary area. Usually surface energy is larger than grain boundary energy, so surface area is eliminated preferentially, resulting in a linear relation between specific surface area and sintered density [19].

Early in sintering, the interface area reduction is largely due to neck growth. The amount of grain boundary area formed determines subsequent grain growth. Late in sintering there is less solid-vapor interface annihilation, so grain growth is the key to

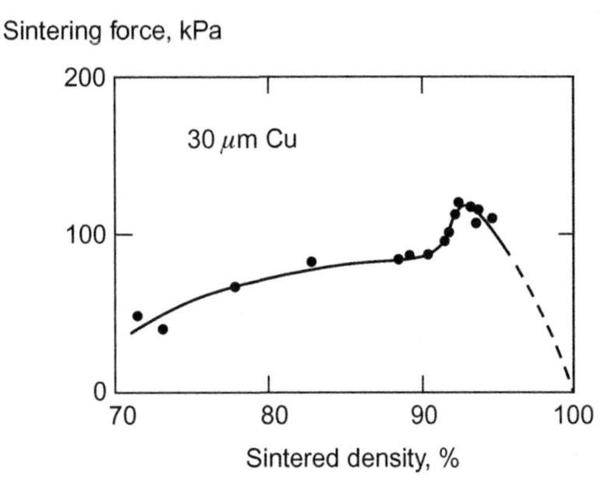

Figure 6.12 The measured sintering force for densification as measured during the sintering of 30 μm copper over the temperature range from 950 to 1150°C (1223 to 1423 K) [18].

interfacial area reduction. The cascade first surface energy converted to grain boundary energy by neck growth, then grain boundary energy is eliminated by grain growth.

The specific surface energy release ΔE_S (units of J/kg) is given by the surface energy γ_{SV} multiplied by the surface area change, as follows:

$$\Delta E_S = \gamma_{SV} S_O \left[\frac{f - f_G}{1 - f_G}\right] \tag{6.21}$$

where f is the sintered fractional density, f_G is the starting green density, and S_O is the starting surface area. At the same time the grain boundary energy increases through grain boundary formation. The specific boundary energy ΔE_G (units of J/kg) depends on the solid-solid interface area and energy γ_{SS}. Early in sintering this can be approximated as follows:

$$\Delta E_G = \frac{3 N_C \gamma_{SS}}{\rho G} \frac{\Delta L}{L_O} \tag{6.22}$$

where $\Delta L/L_O$ is the sintering shrinkage, N_C is the grain coordination number, G is the grain size, and ρ is the theoretical density of the material. This relation divides the solid-solid surface energy in half to split the energy between contacting grains. For intermediate stage sintering with cylindrical pores on the grain edges, the grain boundary energy change is a function of fractional density f, given as follows [20]:

$$\Delta E_G = \gamma_{SS} \frac{3.3}{\rho G} \left[1 - 1.6\sqrt{(1-f)}\right] \tag{6.23}$$

where the grain boundary energy is split between two contacting grains. There are slight variations in the 1.6 constant depending on the grain shape and coordination number.

The densification energy reflects the product of the sintering stress and volumetric strain. The volumetric change is the volume change due to densification divided by the initial volume, $\Delta V/V_O$. Assuming mass conservation, the volumetric strain is equivalent to a function of the green fractional density divided by the sintered fractional density:

$$\frac{\Delta V}{V_O} = 1 - \frac{f_G}{f} \tag{6.24}$$

The sintering stress acts on the compact to induce densification. As shown above, the magnitude of the sintering stress depends on the surface energy, particle size, and neck size. Measurements [21] show that a linear function is best:

$$\sigma = \sigma_O f + c \tag{6.25}$$

where σ_O is the slope and c represents the intercept. Compact strength measurements during sintering find thermal softening is a precursor to rapid densification [22]. For the copper data plotted in Figure 6.12 $\sigma_O = 380$ kPa and $c = -196$ kPa. This means that the system has zero strength at a fractional density of 0.52, near the reported starting green density.

Accordingly, the combination of stress and volumetric strain gives an estimate of the work performed during densification on a per volume basis ΔE_D. To compare this with the surface energy release and grain boundary creation energies, both based on energy per unit mass, requires normalization to the density:

$$\Delta E_D = \frac{1}{\rho f}\left(1 - \frac{f_G}{f}\right)(\sigma_O f + c) \tag{6.26}$$

The energy release by annihilation of surface area, mitigated by the grain boundary creation, gives an estimate of the energy release to perform densification work:

$$\Delta E_D = W[\Delta E_S - \Delta E_G] \tag{6.27}$$

where W is a proportionality factor that accounts for energy loss during densification, representing how efficiently interfacial energy is converted into densification. Other than material parameters, this only depends on fractional density and grain size. It shows that the energy allocated to grain boundaries (ΔE_G) decreases with grain growth, allowing further densification; grain growth releases energy otherwise useful for densification.

For the case of surface transport controlled sintering, necks grow and specific surface area declines, but density remains unchanged. Other models [23–25] include compaction and the grain boundary energy in the surface area determination.

MICROSTRUCTURE CHANGES

This section details the microstructural changes that occur in sintering, focusing on the pore structure and grain structure. These are coupled progressions, evident in pore drag by moving grain boundaries and retarded grain growth in the presence of pores.

Grain Size and Shape

As necks grow between contacting particles there is a concomitant increase in grain boundary area, since a grain boundary forms in each contact plane. Figure 6.13 plots the grain boundary area seen during sintering 48 μm copper powder at 1000°C (1273 K) [26]. One measure of the solid-solid bonding is the contiguity, defined as the fraction of grain perimeter that is solid-solid. Sintered properties such as strength, thermal expansion, and conductivity vary directly with the contiguity.

Grain growth accelerates as full density is approached. The temperatures required to induce sinter bonding also induce grain coarsening. Grain growth involves diffusional exchange across grain boundaries, while densification involves diffusional motion along the brain boundaries. Pores reduce the grain contact area and retard grain boundary migration, so they slow grain growth. As pores are eliminated, an increase in grain boundary area leads to coarsening, simply because there is more interface area, and less retardation. Thus, during sintering the mean grain volume increases linearly with time; the cube of the grain size is proportional to sintering time.

A measure of grain-grain contact during sintering is given by the contiguity C_{SS}—a dimensionless fraction of grain interface that is solid. Contiguity

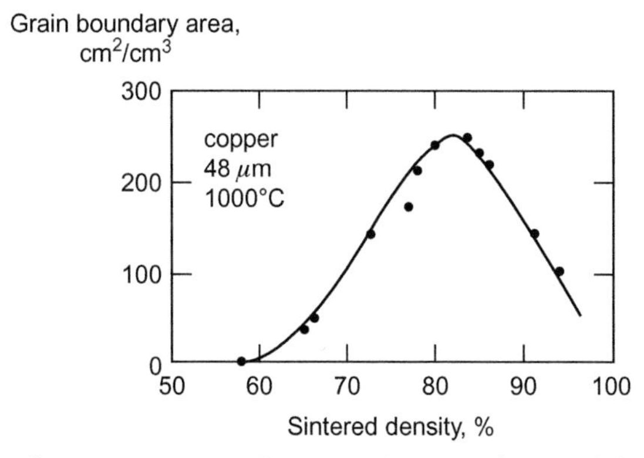

Figure 6.13 Grain boundary area per unit volume as a function of sintered density for 48 μm copper powder sintered at 1000°C (1273 K) [26]. Early neck growth with concomitant formation of grain boundaries at the contacts gives an increase, but grain growth late in sintering reduces the grain boundary area.

varies with fractional porosity $\varepsilon = 1 - f$, where f is the fractional density as follows:

$$C_{SS} = 1 - q\sqrt{\varepsilon} \tag{6.28}$$

where q is a geometric coefficient near 1.5.

One consequence of the pore-boundary coupling is a relation between grain size G and fractional porosity ε (or $1 - f$):

$$G = \frac{\theta \, G_O}{\sqrt{\varepsilon}} \tag{6.29}$$

while G_O is the initial grain size and θ is geometric constant near 0.6. For copper, Figure 6.14 [27] is a plot of grain size versus $1/(1-f)^{1/2}$. This behavior was noted by Bruch [28].

When grain size G is plotted versus fractional density f, grain size increases rapidly as full density reached. Figure 6.15 is a plot for 0.22 μm alumina sintered at 1200 to 1250°C (1473 to 1523 K) for up to 48 h, giving grain size versus fractional density [29]. The relation between fractional density and grain size is valid for pressure-assisted sintering, as evident in Figure 6.16 [30]. This plots grain size versus density during titanium spark sintering.

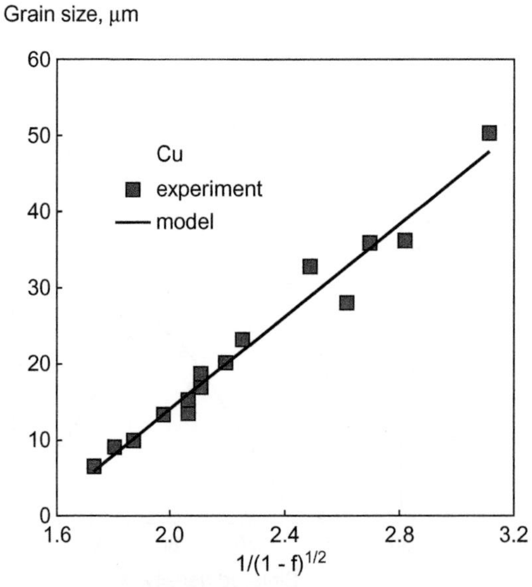

Figure 6.14 Data for copper on the grain size versus the inverse square root of the fractional porosity ($1/\varepsilon^{1/2}$) ranging from 60 to 90% density [27]. The behavior follows the form suggested by Bruch.

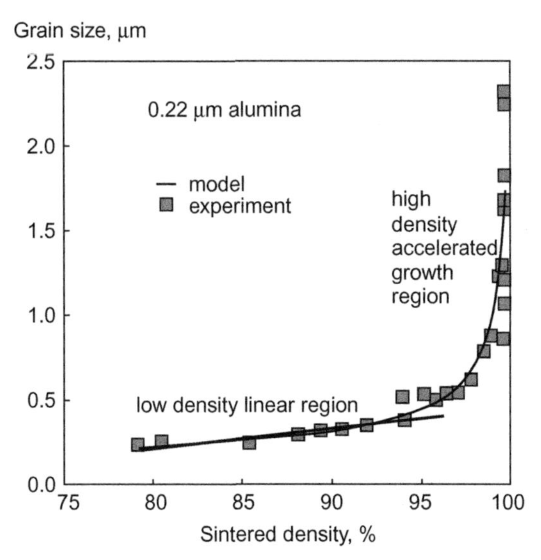

Figure 6.15 Grain size versus sintered density for alumina, illustrating a nearly linear relation at lower densities but more growth as pores are eliminated [29]. The solid line corresponds to Equation 6.29.

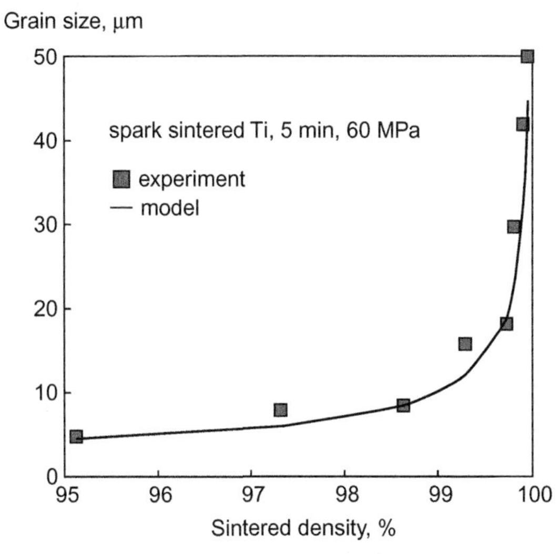

Figure 6.16 Similar to Figure 6.15, this is a plot of grain size versus density for titanium processed under spark sintering conditions at 60 MPa and various temperatures for a hold time of 5 min [30]. The data agree with the model of Equation 6.29.

The sintered grain shape is polygonal, reflecting the grain coordination. Late in sintering that shape converges toward a tetrakaidecahedron with 14 faces. However, the shape is distributed. For example, Figure 6.17 shows the measured distribution in grain contacts for sintered magnesia with a median of 5 [31]. The 14-sided

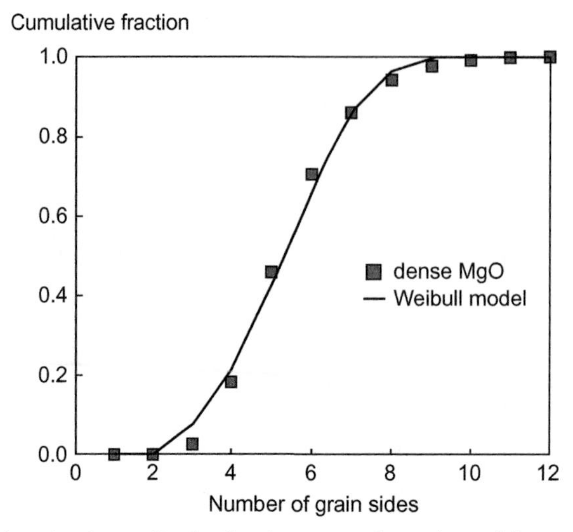

Figure 6.17 Measured grain shape distribution in terms of number of faces per grain for sintered magnesia and the Weibull distribution fit to the grain shape [31].

tetrakaidecahedron has eight hexagonal faces and six square faces, averaging 5.14 edges per face.

The starting particle coordination number is often near seven contacts per grain at 60% initial density, and this increases during sintering. The grain structure coarsens during sintering, converging to self-similar size and shape distributions. Various estimates for the grain coordination number N_C with densification are available, but one that is useful for solid-state sintering is as follows:

$$N_c = 2 + 11f^2 \tag{6.30}$$

It approaches 13 at full density, and the transition from intermediate to final stage sintering occurs with a coordination number from 10 to 12. Regions in a sinter body that start with a low coordination number will grow cracks since they are weak. Thus, from a practical standpoint homogeneous particle packing is beneficial to sintering.

Sintered grain structures are not random. High coordination grains tend to form strings within the microstructure. This is caused by the non-uniform grain boundary energy versus crystal orientation. The low energy cusps lead to preferred orientations during sintering.

Grain shape and grain growth are related. Low coordination grains with fewer faces tend to shrink during coarsening. Since a sphere is the 3-D shape with the lowest surface area per unit volume, the largest grains are more spherical and more stable, while the smaller grains with fewer faces are unstable and disappear during sintering. Larger grains tend toward spherical shapes due to their high coordination numbers. Indeed, grain shape is the best predictor of grain growth. Thus, in sintered materials grain shape and grain size are related. Larger grains have more faces.

Grain Size Distribution

Sintered materials converge to the same grain size distribution shape, that is termed self-similar. Let $F(G)$ be the cumulative fraction of grains with size G. Grain size is measured by the random intercept, equivalent circular diameter, or equivalent spherical diameter. The resulting self-similar grain size follows a distribution as follows:

$$F(G) = 1 - \exp\left(\beta\left(\frac{G}{G_M}\right)^M\right) \tag{6.31}$$

where G_M is the median grain size. The factor $\beta = -\ln 2$ (or -0.6931) ensures $F(G) = 0.5$ when $G = G_M$ at the median size. Fits to this cumulative Weibull grain size model are given in Figure 6.18 for tungsten, zirconia, alumina, and magnesia [20]. The exponent M is near 2 when grain size is measured in two-dimensions and near 3 for three-dimensional grain size data. Unlike a normal distribution, the mode, mean, and median sizes are not the same.

Pore Structure

Early in sintering the pores are angular, reflecting the gaps between particles. Sintering works to reduce the curvature gradients, so the pores round and curvature tends toward a neutral condition. The microstructure moves from a high level of concavity to a neutral or flat structure.

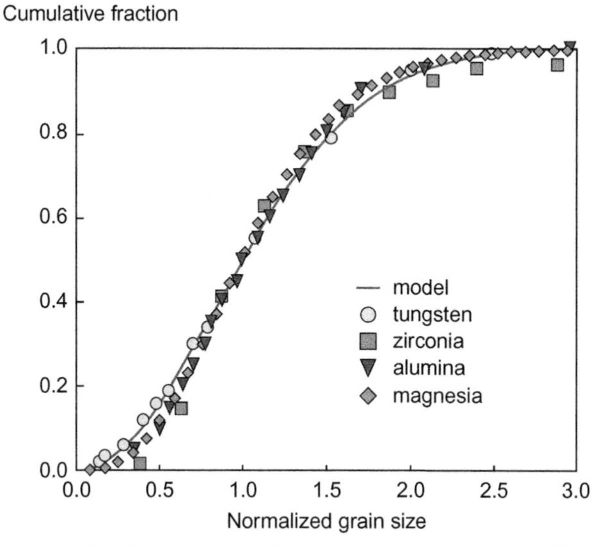

Figure 6.18 Grain size distribution data for sintered tungsten, zirconia, alumina, and magnesia, showing a self-similar Weibull distribution according to Equation 6.31. The grain size is normalized by the median grain size.

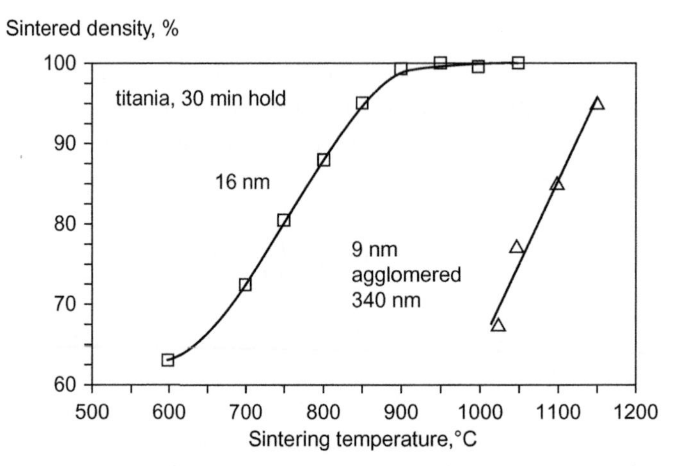

Figure 6.19 Agglomeration hinders sintering due to the large pores that from between the agglomerates. These results plot sintered density versus temperature for 30 min hold for 16 nm and 9 nm powder; the latter powder was agglomerated to 340 nm clusters with a negative impact on sintering [1].

The elimination of curvature gradients leads to a push and pull on the pore structure. The angular pores smoothen, small pores shrink, and large pores grow. The net result is little change in pore size while the porosity decreases. In structures with both large and small pores, the small pores disappear while the large pores remain stable. For example, in experiments with titania (TiO_2), a 16 nm particle without agglomeration sinters to 95% density in 30 min at near 850°C (1123 K) [1]. On the other hand, a smaller agglomerated 9 nm particle giving 340 nm clusters requires over 1150°C (1423 K) to reach the same density, as plotted in Figure 6.19. Agglomeration offsets any advantage of small particle size with respect to sintering.

Pores between agglomerates are larger and grow while pores between particles are small and shrink. This is traced to grain coordination around the pore. As sketched in Figure 6.20, pores with a high number of grains on their surface resist shrinkage. This is why intentional holes and highly detailed features remain in a component after sintering; the high grain coordination resists closure. Meanwhile small pores disappear.

As noted, intermediate stage tubular pores collapse into discrete spherical pores during densification. Measurements show first pore closure at a density of just 85%. Figure 6.21 plots pore closure for urania sintering at 1400°C (1673 K) [32]. As the total porosity decreases, open pores progressively transform into closed pores, giving equal closed and open porosity near 92% density. By about 95% density, all of the pores are closed.

Pore Size Distribution

The pore size depends on the measurement approach. Up to final stage sintering, the tubular pore size is measured by length and diameter. If cross-sectioned, then some of

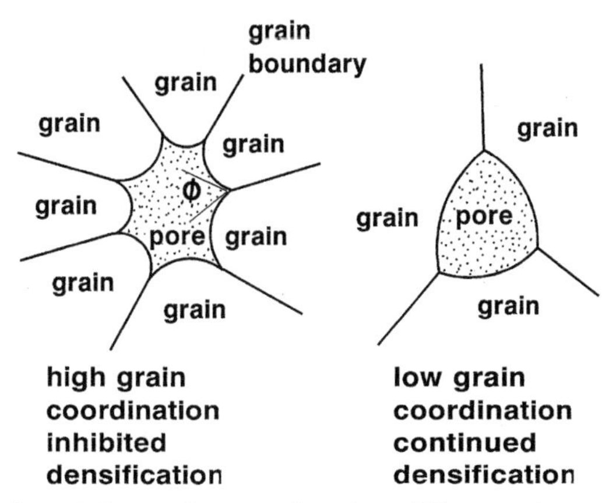

Figure 6.20 The grain size relative to the pore size gives differences in pore shape. Pores with low grain coordination shrink first in sintering, but larger pores resist sintering densification until grain growth reaches a critical condition.

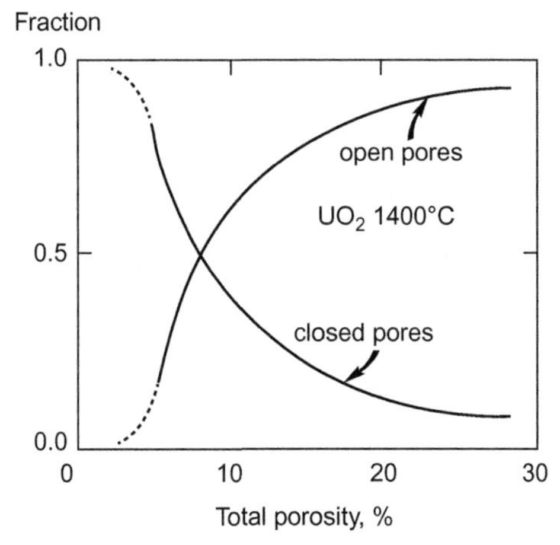

Figure 6.21 Sintering data for urania at 1400°C (1673 K) showing the fraction of open or closed pores versus total porosity, illustrating a change at about 8% porosity [32].

the pores will be captured along their length and others along their diameter. Hence the two-dimensional visualization is not accurate.

The average pore size changes little during sintering, while the porosity decrease and the grain size increases, as illustrated in Figure 6.22 for 4.5 μm iron powder [13]. The number of pores decreases to accommodate densification, but the mean pore size is relatively resistant to change while density increases. However, if the initial pore structure is inhomogeneous, then the structure diverges to form large and stable pores. The large pores between agglomerates contrast with the smaller pores within the agglomerates, as sketched in Figure 6.23. The smaller pores are eliminated early in sintering, but

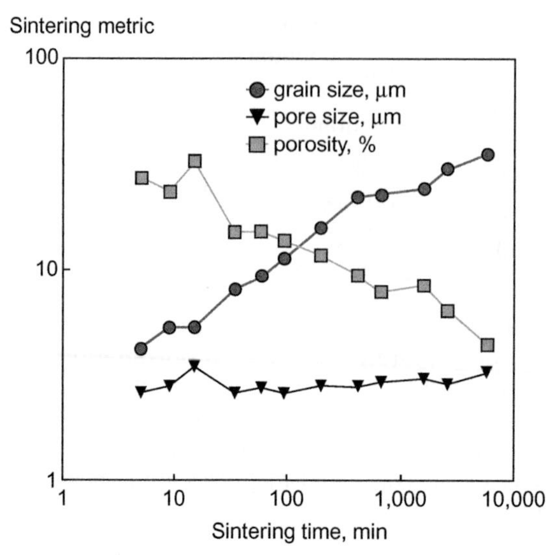

Figure 6.22 Porosity, pore size, and grain size during sintering of 4.5 μm iron powder at 850°C (1123 K) [13]. While the porosity declines grain size increases and pore size remains relatively constant.

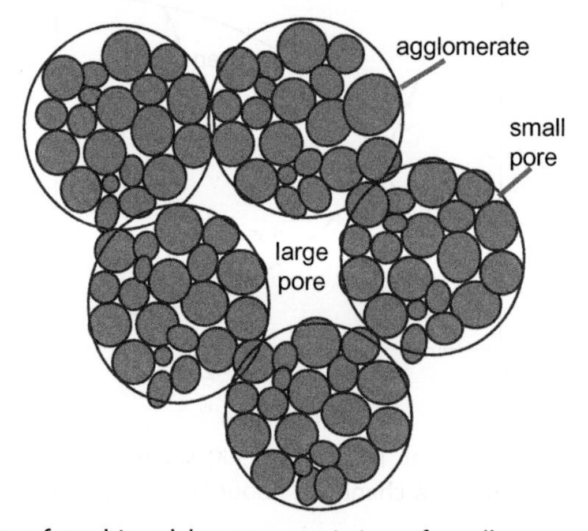

Figure 6.23 Agglomerates form bimodal pores, consisting of small pores within the agglomerates and large pores between the agglomerates. The large pores are resistant to sintering.

the larger inter-agglomerate pores resist sintering. Proper compaction of an agglomerated powder is needed to ensure both pore structures are of similar size.

When sintering in a vacuum, the pore is essentially a collection of vacancies. A small pore has a higher local vacancy concentration, and when heated those vacancies diffuse to grain boundaries, free surfaces, or other interfaces, where they are annihilated. The smaller pores shrink over time but larger pores grow. When gas is trapped in the pores, then gas pressure retards pore elimination. An example of such behavior is evident in

Figure 6.24 [33]. Here the sintered density increases with green density, but once the closed pore condition is reached at 95% density, densification is not improved. At this point the gas pressure in the pore P_G and pore size d are inversely related:

$$P_G = \frac{4\gamma_{SV}}{d} \tag{6.32}$$

with γ_{SV} being the solid-vapor surface energy. The internal gas pressure works against the sintering stress, resulting in less than full densification. Further, swelling can occur since the residual pores coarsen with a concomitant decrease in gas pressure, leading to expansion. Thus, late stage swelling is observed with prolonged sintering, as recorded in

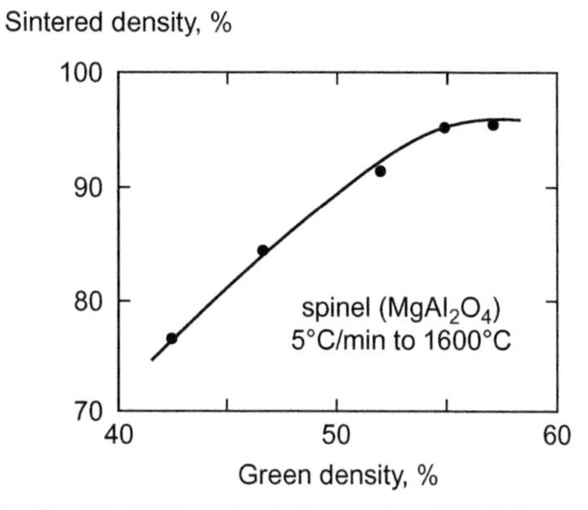

Figure 6.24 Sintered density for $MgAl_2O_4$ spinel sintered at 5°C/min to 1600°C (1873 K) [33]. This plot shows how sintered density increases with green density.

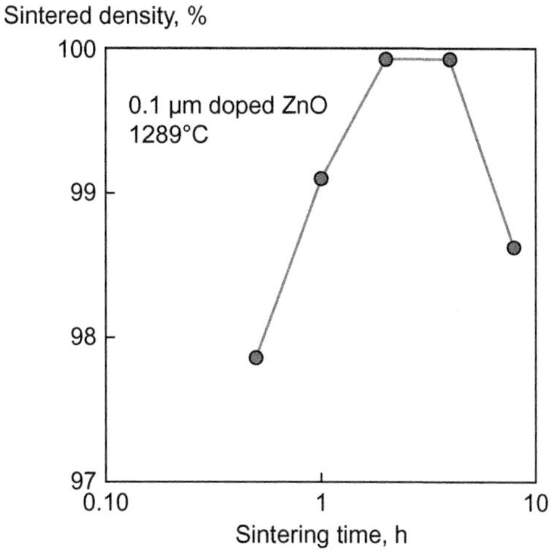

Figure 6.25 Zinc oxide of 0.1 μm particle size sintered for various times at 1289°C (1562 K) [34]. Densification is evident initially, but swelling occurs at longer times.

Figure 6.25 [34]. If the trapped gas is insoluble, then the pores act as pressurized balloons, resisting densification. Higher temperatures inflate the balloons, leading to swelling.

Pores shrink by emitting a vacancy flux annihilated at free surfaces or grain boundaries. At the same time, neighboring pores interact and undergo coarsening by coalescence or diffusion through the solid. All pores can be eliminated by sintering in vacuum, but in atmosphere-based sintering the pores can become stable at less than full density. Atmosphere pressure in the pore decreases as it enlarges, allowing faster pore growth with a curious de-sintering or swelling. In Figure 6.25 the sintered density is plotted versus sintering time, showing swelling at longer times.

Pore attachment to grain boundaries is generally favored during sintering. The focus is to keep the pores coupled to the grain boundaries. A pinning force exists between the pore and the grain boundary, even when they are moving. The magnitude of the pinning force depends on the surface energy and pore shape.

The pore size distribution transforms to become self-similar, in the same way that the grain size distribution and grain shape are self-similar. The cumulative distribution in pore sizes d tends toward a distribution as follows:

$$F(d) = 1 - \exp\left[\beta\left(\frac{d}{d_M}\right)^M\right] \tag{6.33}$$

were d_M is the median pore size. The factor $\beta = -\ln 2$ (or -0.6931) ensures this Weibull cumulative distribution gives a proper fit to the median, that is $F(d) = 0.5$ when $d = d_M$ at the median size. A fit to this distribution is shown in Figure 6.26 for

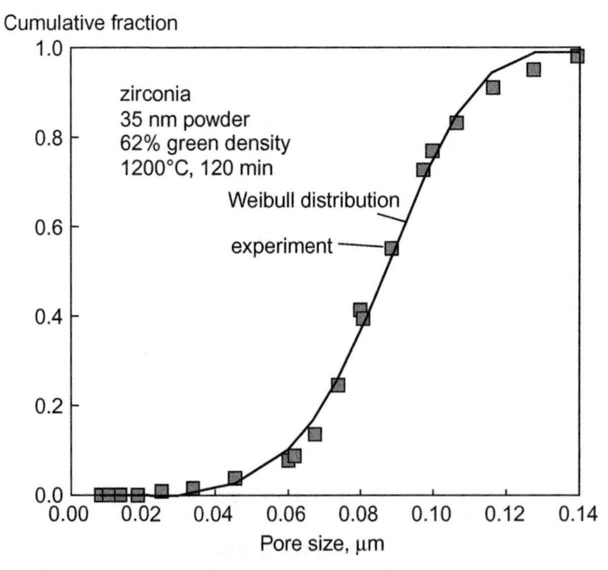

Figure 6.26 The cumulative pore size distribution fit with a Weibull curve. There data are from 35 nm zirconia sintered at 1200°C (1473 K) for 120 min.

Table 6.2 Coordination Effects on Densification and Coarsening during Sintering

	Pore Coordination (number of touching grains)	Particle Coordination (number of touching particles)
densification	low	high
coarsening	high	low

35 nm zirconia sintered at 1200°C (1473 K) for 120 min. In this case the median pore size is 87 nm, about three times the starting pore size.

It is common to observe sintering with simultaneous densification and coarsening. There are sintered products where the lack of densification is desirable. Alternatively, densification is required for structural materials. The balance between densification and coarsening rates is important to determining the sintered properties.

During coarsening the pores grow, while during densification the pores shrink. In many cases this mixture of densification and coarsening leads to growth of the larger pores and shrinkage of the smaller pores. Larger pores grow and regions of high packing density, corresponding to smaller pores, densify. At long times or high temperatures the larger pores eventually densify. However, if the pores are large or coarsen from an internal gas they resist densification. In contrast, when the initial particles are packed to create an ideal monosized pore structure, then sintering densification proves rapid, requiring lower temperatures or less time. This is a reason to avoid particle agglomeration or packing gradients, since the tails of the pore size distribution are critical factors in determining the relative densification and coarsening rates.

As categorized in Table 6.2, the packing coordination number is important in pore shrinkage. Large pores with many neighboring grains grow. Small pores with few neighboring grains shrink. The extreme case is represented by chain-like structures with low particle coordination numbers which coarsen without densification. The combination of small grains and large pores promotes coarsening.

Pore Attachment to Grain Boundaries

Late in sintering, grains coalesce and grow, while grain boundary attachment to pores slows grain growth. Pore accumulation on moving grain boundaries leads to dense zones in the wake of the grain boundary and larger pores accumulated on the moving grain boundary, as sketched in Figure 6.27. This image reproduces the microstructure reported by Burke [35]. Eventually, large pores are unable to remain attached to migrating grain boundaries and become stranded inside the grains. Motion of the grain boundary due to pore drag leads to a bowed grain boundary with a groove, as diagramed in Figure 6.28. At low levels of grain boundary curvature, the increased boundary area is small compared to the area occupied by the pore. But as bowing increases separation becomes favorable. Rapid grain growth occurs once the boundaries break away from the pores, and usually densification halts.

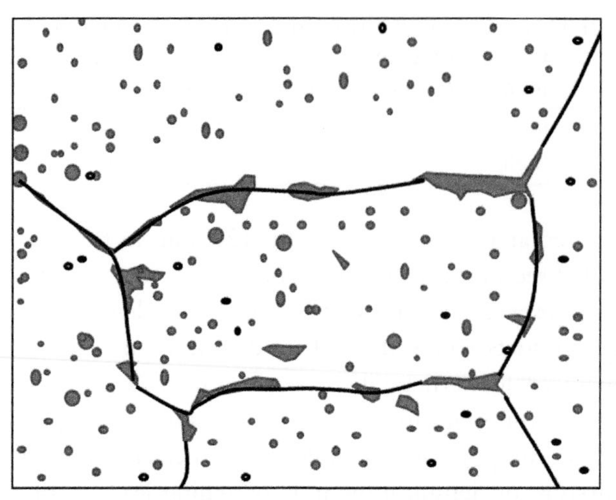

Figure 6.27 The microstructure after sintering as reported by Burke [35], where the migrating grain boundary captures and accumulates pores, leaving a dense region in its wake.

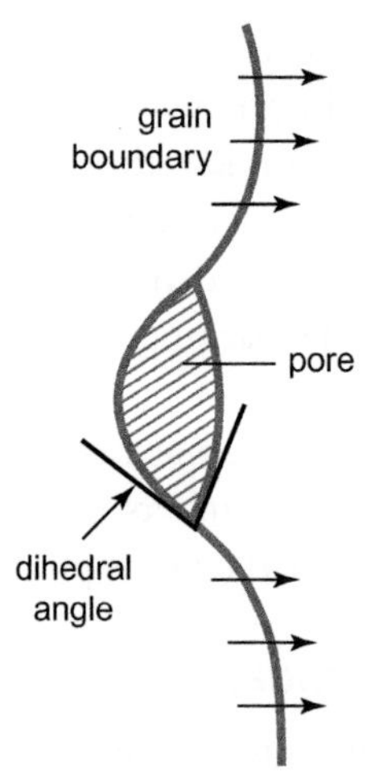

Figure 6.28 Migrating grain boundaries drag pores. The difference in curvature between the front and rear surfaces provides a gradient for mass transport and pore migration.

Pore attachment to the grain boundaries depends on the relative motion of the pores and boundaries. Small pores are easier to either annihilate or move, so the pore size to grain size ratio is a key factor in determining the attachment. A pore and a grain boundary have a binding energy since shared interfaces reduce the overall interfacial energy. As the pore shrinks, the binding energy decreases, as plotted in Figure 6.29. The decreasing binding energy allows for boundary separation from the pore as densification progresses. Grain growth forces the grain boundary to curve,

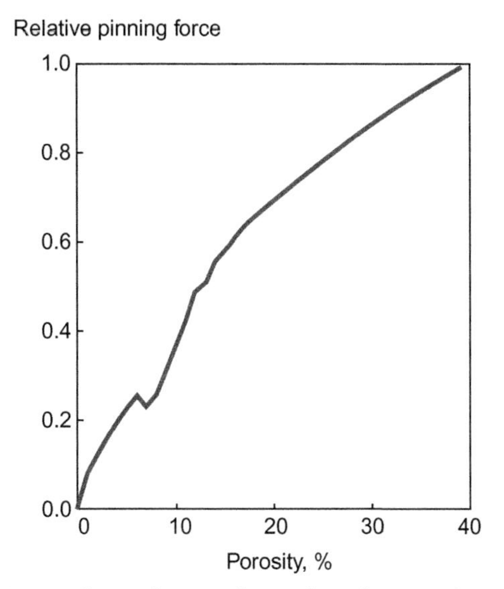

Figure 6.29 The relative pinning force of a pore located on the grain boundary versus the porosity. The bump at 8% porosity corresponds to the change from open to closed pores. As pores are annihilated the attachment diminishes with less than full densification.

leading to increased grain boundary area as the grain boundary bows to sustain contact with a pore. Accordingly, a critical condition is achieved at the point when the boundary breaks away from the pore. In the intermediate stage of sintering with a high porosity, there is a large energy penalty for pore-boundary separation. Densification reduces this penalty. As a consequence, the grain boundary mobility must be controlled to sustain a microstructure conducive to densification.

The result is a relationship between final pores size d, grain size G, and fractional density f, often termed the Zener relation:

$$G = \frac{3d}{8(1-f)} \tag{6.34}$$

A modified form of this relation is observed experimentally [36]:

$$G = \frac{gd}{R(1-f)} \tag{6.35}$$

where the parameter R is the attachment ratio (often measured at 0.7), indicating the fraction of pores attached to grain boundaries and g is a geometric constant near 1.33. This gives the ratio $g/R = 1.9$, while the Zener relation predicts more grain growth retardation. Rearranging this relation predicts the ratio d/G, pore size divided by grain size, is proportional to porosity $\varepsilon = 1 - f$. This is plotted in Figure 6.30 using

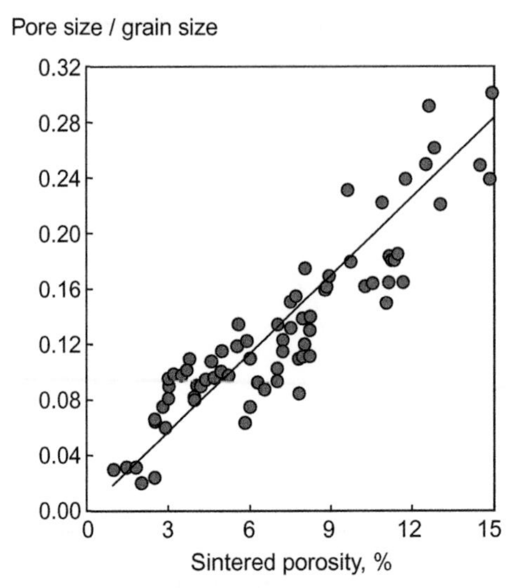

Figure 6.30 Data on the pore size to grain size ratio (*d/G*) during sintering iron, nickel, and copper [37]. Although scattered, the trend is in agreement with a Zener relation.

data for copper, nickel, and iron [37], showing in the final stage of sintering the validity of the Zener relation.

The geometry of a pore depends on the coordination number of contacting grains and the dihedral angle. A pore with a lare number of grains on the surface is stable. On the other hand, pores with a few neighboring grains tend to shrink. Thus, during sintering the larger pores are initially stable, but grain growth eventually makes those same pores unstable. Small pores are the first to annihilate in sintering. Large pores with a high grain coordination number resist densification, but grain growth eventually corrects the situation, assuming there is no simultaneous pore growth.

Grain growth is increasingly active as the pore structure collapses. The pinning effect of the pores diminishes as they shrink and occupy less grain boundary area. Consequently, grain growth dominates late in sintering.

The pore-boundary interaction determines the final sinter density. Additives are used to modify grain growth during sintering. The additives modify grain growth to assist final stage sintering. Some additives retard grain growth, while others accelerate it. This change is evident in the alumina sintering data given in Figure 6.31, where grain size is plotted versus sintered density [38]. All three cases follow a grain size behavior described by Equation 6.29. The undoped alumina is modified by magnesia, which retards grain growth versus iron oxide which accelerates grain growth. If pore-boundary separation occurs early, the sinter density is low and grain growth continues. Many strategies have emerged to control the pore-boundary coupling during sintering. One approach is to make the green structure as homogeneous as possible with all small pores, then pore

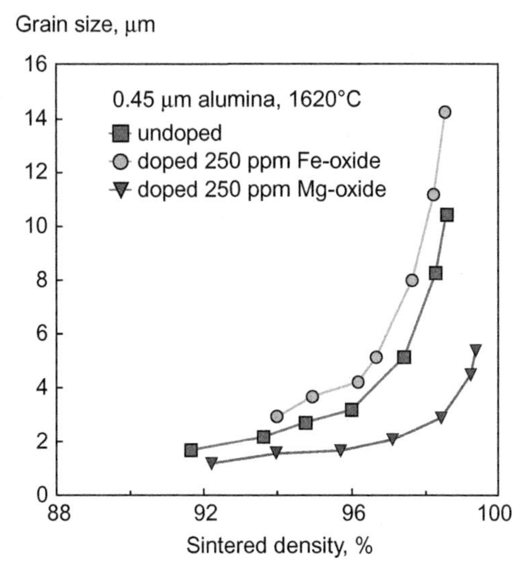

Figure 6.31 Experiments on the sintering of alumina, undoped and doped with two species [38]. The grain size trajectories follow Equation 6.29. The iron oxide additive increases grain growth and the magnesium oxide additive retards grain growth.

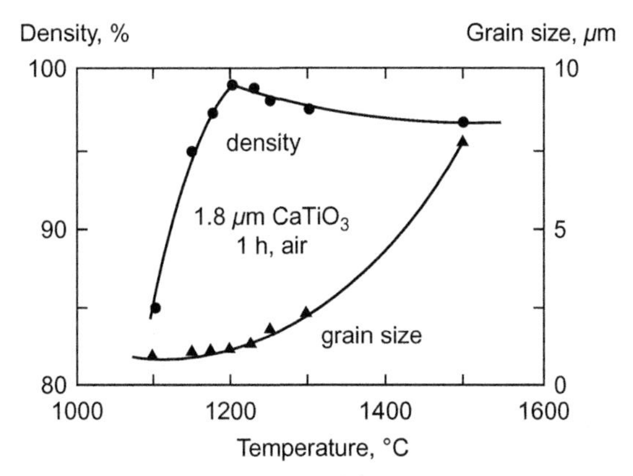

Figure 6.32 Sintered calcium titanate density and grain size versus sintering temperature [40]. Density decreases at higher temperatures, but grain growth continues.

shrinkages occurs rapidly and uniformly, with little grain growth [39]. On the other hand, in cases where pores separate from the boundaries, the pores and grains coarsen independently. The sinter density declines, since the gas pressure decreases as the pores enlarge while the total gas content remains constant. This is evident in Figure 6.32 for sintered calcium titanate [40]. Another control strategy is to reduce the sintering temperature as density increases to a point where grain growth accelerates [41].

MACROSTRUCTURE CHANGES—COMPONENT SIZE AND SHAPE

Neck growth gives a change to the particle spacing, and amplified over millions of necks this results in compact shrinkage or swelling. Swelling occurs during chemical reactions, such as with mixed Ni and Ti sintering to form NiTi. For the more typical systems, sintering results in densification.

One measure of sintering is the compact size. Shrinkage is a dimensional change associated with densification; sintering shrinkage $\Delta L/L_O$ is the change in dimension divided by the initial dimension, where shrinkage a positive change. It is related to the square of the neck size ratio X/D as follows:

$$\frac{\Delta L}{L_O} = \frac{1}{3.6}\left(\frac{X}{D}\right)^2 \tag{6.36}$$

This relation breaks down at large shrinkages, as shown by the glass sintering data in Figure 6.33 [42]. However, for neck size ratios up to 0.5 the relation is accurate.

A related measure is sinter density. If mass is conserved in sintering, then the sintered density f, green density f_G, and shrinkage are related as follows:

$$f = \frac{f_G}{\left(1 - \frac{\Delta L}{L_O}\right)^3} \tag{6.37}$$

The relation is plotted in Figure 6.34 for a range of green densities and for final sintered densities of 100% and 90%. It is common to observe up to 16% dimensional

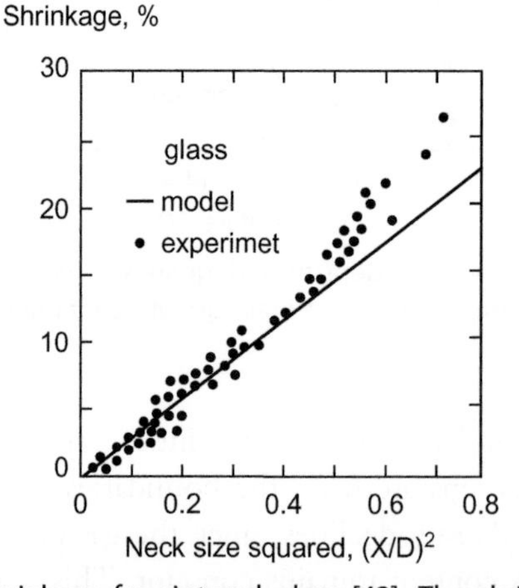

Figure 6.33 Neck size and shrinkage for sintered glass [42]. The shrinkage is plotted versus the square of the neck size ratio. In the initial and intermediate stages of sintering, when X/D remains below 0.5, the relation given by Equation 6.37 is appropriate.

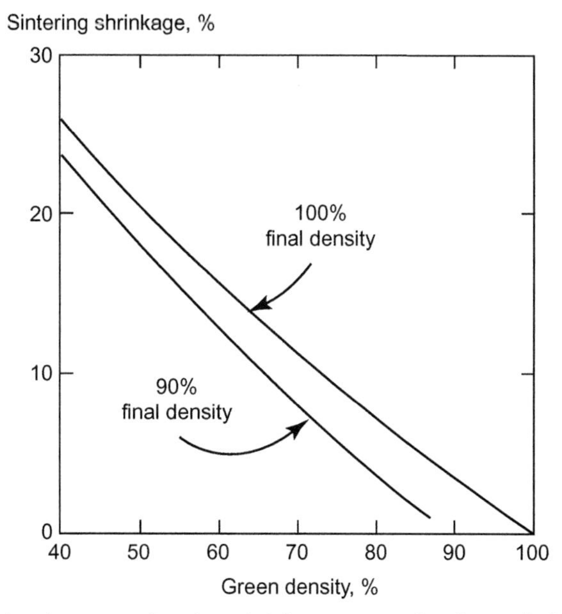

Figure 6.34 The relation between sintering shrinkage, green density, and sintered density, given by Equation 6.38, illustrated for final densities of 90 and 100%.

change in sintering, and even 25% for some ceramics. On the other hand, some ferrous automotive alloys are designed for only 0.1% shrinkage in sintering.

A high compaction pressure increases green density, reducing sintering shrinkage. For example, loose 63 μm copper powder sintered at 1020°C (1293 K) for 120 min shrinks by 4.1%. The same powder compacted at 600 MPa with the same sintering cycle shrinks by 2.3%. One way to rationalize green density changes and sintering shrinkage is to assume constant mass and to rearrange the above equation:

$$\frac{\Delta L}{L_O} = 1 - \left[\frac{f_G}{f}\right]^{1/3} \tag{6.38}$$

Shrinkage is viewed in two extreme ways. Manufacturers of precision components strive to achieve uniform or even no shrinkage during sintering. If shrinkage can be avoided, then pressed compact dimensions are retained, possibly with lower strength due to residual porosity. Shrinkage mandates an oversizing of tooling to bring the final sintered material into acceptable dimensions, but dimensional uniformity is degraded. For some materials, where high densities are mandatory for performance, shrinkage is necessary during sintering. Thus, depending on the material, its ease of compaction, and the required properties, it is possible that shrinkage will be either sought or avoided.

Early sintering densification concepts were developed by Ivensen [43]. He observed that the sintered pore volume remained proportional to the starting pore volume, for a fixed powder and sintering cycle. Thus, the sintered fractional density f varied with the green density f_G:

$$f = C f_G \left(\frac{1-f}{1-f_G} \right) \tag{6.39}$$

For example, in experiments with 1 to 10 μm nickel powder heated to 850°C (1123 K) for 60 min using 38 to 61% green density, the constant $C = 0.285$. A common expression rearranges the sintered and green density terms on each side:

$$\frac{f}{1-f} = C \frac{f_G}{1-f_G} \tag{6.40}$$

Usually dimensional change in sintering is not ideally isotropic, so such relations are only approximations. Several factors induce anisotropic shrinkage—including substrate friction, gravity, and non-uniform heating. Further, the forming process induces density gradients in the starting powder structure. Regions of lower density shrink more than regions of higher density, leading to non-uniform dimensions. This is evident in complex bodies which have thick and thin sections.

Uniform components require repeated fabrication of the same size and shape, upwards to millions of times per day. Size variation between components is a problem. In situations where the sintered density is nearly constant, then according to Equation 6.38 the sintering shrinkage is a function of the cube root of the green density. Figure 6.35 plots this behavior for molybdenum sintered at 1900°C (2173 K) for 8 h, showing a linear relation between shrinkage and the cube root of the green density. Uniform final component size relies on uniform component mass. To form sintered components with reproducible final size requires attention to the mass uniformity in the component shaping process.

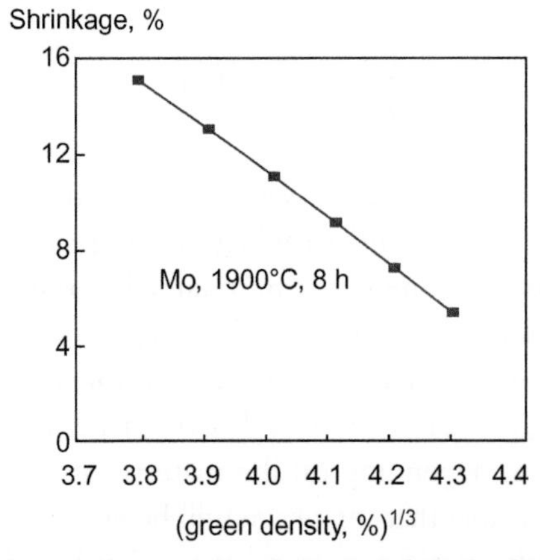

Figure 6.35 Molybdenum sintering data at 1800°C (2073 K) for 8 h, showing shrinkage depends on the cube root of the green density. The data span from 55 to 79% green density.

Sintering works to remove curvature gradients. Up to now the focus has been on microstructure level gradients, but the effect also extends to macroscopic features. Sharp edges tend to round. Indeed, if a material were to be sintered for a very long time, then the shape would be lost since a sphere is the lowest energy final shape.

SURFACE AREA TRAJECTORY

Unlike density, surface area is a sintering monitor applicable to both densification and nondensification mechanisms. Early in sintering the solid-vapor surface area is converted to a grain boundary area. Grain growth accelerates as the grain boundary area increases. Grain boundary area peaks near 85% density, and after that it declines due to grain growth during slow sintering densification [44]. Early in sintering the necks between particles are small, so the added grain boundary area is low. Late in sintering, at lower porosity levels, grain growth removes grain boundary area faster than new grain boundary area is produced by densification. These interface area changes determine the grain size versus the fractional density trajectory, usually giving large grains as full density is approached.

Surface Area Reduction

The specific surface area (measured as area per mass, m^2/g) declines in proportion to the sintered density. This is seen in Figure 6.36 for urania (UO_2) sintering at 1500°C

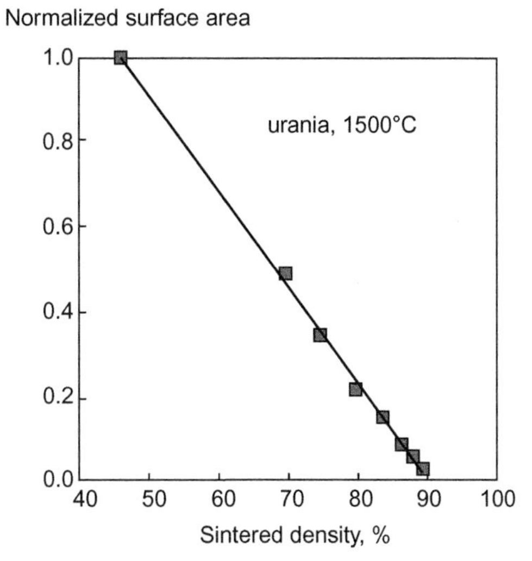

Figure 6.36 An example of the linear surface area relation to sintered density for urania sintering at 1500°C (1773 K) [45]. The surface area is normalized to the initial surface area.

(1773 K) [46]. The specific surface area is normalized to the starting surface area S_O and reaches zero at pore closure near 92% density. In this plot, the specific surface area S_M (solid-vapor) divided by the initial area S_O declines as the fractional sintered density f increases:

$$\frac{S_M}{S_O} = a - bf \tag{6.41}$$

The initial condition gives $S_M = S_O$ at the green density, f_G, so as a first estimate:

$$a = 1 + \frac{b}{f_G} \tag{6.42}$$

The specific surface area reaches zero at a critical density f_C. For gas absorption measurements, this occurs prior to full density. Gas absorption does not measure the area of closed pores, unlike quantitative microscopy which measures both the open and closed pore surface area. Accordingly, f_C ranges from 0.85 to 1.0, giving:

$$f_C = \frac{a}{b} \tag{6.43}$$

and for surface area measured by microscopy f_C is unity, so $a = b$, since it is able to measure closed and open pores. The combination of these two equations gives:

$$b = \frac{f_G}{f_C f_G - 1} \tag{6.44}$$

The tetrakaidecahedronal grain shape gives f_C equal to 92% for surface area measured by gas absorption.

Surface Area and Sintered Density

A linear relation between specific surface area and fractional sintered density is a reflection of neck growth, shrinkage, and packing coordination changes. Consequently the surface area during sintering depends on the neck size ratio X/D and the particle coordination number N_C as follows [46]:

$$\frac{S_M}{S_O} = 1 - 0.26\, N_C \left(\frac{X}{D}\right)^2 + 0.092\, N_C \left(\frac{X}{D}\right)^3 \tag{6.45}$$

These factors relate to the sintered density, so surface area is often expressed as a direct function of sintered density. The specific surface area is S_M, which starts from an initial value of S_O. Figure 6.37 plots the early stage sintering results, showing S_M/S_O as a function of sintered fractional density f starting with fractional green densities of 0.50, 0.55, 0.60, and 0.65. Each line is terminated when the neck size ratio reaches 0.5. Up to this point the correlation between surface area and density is linear.

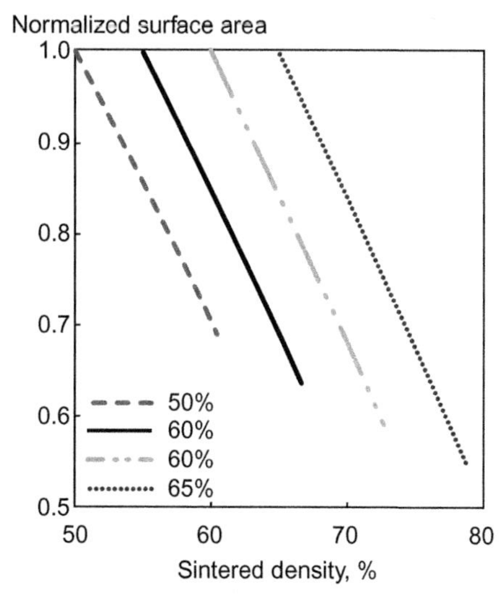

Figure 6.37 Traces of specific surface area versus sintered fractional density for green densities of 50, 60, 60, and 65%, giving nearly linear relations.

Table 6.3 Surface Area versus Sintered Density, Parameter Variation with Green Density*

Fractional Green Density, f_G	*a* Parameter	*b* Parameter
0.50	2.5	3.0
0.55	2.7	3.1
0.60	2.9	3.2
0.62	3.0	3.2
0.65	3.1	3.3

*These data give the following approximate relations:
$a = 0.436 + 4.14 f_G$.
$b = 1.86 + 2.08 f_G$.

Table 6.3 lists the corresponding a and b parameters for the surface area-density relation in Equation 6.41. Typical values are $a = 2.9$ and $b = 3.2$, but they vary slightly with green density. For example, if the initial fractional green density is 0.62, corresponding to loosely packed monosized spheres, then $a = 3.0$ and $b = 3.2$. This combination would indicate pore closure at a density of just under 94%, near that predicted using the tetrakaidecahedron grain shape model, at 92%.

SUMMARY

The geometric progression encountered in sintering tends to follow a similar trajectory for many materials. Many parameters change and converge as surface energy is consumed. Accordingly, the microstructure tracks a path as captured in Figure 6.38,

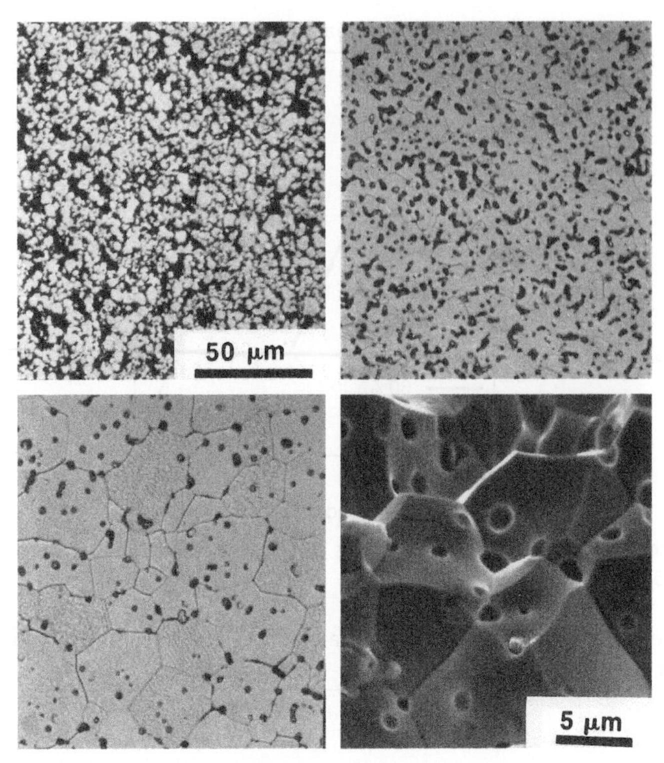

Figure 6.38 Micrographs of sintered tungsten illustrating porosity elimination, reduction in the number of pores, and enlargement of grains during sintering. The fracture image illustrates nearly spherical pores occupying the grain boundaries.

with characteristic shifts in density, grain size, surface area, and pore structure. In these images the black regions are pores and the thin lines evident in the final stage are grain boundaries. The scanning electron micrograph in the lower right corner shows fracture along the grain boundaries for a compact sintered into the final stage. Spherical pores are present on the fractured grain boundaries, a condition that is desirable for continued densification.

Sintering involves a complex interplay of interfacial energies, pore shrinkage and pore growth, particle bonding, and microstructure coarsening. Solid-vapor surface area decreases linearly with sinter density. The relation between grain size and fractional porosity follows an inverse square root relation.

Late in sintering, the surface area is not a useful monitor of sintering, but grain size is important. Grain boundary area increases and then decreases with grain growth. As pores are eliminated the grains increase in mobility and grow, with the annihilation of smaller grains. Accordingly, the peak in the grain boundary area occurs near the onset of final stage sintering, and grain growth accelerates due to inefficient grain boundary pinning by pores.

This chapter emphasizes the following points:

- The solid-vapor surface area declines during sintering.
- For systems that densify the specific surface area decreases linearly with sintered density.
- Neck growth, shrinkage, and coordination number changes track sintering shrinkage.
- Surface area consumed by sintering is initially converted into grain boundary area.
- Grain size distribution and pore size distribution converge to Weibull distributions where only the median size is needed to describe the distribution.
- Grain growth depends on the degree of solid-solid contact.
- Grain boundary area peaks near 85% relative density.
- Grain size tracks with the inverse square root of fractional porosity ($1/\sqrt{\varepsilon}$).

REFERENCES

[1] Z.Z. Fang, H. Wang, Sintering of ultrafine and nanoscale particles, in: Z.Z. Fang (Ed.), Sintering of Advanced Materials, Woodhead Publishing, Oxford, UK, 2010, pp. 434–473.

[2] E.Y. Gutmanas, High-pressure compaction and cold sintering of stainless steel powders, Powder Metall. Inter. 12 (1980) 178–182.

[3] W.S. Slaughter, I. Nettleship, M.D. Lehigh, P.P. Tong, A quantitative analysis of the effect of geometric assumptions in sintering models, Acta Mater. 45 (1997) 5077–5086.

[4] G.C. Kuczynski, Self-diffusion in sintering of metallic particles, Trans. TMS-AIME 185 (1949) 169–178.

[5] R.M. German, Z.A. Munir, Morphology relations during bulk-transport sintering, Metall. Trans. 6A (1975) 2229–2234.

[6] R.M. German, Z.A. Munir, Morphology relations during surface-transport controlled sintering, Metall. Trans. 6B (1975) 289–294.

[7] J. Bernholc, P. Salamon, R.S. Berry, Annealing of fine powders: initial shapes and grain boundary motion, in: P. Jena, B.K. Rao, S.N. Kahanna (Eds.), Physics and Chemistry of Small Clusters, Plenum Press, New York, NY, 1987, pp. 43–48.

[8] M. Upmanyu, G.N. Hassold, A. Kazaryan, E.A. Holm, Y. Wang, B. Patton, et al., Boundary mobility and energy anisotropy effects on microstructural evolution during grain growth, Interface Sci. 10 (2002) 201–216.

[9] C.B. Shumaker, R.M. Fulrath, Initial stages of sintering of copper and nickel, in: G.C. Kuczynski (Ed.), Sintering and Related Phenomena, Plenum Press, New York, NY, 1973, pp. 191–199.

[10] G. Petzow, H.E. Exner, Particle rearrangement in solid state sintering, Z. Metall 67 (1976) 611–618.

[11] O. Lame, D. Bellet, M. Di Michiel, D. Bouvard, Bulk observation of metal powder sintering by X-ray synchrotron microtomography, Acta Mater. 52 (2004) 977–984.

[12] H.E. Exner, Principles of single phase sintering, Rev. Powder Metall. Phys. Ceram. 1 (1979) 7–251.

[13] R. Watanabe, Y. Masuda, Quantitative estimation of structural change in carbonyl iron powder compacts during sintering, Trans. Japan Inst. Met. 13 (1972) 134–139.

[14] R.L. Coble, Sintering crystalline solids. 1. Intermediate and final state diffusion models, J. Appl. Phys. 32 (1961) 787–792.

[15] R.M. German, Analysis of surface diffusion sintering using a morphology model, Sci. Sintering 14 (1982) 13–19.

[16] F.B. Swinkels, M.F. Ashby, Role of surface redistribution in sintering by grain boundary transport, Powder Metall. 23 (1980) 1–7.

[17] A.S. Watwe, R.T. DeHoff, Metric and topological characterization of the advanced stages of loose stack sintering, Metall. Trans. 21A (1990) 2935–2941.
[18] E.H. Aigeltinger, Relating microstructure and sintering force, Inter. J. Powder Metall. Powder Tech. 11 (1975) 195–203.
[19] R.T. DeHoff, R.A. Rummel, H.P. Labuff, F.N. Rhines, The relationship between surface area and density in second-stage sintering of metals, in: H.H. Hausner (Ed.), Modern Developments in Powder Metallurgy, vol. 1, Plenum Press, New York, NY, 1966, pp. 310–331.
[20] R.M. German, Coarsening in sintering – grain shape distribution, grain size distribution, and grain growth kinetics in solid-pore systems, Crit. Rev. Solid State Mater. Sci. 35 (2010) 263–305.
[21] R.A. Gregg, F.N. Rhines, Surface tension and the sintering force in copper, Metall. Trans. 4 (1973) 1365–1374.
[22] R.M. German, Manipulation of strength during sintering as a basis for obtaining rapid densification without distortion, Mater. Trans. 42 (2001) 1400–1410.
[23] F. Amar, J. Bernholc, R.S. Berry, J. Jellinek, P. Salamon, The shapes of first-stage sinters, J. Appl. Phys. 15 (1989) 3219–3225.
[24] J.V. Kumar, The hypothesis of constant relative responses and its application to the sintering process of spherical powders, Solid State Phen. 8 (1989) 125–134.
[25] I. Nettleship, M.D. Lehigh, R. Sampathkumar, Microstructural pathways for the sintering of alumina ceramics, Scripta Mater. 37 (1997) 419–424.
[26] R.T. DeHoff, A. Cell, Model for microstructural evolution during sintering, in: G.C. Kuczynski, A.E. Miller, G.A. Sargent (Eds.), Sintering Heterogeneous Catalysis, Plenum Press, New York, NY, 1984, pp. 23–34.
[27] A.K. Kakar, A.C.D. Chaklader, Deformation theory of hot pressing – yield criterion, Trans. TMS-AIME 242 (1968) 1117–1120.
[28] C.A. Bruch, Sintering kinetics for the high density alumina process, Ceram. Bull. 41 (1962) 799–806.
[29] H.Y. Suzuki, K. Shinozaki, M. Murai, H. Kuroki, Quantitative analysis of microstructure development during sintering of high purity alumina made by high speed centrifugal compaction process, J. Japan Soc. Powder Metall. 45 (1998) 1122–1130.
[30] M. Zadra, F. Casari, L. Girardini, A. Molinari, Microstructure and mechanical properties of CP titanium produced by spark plasma sintering, Powder Metall. 51 (2008) 59–65.
[31] D.A. Aboav, T.G. Langdon, The shape of grains in a polycrystal, Metallog 2 (1969) 171–178.
[32] S.C. Coleman, W. Beere, The sintering of open and closed porosity in UO_2, Phil. Mag. 31 (1975) 1403–1413.
[33] U.C. Oh, Y.S. Chung, D.Y. Kim, D.N. Yoon, Effect of grain growth on pore coalescence during the liquid phase sintering of MgO-$CaMgSiO_4$ systems, J. Amer. Ceram. Soc. 71 (1988) 854–857.
[34] S.I. Nunes, R.C. Bradt, Grain growth of ZnO in ZnO-Bi_2O_3 ceramics with Al_2O_3 additions, J. Amer. Ceram. Soc. 78 (1995) 2469–2475.
[35] J.E. Burke, Recrystallization and sintering in ceramics, in: W.D. Kingery (Ed.), Ceramic Fabrication Processes, John Wiley, New York, NY, 1958, pp. 120–131.
[36] Y. Liu, B.R. Patterson, Particle volume fraction dependence in Zener drag, Scripta Metall. Mater. vol. 29 (1993) 1101–1106.
[37] R. Watanabe, Y. Masuda, Pinning effect of residual pores on the grain growth in porous sintered metals, J. Japan Soc. Powder Metall. 29 (1982) 151–153.
[38] J. Zhao, M.P. Harmer, Sintering of ultra-high purity alumina doped simultaneously with MgO and FeO, J. Amer. Ceram. Soc. 70 (1987) 860–866.
[39] E.A. Barringer, H.K. Bowen, Formation, packing, and sintering of monodisperse TiO_2 powders, J. Amer. Ceram. Soc. 65 (1982) C199–C201.
[40] H. Pickup, The densification and microstructure of calcium titanate, in: A.C.D. Chaklader, J.A. Lund (Eds.), Sintering '91, Trans Tech Publ, Brookfield, VT, 1992, pp. 251–258.
[41] I.W. Chen, X.H. Wang, Sintering dense nanocrystalline ceramics without final stage grain growth, Nature 404 (2000) 168–171.

[42] H.E. Exner, G. Petzow, Shrinkage and rearrangement during sintering of glass spheres, in: G.C. Kuczynski (Ed.), Sintering Catalysis, Plenum Press, New York, NY, 1975, pp. 279–293.
[43] V.A. Ivensen, Densification of Metal Powders during Sintering, Consultants Bureau, New York, 1973.
[44] E.H. Aigeltinger, H.E. Exner, Stereological characterization of the interaction between interfaces and its application to the sintering process, Metall. Trans. 8A (1977) 421–424.
[45] W. Beere, The sintering and morphology of interconnected porosity in UO_2 powder compacts, J. Mater. Sci. 8 (1973) 1717–1724.
[46] R.M. German, Z.A. Munir, Surface area reduction during isothermal sintering, J. Amer. Ceram. Soc. 59 (1976) 379–383.

CHAPTER SEVEN

Thermodynamic and Kinetic Treatments

CURVATURE GRADIENTS AND STRESS

Sinter bonds between particles grow to reduce curvature gradients, surface area, and surface energy. Initially, sintering is dominated by the reduction of curvature gradients [1]. Most of the curvature gradient is eliminated prior to significant densification. Indeed, several materials are fabricated with almost no dimensional change during sintering, yet the component strength increases dramatically due to neck growth. Retained porosity is widely employed in sintered capacitors, bearings, filters, battery electrodes, sound absorbers, permeation devices, ionizers, casting cores, and energy absorption applications. An important point is that densification is not fundamental to sintering, since a wide range of materials and applications involve sintering without densification.

Curvature gradients dominate early stage neck growth between particles [2]. On heating, the atoms move with increased amplitude and move (via random motion) along the curvature gradient. The sinter neck is concave (curved inward) and the particle surface is convex (curved outward). Fundamentally, neck growth is the universal feature associated with sintering. A curved surface is associated with stress, and the concave neck region grows to lower tension while the convex region away from the neck is opposite in stress. Once atomic motion occurs, the atoms randomly move to fill in the neck, thereby reducing the stress and stress gradient. Thus, atoms randomly move into the neck and in doing so the structure becomes more stable.

The scale of the neck and particle determines the magnitude of the stress and stress gradient. For any curved surface the stress varies with the inverse of the curvature. Neck growth eliminates surface area and eliminates the curvature gradient, so naturally the neck growth rate, as a reflection of the sintering rate, slows as the neck enlarges [3]. However, as the neck grows the compact becomes stronger. There are many sintering contacts on each particle, and each contact enlarges to eventually merge into a strong solid.

The temperature needed to induce atomic motion depends on the material and the particle size. Materials with a high chemical stability (measured by a high

Sintering: From Empirical Observations to Scientific Principles
DOI: http://dx.doi.org/10.1016/B978-0-12-401682-8.00007-0

sublimation enthalpy) require higher relative sintering temperatures. Actually, several mass transport mechanisms act during sintering. Their relative role changes as well. For example, when sintering crystalline materials, a grain boundary grows at the interface between the particles, so grain boundary area increases while surface area decreases, thus grain boundary diffusion overtakes surface diffusion.

Transport mechanisms detail the paths by which the atoms move; for solid-state sintering the candidates include surface diffusion, volume diffusion, grain boundary diffusion, viscous flow, plastic flow, and vapor transport from solid surfaces [4,5]. In liquid phase sintering, the dominant process is diffusion through the liquid. Coupled with the mass transport paths are the various geometric stages traversed during sintering. The first stage occurs when particles come into contact with each other, since there is a weak cohesive bond at the contacts. Initial stage sintering occurs during heating and is characterized by neck growth, often without densification. The actual neck volume in the initial stages is small, so substantial neck growth occurs with a relatively small amount of mass. In the intermediate stage of sintering, the pore structure becomes less angular and develops an interconnected tubular nature. The concomitant reduction in curvature and surface area results in slower sintering. It is common for grain growth to occur in the latter portion of the intermediate stage of sintering, giving a larger average grain size and fewer grains.

An open pore network is geometrically unstable when the porosity reduces to about 8% (92% density), because the pores are elongated due to grain growth while shrinking in diameter [6,7]. This is analogous to the breakup event that changes a long, thin stream of water into discrete droplets. The somewhat cylindrical pores collapse and pinch-off into lenticular or spherical pores, which are less effective in slowing grain growth. The appearance of these isolated pores indicates the final stage of sintering with slow densification. Gas in the pores limits densification; accordingly, vacuum produces the highest final density.

There is no clear distinction between sintering stages, since the geometric progression varies from point to point in the microstructure. Clusters of small particles reach final stage sintering first, while regions of large particles might still be in the initial stage. Initial stage sintering operates while there are large curvature gradients, necessarily with smaller neck sizes. The neck size ratio (neck diameter divided by the particle diameter) is less than 0.3, shrinkage is low (less than 3%), and the grain size is about the same as the initial particle size. The surface area is still at least 50% of the original value. In the intermediate stage, the pores are smoother and the density is usually between 70 and 92% of theoretical. Grain growth accelerates as more grain contacts form, so late in the intermediate stage the grains enlarge to become much larger than the initial particles. Pore growth is also observed [8]. By the final stage of sintering, the pores are spherical and closed, grain growth is evident, and total porosity is less than 8% [9,10].

This chapter emphasizes solid-state sintering and the combinations of driving force, mass transport mechanisms, and sintering stages that describe sintering.

ATMOSPHERIC REACTIONS

Swelling during sintering is often an unexpected problem that occurs late in the sintering cycle. It occurs for a wide variety of materials—oxide ceramics, copper, steels, tungsten, cermets, and titanium-nickel. Swelling is frequently caused by residual gas in the final stage closed pores. As an example of this type of reaction, consider the sintering of copper in a hydrogen atmosphere. Oxygen is soluble in copper and is a common impurity. The dissolved oxygen reacts with the sintering atmosphere to produce water vapor. Unlike oxygen, steam is insoluble in copper so it fills the pores as a reaction product. As oxygen diffuses to the pores, the reaction progresses and the material bloats and in some cases blisters. Figure 7.1 is a photograph of sintered tensile samples with considerable blistering. This swelling is often associated with chemical reactions, such as copper melt spreading on iron grain boundaries (Fe-Cu-C alloys), tin dissolving into copper (Cu-Sn), nickel reacting with titanium (Ti-Ni), or unbalanced diffusion in mixed powders (Fe-Al).

Depending on the retained gases in the pores, possible sintering densification trajectories arise as follows:

- Sintering in vacuum—there is no impediment to full densification other than reaction vapors given off by the material.
- Sintering in a soluble atmosphere—the material permeates gas out of the pores so there is only a small pressure resisting densification, this depends on gas solubility at the sintering temperature.

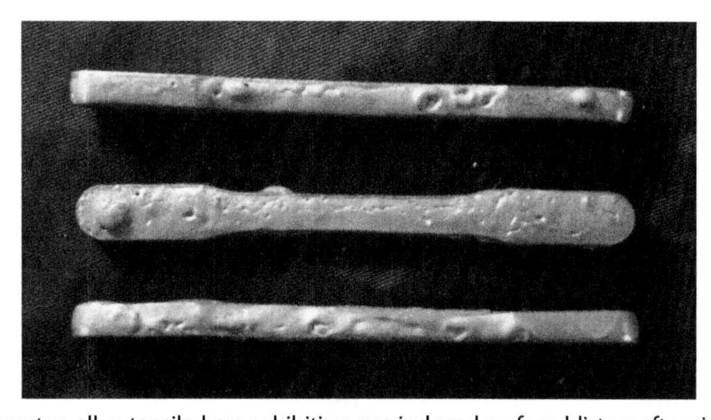

Figure 7.1 Tungsten alloy tensile bars exhibiting gas induced surface blisters after sintering.

- Sintering in a low solubility or insoluble atmosphere—the gas accumulates in the residual pores and resists densification as the pressure increases, possibly leading to swelling.
- Dissolved species react—gas is produced inside the pores, the reaction continues to increase pore pressure with insoluble products, as might be seen by the generation of carbon dioxide or steam.

Further, pore coarsening during sintering contributes to swelling. Both pore size and grain size increase due to coalescence and diffusion. Since the pore pressure decreases as the pore size increases, the gas takes up more volume and causes swelling.

Assume that the sintering pores are pinched closed and exist as spherical voids located on the grain corners. Closed pores exist at densities over about 92% and are initially assumed to be spherical. Gas in the pores increases in pressure as reactions produce vapor while at the same time densification is reducing the porosity. The competing gas production and pore shrinkage reach a point where the pore gas pressure P equals the capillary pressure from the solid-vapor surface energy γ_{SV};

$$P = \frac{4\gamma_{SV}}{d} \tag{7.1}$$

where d is the pore diameter. Pore pressure results in incomplete densification.

Delayed swelling occurs from gas in the pores [11–13]. Based on the tetrakaidecahedron grain shape the effective grain diameter G and pore diameter d are related. Pores coarsen as the number of pores declines [14]. The grain volume V_G (solid and pore) is:

$$V_G = \frac{\pi}{6} G^3 \tag{7.2}$$

The volume of each spherical pore is V_P:

$$V_P = \frac{\pi}{6} d^3 \tag{7.3}$$

Since there are 24 pores on the corners of the grain structure, each shared by four grains, the structure has six pores per grain, so the fractional porosity ε is:

$$\varepsilon = 6\left(\frac{d}{G}\right)^3 \tag{7.4}$$

Rearranging gives the pore size d as a function of porosity and grain size as follows:

$$d = G\left[\frac{\varepsilon}{6}\right]^{1/3} \tag{7.5}$$

Grain growth is active during sintering, leading to a pore size increase as the porosity declines. Since growth is typical during sintering, the sintered grain size G depends on the starting grain size G_O and fractional porosity ε as follows:

$$G = \frac{\theta\, G_O}{\sqrt{\varepsilon}} \tag{7.6}$$

Usually θ is about 0.6, but it depends on the processing cycle, reflecting pore attachment to the grain boundary during grain growth. The pore size as a function of porosity is:

$$d = G_O \frac{\theta}{6^{1/3}} \varepsilon^{-1/6} \tag{7.7}$$

This low sensitivity to porosity, reflected by the inverse sixth root, predicts small pore size changes during latter stage sintering—the pore size at 4% porosity is 12% larger than at pore closure, 26% larger at 2% porosity, and 41% larger at 1% porosity. The number of pores diminishes at the same time. In experiments on iron [15], the pore size at 7.8% porosity (point of pore closure) is 2.94 μm and at 4.4% porosity the pore size is 3.26 μm; the pore size enlarged by 11% while the porosity dropped by 44%; the predicted pore size ratio would be $(0.078/0.044)^{1/6} = 1.10$, nearly the same as measured experimentally.

If the gas mass is conserved, then porosity ε and pore pressure P are related [16]:

$$\varepsilon P = C \tag{7.8}$$

where C is a constant. Sintering densification decreases the porosity with a concomitant increase in pore gas pressure if the gas is insoluble in the material. Rearranging, the relation between ε porosity and pore pressure P is predicted to be as follows:

$$\frac{\varepsilon}{\varepsilon_C} = \frac{P_C}{P} \tag{7.9}$$

where P_C is the atmosphere pressure in the pores at pore closure and ε_C is the porosity at pore closure. Inserting the relation between pore size and pore pressure gives:

$$\varepsilon^{2/3} = \frac{\varepsilon_C P_C}{4\gamma_{SV} 6^{1/3}} G \tag{7.10}$$

This indicates that once the pores close and trap gas, grain growth results in a porosity increase seen as swelling.

Gas trapped in pores retards densification and induces swelling as the pores and grains coarsen. Capillarity from surface energy drives pore shrinkage, but internal gas pressure halts and reverses densification. The amount of gas needed to halt densification is small. For example, if 5 μm pores exist at pore closure during sintering at 1300 K in one atmosphere at 8% porosity, then sintering shrinkage terminates when the collapsed pores are pressurized to give 0.6% porosity. For copper at this temperature, this corresponds to about 12 μg per cubic centimeter, or about 1.3 ppm oxygen. Sintering in hydrogen induces reaction with trace oxygen to form steam. With just parts per million of residual oxygen, this steam is sufficient to resist full densification. For most of the following discussion on the sintering mechanisms and stages, the general treatment will ignore trapped gases; however, realize that gas reactions are a major influence on behavior.

MASS TRANSPORT MECHANISMS

Transport mechanisms determine how mass flows in response to the driving force for sintering. There are two different types of mechanisms, and both contribute to neck growth. However, only bulk transport processes give densification. Surface transport mechanisms do not give densification, since mass is repositioned on the pore surface to lower surface area and removing curvature gradients. Bulk transport mechanisms move mass from inside the solid to deposit on the pores. Often the mechanisms operate in collaboration.

The differences in the transport mechanisms are sketched in Figure 7.2. Note the shrinkage for the bulk transport options, where the spheres move together as mass

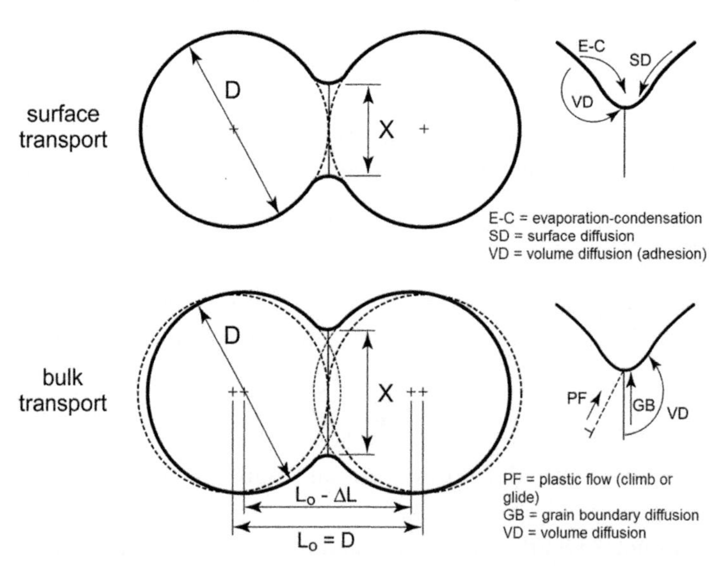

Figure 7.2 Neck growth in sintering models for spheres of diameter D. The upper drawing shows neck growth measured by the neck diameter X via surface transport mechanisms that do not produce shrinkage. The lower drawing is for bulk transport by mechanisms that move mass from between the particles to allow densification.

moves from the interior to the surface. The mass transport pathways are marked near the neck for several pathways. Not shown is the viscous flow process, applicable to amorphous materials. To understand the mathematical treatments, it is typical to assume the pores are large accumulations of vacancies. Classic treatments of sintering mechanisms examine the motion of vacancies as a basis for understanding pore elimination.

Several mechanisms are envisioned. Mass flow is represented as vacancy and atom exchanges, where the atoms move along particle surfaces (surface diffusion), across pore spaces (evaporation-condensation), along grain boundaries (grain boundary diffusion), and through the lattice interior (viscous flow or volume diffusion). Also, the dislocation structure plays a role in plastic flow and dislocation climb. Additionally, vacancies migrate between pores, leading to the growth of larger pores while the smaller pores shrink.

Surface transport processes produce neck growth without a change in particle spacing (no shrinkage or densification). Mass flow originates and terminates at the particle surface. Surface diffusion is commonly the dominant sintering mechanism early in sintering while a large surface area is still present. Also, the surface diffusion activation energy is lower, so surface atoms start moving at a lower temperature versus interior atoms [17]. Evaporation-condensation is not a common process unless the material has a high vapor pressure at the sintering temperature.

While both surface and bulk transport processes promote neck growth, as noted the large difference is in shrinkage. Densification occurs by bulk transport sintering since the mass that grows the sinter neck comes from inside the body. Bulk transport mechanisms include volume diffusion, grain boundary diffusion, plastic flow, dislocation climb, and viscous flow. Plastic flow is important during heating before the dislocation population anneals out of the material. It is especially important in compacted powders when the initial dislocation density is high from work hardening in compaction. Surface energy is generally insufficient to generate new dislocations, so sintering corresponds to a declining dislocation density and declining role from plastic flow [18].

Amorphous materials, glasses and polymers, sinter by viscous flow, where the particles coalesce at a rate that depends on the particle size and material viscosity [19]. Viscous flow is also possible for metals when liquid phases form on the grain boundaries [20].

Grain boundary diffusion is fairly important to densification for most crystalline materials, and generally dominates densification in sintering. The junction of two grains is a grain boundary that is defective in bonding, thereby providing a pathway for rapid diffusion. With sufficient grain boundary area, grain boundary diffusion dominates sintering. Grain growth and elimination of grain boundaries is detrimental to sintering [21].

Volume diffusion operates at higher temperatures and is somewhat active in sintering. It is one of the possible bulk transport processes, but is usually only active at the highest temperatures.

The sinter bond between the contacting particles is the critical region. It is the point where atoms deposit to reduce surface energy. The measures of sintering can almost always be related back to how the mass transport processes interact to drive neck growth and pore changes. The sintering stages then assess the geometric progression [22]. Models for sintering mostly treat specific combinations of the sintering stage and mass transport mechanism. This is a confusing partition of mechanisms and stages. Here the mechanisms are introduced first, and subsequently attention is directed to the stages. However, sintering involves several mechanisms operating simultaneously over overlapping stages.

Sintering models assume monosized spheres in point contact under isothermal conditions. Most sintering involves non-spherical powders, wide particle size distributions, and compacted powder. For large size differences, the small particles absorb into the large particles with the disappearance of the interparticle grain boundary. Compaction deforms the particles, collapses large pores, and may introduce dislocations. Further, sintering involves several concurrent mechanisms working in cooperation. For now each mechanism will be treated in isolation

Viscous Flow

By ignoring grain boundaries, the first model for neck growth introduced the concept of neck growth over time. Frenkel [23] assumed two amorphous spheres of diameter D in contact, growing a sinter bond of diameter X. This is seen during the sintering of glass spheres or many polymers. Since there was no grain boundary, the neck is smooth. Amorphous materials exhibit a decreasing viscosity as temperature increases. Under the action of the capillary stress from the concave neck, mass flows to grow the sinter bond. Thus, at high temperatures, glass and polymer powders densify by viscous flow in response to the sintering stress. Sintering rates increase as temperature increases. If an external stress is applied, then the rate of sintering increases in proportion to the applied stress.

Over a limited temperature range, the viscosity η varies with temperature as:

$$\eta = \eta_O \, exp\left[\frac{Q}{R\,T}\right] \tag{7.11}$$

where Q is an activation energy, η_O is a pre-exponential coefficient, T is the absolute temperature, and R is the gas constant. This is not the only accepted form for the viscosity temperature dependence, but generally gives a good fit to experimental data.

Frenkel's viscous flow sintering model assumed surface energy dissipation balanced densification. Subsequent models clarified the model, giving the neck size ratio as a function of sintering time t:

$$\left(\frac{X}{D}\right)^2 = \frac{3\,\gamma\, t}{D\,\eta} \tag{7.12}$$

where γ is the surface energy. Since higher temperatures decrease the viscosity, there is a progressive increase in neck growth rate with higher temperatures. Figure 7.3 plots neck size versus square-root time behavior for soda-lime glass (window glass) spheres sintering at 750°C (1023 K) [24]. This example shows the generic behavior of the neck size ratio and its dependence on particle size, time, surface energy, and a mass flow.

Sintering shrinkage is related to the square of the neck size for bulk transport sintering, so as sphere centers approach there is shrinkage. Initially shrinkage tracks neck growth with a square relation, giving a linear relation between shrinkage and isothermal hold time [25]. An example is plotted Figure 7.4 for a soda-lime glass [26]. For the first 8% sintering shrinkage the behavior is linear with time, but as sintering progresses the driving force declines and the rate of sintering naturally decays.

Amorphous materials lack grain boundaries, so as neck growth proceeds the amorphous materials achieves a zero curvature condition where the convex and concave radii are equal, but opposite in sign. This occurs at a neck size ratio of approximately $X/D = 2/3$ or about 11% linear shrinkage ($\Delta L/L_O$), leading to greatly reduced sintering as densification advances. This is evident in the plot of shrinkage versus time, and after 40 min densification slows. As a practical consequence, low packing density

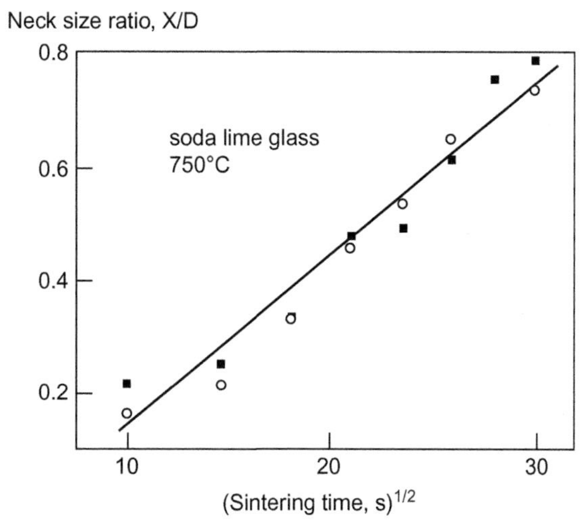

Figure 7.3 Neck size ratio (X/D) plotted versus the square-root of the sintering time to illustrate behavior characteristic of viscous sintering [24].

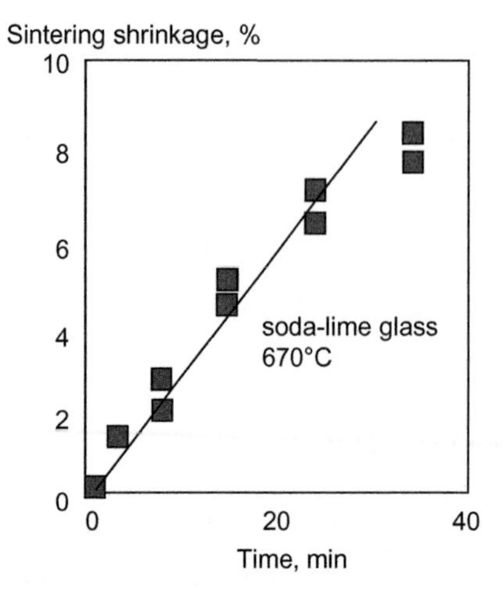

Figure 7.4 Sintering shrinkage data for glass powder during isothermal sintering at 670°C (943 K) [26]. The linear behavior is expected for early stage viscous flow controlled sintering.

structures cannot fully densify by viscous flow since the curvature gradients dissipate. An external pressure is effective in correcting this difficulty [27]. As an alternative view, Zagar [28] derived an isothermal densification model for viscous flow:

$$log\left[\frac{\varepsilon}{\varepsilon_O}\right] = -\frac{\gamma\, t}{D\, \eta} \tag{7.13}$$

where ε is the porosity after sintering and ε_O is the starting porosity.

In addition, curvature is associated with the initial pore structure and usually the particle structure is not ideally homogeneous. As a consequence, the inhomogeneities result in a torque that induces rearrangement in the coordination and pore structure, and both influence sintering [29]. The capillary forces on the particles induce rearrangement; this is especially true for milled glass particles with sharp corners or flat faces. In glass sintering the large pores grow while the small pores shrink, leading to stable residual pores.

Early sintering models attempted to equate viscous flow with diffusional creep. It was based on a conceptual link between diffusion $D_V$and viscosity η as follows:

$$\eta = \frac{k\, T}{\delta\, D_V} \tag{7.14}$$

with k equal to Boltzmann's constant, T being the absolute temperature, and δ reflecting a dimension typically taken as the atomic size. Attempts to create a universal viscosity concept for sintering permeate computer modeling efforts, although the complexity of the situation advises against this assumption [30]. However, the densification process does not typically take place by volume diffusion, so such a simplification suffers from a lack of microstructural reality. Further assumption of a purely viscous analog to sintering fails to properly embed non-densification surface transport effects. Even, so for several years it provided a means to model sintering densification.

Surface Diffusion

Crystal surfaces consist of defects that include extra atoms, surface vacancies, terraces, ledges, kinks, and adsorbed atoms. Although the surface might appear smooth, at the atomic scale it is quite defective, as is schematically illustrated in Figure 7.5. Atomic motion occurs between defects, for example the atom on a terrace corner (kink) might jump to fill a nearby surface vacancy. This is surface diffusion—the motion of atoms (or ions or molecules) between surface sites.

Three steps are involved in a typical surface diffusion event. In the first step an atom breaks existing bonds, typically from a surface kink. This source is on the pore surface. Once dislodged, the atom tumbles across the pore surface via random motion. Usually the jump is a fast step. Finally, the atom attaches at a new surface site, possibly again at a surface vacancy or kink. The attachment site is an atomic sink, also located on the pore surface. Since the atoms are only repositioning to create a smooth surface, there is no substantial center motion between particles, and thus no shrinkage [31]. Although atomic motion is random, atoms tend to migrate from convex to concave surfaces due to differences in defect concentrations. The result is a reduction in curvature—effectively rounding the pores. The neck between particles is particularly

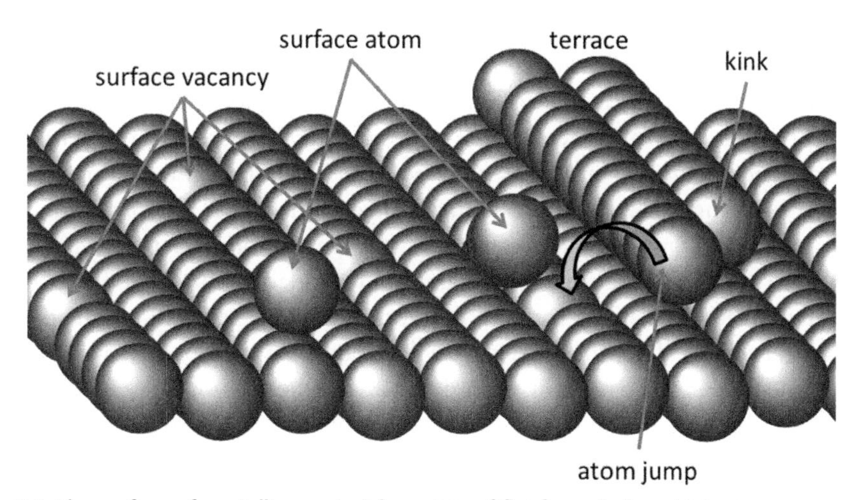

Figure 7.5 The surface of crystalline material consists of flat faces, ledges, kinks, terraces, vacancies, and additional atoms. Surface diffusion occurs by atoms moving between sites, such as the jump from a kink to a surface vacancy.

stressed, so is a preferred sink for atoms. Atomic deposition in the neck grows the interparticle bond and forms a grain boundary where the two particles meet. That grain boundary is important to subsequent sintering. Later, surface diffusion acts in concert with grain boundary diffusion in a cooperative manner [32]. Densification occurs by diffusion in the grain boundary while surface diffusion redistributes mass to smooth the pore.

The product of the population of defective sites (P_{defect}) and the probability of motion (P_{motion}) between sites gives an approximation for the net diffusion mobility:

$$M = P_{defect} P_{motion} \tag{7.15}$$

Both probabilities are thermally activated, so they exhibit similar temperature effects:

$$P = P_O \exp\left[-\frac{Q}{R\,T}\right] \tag{7.16}$$

where Q is an activation energy, R is the gas constant, T is the absolute temperature, and P_O is a material constant. A further refinement is to recognize the anisotropic character of surface diffusion, but for sintering that is ignored since all crystal orientations occur in a powder ensemble. The flux increases with temperature. Since defect creation and atomic motion are tied to temperature, the protocol is to lump both into one activation energy, Q_S with one pre-exponential frequency factor D_{OS}, so the rate of surface diffusion is given as:

$$D_S = D_{OS} \exp\left[-\frac{Q_S}{R\,T}\right] \tag{7.17}$$

The populations of source sites and sink, and the ease of motion all determine the surface diffusion rate. Curvatures determine the concentration C of sources and sinks relative to the equilibrium concentration C_O; but most important to sintering is the change in concentration (curvature) with position over the local geometry. At the base of the sintering neck, the concave radius is dominant, but a short distance away the particle convex radius is dominant. The change in concentration with distance and curvature gradient drives initial neck growth by surface diffusion [33].

High temperatures increase surface diffusion, but as surface area is lost its relative contribution declines. The sinter neck growth depends on the volume of atoms arriving at the neck surface, which in turn depends on the diffusion area and diffusion flux. In turn the flux depends on the curvature gradient and surface diffusivity. A combination of all factors leads to an approximation for the neck size versus time, temperature, and particle size for surface diffusion controlled sintering. Surface diffusion controlled sintering was first treated by Kuczynski [34] and subsequently was a favorite of other investigators [2,35–38].

At the particle level, crystal orientation influences surface diffusion controlled sintering. However, usually an average rate of transport is assumed without attention to specific crystal orientations. Adsorbed species influence surface diffusion rates, especially at low temperatures, but sintering usually occurs at temperatures where adsorbed species evaporate and no longer play a role.

Surface diffusion is most active during heating to the sintering temperature. The activation energy for surface diffusion is usually low compared to that for other mass transport processes. Consequently, it initiates at lower temperatures and is dominant while there is a high surface area and little grain boundary area (the latter increases as interparticle bonds grow). As surface area is consumed, surface diffusion naturally declines in importance.

Surface diffusion dominates early sintering for a broad array of materials. It depends on many factors, and is sensitive to the initial surface area (particle size), surface impurities, and temperature. Metals such as Fe, Ni, Ag, Cu, Cu, and Pd exhibit early sintering by surface diffusion, and many ceramics are especially prone to surface diffusion controlled sintering, including covalent ceramics such as SiC, Al_2O_3, MgO, FeO, and TiO_2. As sintering occurs the loss of surface area reduces the relative role of surface diffusion.

Volume Diffusion

In a crystalline material, volume diffusion involves atom exchange with vacancies, also termed lattice diffusion. At any temperature there is an equilibrium population of vacancies. The application of heat induces atomic motion, which is at times sufficient to jump into a vacancy. A result is a position change between the atom and vacancy as sketched in Figure 7.6. In sintering, pores are assumed to be large vacancy clusters, emitting vacancies into the surrounding solid with a counter flow of atoms into the pores. This is effectively volume diffusion during sintering. There is also a flow of vacancies between pores, leading to an increase in the median pore size as porosity decreases. Early sintering concepts assumed volume diffusion resulted in an effective viscosity.

Besides atoms, the diffusing species might be ions (for example Cl^- in salt, NaCl) or water molecules in ice. The number of atoms is preserved during volume diffusion, but vacancies are created and annihilated at pores, grain boundaries, interfaces, and dislocations. Atomic motion takes place from a source to a sink—effectively mountain tops are eroded away to fill in valleys. The same is true in sintering—mass is moved from convex outward curved surfaces to concave surfaces, namely the necks between particles.

A vacancy is created when an atom jumps from the crystal onto the pore surface. The resulting vacancy moves by a sequence of random position changes by atoms

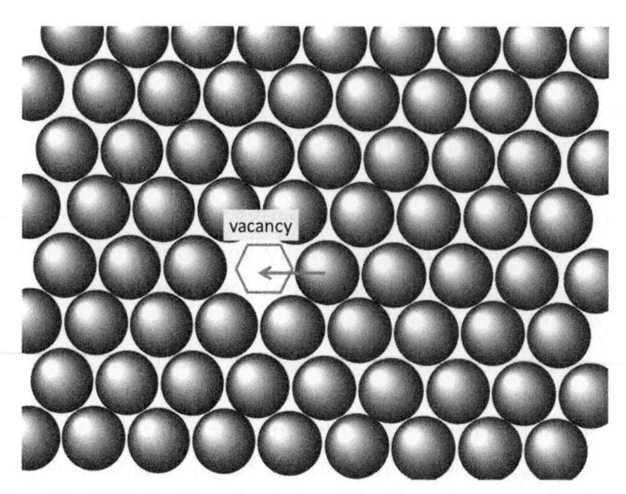

Figure 7.6 Two-dimensional illustration of volume diffusion, where a vacancy and atom exchange positions. During sintering it is typical for each atom to change position about six times per second.

inside the crystal. At the sintering temperature, the rate of atomic vibration is 10^{14} oscillations per second. Only a few of those oscillations lead to an atom exchange with a vacancy, yet every atom moves to a new position about every 0.1 s at typical sintering temperatures. Thus, constant motion and exchange exists between atoms and vacancies. Treating pores as large collections of vacancies which permeate though the crystal structure leads to models for sintering by volume diffusion. Vacancies are then annihilated inside the body, at dislocations, grain boundaries, surfaces, or interfaces. Grain boundaries are efficient vacancy sinks, acting in cooperation with grain rotation to consume vacancies [39].

Volume diffusion sintering involves the motion of vacancies along paths as sketched in Figure 7.7. One is from the neck surface, through the particle interior, with subsequent emergence at the particle surface, resulting in mass deposition at the neck surface. This is effectively transport from a surface to a surface so there is no densification or shrinkage. It is termed volume diffusion adhesion to distinguish it from the more important densification process; it is a surface transport process since there is no densification. Although treated theoretically, there is little evidence for this occurring in sintering cycles.

The second path is termed volume diffusion densification. It is a bulk transport process that involves vacancy flow to the interparticle grain boundary from the neck surface. Of course this requires neck growth and the emergence of a grain boundary, typically by surface diffusion. Volume diffusion transports mass from the grain boundary to the pore, effectively moving vacancies from pores to annihilation at the defective region of the grain boundary. Densification and shrinkage occur since layers of

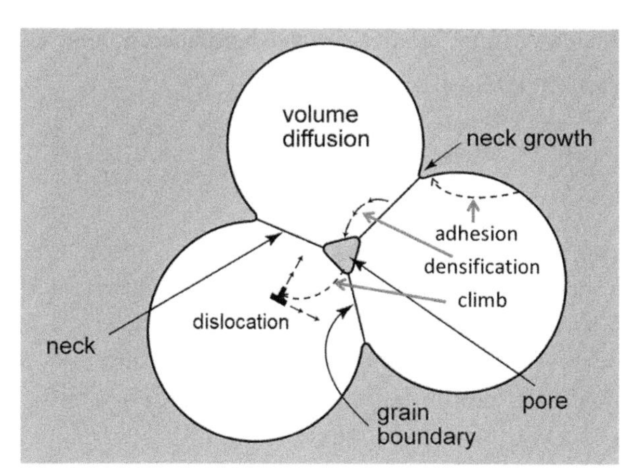

Figure 7.7 Three particles in cross-section with various processes sketched, including adhesion (vacancy source and sink are the pore surface), densification (vacancy source is pore and sink is grain boundary), and dislocation climb (vacancy source is pore and sink is dislocation).

atoms are removed along the particle contacts and repositioned on the pore surface. Consequently, the particle centers approach as the sinter bond grows. Cooperative vacancy annihilation by grain boundary accommodation, rotation, or sliding is required [40]. Two-phase interfaces are effective vacancy annihilation sites in multiphase materials. As a consequence, sintering halts when grain boundaries or interfaces disappear.

In the third volume diffusion option, vacancies interact with the dislocations, diffusing from the pores to annihilate via dislocation climb. In hot stage transmission microscopy, this process is seen during heating [41]. It is enhanced by increasing the dislocation density prior to sintering [42]. Since dislocations are not conserved during sintering, volume diffusion coupled to dislocation climb diminishes in importance during heating. The vacancy path is in the opposite direction to the atomic flux in each case.

The rate of volume diffusion depends on three factors—temperature, composition, and stress. The stress arises from surface curvature or external pressure. The temperature role is a combination of vacancy creation and vacancy-atomic motion. Both depend on temperature. In compounds of two or more atomic species, stoichiometry is an additional factor. Slight changes in composition induce vacancy population changes, thereby affecting volume diffusion. This is especially true in ionic solids. Stress changes the equilibrium vacancy population. Surfaces under compression tend to have fewer vacancies to reduce volume, and surfaces under tension have an excess of vacancies to increase volume and relax the stress. Curvature gradients give vacancy concentration gradients that induce volume diffusion, with eventual removal of the gradients and curvature differences.

The vacancy concentration C under a curved surface depends on the curvature of the two perpendicular radii of curvature for that surface:

$$C = C_O \exp\left[-\frac{Q_{VF}}{R\,T}\right]\left\{1 - \frac{\gamma\Omega}{R\,T}\left(\frac{1}{R_1} + \frac{1}{R_2}\right)\right\} \tag{7.18}$$

where C_O is the pre-exponential vacancy concentration term, Q_{VF} is the activation energy for vacancy formation (related to melting temperature for the material), R is the gas constant, T is the absolute temperature, γ is the surface energy, and Ω is the molar volume. The more curved a surface, then the smaller are R_1 and R_2 so the greater the vacancy concentration departs from equilibrium as reflected by a flat surface where R_1 and R_2 are infinite. An alternative is to determine the effective stress gradient corresponding to the concentration gradient.

Sinter bond growth by volume diffusion reflects a sequential exchange of atom and vacancy positions. For a concave surface, the vacancy concentration is higher than equilibrium, while for a convex surface the vacancy concentration is below equilibrium. The particle surface is convex and the sinter bond is concave, thus there is a vacancy concentration gradient between the two. Mass flow acts to remove that concentration gradient, just as water flows downhill in response to a gravitational gradient. The result is vacancy flow from the neck or atomic flow into the neck. At temperatures where atoms and vacancies move, the vacancy concentration gradient induces volume diffusion controlled sintering. Fick's first law is used to explain the sintering rate:

$$J = -D_V \frac{\partial C}{\partial x} \tag{7.19}$$

where J is the flux in terms of atoms (or vacancies) per unit area per unit time, D_V is the volume diffusivity, ∂C is the vacancy concentration change over a distance ∂x. Immediately the role of particle size is apparent, since smaller particles correspond to shorter distances. As noted, volume diffusion depends on the number of vacancies and the mobility of the vacancies, and both depend on composition and temperature. Like other expressions:

$$D_V = D_{VO} \exp\left[-\frac{Q_V}{R\,T}\right] \tag{7.20}$$

where D_{VO} is the pre-exponential frequency factor, Q_V is the activation energy, R is the gas constant, and T is the absolute temperature.

Stoichiometry additionally influences the vacancy population, especially in ionic materials. The total flux by volume diffusion is the combined action of the thermally induced vacancies and any excess vacancies occurring from a change in stoichiometry [43]. If the excessive ionic vacancies are associated with the slow moving species, then accelerated sintering results. The stoichiometric effect is accessible through the original compound formulation or through the process atmosphere or by chemical additions. For example in sintering urania, a hyper-stoichiometric oxygen level containing slightly more oxygen (2.02 oxygen ions for each uranium ion) gives the highest sintered density.

Late in sintering, pores are smooth, nearly spherical and are conceptualized as vacancy collections. A difference in size between neighboring pores contributes to a vacancy concentration gradient. Consequently, large pores accumulate vacancies while small pores are vacancy sources, leading to progressive large pore growth and small pore elimination [44]. For a high sintered density it is important to avoid trapped atmosphere in the pores and to sustain vacancy annihilation sites, such as grain boundaries, to avoid pore coarsening during sintering. Thus, attention is directed toward grain growth control and the coupling of pores to grain boundaries to achieve full densification.

Although volume diffusion is active at high temperatures, often it is a minor contributor to sintering. This is especially true for small powders with a high surface area. Surface diffusion is typically more active early in heating. Thus, the common situation is for early neck growth by surface diffusion and later densification by grain boundary diffusion [45,46]. However, volume diffusion controls the sintering of several stoichiometric compounds, including beryllium oxide (BeO), calcium oxide (CaO), cerium oxide (CeO_2), chromium oxide (Cr_2O_3), copper oxide (CuO), titanium oxide (TiO_2), uranium oxide (UO_2), yttrium oxide (Y_2O_3), and zirconia (ZrO_2) [47].

Grain Boundary Diffusion

Grain boundary diffusion is important for sintering densification. It dominates the sintering of most metals and many compounds. Grain boundaries form within the neck between individual particles as a consequence of random grain contacts leading to misaligned crystals. A grain boundary is essentially a collection of repeated misorientation steps. The defective character of the grain boundary allows mass flow along this interface with an activation energy that is usually intermediate between that for surface diffusion and volume diffusion. Although the grain boundary is rather narrow, probably only five atoms wide, it is an active transport path that promotes densification [48].

Early sintering studies learned that sintering was enabled by grain boundaries [49,50]. Indeed, it is commonly considered to be the dominant process for metals. Usually grain boundary diffusion acts in cooperation with other transport processes, and this often confuses studies seeking to find a single dominant process [51–53]. Once the various processes are isolated, heating cycles are manipulated to maximize one event over another, such as by two-stage heating to minimized surface diffusion while enhancing grain boundary diffusion.

Grain boundary diffusion depends on grain boundary area per unit volume. As surface area is consumed and surface diffusion declines in importance, the simultaneous emergence of new grain boundaries increases the role of grain boundary diffusion. Indeed the grain boundary area peaks during intermediate stage sintering. Since powder compacts are composed of large numbers of grain boundaries, it is reasonable to ignore differences in diffusion rates with orientation and assume average behavior.

A theoretical description of grain boundary diffusion controlled sintering came from Coble [54,55]. Subsequent models [56–59] applied refined concepts to shrinkage and densification. In sintering by grain boundary diffusion, mass is removed from the interparticle grain boundary and deposited on the neck surface. Pores act to emit vacancies. Then the interparticle grain boundaries, and internal grain boundaries in the particles, act as vacancy annihilation sites. The mass flows along the grain boundary to deposit on the neck surface. Without surface diffusion, a hillock forms, similar to an ant hill [60]. Typically surface diffusion operates to redistribute the mass over the neck surface in a cooperative process.

Grain boundary diffusion is sensitive to impurities, crystal orientation, and temperature. The diffusion rate depends on an Arrhenius relation with exponential temperature dependence. Normally the grain boundary width is unknown, so a nominal value of 5 to 10 atom diameters is assumed. This gives the thermally activated grain boundary diffusivity in m^3/s as the diffusivity times and the nominal grain boundary width $\delta\ D_B$ as follows:

$$\delta D_B = \delta D_{BO} \exp\left[-\frac{Q_B}{R\ T}\right] \tag{7.21}$$

where Q_B is the activation energy and D_{BO} is the frequency pre-exponential. Thus, unlike other diffusivities, where the units are m^2/s, grain boundary diffusion includes the boundary width to give diffusivity as m^3/s.

Grain growth and grain boundary diffusion involve similar atomic jumps during sintering. Atoms moving across the boundary cause grain size changes, while atoms moving along the boundary cause densification. It is no surprise that grain growth and densification have very similar dependences. Further, it is difficult to separate the two events. Thus, as sintering progresses, grain size enlargement seems to go hand in hand.

Dislocation Motion—Climb and Slip

Plastic flow, the motion of dislocations under stress, is an elusive aspect of sintering theory. Multiple experiments and calculations provide evidence that dislocation motion occurs in response to the sintering stress. Depending on the experimental design, the evidence tends to favor a transient contribution that decays as the dislocation structure is annealed to a low concentration after the initial heating period.

Two roles are recognized:

- Dislocation climb due to vacancy absorption [18].
- Dislocation glide due to surface stresses [61].

Dislocation glide occurs when the sintering stress exceeds the flow stress of the material at the sintering temperature. The consensus is that dislocations participate in sintering during heating, especially if the powder has a high initial dislocation density, such as from compaction. These results are illustrated by experiments using deformed iron powder with a high initial dislocation density [62]. When compared to the sintering of the powder after annealing, the cold worked powder exhibited more rapid sintering.

Dislocations interact with vacancies during sintering via dislocation climb; vacancies from the pores are absorbed by the dislocations (effectively the dislocation emits atoms as it moves to a parallel slip plane). Densification occurs by volume diffusion, but nearby dislocations act as vacancy sinks close to the neck. Accordingly, the dislocation population declines as dislocations are removed, halting the process. Schatt and Friedrich [63] demonstrated densification rate (change in porosity ε divided by the change in time t) improvements because of dislocation climb, with the rate of pore elimination given as:

$$\frac{d\varepsilon}{dt} = \frac{\sigma\Omega\, D_V}{R\, T\, \lambda^2} \tag{7.22}$$

where σ is the surface stress from the pore curvature, Ω is the atomic volume, D_V is the diffusivity, R is the gas constant, T is the absolute temperature, and λ is the average spacing between dislocations. Accordingly, the volume diffusion rate of sintering is increased nearly 100-fold by the dislocations accumulated in the neck region, both from plastic flow prior to sintering and from surface stresses.

Finite element analysis shows the early sintering stress is sufficient to exceed the flow stress [64]. The maximum shear stress is near the neck surface. As the neck grows, the shear stress declines and falls below the flow stress for the material. At this point dislocation participation in densification becomes ineffective. Thus, plastic flow is important early in sintering when compaction stresses and thermal stresses add to the sintering stress. But the dislocation density declines during heating and the process is not generally effective in late stage sintering. The process of plastic flow is favored by rapid heating, small particles, and pressure-assisted cycles.

In polymorphic materials, dislocation motion is also induced by phase transformations during heating. Direct observation of plastic flow during sintering is found in Al_2O_3, Ag, CaF_2, CoO, Cu, Fe, MgO, NaCl, Ni, Pb, ThO_2, Ti, W, and Zn during heating, but not under isothermal conditions.

Evaporation-Condensation

Vapor transport during sintering repositions atoms from pore surfaces into the neck region, without densification [25]. The net result is a reduction in the total surface area as necks grow between touching particles. Since the source and sink for the atoms is on the free surfaces there is no change in the distance between particle centers. The fraction of atoms on surface sites declines as surface area is consumed during sintering. The vapor pressure P depends on the absolute temperature T as:

$$P = P_O \exp\left[-\frac{Q}{R\,T}\right] \tag{7.23}$$

where P_O is a pre-exponential constant, Q is the activation energy for evaporation, and R is the gas constant. Vapor pressure increases with temperature, resulting in more flux to smooth pores and removal of surface area. In a porous material during sintering, evaporation occurs preferentially from convex particle surfaces. Preferential deposition occurs at concave necks due to a slightly reduced vapor pressure.

Usually evaporation-condensation is not significant in sintering, due to the low vapor pressure of most materials at the sintering temperature. Thus, it is often ignored. Alternatively, vapor transport occurs with small powders of high vapor pressure materials, such as NaCl, PbO, TiO_2, H_2O, Si_3N_4, BN, and ZrO_2. Generally, materials that lose weight during sintering tend to also exhibit vapor transport.

Atmosphere additions can promote sintering by evaporation-condensation. In tungsten sintering, evaporation-condensation varies with the oxygen, water, and halide concentrations. In many cases halides are effective in inducing vapor transport [64–66]. Figure 7.8 illustrates the marked change in zirconia (ZrO_2) sintering that becomes possible by doping the atmosphere with HCl [66]. Sintering in air gives

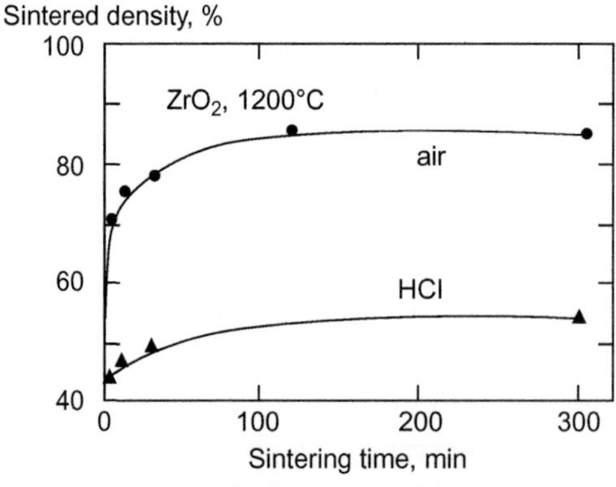

Figure 7.8 Zirconia sintering data showing sintered density versus sintering time at 1200°C (1473 K) in air and in air doped with HCl to promote evaporation-condensation versus densification [66].

sintering densification, while sintering in the HCl doped atmosphere shifts to evaporation-condensation. This latter process consumes the sintering potential of the system without contributing to densification.

Necks growth reaches a limited size dictated by the surface energies, as evident by the dihedral angle ϕ. Consequently, the neck size is limited by the grain size as follows:

$$X = G \sin\left(\frac{\phi}{2}\right) \tag{7.24}$$

Once vapor phase transport achieves a neck size ratio X/G defined by the dihedral angle, further neck growth depends grain growth. Depending on the transport process, grain size enlargement typically depends on the cube root of the sintering time.

Late in sintering, the pores become lenticular in shape due to the dihedral angle. Grain growth induces grain boundary migration, with the slower moving pores remaining attached but exerting a drag force on the grain boundary. Final stage densification depends on minimized grain growth and attachment of the pores to the grain boundaries. Vapor transport provides one means for this process to take place. Unfortunately, it is common for the grain boundary to separate from the pore, leaving stranded spherical pores in the grain interior.

KINETIC RELATIONS

Several mass transport mechanisms exist, categorized as surface transport or bulk transport depending on the mass source. The latter are responsible for densification, but both contribute to particle bonding. Surface diffusion is the most common surface transport process and grain boundary diffusion is the most common bulk transport process. For most materials, vapor phase transport contributions are small and are ignored. In reactive atmospheres (including hydrogen, oxygen, halides, and water) a high vapor pressure induces surface area loss during sintering without densification.

Plastic flow is favored during heating to the sintering temperature and involves both dislocation climb (via vacancy annihilation) and possibly dislocation glide. It is only a transient process that is not sustainable during isothermal sintering.

For amorphous materials, sintering takes place by viscous flow. Viscosity declines as temperature increases. Bonding progresses until the curvature gradients are eliminated, since there is no grain boundary or dihedral angle. If the initial packing density is high, then growing necks overlap prior to elimination of the curvature gradients, making full density possible. At low green densities, densification may terminate at less than full density. Small particles, high green densities, high temperatures, and long times improve densification.

Crystalline materials densify during sintering by bulk transport processes, predominantly grain boundary diffusion [21,47]. Vacancies emitted by the pores diffuse through the lattice or along grain boundaries, resulting pore filling. Vacancies are annihilated at grain boundaries, dislocations, phase boundaries, or other interfaces. The emission of vacancies is high under sharply curved concavities, giving rapid early sintering. Although volume diffusion is active, the lower activation energy for grain boundary diffusion and the presence of segregated species on grain boundaries often makes the grain boundary a preferred densification mechanism. Diffusion in ionic compounds varies with stoichiometry, so small doping or composition changes prove significant to the observed sintering rate. In a sense, the dominance of liquid phase sintering is simply an example of taking the grain boundary segregation to an extreme—a liquid film is a very fast transport path, more so than a grain boundary.

In general, surface diffusion is initially active; however, densification depends on grain boundary diffusion [67,68]. The interparticle grain boundary is an effective path for mass flow to the neck, especially in light of the lower activation energy in comparison with volume diffusion. Contamination of the grain boundary, intentional in some cases, dramatically changes the sintering rate—for example, activated sintering can enhance sintering by 100-fold or more [69].

The annihilation of vacancies during sintering requires cooperative events—dislocation climb and grain boundary rotation. Additionally, grain boundary diffusion requires cooperative surface diffusion to avoid building up hillocks where the grain boundary emerges at a surface. Thus, sintering involves multiple, parallel mechanisms where the dominant mechanism changes as the microstructure transforms from particles to a solid [70]. Computer simulations provide a means to deduce these mechanism shifts.

As detailed in Chapter 6, the sintering stages reflect the geometric progression from a loose powder to a dense product. In the following sections, the transport mechanisms are overlaid on the sintering stages to detail the kinetics of sintering.

Initial Stage

The curvature gradient at this stage shifts from convex to concave over a distance that is a fraction of the particle size. This gradient drives initial stage sintering. Several mass transport paths are possible, often with contributions coming from all of them at the same time. Fick's first law is invoked to determine the way that the mass flux depends on the curvature gradient. At each point in the sintering microstructure, atomic motion responds to the atomic mobility to set the rate of sintering; neck growth depends on the net (arrival rate less departure rate) mass accumulation in the sinter bond:

$$\frac{dV}{dt} = J\,A\,\Omega \tag{7.25}$$

where J is the atomic flux, A is the bond area over which the mass is distributed, and Ω is the volume of a single atom or molecule. Curvature gradients direct mass flow. Atoms are removed or deposited to change the neck size and shape. In turn the curvature gradient progressively relaxes and the flux declines. High temperatures promote faster mass transport, contributing to faster neck growth. Many concurrent processes react to the same driving force. Consequently, accurate calculations of sintering rates rely on numerical techniques.

Simplified sintering models are used to approximate the sintering behavior for a single transport mechanism. Most common are neck size and shrinkage models. For neck growth, the neck size ratio X/D (neck diameter divided by particle diameter) is given as a function of sintering time t under isothermal (constant temperature T) conditions:

$$\left(\frac{X}{D}\right)^n = \frac{B\,t}{D^m} \tag{7.26}$$

Here B is a term made up of material and geometric constants, as specified in Table 7.1. This table also lists the typical values for the exponents n and m. The particle size dependence m is known as the Herring scaling law exponent [71]. Although given as integers, the exponents change with the degree of sintering. For example,

Table 7.1 Initial Stage Sintering Equations
$(X/D)^n = B\,t\,/\,D^m$

Mechanism	n	m	B
viscous flow	2	1	$3\,\gamma\,/\,(\eta)$
plastic flow	2	1	$9\,\pi\,\gamma\,b\,D_V\,/\,(R\,T)$
evaporation-condensation	3	2	$(3P\,\gamma\,/\,\rho^2)\,(\pi/2)^{1/2}\,(M\,/\,(R\,T))^{3/2}$
volume diffusion	5	3	$80\,D_V\,\gamma\,\Omega\,/\,(R\,T)$
grain boundary diffusion	6	4	$20\,\delta\,D_B\,\gamma\,\Omega\,/\,(R\,T)$
surface diffusion	7	4	$56\,D_S\,\gamma\,\Omega^{4/3}\,/\,(R\,T)$

symbols

γ = surface energy, J/m^2
η = viscosity, Pa · s
b = Burgers vector, m
R = gas constant, J/(mol K)
T = absolute temperature, K
ρ = theoretical density, kg/m^3
δ = grain boundary width, m
D_V = volume diffusivity, m^2/s
D_S = surface diffusivity, m^2/s
D_B = grain boundary diffusivity, m^2/s
P = vapor pressure, Pa
M = molecular weight, kg/mol
Ω = atomic volume, m^3/mol

some models for surface diffusion sintering give n values up to 7.5. The parameter B embeds material properties such as diffusion, so it depends on temperature:

$$B = B_O \exp\left[-\frac{Q}{R\,T}\right] \tag{7.27}$$

where R is the gas constant, T is the absolute temperature, and B_O consists of material parameters such as surface energy and atomic size, as detailed in Table 7.1.

The integral neck size model represented by Equations 7.26 and 7.27 is reasonable to the end of initial stage sintering. The diffusion coefficient embedded in the parameter B introduces significant temperature sensitivity. The frequency factor and activation energy for common materials are available in several tabulations. However, this conceptualization is only approximate, with a typical error of 10 to 20% when compared to numerical solutions, and since most powders have a size distribution, variations in neck size coexist naturally during sintering.

Although only approximate, the isothermal neck growth model illustrates some key factors. Smaller particles sinter more rapidly. Surface diffusion and grain boundary diffusion have the highest sensitivity to particle size, so they are enhanced relative to the other processes by smaller particles. Temperature appears in an exponential term, so small temperature changes have a large effect. Finally, time diminishes in importance as the curvature gradients are relaxed.

Bulk transport processes decrease the interparticle spacing (shrinkage) while also contributing to neck growth, resulting in compact shrinkage. Densification also induces new contacts as particle centers approach one another, giving delayed onset of neck growth. Since it is relatively easy to monitor compact dimensional change as a function of time, temperature, or particle size, shrinkage is a favorite monitor of initial stage sintering. Accordingly, the time-dependent, isothermal shrinkage behavior is as follows:

$$\left(\frac{\Delta L}{L_O}\right)^{n/2} = \frac{B\,t}{D^m} \tag{7.28}$$

with sintering shrinkage $\Delta L/L_O$ normalized to the initial length. Shrinkage is a negative value, but usually the sign is ignored. Such behavior is plotted in Figure 7.9 [25] as a log-log plot of shrinkage versus time for two temperatures. The slope is $n/2 = 2.5$, corresponding to volume diffusion.

Dimensional change is associated with densification or density changes, so it is useful for monitoring the property gains. Dimensional change is a bulk measurement versus neck size and requires repeated microscopic measures. There are two very different views on shrinkage during sintering:

- Some manufacturers of precision components strive to achieve no shrinkage during sintering, so that final dimensions are used to design shaping tools. Densification is obtained by using high forming pressures prior to sintering. This

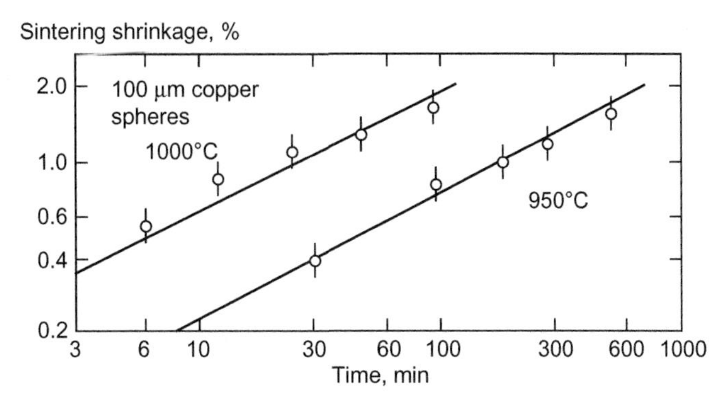

Figure 7.9 Sintering shrinkage for copper spheres at two temperatures [25]. This log shrinkage versus log time plot is predicted by Equation 7.28.

is common in forming ferrous automotive engine components, such as oil pumps, linkages, pulleys, and timing gears using Fe-Cu-Ni-C steels.

- Other manufactures plan for significant dimensional change and incorporate that into the tooling, anticipating 15 to 25% shrinkage. This is common in hard powders that cannot be compacted to a high green density, such as cemented carbide cutting tools, mining tips, and wire drawing dies formed from WC-Co.

If shrinkage is avoided, the pressed compact dimensions are retained in the sintered compact, although properties such as strength are degraded by residual porosity. Shrinkage mandates an oversizing of tooling to bring the final sintered material to acceptable dimensions. For some materials, where high densities are mandatory for performance, shrinkage is necessary during sintering. For example, this is the case in surgical tools where contamination by blood trapped in pores within the part is unacceptable. Thus, depending on the material, its ease of compaction, and the required properties, shrinkage can either be sought or avoided in sintering.

Surface area is another means of tracking sintering. It too is a bulk property and is applicable to small powders. Surface area loss depends on the neck size ratio and particle packing coordination, since each contact contributes concurrently to the loss of surface area. In the early stage of sintering, the surface area reduction parameter $\Delta S/S_O$, similar to shrinkage, is used to track sintering versus time t:

$$\left(\frac{\Delta S}{S_O}\right)^V = Ct \tag{7.29}$$

where $\Delta S/S_O$ is the change in surface area divided by the initial surface area, C is a kinetic term proportional to B, so it includes mass transport parameters and other factors as previously given in Table 7.1. The exponent V is approximately equal to $n/2$. To illustrate surface area reduction, a log-log data plot is given in Figure 7.10 for 100

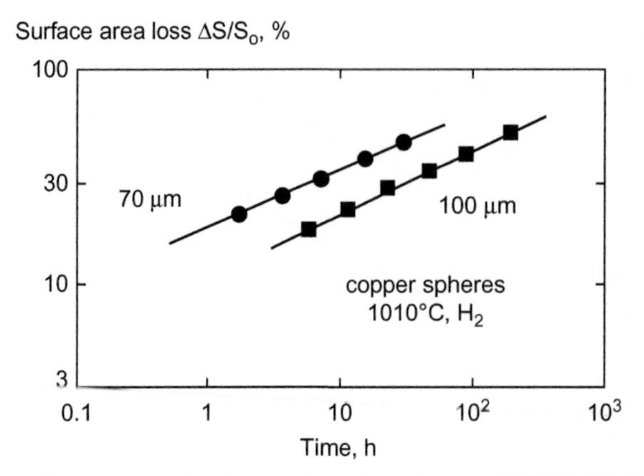

Figure 7.10 Copper surface area reduction during sintering at 1010°C (1283 K) for two particle sizes plotted on a log-log basis.

and 70 μm copper spheres sintered at 1010°C (1283 K). As expected, the behavior shows progressive annihilation of surface area with isothermal sintering time in accordance with Equation 7.29.

Similar laws apply to non-spherical particles [72]. Indeed, models exist for a range of geometries—sphere on plate, wire on plate, wire on wire, needle on plate, knife edge to knife edge, knife edge on plate, and needle tip to needle tip. These forms are similar to those presented here. Thus, several measures are possible and some keys are as follows:

- Initial sintering is comparatively rapid.
- Curvature gradients from convex to concave surfaces are the driving force.
- Initial mass flow is dominated by surface diffusion.
- Initial neck growth occurs without significant shrinkage.
- The elimination of surface area shifts dominance to bulk transport processes.
- With neck growth grain boundary diffusion increases in importance.
- Neck growth is the fundamental measure of initial stage sintering.

Intermediate Stage

Initial stage sintering is active for the early portion of neck growth while sharp curvature gradients are evident in the microstructure, but there is only a minor level of shrinkage or densification. Neck growth eliminates curvature gradients. Intermediate stage sintering further rounds the pores by fusing neighboring necks. In three dimensions, the pores progress toward a tube-like structure as the convex surfaces are eliminated. This is the intermediate stage of sintering, when necks merge to produce a wormhole structure in the sintering body. The growing neck, that was the focus of

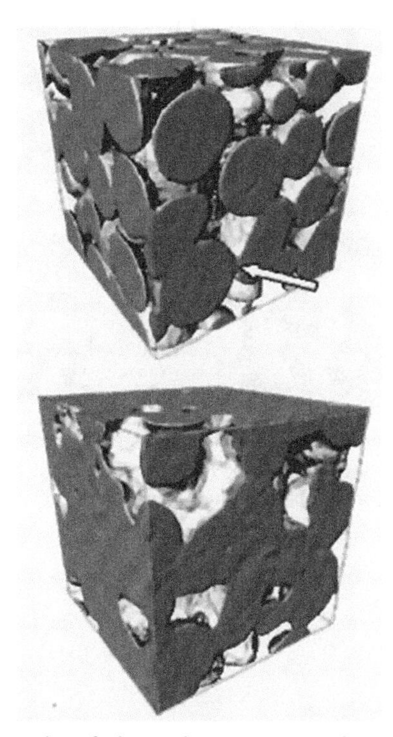

Figure 7.11 Computed tomography of glass sphere sintering after 10 min (top) and 190 min (bottom) at 720°C (993 K) [73]. Neck growth and densification are accompanied by loss of curvature and pore coarsening.

initial stage sintering, is less evident so attention shifts to the pore-grain structure. Figure 7.11 shows computed tomography images of a glass powder compact sintered for 10 min (top) and 190 min (bottom), showing several changes in the pores [73]. The number of pores decrease, the curvature decreases, and the pore size increases while the structure densifies.

Densification, if it occurs, is most evident in the intermediate stage. The mixture of concave and convex surfaces is evident in the two-dimensional microstructure shown in Figure 7.12. Sintering removes curvature and surface area, as evident for the same material sintered for longer in Figure 7.13. The first of these images corresponds to the start of the intermediate stage and the latter image corresponds to the end of the intermediate stage. During this transition, properties such as strength significantly improve due to simultaneous pore rounding and pore elimination. The driving force is the reduction in surface energy, since the curvature gradients have been reduced. A differentiation from the final stage of sintering is in how the pore space is interconnected.

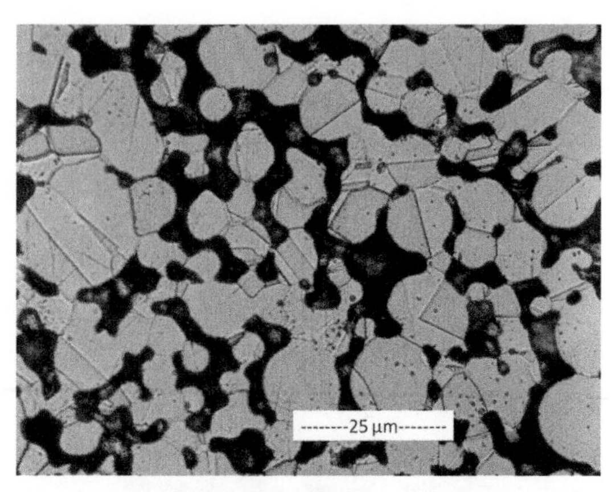

Figure 7.12 A cross-sectional micrograph of a stainless steel powder at the transition from the initial to intermediate stage sintering. The black regions are pores. Neck growth during the initial stage occurs with a mixture of concave and convex features.

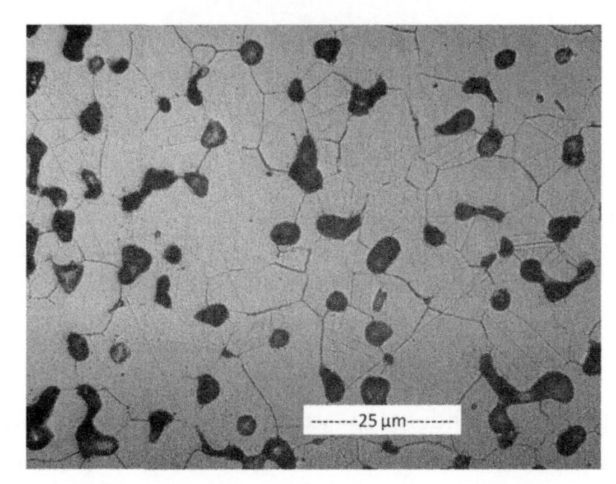

Figure 7.13 A cross-sectional micrograph is for the same stainless steel powder shown in Figure 7.12, sintered to a higher density with reduced pore space, rounder pores, and enlarged grains.

Sintering calculations place the cylindrical pores on the grain edges. Cylindrical pores located on grain edges as sketched by the idealized geometry are shown in Chapter 6. A segment of the pore structure around the neck is sketched in Figure 7.14. The part labeled grain boundary sits at the interparticle neck, and the pore structure forms a network on the grain edges. The particle size distribution in the starting material leads to a pore size distribution in the intermediate stage. Similar to the grain size distribution, the pore size distribution becomes self-similar in sintering. This means the simultaneous pore shrinkage and pore coarsening events reach a balance. Late in sintering, as full density is approached, the grain shape approaches a

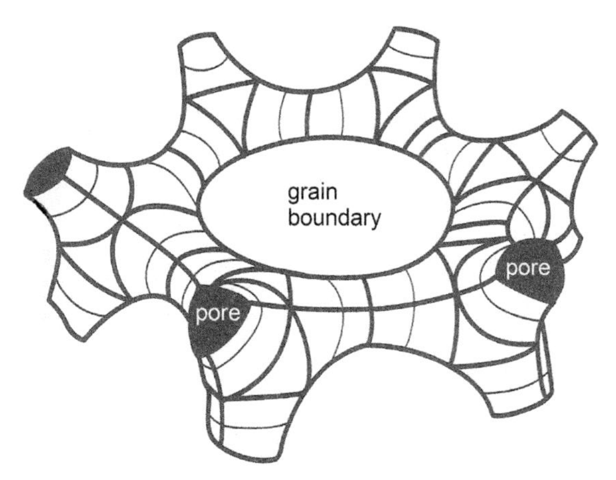

Figure 7.14 Graphical representation of the pore structure in intermediate stage sintering. Necks containing grain boundaries bond the grains while tubular pores are around the necks.

12 or 14 sided polygon with cylindrical pores stretched on the grain edges. In two-dimensional cross-sections the connectivity of the pore space and the tetrakaidecahedron grain shape are not evident.

Pores and grain boundaries are the prime focus in intermediate stage sintering. During sintering of this pore-grain geometry, the approximate relation between the pore diameter d, grain size G, and fractional density f is given by:

$$f = 1 - \pi \left(\frac{d}{G} \right)^2 \tag{7.30}$$

If pore size does not change, as is often the case, this results in grain size varying with the inverse square-root of the porosity. Densification depends on both changes in pore size and grain size. Typically, pores remain attached to the grain boundaries through much of the intermediate stage. The energy penalty associated with pore-boundary separation is high as long as the dihedral angle is below 120°. Consequently, the densification rate depends on diffusion from the pore:

$$\frac{df}{dt} = J\, A\, N\, \Omega \tag{7.31}$$

where N is the number of pores per unit volume, A is the pore area, J is the diffusion flux per unit area per unit time, and Ω is the atomic volume.

To calculate the diffusion flux using Fick's first law, the concentration gradient between the grain boundary and pore surface is determined. As noted earlier, the vacancy concentration C varies with curvature. The change in concentration divided by the change in position gives the concentration gradient that drives diffusion. In the intermediate stage curvature depends on the pore diameter d, the most curved aspect of the microstructure.

The vacancy sink is at the center of the grain boundary where an equilibrium vacancy concentration exists that is equal to C_O. The pore surface has a higher vacancy concentration. The distance from the vacancy source to the vacancy sink is approximately $G/6$, where G is the grain size. The diffusion flux depends on the gradient, the change in concentration ($\Delta C = C - C_O$) and the distance ($G/6$) times the diffusivity D_V (assuming volume diffusion). The area A over which mass flows also depends on the pore size d and grain size G, estimated as $A = \pi\ dG/3$. Consequently, the densification rate df/dt is calculated as:

$$\frac{df}{dt} = \frac{g\,\gamma_{SV}\Omega D_V}{R\,T\,G^3} \tag{7.32}$$

where f is the fractional density, t is the time, g is near 5 (depending on geometric assumptions such as grain shape and how grain size is measured—intercept, area, or volume). Due to the inverse cube effect, smaller grains significantly aid densification, in part because smaller grains are associated with more sharply curved pores.

For illustration, assume densification by volume diffusion. Each grain face is shared by two grains. During intermediate stage sintering, grain growth leads to enlarged grains and slower densification. The characteristic grain volume increases linearly with time, where K is a rate parameter that depends on the material, temperature, and impurities.

The grain growth rate parameter is thermally activated, with an activation energy that reflects diffusion events across the grain boundary. Often it is similar to the activation energy for grain boundary diffusion, but is modified by pore mobility and impurity drag effects [74–76]. The combined factors give a fractional density change law [9,77]:

$$f = f_I + B_I\ ln\left(\frac{t}{t_I}\right) \tag{7.33}$$

where f is the fractional sintered density, f_I is the fractional density at the beginning of the intermediate stage (typically about 0.7), B_I is a rate term, t is the total sintering time, and t_I is the time for the onset of the intermediate stage. The rate term B_I includes surface energy, diffusion, atomic volume, pore curvature, and temperature. This relation predicts that sintered density is proportional to the logarithm of sintering time.

A similar form holds for densification by grain boundary diffusion control. There is a strong role played by grain boundaries. Smaller grains increase the curvature, make diffusion distances smaller, and retard grain growth. Temperature is a major factor as evident in the data shown in Figure 7.15 for 45 μm copper powder [78]. Here

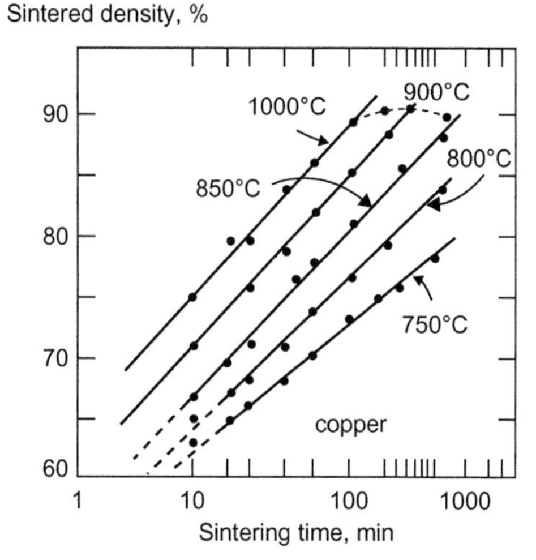

Figure 7.15 Sintered density at various temperatures during intermediate stage sintering for copper [78].

the density is given versus the logarithm of sintering time for temperatures between 750 and 1000°C (1023 and 1273 K); the densification rate increases with temperature.

Other intermediate stage sintering models exist, with variations in the assumed curvature gradient, grain growth rate, vacancy equilibrium, or diffusion path [79,80]. Some models treat densification in terms of simultaneous processes and include surface area and pore curvature [81]. Typically, experimental densification rates are faster than predicted, because the models assume uniform microstructures. An important attribute of such models is an inverse relation between the sintering rate and the grain size.

Substantial surface area reduction occurs in the intermediate stage, and the rate of surface area loss depends on the surface area:

$$\frac{dS}{dt} = S^{\alpha} \tag{7.34}$$

where dS/dt is the surface area loss rate, α depends on the transport mechanism, and S is the remaining surface area. The open surface area is consumed by the end of intermediate stage sintering. The only remaining pore space is associated with closed pores.

Grain size enlarges as the pore structure collapses. The median grain volume increases linearly with time. As a demonstration of this behavior, 900°C (1173 K) copper sintering data are plotted as grain size cubed versus hold time in Figure 7.16 [82]. The individual measures during densification are shown as symbols giving an excellent fit to the cubic law.

Pinning of grain boundaries by pores slows grain growth, but this effect diminishes as porosity declines. As necks enlarge the grain boundary increases, but as grains grow

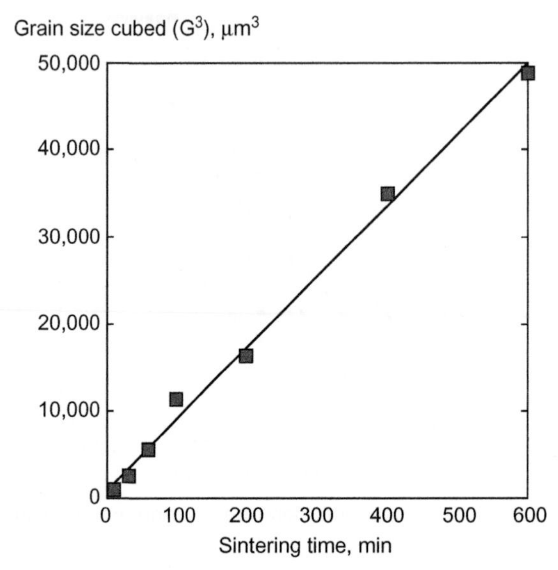

Figure 7.16 Grain growth during sintering for copper sintered at 900°C (1173 K) [82]. The grain volume (cube of the grain size) increases linearly with sintering time.

the grain boundary area decreases. The grain boundary area peaks at approximately 85% density, after which grain growth accelerates.

It is most desirable for the pores to remain attached to the grain boundaries to sustain densification and retard grain growth. Densification depends on volume and grain boundary diffusion. Pores located on grain boundaries disappear faster than isolated pores, due to the efficiency of grain boundaries as vacancy annihilation sites. At the same time the grain boundaries provide transport between pores, allowing pore coarsening.

Surface transport is active during intermediate stage sintering, providing pore rounding and assisting with pore migration during grain growth. However, as with the initial stage, the surface transport processes that rearrange atoms on the pore surfaces do not contribute to densification or shrinkage.

Final Stage

Final stage sintering is characterized by a declining rate of densification and an accelerating rate of microstructure coarsening. The pores are closed, and usually decrease in number but increase in average size during the final stage. There is not a dramatic shift to closed pores; rather a gradual transition occurs at the onset of the final stage. The closed pores form as pore diameter decreases and grain growth induces a stretching of the pore length to close the pores. Details of this condition were presented in

Figure 7.17 Fracture surface of spark sintered tantalum carbide, showing pores located on grain boundaries as expected in final stage sintering.

Chapter 6. This should occur with about 8% porosity. Microstructure variations result in a transition progressively from about 15% porosity to 5% porosity. Initially the pores are spherical, but grain boundary energy usually results in lenticular shaped pores.

Sintering is much slower in the final stage, since curvature gradients and surface energy are both much reduced. Coarsening of the pores and grains impedes densification. Initially the pores attach to grain corners, giving an idealized microstructure. The scanning electron micrograph fracture surface in Figure 7.17 shows nearly spherical pores on the grain boundaries, many evidencing the anticipated lenticular shape.

During sintering, the pores on grain boundaries absorb atoms and emit vacancies. The fractional density, porosity, pore size, and grain size are related. The pore is bound to the grain boundary since the pore reduces the total grain boundary area. Effectively, this is a pore-boundary binding energy, and it increases with porosity and pore size. As densification proceeds, the binding energy between pores and grain boundaries declines, eventually allowing grain growth with stranded pores left behind.

Everything depends on the relative rates for densification, grain growth, pore shrinkage or growth, and pore migration. Often the activation energies are similar, especially for densification and grain growth. Thus, manipulation is difficult if it uses only temperature adjustments. Rapid grain growth tends to strand the pores away from the grain boundaries, resulting in slow densification and residual porosity. Alternatively, high pore mobility keeps the pore-boundary structure together, leading to rapid convergence to full density. Small pores are most helpful, and they often migrate by surface diffusion or evaporation-condensation. But, usually grain boundary mobility is larger than the pore mobility, leading to pore drag and retarded grain growth during final stage sintering, as long as the pores remain attached to the grain boundaries.

One means of predicting the propensity for full densification is via comparison of the surface and grain boundary diffusion rates:

$$\Gamma = \frac{D_S \gamma_{SS}}{300\ D_B \gamma_{SV}} \tag{7.35}$$

where D_S is the surface diffusion rate, D_B is the grain boundary diffusion rate, γ_{SS} is the grain boundary energy, and γ_{SV} is the solid-vapor surface energy. When Γ is less than 1, full density is possible. Densification is incomplete when the grain size increases quickly.

A modified final stage densification model includes the retardation from gas trapped in the pores [67,83]. If no external pressure is applied, then densification primarily depends on grain boundary diffusion:

$$\frac{df}{dt} = \frac{a\,\Omega\,\delta\,D_B}{R\,T\,G^3}\left(\frac{4\gamma_{SV}}{d} - P\right) \tag{7.36}$$

where f is the fractional density, t is the isothermal hold time, a is a geometric constant equal to 5, Ω is the atomic volume, δ is the grain boundary width (assumed to be five times the atomic size), D_B is the diffusivity, R is the gas constant, T is the absolute temperature, G is the grain size, γ_{SV} is the solid-vapor surface energy, d is the pore size, and P is the pore gas pressure. Normally the grain boundary width is included in the diffusion frequency factor:

$$\delta D_B = \delta D_{Bo}\ exp\left[-\frac{Q_B}{R\,T}\right] \tag{7.37}$$

where Q_B is the grain boundary diffusion activation energy, R is the gas constant, and D_{BO} is the frequency factor for grain boundary diffusion.

An external pressure (from hot pressing, hot isostatic pressing, sinter forging, or spark sintering) results in an additional term which is proportion to the external pressure as it is amplified by the porosity.

Late in sintering, after say an hour, the pore dihedral angle between the grain boundary and pores causes the pores to become lenticular in shape. Grain growth depends on the relative attachment and mobility of the pores. Large pores are unable to remain attached to moving grain boundaries and become stranded in the grain's interior. Rapid grain growth occurs once the grain boundaries break away from the pores. Successful sintering achieves pore attachment on the boundaries, by either retarding grain growth or enhancing pore mobility. It is beneficial to have a small pore size compared to the grain size.

PROCESSING VARIABLES

Several adjustable processing variables impact sintering. The relative role of temperature, time, particle size, and pressure depends on the material. Some materials are dominated by surface diffusion while others rely on mostly grain boundary diffusion. Early sintering models assumed only one mechanism operated at a time. Sintering is a multiple mechanism process where mass fluxes are summed to determine the instantaneous rate [81,84].

In assessing the rate of sintering, several independent parameters are encountered:

- material
- particle size
- green density
- heating rate
- peak temperature
- hold time.

The sintering response involves a combination of interacting factors and multiple transport events. The complexity of the interactions, mass transport events, sintering stages, and microstructure changes is best treated using computer simulations. Initially the predictions were made in the form of sintering maps, but more recently they are conceived in terms of master sintering curves. Both are treated in Chapter 14. Fundamentally, each approach relies on understanding how each mechanism responds to the process variables, building up by summation the net impact of the individual contributions.

Temperature

Temperature is the most important sintering variable. Sintering models involve thermally activated events, so any measure of sintering Y (for example neck size ratio, surface area reduction, or shrinkage) depends on temperature via the exponential Arrhenius relation:

$$Y^n = \frac{C}{T}\, exp\left[-\frac{Q}{R\,T}\right] \tag{7.38}$$

where T is the absolute temperature, Q is an activation energy that relates to the melting temperature of the material, R is the gas constant, and C is a cluster of material and geometric constants. The pre-exponential temperature term is relatively weak compared to the exponential temperature term. The parameter Y is one of the several measures of sintering, such as shrinkage, density, densification, surface area change, or neck size, each with an appropriate time exponent n. This behavior is seen in constant

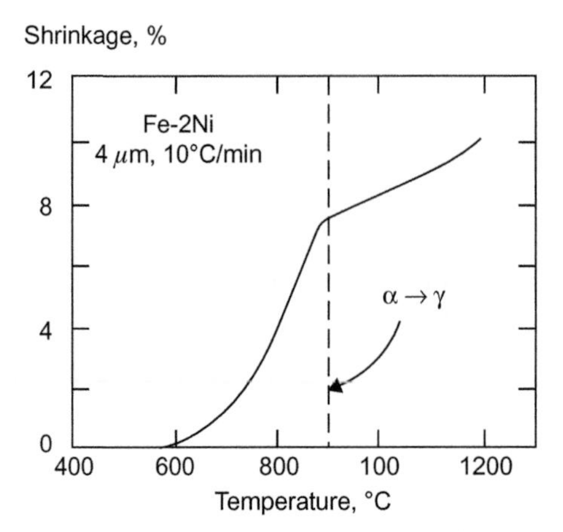

Figure 7.18 Sintering shrinkage for Fe-2Ni during constant rate heating. Shrinkage slows with the phase transform from alpha to gamma.

heating rate experiments, where the metric shows a steady increase with temperature. An example plot is given in Figure 7.18 for a mixture of 4 μm iron and nickel powders (corresponding to Fe-2Ni) heated at 10°C/min in hydrogen. The sintering shrinkage, as measured by dilatometry, is plotted versus temperature. Sintering behavior shifts when the iron goes through a phase transition at 910°C (1183 K). The lower temperature α phase (body-centered cubic) exhibits higher diffusion rates versus the γ phase (face-centered cubic), leading to a dramatic reduction in the shrinkage rate.

Over a narrow temperature range the pre-exponential temperature in Equation 7.38 is ignored and the logarithmic form gives:

$$\ln(Y^n) \approx \ln A - \frac{Q}{R\,T} \tag{7.39}$$

A plot of $\ln(Y)$ versus $1/T$ gives a straight line with slope proportional to Q/nR. The slope includes the time exponent which in turn depends on the mechanism and measurement; for example, if sintering is measured by the neck size ratio X/D and is controlled by volume diffusion, then $n = 5$. Two example plots are given in Figure 7.19 for surface area reduction and shrinkage during constant rate heating for alumina [67]. Knowing time dependence from parallel isothermal experiments allows extraction of the activation energy. This is critical to understanding how temperature changes influence sintering.

Time

Sintering time is a mixed variable, since more hold time at the peak temperature increases sintering, but longer times contribute to rapid microstructural coarsening. A longer sintering time typically has less impact than a higher sintering temperature.

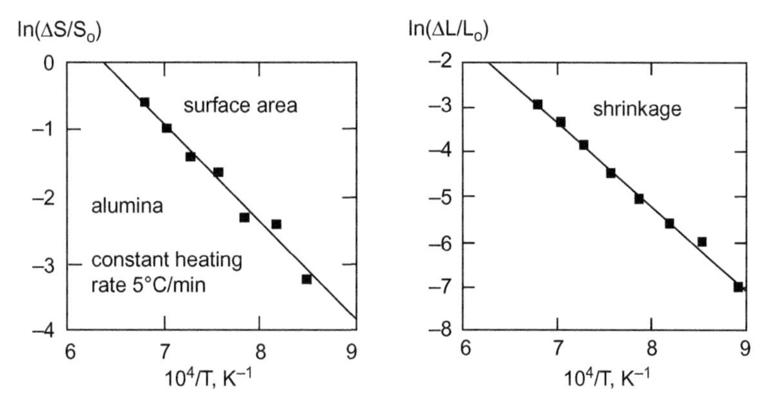

Figure 7.19 Logarithmic surface area reduction and shrinkage data obtained for alumina during constant heating rate sintering [67]. Both parameters are plotted versus the inverse absolute temperature for extraction of apparent activation energies.

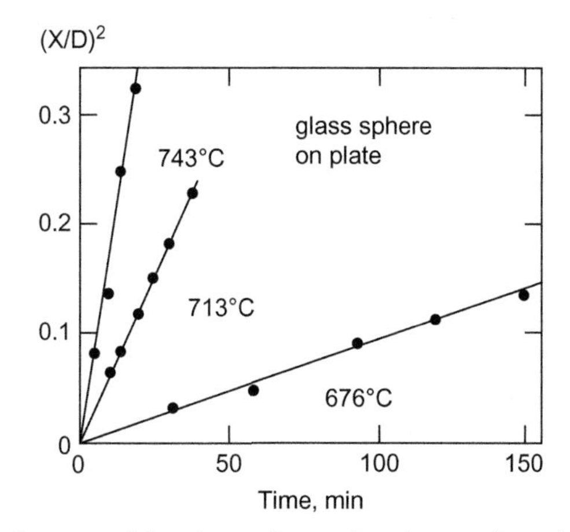

Figure 7.20 Neck size ratio squared for glass spheres sintering to glass plate at three temperatures [85]. The behavior agrees with viscous flow controlled sintering.

This comparison is evident in Figure 7.20, in which the square of the neck size is plotted versus time for three temperatures [85]. Typically time is a power law factor:

$$rate = t^z \tag{7.40}$$

The exponent z ranges from 0.1 to 0.5 and is related to the time exponents given in Table 7.1. Accordingly, since sintering rates are dependent on other factors such as grain size, microstructure coarsening is an important factor that offsets gains envisioned from long sintering times; effectively long sintering times are often counterproductive and industrial sintering times might be as short as 10 min at the peak temperature.

Heating Rate

Heating rate is effectively a combination of time and temperature into a single parameter. A fast heating rate often suppresses surface diffusion, thereby inducing more sintering densification at high temperatures. On the other hand, rapid heating reduces the time at temperature to reduce net shrinkage. At the same time, grain growth is reduced by fast heating. Normally, slow heating at rates of 2 to 10°C/min are preferred, since there is less component warpage from non-uniform heat transfer, but for small and thin sections faster heating is possible using some of the novel heating techniques.

Particle Size

The median particle size is the most important powder characteristic. Usually the particle size is based on the volume or mass of particles, but for microscopy or laser scattering, the number of particles is measured. The median size corresponds to the size where half the particles are smaller and half are larger.

Experimental results where nickel of different particles sizes was sintered at a constant heating rate of 5°C/min are plotted in Figure 7.21 [86]. Sintered density is plotted against the temperature for powders of 0.05, 5, and 50 μm. As the particle size decreases there is a shift to a lower sintering onset temperature and more densification at all temperatures.

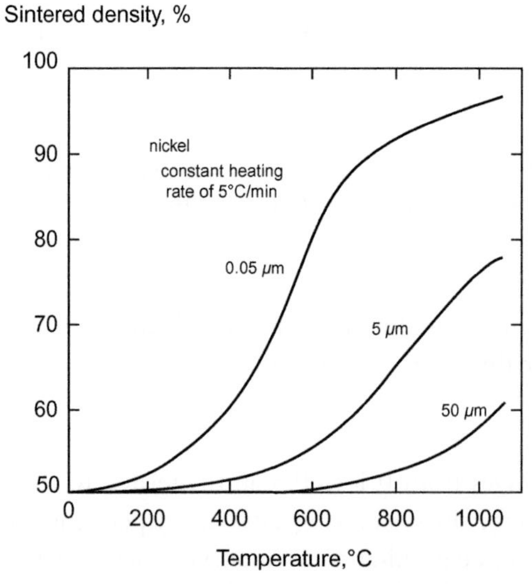

Figure 7.21 Constant hearing rate sintering for three nickel particle sizes, giving density versus temperature [86].

Other factors related to the powder have an impact on sintering, and one that is often ignored is the strain energy stored in the powder. A milled powder releases stored strain (dislocation entanglements) during heating, giving a boost to the sintering rate. This is a transient, since sintering anneals and relaxes the defects and excess dislocations, but the effect is measurable and an advantage in sintering [87,88]. Likewise, constraints against sintering retard densification [89]. Thus, composites, while strong after sintering, prove difficult to sinter.

Green Density

Usually green density is adjustable over a narrow range prior to sintering. Too low a green density and the material is not easily handled. Usually the vibrated powder density (tap density) is the lowest possible starting point. Compaction to very high pressures, essentially 100% density, is demonstrated [90] but not often used due to practical limitations in tooling and press capacity. Green density is a factor in the sintering response, generally giving higher sintered properties as the green density increases—as is well demonstrated in a host of materials. Simply, a high green density gives more particle-particle contacts (more sinter necks form and grow) and smaller pore size (higher curvature), both factors driving faster sintering.

SUMMARY

Sintering involves both densification and coarsening. Although many materials densify as part of sintering, not all sintering involves densification. Many materials form sinter bonds by surface diffusion or evaporation-condensation without densification. Some sintered products intentionally avoid densification, but densification is desirable in structural materials since residual pores lower properties. Thus, the balance between densification and coarsening determines microstructure and properties. Even when there is no net dimensional change, sintering produces surface area reduction, grain size enlargement, and compact strengthening, and might have attendant changes in the pore structure.

The pore size is a measure of the sintered microstructure. Pores grow due to coarsening but shrink under the action of densification. This mixed trajectory usually grows the larger pores while simultaneously shrinking the smaller ones. At long times or high temperatures even the larger pores eventually disappear. However, several situations arise to stabilize the pores to prevent full densification, sometimes resulting in compact swelling at long sintering times.

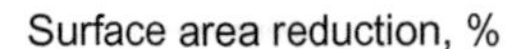

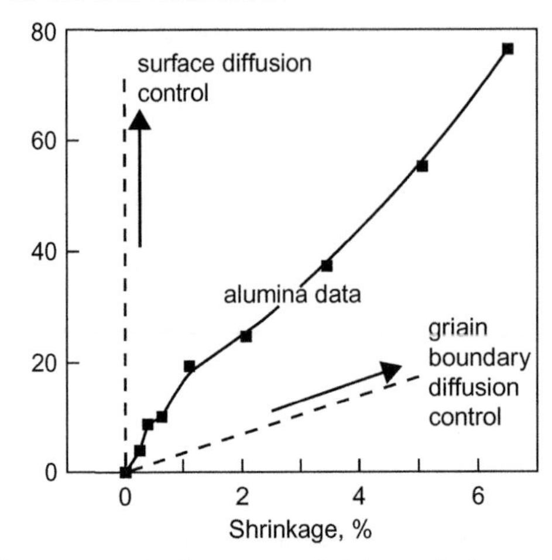

Figure 7.22 A plot of surface area reduction versus shrinkage during constant heating rate sintering for 0.14 μm alumina [67]. Two ideal pathways are shown, one for surface diffusion control and one for grain boundary diffusion control. The sintering data take on an intermediate behavior, indicating both processes contribute to sintering.

If the particles are packed homogeneously to create an ideal monosized pore structure, then sintering is rapid, requiring lower sintering temperatures or shorter sintering times. This is a reason to avoid particle agglomeration or packing gradients, since the very large and very small parts of the pore size distribution are critical in determining densification and coarsening rates.

This chapter has outlined the classic mixture of sintering stages, mass transport processes, and microstructure monitors used to track sintering. As discussed in Chapter 14, the complexity of these events are typically addressed using computer simulations, in which the simultaneous, multiple events are treated in iterative time advance simulations. However, this chapter is necessary to capture the underlying concepts behind those simulations. As a demonstration of the sintering pathway actually taken in sintering, Figure 7.22 plots the surface area reduction versus the sintering shrinkage for 0.14 μm alumina during heating at 5°C/min [67]. If sintering took place totally by surface diffusion, there would be no shrinkage and the data would track the vertical line labeled surface diffusion control. On the other hand, if sintering took place entirely by grain boundary diffusion control, there would be both surface area reduction and shrinkage, tracking along the line labeled grain boundary diffusion control. The experimental data follow an intermediate path, attributed to simultaneous contributions from both mechanisms. Such multiple mechanism character is frequently encountered in sintering, and requires computer simulations to properly handle the solutions.

REFERENCES

[1] H. Djohari, J.I. Martinez-Herrera, J.J. Derby, Transport mechanisms and densification during sintering I, viscous flow versus vacancy diffusion, Chem. Eng. Sci. 64 (2009) 3799–3809.
[2] P. Schwed, Surface diffusion in sintering of spheres on planes, Trans. TMS-AIME 191 (1951) 245–246.
[3] A. Shimosaka, Y. Ueda, Y. Shirakawa, J. Hidaka, Sintering mechanism of two spheres forming a homogeneous solid solubility neck, Kona (21) (2003) 219–233.
[4] F. Thummler, W. Thomma, The sintering process, Metall. Rev. 12 (1967) 69–108.
[5] R.M. German, Thermodynamics of sintering, in: Z.Z. Fang (Ed.), Sintering of Advanced Materials, Woodhead Publishing, Oxford, UK, 2010, pp. 3–32.
[6] H. Kuroki, H.Y. Suzuki, G. Han, K. Shinozaki, Effect of pore morphology on driving force for pore closure in sintered materials, Sci. Sintering 32 (2000) 69–72.
[7] C.F. Yen, R.L. Coble, Spheroidization of tubular voids in alumina crystals at high temperatures, J. Amer. Ceram. Soc. 55 (1972) 507–509.
[8] A.P. Greenough, Grain boundaries and sintering, Nature 166 (1950) 904–905.
[9] R.L. Coble, Sintering crystalline solids. 1. Intermediate and final state diffusion models, J. Appl. Phys. 32 (1961) 787–792.
[10] J.W. Noh, S.S. Kim, K.S. Churn, Collapse of interconnected open pores in solid-state sintering of W-Ni, Metall. Trans. 23A (1992) 2141–2145.
[11] I. Amato, The effect of gas trapped within pores during sintering and density regression of ceramic bodies, Mater. Sci. Eng. 7 (1971) 49–53.
[12] F.N. Rhines, C.E. Birchenall, L.A. Hughes, Behavior of pores during the sintering of copper compacts, Trans. TMS-AIME 188 (1950) 378–388.
[13] Y.J. Lin, K.S. Hwang, Swelling of copper powders during sintering of heat pipes in hydrogen containing atmospheres, Mater. Trans. 51 (2010) 2251–2258.
[14] T.K. Gupta, R.L. Coble, Sintering of ZnO: 2, density decrease and pore growth during the final stage of the process, J. Amer. Ceram. Soc. 51 (1968) 525–528.
[15] R. Watanabe, Y. Masuda, Quantitative estimation of structural change in carbonyl iron powder compacts during sintering, Trans. Japan Inst. Met. 13 (1972) 134–139.
[16] R.M. German, K.S. Churn, Sintering atmosphere effects on the ductility of W-Ni-Fe heavy metals, Metall. Trans. 15A (1984) 747–754.
[17] W.R. Rao, I.B. Cutler, Initial sintering and surface diffusion in Al_2O_3, J. Amer. Ceram. Soc. 55 (1972) 170–171.
[18] Y.I. Boiko, Y.E. Geguzin, V.G. Kononeknko, F. Friedrich, W. Schatt, Theory and technology of sintering, thermal, and chemicothermal treatment processes, Powder Metall. Metal Ceram. (1980) 675–682 vol. 19.
[19] A.R. Boccaccini, E.A. Olevsky, Anisotropic shrinkage during sintering of glass-powder compacts under uniaxial stresses – qualitative assessment of experimental evidence, Metall. Mater. Trans. 28A (1997) 2397–2404.
[20] R.M. German, Supersolidus liquid-phase sintering of prealloyed powders, Metall. Mater. Trans. 28A (1997) 1553–1567.
[21] S.J.L. Kang, Sintering densification, Grain Growth, and Microstructure, Elsevier Butterworth-Heinemann, Oxford, UK, 2005.
[22] C.A. Handwerker, J.E. Blendell, R.L. Coble, Sintering of ceramics, in: D.P. Uskokovic, H. Palmour, R.M. Spriggs (Eds.), Science of Sintering, Plenum Press, New York, NY, 1980, pp. 3–37.
[23] J. Frenkel, Viscous flow of crystalline bodies under the action of surface tension, J. Phys. 9 (1945) 385–391.
[24] M. Godinho, E. Longo, E.R. Leite, R. Aguiar, *In Situ* observation of glass particle sintering, J. Chem. Edu. 83 (2006) 410–413.
[25] W.D. Kingery, M. Berg, Study of the initial stages of sintering solids by viscous flow, evaporation-condensation, and self-diffusion, J. Appl. Phys. 26 (1955) 1205–1212.
[26] A.R. Boccaccini, R. Kramer, Experimental verification of a stereology-based equation for the shrinkage of glass powder compacts during sintering, Glass Tech. 36 (1995) 95–97.

[27] M.N. Rahaman, L.C. De Johghe, Sintering of spherical glass powder under a uniaxial pressure, J. Amer. Ceram. Soc. 73 (1990) 707–712.

[28] L. Zagar, Theoretical aspects of sintering glass powders, in: M.M. Ristic (Ed.), Sintering – New Developments, Elsevier Scientific, New York, NY, 1979, pp. 57–64.

[29] M. Nothe, K. Pischang, P. Ponizil, B. Kieback, J. Ohser: Study of particle rearrangement during sintering process by microfocus computer tomograph (micro-ct); Proceedings PM2004 Powder Metallurgy World Congress, vol. 2, European Powder Metallurgy Association, Shrewsbury, UK, 2004, pp. 221–226.

[30] M.W. Reiterer, K.G. Ewsuk, J.G. Arguello, An arrhenius type viscosity function to model sintering using the Skorohod-Olevsky viscous sintering model within finite element code, J. Amer. Ceram. Soc. 89 (2006) 1930–1935.

[31] R.T. DeHoff, A general theory of microstructural evolution by surface diffusion, Sci. Sintering 16 (1984) 97–104.

[32] F.B. Swinkels, M.F. Ashby, Role of surface redistribution in sintering by grain boundary transport, Powder Metall. 23 (1980) 1–7.

[33] A. Moitra, S. Kim, S.G. Kim, S.J. Park, R.M. German, Investigation on sintering mechanism of nanoscale powder based on atomistic simulation, Acta Mater. 58 (2010) 3939–3951.

[34] G.C. Kuczynski, Self-diffusion in sintering of metallic particles, Trans. TMS-AIME 185 (1949) 169–178.

[35] N. Cabrera, Note on surface diffusion in sintering of metallic particles, Trans. TMS-AIME 188 (1950) 667–668.

[36] J.G.R. Rockland, On the rate equation for sintering by surface diffusion, Acta Metall. 14 (1966) 1273–1279.

[37] B.Y. Pines, A.F. Sirenko, Sintering kinetics of wire and spherical granules by surface diffusion, Fizik Metal. Metall. 28 (1969) 832–836.

[38] J.W. Bullard, Digital image based models of two dimensional microstructural evolution by surface diffusion and vapor transport, J. Appl. Phys. 81 (1997) 159–168.

[39] S. Arcidiacono, N.R. Bieri, D. Poulikakos, C.P. Grigoropoulos, On the coalescence of gold nanoparticles, J. Multi. Flow 30 (2004) 979–994.

[40] A.P. Sutton, R.W. Balluffi, General aspects of interfaces as sources/sinks, Interfaces in Crystalline Materials, Clearndon Press, Oxford, UK, 1995, pp. 599–621

[41] L.L. Hall, C.S. Morgan, Observation of dislocations occurring during sintering, J. Amer. Ceram. Soc. 54 (1971) 55.

[42] C.C. Fatino, J.S. Hirschhorn, Effect of strain on the loose sintering of stainless-steel powder, Trans. TMS-AIME 239 (1967) 1499–1504.

[43] M.J. Bannister, W.J. Buykx, The sintering mechanism in UO_{2+X}, J. Nucl. Mater. 64 (1977) 57–65.

[44] A.J. Markworth, On the coarsening of gas-filled pores in solids, Metall. Trans. 4 (1973) 2651–2656.

[45] H. Ichinose, H. Igarashi: Contributions of Volume and Surface Diffusion in High Temperature Sintering of Copper; Proceedings 1993 Powder Metallurgy World Congress, Part 1, Y. Bando and K. Kosuge (Eds.), Japan Society Powder and Powder Metallurgy, Kyoto, Japan, 1993, pp. 349–352.

[46] J.C. Wang, Analysis of early stage sintering with simultaneous surface and volume diffusion, Metall. Trans. 21A (1990) 305–312.

[47] R.M. German, Sintering Theory and Practice, Wiley-Interscience, New York, NY, 1996.

[48] G.C. Kuczynski, The mechanism of densification during sintering of metallic particles, Acta Metall. 4 (1956) 58–61.

[49] L. Seigle, Role of grain boundaries in sintering, in: W.D. Kingery (Ed.), Kinetics of High-Temperature Processes, John Wiley, New York, NY, 1959, pp. 172–178.

[50] M. Tikkanen, The part of volume and grain boundary diffusion in the sintering of one-phase metallic systems, Plansee. Pulvermet. 11 (1963) 70–81.

[51] W. Zhang, I. Gladwell, Sintering of two particles by surface and grain boundary diffusion – a three dimensional model and numerical study, Comp. Mater. Sci. 12 (1998) 84–104.

[52] Z. He, J. Ma, Constitutive modeling of alumina sintering: grain size effect on dominant densification mechanism, Comp. Mater. Sci. 32 (2005) 196–202.
[53] J. Svoboda, H. Riedel, Quasi-equilibrium sintering for coupled grain boundary and surface diffusion, Acta Metall. Mater. 43 (1995) 499–506.
[54] R.L. Coble, Initial sintering of alumina and hematite, J. Amer. Ceram. Soc. 41 (1958) 55–62.
[55] W.S. Coblenz, J.M. Dynys, R.M. Cannon, R.L. Coble, Initial stage solid state sintering models, a critical analysis and assessment, in: G.C. Kuczynski (Ed.), Sintering Processes, Plenum Press, New York, NY, 1980, pp. 141–157.
[56] D.L. Johnson, A general model for the intermediate stage of sintering, J. Amer. Ceram. Soc. 53 (1970) 574–577.
[57] K. Breitkreutz, K. Haedecke, Method for determining activation energies of shrinkage processes during sintering, Part 1; theoretical model for superimposed processes, Powder Metall. Inter. 14 (1982) 160–163.
[58] G.N. Hassold, I.W. Chen, D.J. Srolovitz, Computer simulation of final stage sintering: I, model, kinetics, and microstructure, J. Amer. Ceram. Soc. 73 (1990) 2857–2864.
[59] H.N. Chng, J. Pan, Cubic spline elements for modelling microstructural evolution of materials controlled by solid-state diffusion and grain boundary migration, J. Comp. Phys. 196 (2004) 724–750.
[60] E.E. Adams, D.A. Miller, R.L. Brown, Grain boundary ridge on sintered bonds between ice crystals, J. Appl. Phys. 90 (2001) 5782–5785.
[61] H. Zhu, R.S. Averback, Sintering processes of two nanoparticles: a study by molecular dynamics simulations, Phil. Mag. Lett. 73 (1996) 27–33.
[62] S. Siegel, A. Elshabiny, W. Hermel, Sinterbeschleunigung durch mechanisches Aktivieren von Eisenpulver, Z. Metall. 75 (1984) 911–915.
[63] W. Schatt, E. Friedrich, Sintering as a result of defect structure, Cryst. Res. Tech. 17 (1982) 1061–1070.
[64] R.D. McIntyre, The effect of HCl-H_2 sintering atmospheres on the properties of compacted iron powder, Trans. Quart. Amer. Soc. Met. 57 (1964) 351–354.
[65] D.W. Readey, Vapor transport and sintering, in: C.A. Handwerker, J.E. Blendell, W. Kaysser (Eds.), Sintering of Advanced Ceramics, Amer. Ceramic Society, Westerville, OH, 1990, pp. 86–110.
[66] M.J. Readey, D.W. Readey, Sintering of ZrO_2 in HCl atmospheres, J. Amer. Ceram. Soc. 69 (1986) 580–582.
[67] S.H. Hillman, R.M. German, Constant heating rate analysis of simultaneous sintering mechanisms in alumina, J. Mater. Sci. 27 (1992) 2641–2648.
[68] J. Pan, A.C.F Cocks, S. Kucherenko, Finite element formulation of coupled grain boundary and surface diffusion with grain boundary migration, Proc. Royal Soc. London A 453 (1997) 2161–2184.
[69] R.M. German, Z.A. Munir, Enhanced low-temperature sintering of tungsten, Metall. Trans. 7A (1976) 1873–1877.
[70] P. Redanz, R.M. Mcmeeking, Sintering of a BCC structure of spherical particles of equal and different sizes, Phil. Mag. 83 (2003) 2693–2714.
[71] C. Herring, Effect of change of scale on sintering phenomena, J. Appl. Phys. 21 (1950) 301–303.
[72] P.M. Raj, W.R. Cannon, 2-D sintering of oriented ellipses by grain boundary and surface diffusion: a numerical study of the shrinkage anisotropy, in: R.M. German, G.L. Messing, R.G. Cornwall (Eds.), Sintering Science and Technology, The Pennsylvania State University, State College, PA, 2000, pp. 393–398.
[73] D. Bernard, D. Gendron, J.M. Heitz, J.M Heintz, S. Bordere, J. Etourneau, First direct 3d visualisation of microstructural evolutions during sintering through X-ray computed micro tomography, Acta Mater. 53 (2005) 121–128.
[74] A.H. Heuer, The role of MgO in the sintering of alumina, J. Amer. Ceram. Soc. 62 (1979) 317–318.
[75] T.W. Sone, J.H. Han, S.H. Hong, D.Y. Kim, Effect of surface impurities on the microstructure development during sintering of alumina, J. Amer. Ceram. Soc. 84 (2001) 1386–1388.
[76] L. Montanaro, J.M. Tulliani, C. Perrot, A. Negro, Sintering of industrial mullites, J. Europ. Ceram. Soc. 17 (1997) 1715–1723.

[77] R.L. Coble, Intermediate-stage sintering: modification and correction of a lattice-diffusion model, J. Appl. Phys. 36 (1965) 2327.
[78] R.L. Coble, T.K. Gupta, Intermediate stage sintering, in: G.C. Kuczynski, N.A. Hooton, C.F. Gibbon (Eds.), Sintering and Related Phenomena, Gordon and Breach, New York, NY, 1967, pp. 423−441.
[79] W. Beere, The second stage sintering kinetics of powder compacts, Acta Metall. 23 (1975) 139−145.
[80] W. Beere, The intermediate stage of sintering, Met. Sci. J. 10 (1976) 294−296.
[81] J.D. Hansen, R.R. Rusin, M.H. Teng, D.L. Johnson, Combined-stage sintering model, J. Amer. Ceram. Soc. 75 (1992) 1129−1135.
[82] A.K. Kakar, A.C.D. Chaklader, Deformation theory of hot pressing − yield criterion, Trans. TMS-AIME 242 (1968) 1117−1120.
[83] A.J. Markworth, On the volume diffusion controlled final stage densification of a porous solid, Scripta Metall. 6 (1972) 957−960.
[84] D.L. Johnson, Sintering kinetics for combined volume, grain boundary and surface diffusion, Phys. Sintering 1 (1969) B1−B22.
[85] G.C. Kuczynski, Study of the sintering of glass, J. Appl. Phys. 20 (1949) 1160−1163.
[86] R.A. Andrievski, Compaction and sintering of ultrafine powders, Inter. J. Powder Metall. 30 (1994) 59−66.
[87] Q. Wei, H.T. Zhang, B.E. Schuster, K.T. Ramesh, R.Z. Valiev, L.J. Kecskes, et al., Microstructure and mechanical properties of superstrong nanocrystalline tungsten processed by high pressure torsion, Acta Mater. 54 (2006) 4079−4089.
[88] C.C. Fatino, J.S. Hirschhorn, Effect of strain on the loose sintering of stainless-steel powder, Trans. TMS-AIME 239 (1967) 1499−1504.
[89] A.F. Whitehouse, T.W. Clyne, Cavity formation during tensile straining of particulate and short fiber metal matrix composites, Acta Metall. Mater. 41 (1993) 1701−1711.
[90] E.Y. Gutmanas, High-pressure compaction and cold sintering of stainless steel powders, Powder Metall. Inter. 12 (1980) 178−182.

CHAPTER EIGHT

Microstructure Coarsening

INTRODUCTION

Characteristics

Microstructure coarsening is inherent to sintering. Coarsening happens because of energy differences across interfaces—there is a small energy difference between large and small microstructural features, such as grains or pores. The energy per unit volume varies with the inverse of the size, so small pores or small grains have a tendency to coalesce into larger pores or grains. The interface between the microstructural features might be a grain boundary, a liquid film, a second phase, or a pore. Typically the mean (or median) feature size is tracked during sintering. The greatest attention is devoted to grain growth with secondary attention being given to pore growth [1−9].

For grain growth, the focus is on the crystal grains inside the sintering body. Particles exist prior to sintering. They are discreet solids that are able to flow during shaping. On the other hand, grains are solid regions with a specific crystal orientation. After sintering the structure consists of grains bonded to one another. Sintered particles become grains, and characteristically grains enlarge to reduce the interfacial energy. Consequently the number of grains declines as the grain size increases. One exception is in liquid phase sintering, where a solid dissolves in a newly formed liquid to reach the solubility limit, giving a slight initial reduction in the solid size.

During sintering the median or mean grain size G_M increases and the number of grains N_G decreases:

$$N_G G_M^3 = \text{constant} \tag{8.1}$$

Thus, the terminal condition in sintering is one large grain, but such a condition would require years of sintering.

For pores, the compressibility of the vapor phase means that volume is not conserved. Indeed pore coarsening during final stage sintering often produces swelling [10].

Grain boundary energy is lower than surface energy. Thus, sintering initially converts surface area to grain boundary area by growing interparticle necks, and then grain boundary area is eliminated by coarsening. The large grains consume neighboring small grains, just as big fish consume small fish.

Sintering: From Empirical Observations to Scientific Principles
DOI: http://dx.doi.org/10.1016/B978-0-12-401682-8.00008-2

One mechanism for growth is direct absorption of one grain into a neighboring grain via coalescence. This preferentially occurs when two grains contact with a low degree of crystallographic misorientation [11–17]. Otherwise, grain rotation is required to align the crystals, thereby removing the grain boundary and allowing the grains to fuse. Figure 8.1 sketches such an event. The initial grain contact is random with respect to crystal orientation. A grain boundary forms in the neck due to the differing crystal orientations. Subsequently the grain boundary is eliminated if the grains rotate into crystal alignment [18–20]. Figure 8.2 is a scanning electron micrograph showing several coalescence events captured during the sintering of a W-Ni alloy [21]. Because grain size is not uniform, several situations arise to favor coalescence within a microstructure, usually with smaller grains melding into larger grains.

Coarsening also occurs by atomic exchange across interfaces—grain boundaries, liquid films, or vapor phases. These diffusion mechanisms, in which larger grains grow at the expense of smaller grains, are sketched in Figure 8.3. Besides transport across grain boundaries, other pathways such as surface diffusion, transport in a second phase, or vapor phase transport are observed. The larger grain grows to produce significantly fewer grains. Of course, a grain might be growing at one interface with a smaller grain while shrinking at another interface with a larger grain.

Since the grain volume is proportional to the grain size cubed, a two-fold grain size increase corresponds to an eight-fold reduction in the number of grains. In parallel, changes occur in the pores; the porosity and the number of pores usually decrease, but the pore size tends to grow in proportion to the grain size. Unlike grain volume,

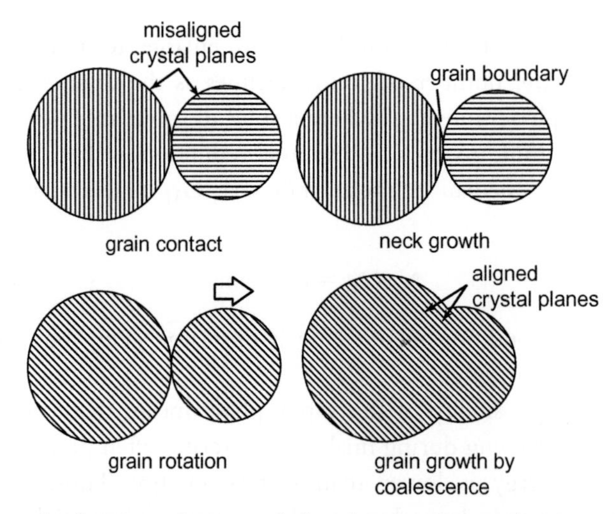

Figure 8.1 A schematic of grain coalescence during sintering. Two grains come into contact during sintering and grow a bond with a grain boundary at the point of contact. That grain boundary is eliminated by grain rotation, typically by the smaller grain, with melding together to form a single grain.

which is conserved during coarsening, pore volume is not conserved. Large pores have lower pressure than small pores, so when two small pores coalesce the resulting pore is lower in pressure and larger in total volume. Thus, the pore size trajectory is complicated by simultaneous pore annihilation and pore growth processes [22–27]. Coarsening also results in grain shape changes and second phase enlargement. One critical aspect of sintering is the pore interaction with the grain boundaries; both move and change size within the structure. It is desirable for the pores to attach and move with grain boundaries [28,29].

Figure 8.2 Scanning electron micrograph of a W-Ni liquid phase sintered sample with several examples of grain coalescence with no grain boundary at the point of coalescence.

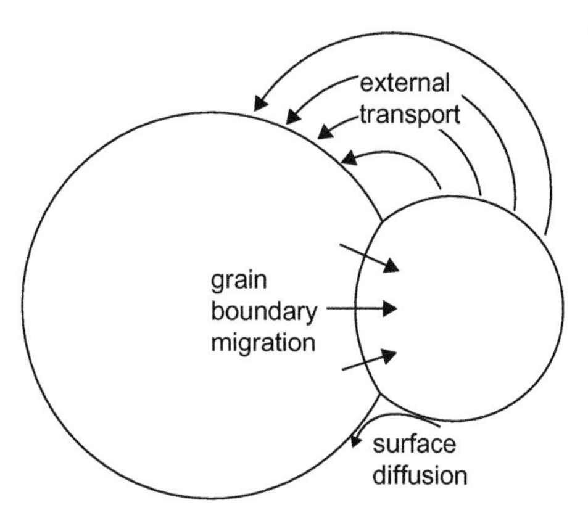

Figure 8.3 Diffusion controlled grain coarsening occurs by mass transfer from the smaller grain to the larger grain, leading to elimination of the smaller grain. This sketch shows grain boundary migration, surface diffusion, and transport via vapor or liquid phases as possible mechanisms.

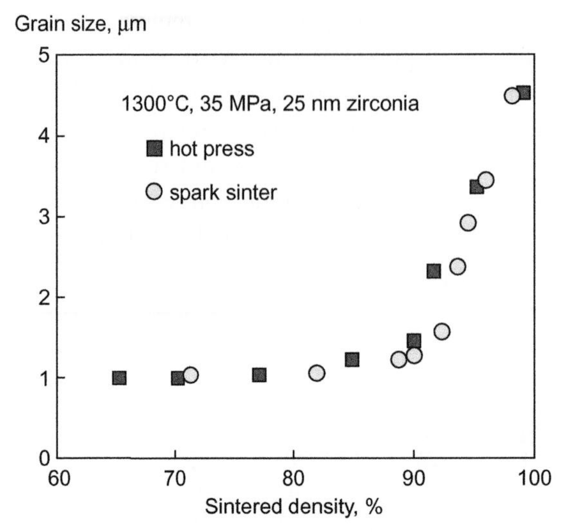

Figure 8.4 The coarsening trajectory for hot pressing and spark sintering 25 nm zirconia [30]. There is little variation with the consolidation technique since the microstructure pathway is inherent to sintering densification.

Importance

Mathematical treatment of coarsening dates from the 1890s. Wilhelm Ostwald (1909 Nobel Prize) modeled colloidal suspension size change over time. In sintering, the conditions responsible for bonding and densification also induce microstructure coarsening. But the process is very different from that of a suspension. In Ostwald ripening there is no densification and no solid contact. By contrast, sintering involves solid bonding and densification without volume conservation. As sintering densification progresses, the grain contact area increases. Thus, grain size changes with sintered density. An example of the behavior common to sintering is given in Figure 8.4 for zirconia consolidated by hot pressing and spark sintering [30]. Grain size enlarges with density following a trajectory that is independent of the consolidation route.

Microstructure is second only to density in determining sintered properties. Residual pores reduce strength by reducing the load bearing cross-sectional area, but grain size is also a contributing factor. For example, sintered alumina exhibits an eight-fold strength decrease when the grain size increases from 0.8 μm to 6.9 μm. While sinter bonding and densification lead to property improvements, coarsening degrades many properties. Accordingly, sintered properties pass through a maximum and decline with excessive sintering. Some sintering is good, but it is possible to sinter

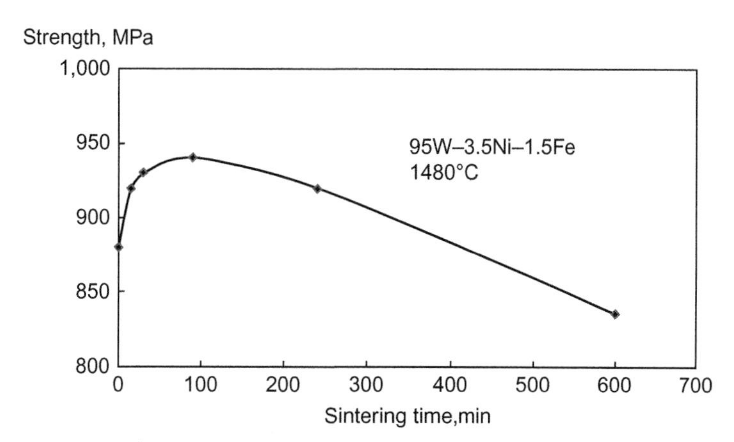

Figure 8.5 Strength data for a sintered tungsten alloy showing optimized strength at a shorter sintering time because microstructure coarsening reduces properties [31].

too much. Figure 8.5 gives an example for a 95 wt.% tungsten alloy sintered at 1480°C (1753 K) [31]. Its strength peaks after 90 min of sintering, when the alloy reaches 99.8% density. Grain coarsening then occurred, going from 30 to 45 μm between 90 to 600 min. Likewise, ductility is highest at 22% elongation after 90 min, declining to 14% after 600 min. Densification and grain growth compete; at short sintering times densification improves mechanical properties, but at long sintering times microstructural coarsening has the opposite effect.

Interactions

The microstructure coarsens and converges to a self-similar character during sintering. For a given material this means the microstructure "looks" the same, except for a progressive increase in size over time. In simple terms, the structure retains a common morphological arrangement. For school students the normal distribution is a common basis for assigning grades. It is a two-parameter distribution that requires information on the mean and standard deviation. In sintering, the grain size distribution likewise tracks the self-similar character. The mean size increases, but the normalized dispersion around the mean remains the same. This means that if the coarsening rate is known, then the distribution is predicable.

Grain growth depends on temperature, time, and composition. Growth largely occurs by diffusion across boundaries. Pores on grain boundaries represent missing interface area that usually does not participate in grain coarsening. Further, grain growth reduces grain boundary interaction.

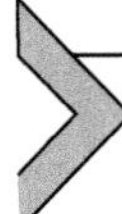

GRAIN COARSENING

Grain Growth Rate

During sintering, additives influence grain growth:

- additives that form solid solutions usually increase grain growth
- additives that form second phases usually retard grain growth.

This reflects the additive role on mass transport rates. Linked to the sintered grain size is the grain growth rate parameter. It is sensitive to temperature and composition.

Coarsening causes the mean grain volume to change linearly with the sintering time. For a grain of size G, the size change rate dG/dt, depends on its surface area over which transport occurs (proportional to the grain area G^2), and the size difference normalized to the mean size ($\Delta G/G_M$ where $\Delta G = G - G_M$) as follows [32]:

$$\frac{dG}{dt} = \frac{h\,K\,\Delta G}{G^2 G_M} \tag{8.2}$$

where h is a geometric constant and K is the rate parameter with units of m^3/s. The rate parameter is a combination of atomic volume, transport rates, and composition terms. In this conceptualization, the size change for an individual grain relates to the mean grain size in what is termed the mean field theory. In this accounting, a grain smaller than the median size is shrinking (negative growth), while a grain larger than the median size is growing. Observations show that the grain size change depends on the size of neighboring grains [33]. Depending on the surrounding grain sizes, grains smaller than the mean size sometimes grow and grains larger than the mean size sometimes shrink.

Integration gives the mean grain volume as a function of time. The grain size G increases with time t as follows:

$$G^N = G_O^N + K\,t \tag{8.3}$$

with the initial grain size G_O corresponding to time $t = 0$, with the rate parameter K given in $\mu m^3/s$. Usually $N = 3$ for sintering, so the grain volume is proportional to time. An example is given by the grain size data for copper sintering plotted in Figure 8.6 [34]. This log-log plot gives a slope of one-third indicated by the solid line.

The grain growth rate parameter depends on temperature as follows:

$$K = K_O \exp\left[-\frac{Q_{GG}}{R\,T}\right] \tag{8.4}$$

where R is the gas constant, T is the absolute temperature, and K_O is the frequency factor. It is similar to diffusion since grain growth also depends on atomic transport.

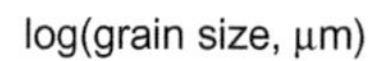

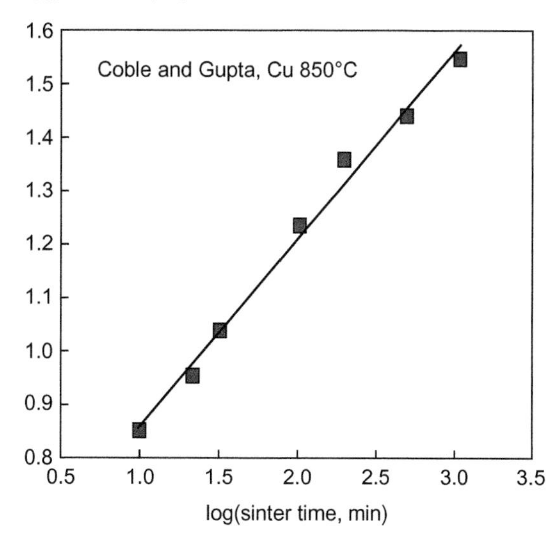

Figure 8.6 Isothermal sintering of copper giving grain size versus sintering time on a logarithmic scaling [34]. The plot corresponds to the cubic grain growth law typical to sintering.

As such, the activation energy for grain growth Q_{GG} is similar to the activation energy for sintering. This implies coarsening and sintering go hand in hand. However, sintered density reaches a terminal value first while grain growth continues for a very long time before reaching one grain.

Equations 8.3 and 8.4 provide a means of extracting the activation energy for grain growth. One approach is to solve for the time required to reach a given grain size (say 50 μm) at different temperatures. Figure 8.7 plots the logarithmic time versus inverse absolute temperature for MgO during coarsening [35]. The activation energy is 260 kJ/mol for both the 1% and 2% added vanadium contents, but the time to reach a 50 μm grain size is different. Both systems exhibit a linear dependence of the mean grain volume on sintering time.

Grain growth is slower for systems with two different solid phases [36,37]. This is because of reduced grain contacts, similar to the way that pores hinder coarsening.

A few materials differ from the G^3 proportional to time relation. A square law of $G^2 \approx t$ is associated with anisotropic surface energies. Flat faced grains indicate grain growth throttled by a limited population of surface sites for growth. For example, in WC-Co sintering initial grain growth is rapid, so to control grain growth additives are used to form polygonal grains with slower grain growth [38–41].

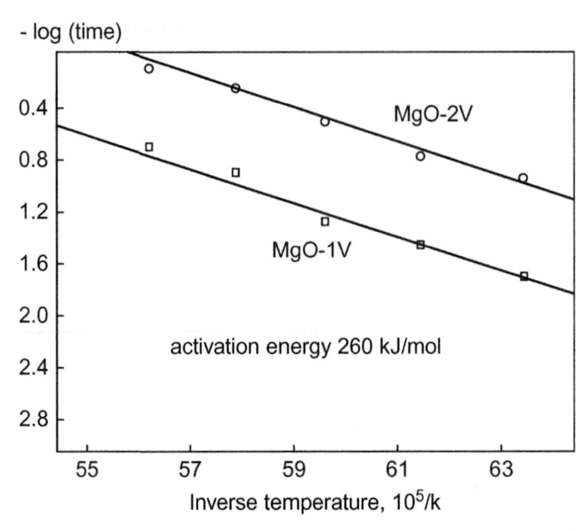

Figure 8.7 Extraction of the grain growth activation energy for two magnesia compositions, where the logarithm of the time to reach a grain size of 50 μm is plotted versus the inverse absolute temperature, giving 260 kJ/mol [35].

Porosity, Liquid, Vapor Effects

Grain growth accelerates as pores are eliminated. Of course, if high pressures are applied then densification events dominate. Thus, ultrahigh pressure consolidation is one means to reduce sintering temperature and impede grain growth.

Grain growth is sensitive to additives, such as liquids, active vapors, and second phases. As noted earlier, grain growth is slower for systems with two different solid phases. Even so the grain size behavior tracks against the inverse square root of the porosity, as evident in the three alumina systems in Figure 8.8 [42]. The data are for pure alumina, MgO doped alumina which retards grain growth during sintering, and FeO doped alumina which accelerates grain growth during sintering. Although different in slope, the three systems exhibit grain growth which is linked to porosity.

Pores interfere with coarsening because they hinder mass transport. During heating, there is ample porosity and little solid bonding, so grain growth is slow because there is little interface area. At the end of sintering the opposite occurs; the increased solid-solid interface area associated with densification enables more coarsening. Diffusion rates might be high during early sintering, but until the solid bonds are formed grain growth is slow.

An underlying aspect of coarsening is the rate parameter. A generalized case corresponds to solid, liquid, and vapor phases coexisting during sintering. Usually one interface dominates coarsening. Grain size enlargement is proportional to the rate parameter K. In turn, there is a concurrent shift in the rate parameter as the interfacial

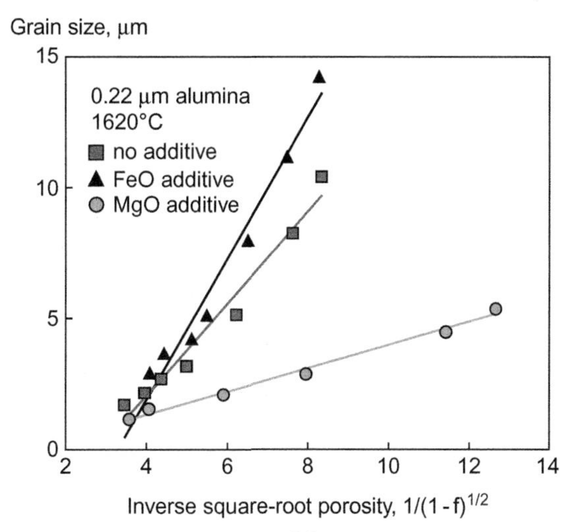

Figure 8.8 Three alumina compositions compared for grain size versus inverse square root of the fractional porosity [42]. The additives retard (MgO) or accelerate (FeO) grain growth, but the porosity role remains similar.

area changes. It is wrong to think of K as a constant, since it varies during sintering. The effective rate parameter depends on the various contributions [43]:

$$K = C_{SS}K_{SS} + C_{SL}K_{SL} + C_{SV}K_{SV} \quad (8.5)$$

In this model, C represents the contiguity or fractional interface area (solid-solid or SS, solid-liquid or SL, or solid-vapor or SV) and each interface has a corresponding rate parameter. The fractional interface areas are bounded by $C_{SS} + C_{SL} + C_{SV} = 1$, but the terms change during sintering. Pores reduce grain-grain contact, so grain growth is slow early in sintering because of the reduced grain-grain contact area (low C_{SS}). The grain boundary area increases as necks grow to convert solid-vapor surface area into grain boundary area. Accordingly, early in sintering C_{SS} is nearly a linear function of fractional density.

In liquid phase sintering, as long as the solid is soluble in the liquid, the solid-liquid parameter K_{SL} is the dominant path for coarsening. Then solid-solid parameter K_{SS} is usually small, and is ignored. This assumes no significant vapor phase transport, although sintering in halide atmospheres makes vapor transport dominant [44–47]. Accordingly, in liquid phase sintering the grain growth rate parameter is normalized to the Ostwald ripening rate (K_O) and given as:

$$K_O = \frac{q\, D_S S\, \Omega\, \gamma_{SL}}{RT} \quad (8.6)$$

where the geometric constant $q \approx 7$, D_S is the solid diffusivity in the liquid (Arrhenius temperature dependence), S is the solubility of the solid in the liquid, Ω is the solid molar volume, γ_{SL} is the solid-liquid surface energy, R is the gas constant, and T is the absolute temperature. The corresponding grain growth rate parameter depends on the inverse of the liquid volume fraction raised to the 2/3 power. Ignoring pores, the solid-liquid grain growth rate parameter is given as [43]:

$$K_{SL} = \frac{K_O}{(1 - V_S)^{2/3}} \tag{8.7}$$

where V_S is the solid volume fraction. The solid volume fraction is given as $V_S = 1 - V_L$ for zero porosity, where V_L is the volume fraction of liquid. Note that as the liquid content changes from 5 to 15 vol.% the rate parameter more than doubles. Figure 8.9 plots the rate parameter for several systems based on Equation 8.7.

One effect of the liquid content on coarsening is observable by a grain size gradient in the gravitational direction. The grain size at the top of a component is smaller than at the bottom due to solid-liquid separation induced by density differences (assuming the solid is denser than the liquid) [48–50]. This occurs because gravity stratifies liquid content, in turn changing the coarsening rate parameter with position.

For cases of liquid phase sintering in which the solid is not soluble in the liquid, coarsening still occurs. In this case the rate parameter for solid-liquid transport is zero, but the solid-solid contribution is not zero and allows coarsening [51].

Vapor transport as a coarsening mechanism is often correctly ignored during sintering. In some situations it is active and contributes significantly to coarsening

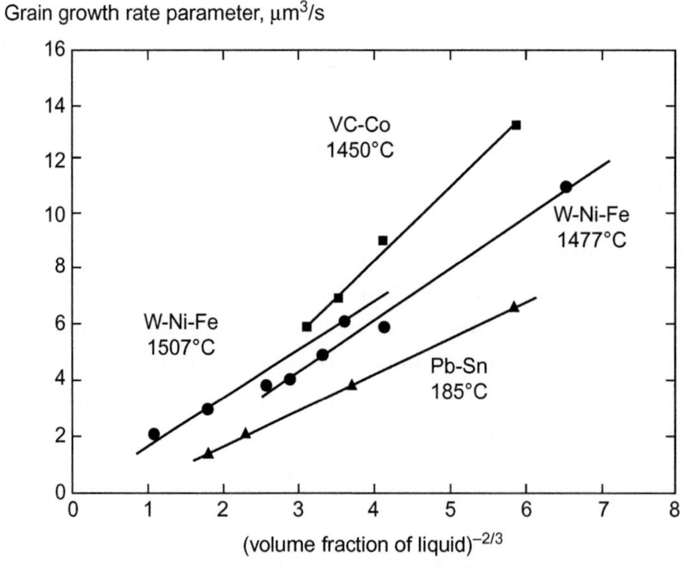

Figure 8.9 Grain growth rate parameter for various liquid phase sintering materials expressing the dependence on the liquid volume fraction.

[44–47,52–54]. Vapor collects in the pores as densification progresses. For insoluble vapors, the pore pressure increases in the final stage of sintering, to the point where pore shrinkage terminates at less than full density. Stable pores slow grain growth, but also can coalesce or coarsen to induce swelling.

A high vapor pressure, such as with materials containing lead oxide or zinc, or active vapor complexes, such as halides, results in significantly increased vapor transport. As a consequence there is a vapor-solid interface contribution to coarsening during sintering, even as densification halts. Pore motion by vapor transport enables grain growth. For most materials the pores are inactive with respect to coarsening, so even though the solid-vapor contact term C_{SV} is large, the small solid-vapor rate parameter, K_{SV}, gives a process dominated by the solid-solid transport mechanism. With an active vapor species in the pores the vapor contribution is significant, enabling grain coarsening and pore coarsening even when densification is slow [55].

When vapor transport is dominant, grain coarsening occurs without densification. For example, in air sintering 0.2 μm titania (TiO_2) for 100 min at 1200°C (1573 K) the sintered density is 76% of theoretical. Parallel sintering in HCl vapor gave 45% density (essentially the green density); however, grain coarsening gave a 6 μm grain size versus the 1 μm grain size from air sintering. Vapor transport did not contribute to densification, but did accentuate coarsening. In such cases a mass loss is one indication of vapor induced coarsening. Grain volume increases linearly with time, but density remains low.

Pore migration occurs with an active pore vapor, allowing evaporation from a concave surface and condensation on a flatter surface. The pores are bowed to enable evaporation from one surface and condensation on the opposite surface.

Stable Grain Shape

Early coarsening models assumed spherical grains. In reality the grains are rounded polygons with flat contact faces. For example, Figure 8.10 is a scanning electron micrograph of zinc oxide grains extracted from a sintered microstructure. It is a mixture of flat grain contacts corresponding to grain boundaries and rounded edges. During coarsening, shrinking grains are smaller and rounded while growing grains are larger and polygonal.

Grain growth ensures that the grain size distribution and grain shape distribution approach self-similar characteristics. The larger grains have more faces, so grain size and grain shape are correlated [56–58]. Grain shape is distributed, just as the grain size is distributed. Smaller grains have fewer faces and larger grains have more faces. Within a grain shape class, where all grains have the same number of faces N, the cumulative distribution of grain sizes follows a Weibull distribution. For example all 10-faced grains have a median size and a distribution about that median. The larger

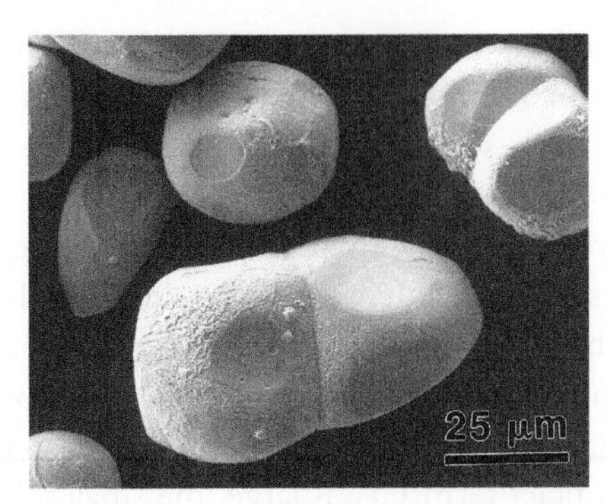

Figure 8.10 Scanning electron micrograph of ZnO grains after sintering. The flat faces represent grain boundaries between the grains.

the number of faces N, the larger is the median size for grains with N faces. The two-dimensional median size of grains with N faces follows a simple relation [59]:

$$L_N = L_{50}(N - 2) \tag{8.8}$$

where L_{50} is the median intercept grain size for all grains. Smaller grains have fewer contacting grains because they are shrinking. For three-dimensional microstructures, the relation between median grain size G_{50} and grain coordination N is [60]:

$$G_N = G_{50}\left(\frac{N}{N_{50}}\right)^2 \tag{8.9}$$

where G_N is the average grain size for grains with a three-dimensional coordination of N (versus Equation 8.8 for two-dimensional analysis). For two-dimensional measurements, the median grain size is expressed as a linear function of the grain faces.

The grain shape distribution is self-similar and, like the grain size, it fits an exponential cumulative distribution as follows:

$$F(N) = 1 - \exp\left[-0.7\left(\frac{N}{N_{50}}\right)^M\right] \tag{8.10}$$

where in two dimensions $F(N)$ is the cumulative fraction of grains with up to N faces and N_{50} is the median number of grain faces, typically about six as full density is approached.

During sintering the grain contacts enlarge to a limit which is set by the dihedral angle, where the diameters of the flat contacts between grains are defined by the neck diameter X, which are evident as the ovals on the grain surfaces in Figure 8.11. The bond size is proportional to the grain diameter G as follows:

$$X = G \sin\left[\frac{\phi}{2}\right] \tag{8.11}$$

Figure 8.11 Fracture surface showing neck size and how it reflects the grain size and dihedral angle.

with ϕ being the dihedral angle determined by the interfacial energies. The neck size grows in tandem with the grain size, while the grain shape depends on the solid density.

Stable Distributions

The sintered grain size distribution converges to the same form independent of the starting particle size distribution. This is termed self-similar [61,62]. Early coarsening models predicted a narrow grain size distribution, but subsequent models which include coalescence and other coarsening events better fit to experimental results [43,63].

As already noted, during grain coarsening the grain sizes track to a predictable size distribution. Cumulative grain intercept distributions find that the largest grains are about three-fold larger than the median size. This is in disagreement with all of the Ostwald ripening theories [64,65]. Chapter 6 details the geometric distributions for sintered materials.

PORE STRUCTURE CHANGES

Whenever there is a feature size difference within a microstructure, the concomitant difference in energy with location induces coarsening. Pores exhibit a size distribution, so they coarsen during sintering, but are annihilated by densification. After an initial transient the pores take on self-similar distribution characteristics. The pore structure change depends on the relative pore size. When sintering in a vacuum, the pore is a collection of vacancies with no impediment to collapse. Small pores generate higher local vacancy concentrations, and when heated those vacancies diffuse to grain boundaries, free surfaces, or interfaces, where they are annihilated. Hence, pores near free surfaces disappear while internal pores coarsen. Consequently, density is not uniform within the sintered component.

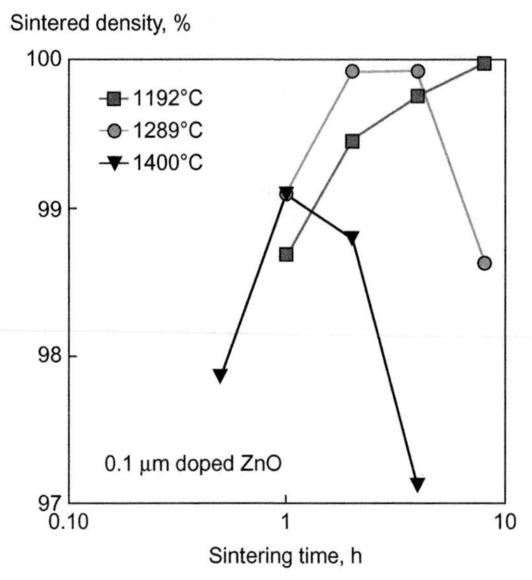

Figure 8.12 Sintered density of 100 nm zinc oxide processed for various times and temperatures, mapping the conditions that produce swelling at higher temperatures [66].

Gas trapped in the pores complicates the situation, since the trapped gas swells on heating. The greatest difficulty arises from an insoluble atmosphere. In this case the pores act as pressurized balloons, resisting densification. Higher temperatures enlarge the balloons, leading to component swelling. A trapped atmosphere that is soluble in the solid allows the small pores to shrink, while the large pores grow. Pore coarsening leads to reduced gas pressure in the pores and swelling, as evident by the density versus time data plotted earlier in Figure 8.12; here the inherent ZnO vapor pressure increased porosity [66].

Agglomerated powders have small pores inside the agglomerates and large pores between the agglomerates. These bimodal pore structures are difficult to sinter. The small pores disappear early but the large pores grow [67].

Pores spheroidize at elevated temperatures. In ductile materials, the pores are initially flattened by compaction, but on heating they spheroidize accompanied by shrinkage perpendicular to the compression axis, but swelling along the compression axis. Component distortion results from this differential shrinkage.

COARSENING INTERACTIONS

A small grain size represents stored energy in the form of interface area per unit volume. Grain growth reduces the grain boundary energy. The motion of a grain boundary depends on the rate of diffusion across the boundary. Simultaneously pores attached to the grain boundaries exhibit several possible trajectories:

- Pores retard grain growth, causing boundary bowing due to pore drag.
- Pores with high mobility move and coarsen with grain boundaries migration.
- Grain boundaries break away from the pores, leaving pores isolated in the grains.

Early in sintering the pores hinder grain growth. Grain boundary motion associated with grain growth leads to pore drag by the moving grain boundary. In response to the drag, the pores move by volume diffusion, surface diffusion, or evaporation-condensation [68]. With pore elimination, the rate of grain growth increases to a point where the boundaries eventually break away from the pores. Pores on the grain boundaries shrink faster than pores inside the grain [69]. Thus, avoiding grain boundary separation from the pores is a target manipulation during heating.

Large pores are unable to move with grain boundaries, so they become stranded. An alternative view is to consider the number of grains surrounding a pore; a relatively small pore compared to the grain size is desirable. Pore mobility depends on transport processes across and around the pore, mostly surface diffusion and evaporation-condensation, while grain boundary mobility depends on transport across the boundary. Both are thermally activated and have similar, but not the same activation energies. Once grain growth starts, large grains sweep through the microstructure, leaving pores stranded in their wake as porosity declines below about 5%.

Realizing that grain growth and pore attachment are critical to densification leads to techniques to control grain boundary and pore mobility. Additives are one common means to control mobility, especially if the additive segregates to the grain boundary. The addition of 0.1 wt.% MgO to Al_2O_3 is one example. Magnesia slows grain growth. Alternatively, the addition of FeO to Al_2O_3 has the opposite effect. The critical condition for the MgO benefit depends on the CaO impurity content. Calcia segregates to grain boundaries and forms a viscous grain boundary phase. When the MgO:CaO ratio is over a critical value, CaO segregation is avoided, enabling full density [70].

Nanoscale powders provide a means of lowering sintering temperatures. For example, with tungsten, first sintering is detected at 825°C (1098 K) for 31 nm powder, 725°C (998 K) for 16 nm powder, and 625°C (798 K) for 9 nm powder. A progressively lower rate of grain growth comes with this temperature reduction, so nanoscale powders offer some promise of new property combinations. A change to a new, smaller particle size D_N is accommodated by a change to a new, lower temperature (absolute) T_N to attain equivalent sintering as with an old temperature of T_O for particle size D_O as follows:

$$T_N = \frac{1}{\frac{1}{T_O} - \frac{R\,m}{Q} \ln\left(\frac{D_N}{D_O}\right)} \tag{8.12}$$

where R is the gas constant, Q is the activation energy, and m is a mechanism dependent particle size constant typically near 3 to 4. A smaller particle size lowers the sintering temperature to obtain an equivalent degree of sintering, but with less grain growth because the grain growth rate parameter is smaller at lower temperatures.

A natural affinity of pores to grain boundaries occurs since the configurational energy is lower. At high sintered densities pores cluster with the largest grains. The probability of pore-grain boundary attachment is up to 5.7 times that for random contact. Consequently, the relation between grain size G, pore diameter d, and fractional density f is as follows:

$$\frac{G}{d} = \frac{K}{\psi(1-f)} \tag{8.13}$$

where ψ represents the ratio of attached pores to randomly placed pores, and K is a geometric constant. Values of ψ range from 1.7 to 5.7 for various materials and are essentially constant during sintering.

Curiously, sintering densification results in an increasing pore size [71,72]. Since the densification rate depends on the inverse of the pore size, enlarged pores slow densification. Usually the grain size to pore size ratio needs to be larger than 2 to sustain densification. Thus, agglomerated powders with broad pore size distributions are more difficult to sinter. Smaller pores, higher green densities, and narrow pore size distributions enable rapid sintering densification; consequently, narrow particle size distributions prove easier to sinter [73].

The small grain size and low porosity solution for pore mobility M_P is as follows:

$$M_P = \frac{\pi\, \delta\, \Omega\, D_S}{4RT} \frac{S_{SV} S_{GB}}{(1-f)^2} \tag{8.14}$$

where Ω is the atomic volume, δ is the grain boundary width, D_S is the surface diffusivity, R is the gas constant, T is the absolute temperature, f is the fractional density, and S is the surface or interface area for SV = solid-vapor and GB = grain boundary. In intermediate stage sintering, the shape factor B is about 70. Pore-boundary separation occurs when the pore size is increasing because there are fewer pores to slow grain growth. The resistance to grain growth offered by the pores decreases with densification. The natural variation of grain size and pore size in the microstructure leads to a range of breakaway conditions, so pore-boundary separation is not uniform within the microstructure.

SUMMARY

Modified coarsening rates occur when two solid phases coexist. The second phases might be dispersoids, reinforcing fibers, or insoluble liquids. Insoluble phases

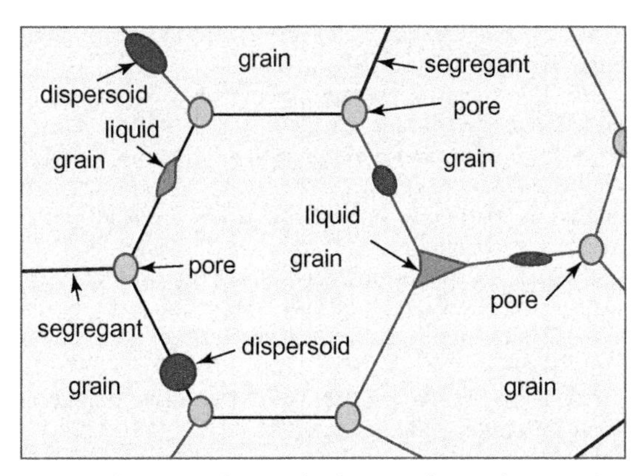

Figure 8.13 A schematic to illustrate the multiple interfaces that might impact microstructure coarsening during sintering. Shown on this sketch are pores, segregated species, dispersoids, and liquid films. A treatment of coarsening is difficult since so many interactions must be understood.

retard grain growth since the grain boundaries area increases on breaking away from the second phase. To illustrate the principle, the schematic microstructure in Figure 8.13 includes pores, liquid pools, dispersoids, and segregants at grain contacts.

Coarsening is avoidable under ideal conditions. Monosized spherical powders packed in an ideal array provide a means to avoid pore and grain coarsening [74,75]. Otherwise, microstructure imbalances initiate coarsening, wherein large features consume small features—the rich get richer.

The grain size to pore size ratio is an important control parameter. Large pores grow, so defects introduced in the component forming process are near impossible to sinter away and require pressure-assisted sintering. On the other hand, small uniform pores (compared to the particle size) are desirable and enable full density. The best measure of coarsening is via sintered properties. A property decline with longer sintering times or higher sintering temperatures indicates microstructure coarsening.

REFERENCES

[1] G.F. Bolling, Remarks on sintering kinetics, J. Amer. Ceram. Soc. 48 (1965) 168–169.

[2] A. Mocellin, W.D. Kingery, Microstructural changes during heat treatment of sintered Al_2O_3, J. Amer. Ceram. Soc. 56 (1973) 309–314.

[3] J. Gurland, Observations on the structure and sintering mechanism of cemented carbides, Trans. TMS-AIME 215 (1959) 601–608.

[4] J.A. Varela, O.J. Whittemore, M.J. Ball, Structural evolution during the sintering of SnO_2 and SnO_2–2 mole % CuO, in: G.C. Kuczynski, D.P. Uskokovic, H. Palmour, M.M. Ristic (Eds.), Sintering '85, Plenum Press, New York, NY, 1987, pp. 259–268.

[5] A. Xu, A.A. Solomon, The effects of grain growth on the intergranular porosity distribution in hot pressed and swelled UO_2, in: J.A. Pask, A.G. Evans (Eds.), Ceramic Microstructures '86, Plenum Press, New York, NY, 1988, pp. 509–518.

[6] A.V. Laptev, S.S. Ponomarev, L.F. Ochkas, Solid-phase consolidation of fine-grained WC-16% Co hardmetal, J. Adv. Mater. 33 (3) (2001) 42–51.

[7] H. Han, P.Q. Mantas, A.M.R. Senos, Sintering kinetics of undoped and Mn-doped zinc oxide in the intermediate stage, J. Amer. Ceram. Soc. 88 (2005) 1773–1778.

[8] D.J. Chen, M.J. Mayo, Densification and grain growth of ultrafine 3 mol.% Y_2O_3-ZrO_2 ceramics, Nano. Mater. 2 (1993) 469–478.

[9] J.L. Shi, Relation between coarsening and densification in solid-state sintering of ceramics: experimental test on superfine zirconia powder compacts, J. Mater. Res. 14 (1999) 1389–1397.

[10] Y.J. Lin, K.S. Hwang, Swelling of copper powders during sintering of heat pipes in hydrogen containing atmospheres, Mater. Trans. 51 (2010) 2251–2258.

[11] E.G. Zukas, H. Sheinberg, Sintering mechanisms in the 95% W – 3.5% Ni – 1.5% Fe composite, Powder Tech. 13 (1976) 85–96.

[12] H.Y. Kim, S.H. Lee, H.G. Kim, J.H. Ryu, H.M. Lee, Molecular dynamic simulation of coalescence between silver and palladium clusters, Mater. Trans. 48 (2007) 455–459.

[13] K.Z.Y. Yen, T.K. Chaki, A dynamic simulation of particle rearrangement in powder packings with realistic interactions, J. Appl. Phys. 71 (1992) 3164–3173.

[14] S. Arcidiacono, N.R. Bieri, D. Poulikakos, C.P. Grigoropoulos, On the coalescence of gold nanoparticles, Inter. J. Multi. Flow 30 (2004) 979–994.

[15] M. Yeadon, J.C. Yang, R.S. Averback, J.W. Bullard, J.M. Gibson, Sintering of silver and copper nanoparticles on (001) copper observed by *in situ* ultrahigh vacuum transmission electron microscopy, Nano. Mater. 10 (1998) 731–739.

[16] S. Takajo, W.A. Kaysser, G. Petzow, Analysis of particle growth by coalescence during liquid phase sintering, Acta Metall. 32 (1984) 107–113.

[17] G. Petzow, S. Takajo, W.A. Kaysser, Application of quantitative metallography to the analysis of grain growth during liquid phase sintering, in: J.L. McCall, J.H. Steele (Eds.), Practical Applications of Quantitative Metallography, Amer. Society for Testing and Materials, Philadelphia, PA, 1984, pp. 29–40.

[18] L. Froschauer, R.M. Fulrath, Direct observation of liquid-phase sintering in the system iron-copper, J. Mater. Sci. 10 (1975) 2146–2155.

[19] L. Froschauer, R.M. Fulrath, Direct observation of liquid-phase sintering in the system tungsten carbide-cobalt, J. Mater. Sci. 11 (1976) 142–149.

[20] W.J. Huppmann, R. Riegger, Modeling of rearrangement processes in liquid phase sintering, Acta Metall. 23 (1975) 965–971.

[21] Y. Liu, R.G. Iacocca, J.L. Johnson, R.M. German, S. Kohara, Microstructural anomalies in a W-Ni alloy liquid phase sintered under microgravity conditions, Metall. Mater. Trans. 26A (1995) 2484–2486.

[22] L.J. Perryman, P.J. Goodhew, A description of the migration and growth of cavities, Acta Metall. 36 (1988) 2685–2692.

[23] C.V. Santilli, S.H. Pulcinelli, J.A. Varela, J.P. Bonnet, Effect of green compact pore size distribution on the sintering of alpha-Fe_2O_3, in: D.P. Uskokovic, H. Palmour, R.M. Spriggs (Eds.), Science of Sintering, Plenum Press, New York, NY, 1989, pp. 519–527.

[24] N. Wade, M. Ohara, O. Suda, Heterogeneous sintering behavior of injection molded SUS304 steel, J. Japan Soc. Powder Powder Metall. 40 (1993) 384–387.

[25] R.M. German, M. Bulger, The effects of bimodal particle size distribution on sintering of powder injection molded compacts, Solid State Phen. 25 (1992) 55–61.

[26] N. Shinohara, M. Okumiya, T. Hotta, K. Nakahira, M. Naito, K. Uematsu, Morphological changes in process related large pores of granular compacted and sintered alumina, J. Amer. Ceram. Soc. 83 (2000) 1633–1640.

[27] I.M. Robertson, G.B. Schaffer, Suitability of nickel as alloying element in titanium sintered in solid state, Powder Metall. 52 (2009) 225–232.

[28] Y. Liu, B.R. Patterson, Frequency of pore location in sintered Al_2O_3, J. Amer. Ceram. Soc. 75 (1992) 2599–2600.

[29] B.R. Patterson, Y. Liu, Quantification of grain boundary-pore contact during sintering, J. Amer. Ceram. Soc. 73 (1990) 3703–3705.

[30] J. Langer, M.J. Hoffmann, O. Guillon, Electric field-assisted sintering in comparison with the hot pressing of yttria stabilized zirconia, J. Amer. Ceram. Soc. 94 (2010) 24−31.
[31] R.M. German, A. Bose, S.S. Mani, Sintering time and atmosphere influences on the microstructure and mechanical properties of tungsten heavy alloys, Metall. Trans. 23A (1992) 211−219.
[32] M. Hillert, On the theory of normal and abnormal grain growth, Acta Metall. 13 (1965) 227−238.
[33] W.J. Boettinger, P.W. Voorhees, R.C. Dobbyn, H.E. Burdette, A study of the coarsening of liquid-solid mixtures using synchrotron radiation microradiography, Metall. Trans. 18A (1987) 487−490.
[34] R.L. Coble, T.K. Gupta, Intermediate stage sintering, in: G.C. Kuczynski, N.A. Hooton, C.F. Gibbon (Eds.), Sintering and Related Phenomena, Gordon and Breach, New York, NY, 1967, pp. 423−441.
[35] G.C. Nicholson, Grain growth in magnesium oxide containing a liquid phase, J. Amer. Ceram. Soc. 48 (1965) 525−528.
[36] H. Gleiter, B. Chalmers, Grain boundary melting, High-Angle Grain Boundaries, Pergamon Press, Oxford, UK, 1972, pp. 113−126.
[37] D.M. Owen, A.H. Chokshi, An evaluation of the densification characteristics of nanocrystalline materials, Nano. Mater. 2 (1993) 181−187.
[38] T. Tanase, Development of high performance carbide tools, J. Japan Soc. Powder Powder Metall. 54 (2007) 243−250.
[39] P. Maheshwari, Z.Z. Fang, H.Y. Sohn, Early Stage sintering densification and grain growth of nanosized WC-Co powders, Inter. J. Powder Metall. 43 (2) (2007) 41−47.
[40] H.R. Lee, D.J. Kim, N.M. Hwang, D.Y. Kim, Role of vanadium carbide additive during sintering of WC-Co: mechanism of grain growth inhibition, J. Amer. Ceram. Soc. 86 (2003) 152−154.
[41] A. Adorjan, W.D. Schubert, A. Schon, A. Bock, B. Zeiler, WC grain growth during the early stages of sintering, Inter. J. Refract. Met. Hard Mater. 24 (2006) 365−373.
[42] H.Y. Suzuki, K. Shinozaki, M. Murai, H. Kuroki, Quantitative analysis of microstructure development during sintering of high purity alumina made by high speed centrifugal compaction process, J. Japan Soc. Powder Powder Metall. 45 (1998) 1122−1130.
[43] P. Lu, R.M. German, Multiple grain growth events in liquid phase sintering, J. Mater. Sci. 36 (2001) 3385−3394.
[44] R.D. McIntyre, The effect of HCl-H_2 sintering atmospheres on the properties of compacted iron powder, Trans. Quart. Amer. Soc. Met. 57 (1964) 351−354.
[45] R.D. McIntyre, The effect of HCl-H_2 sintering atmospheres on properties of compacted tungsten powder, Trans. Quart. Amer. Soc. Met. 56 (1963) 468−476.
[46] M.J. Readey, D.W. Readey, Sintering TiO_2 in HCl atmospheres, J. Amer. Ceram. Soc. 70 (1987) C358−C361.
[47] T. Quadir, D.W. Readey, Microstructure development of zinc oxide in hydrogen, J. Amer. Ceram. Soc. 72 (1989) 297−302.
[48] C.M. Kipphut, A. Bose, S. Farooq, R.M. German, Gravity and configurational energy induced microstructural changes in liquid phase sintering, Metall. Trans. 19A (1988) 1905−1913.
[49] J.L. Johnson, A. Upadhyaya, R.M. German, Microstructural effects on distortion and solid-liquid segregation during liquid phase sintering under microgravity conditions, Metall. Mater. Trans. 29B (1998) 857−866.
[50] J.L. Johnson, L.G. Campbell, S.J. Park, R.M. German, Grain growth in dilute tungsten heavy alloys during liquid phase sintering under microgravity conditions, Metall. Mater. Trans. 40A (2009) 426−437.
[51] J.L. Johnson, R.M. German, Solid-state contributions to densification during liquid phase sintering, Metall. Mater. Trans. 27B (1996) 901−909.
[52] M.J. Readey, D.W. Readey, Sintering of ZrO_2 in HCl atmospheres, J. Amer. Ceram. Soc. 69 (1986) 580−582.
[53] P. Wynblatt, N.A. Gjostein, Particle growth in model supported metal catalysts − 1. Theory, Acta Metall. 24 (1976) 1165−1174.
[54] I.H. Moon, J.H. Kim, M.J. Suk, K.M. Lee, J.K. Lee, Observation of W particle growth in a W powder compact during sintering in a non-reducing atmosphere, J. Mater. Syn. Proc. 1 (1993) 309−315.

[55] D.W. Readey, D.J. Aldrich, M.A. Ritland, Vapor transport and sintering, in: R.M. German, G.L. Messing, R.G. Cornwall (Eds.), Sintering Technology, Marcel Dekker, New York, NY, 1996, pp. 53–60.

[56] D. Weaire, Some remarks on the arrangement of grains in a polycrystal, Metallog 7 (1974) 157–160.

[57] D.A. Aboav, The arrangement of grains in a polycrystal, Metallog 3 (1970) 383–390.

[58] M. Blanc, A. Mocellin, Grain coordination in plane sections of polycrystals, Acta Metall. 27 (1979) 1231–1237.

[59] D.A. Aboav, T.G. Langdon, The shape of grains in a polycrystal, Metallog 2 (1969) 171–178.

[60] A. Tewari, A.M. Gokhale, R.M. German, Effect of gravity on three-dimensional coordination number distribution in liquid phase sintered microstructures, Acta Mater. 47 (1999) 3721–3734.

[61] Z. Fang, B.R. Patterson, M.E. Turner, Influence of particle size distribution on coarsening, Acta Metall. Mater. 40 (1992) 713–722.

[62] Z. Fang, B.R. Patterson, Experimental investigation of particle size distribution influence on diffusion controlled coarsening, Acta Metall. Mater. 41 (1993) 2017–2024.

[63] R.M. German, Coarsening in sintering – grain shape distribution, grain size distribution, and grain growth kinetics in solid-pore systems, Crit. Rev. Solid State Mater. Sci. 35 (2010) 263–305.

[64] R.T. DeHoff, A geometrically general theory of diffusion controlled coarsening, Acta Metall. Mater. 39 (1991) 2349–2360.

[65] L. Zeng, B.R. Patterson, Growth path envelope analysis of ostwald ripening, Advances in Powder Metallurgy and Particulate Materials, vol. 2, Metal Powder Industries Federation, Princeton, NJ, 1993, pp. 195–202.

[66] T. Senda, R.C. Bradt, Grain growth of zinc oxide during the sintering of zinc oxide – antimony oxide ceramics, J. Amer. Ceram. Soc. 74 (1991) 1296–1302.

[67] W.H. Rhodes, Agglomerate and particle size effects on sintering yttria-stabilized zirconia, J. Amer. Ceram. Soc. 64 (1981) 19–22.

[68] W. Villanueva, G. Amberg, Some generic capillary driven flows, Inter. J. Multi. Flow 32 (2006) 1072–1086.

[69] J.E. Burke, Role of grain boundaries in sintering, J. Amer. Ceram. Soc. 40 (1957) 80–85.

[70] S.I. Bae, S. Baik, Critical concentration of MgO for the prevention of abnormal grain growth in alumina, J. Amer. Ceram. Soc. 77 (1994) 2499–2504.

[71] R. Watanabe, Y. Masuda, Pinning effect of residual pores on the grain growth in porous sintered metals, J. Japan Soc. Powder Metall. 29 (1982) 151–153.

[72] J. Zheng, J.S. Reed, Effects of particle packing characteristics on solid-state sintering, J. Amer. Ceram. Soc. 72 (1989) 810–817.

[73] T.S. Yeh, M.D. Sacks, Effect of green microstructure on sintering of alumina, in: C.A. Handwerker, J.E. Blendell, W.A. Kaysser (Eds.), Sintering of Advanced Ceramics, American Ceramic Society, Westerville, OH, 1990, pp. 309–331.

[74] E.A. Barringer, R. Brook, H.K. Bowen, The sintering of monodisperse TiO_2, in: G.C. Kuczynski, A.E. Miller, G.A. Sargent (Eds.), Sintering and Heterogeneous Catalysis, Plenum Press, New York, NY, 1984, pp. 1–21.

[75] E.A. Barringer, H.K. Bowen, Formation, packing, and sintering of monodisperse TiO_2 powders, J. Amer. Ceram. Soc. 65 (1982) C199–C201.

CHAPTER NINE

Sintering With a Liquid Phase

OVERVIEW

Most industrial sintering involves the formation of a liquid during the heating cycle. When properly engineered, the liquid rapidly bonds the grains. Liquid phase sintering typically starts with mixed particles, of which one remains solid while the other forms a liquid. The technology arose before any basic understanding of the process, and is widely used in everything from bronze bearings to aluminum nitride heat sinks. The dominant use is in the consolidation of hard materials, such as TiC-Ni and WC-Co. A critical advantage derives from diffusion in a liquid, which is often hundreds to thousands of times faster than solid state diffusion. Most of the products are sintered to full density, so this technique is a favorite for sintering materials otherwise difficult to fabricate.

Persistent liquid phase sintering implies the liquid remains in existence as long as the peak temperature is held. An example sintered microstructure is given in Figure 9.1. This is a stainless steel sintered with a boron addition that formed the liquid phase between grains. It is used for hardware applications, such as valves and fittings for industrial gas systems. The structure consists of solid grains with sinter necks. The solidified liquid phase fills gaps between the grains, so the structure has no porosity. Several events occur to remove porosity and bond the solid into a sintered skeleton.

Liquid phase sintering variants include persistent, transient, reactive, and supersolidus approaches. They differ in a few characteristics, namely:

- persistent – mixed powders, solid soluble in liquid, liquid low solubility in solid; example WC-Co to form metal cutting tools
- transient – mixed powders, liquid soluble in solid, liquid disappears after forming; example Cu-Sn to form porous bronze bearings
- reactive – mixed powders, highly exothermic reaction, stoichiometric product; example $MoSi_2$ for use as furnace heating elements
- supersolidus – prealloyed powder, solid soluble in liquid, liquid low solubility in solid; example tool steels for wear applications.

Each variant is applied commercially, and estimates are that 90% of sintered commercial products involve formation of a liquid during sintering.

Sintering: From Empirical Observations to Scientific Principles
DOI: http://dx.doi.org/10.1016/B978-0-12-401682-8.00009-4

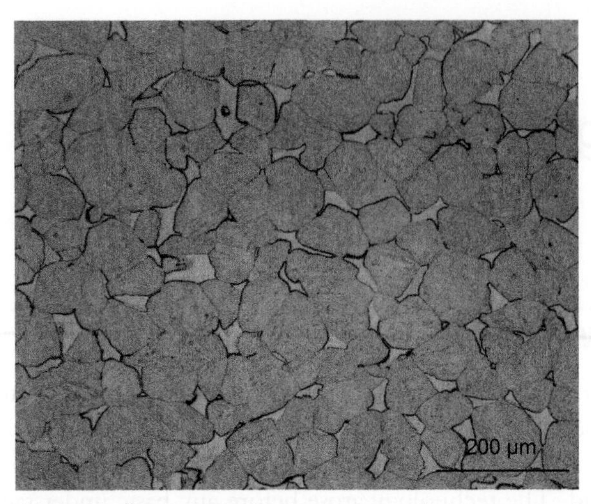

Figure 9.1 An example liquid phase sintered microstructure for a vacuum sintered stainless steel doped with boron to form a liquid phase during sintering. The solidified liquid has filled the gaps between the solid grains which are bonded to each other.

CONCEPTUAL DEVELOPMENTS

Early sintered ceramics were made from a mixture of crystalline oxide grains and a glassy matrix. The glass softened at high temperature to form a viscous liquid that enabled densification of the mixture. Sometimes this is termed viscous phase sintering, since densification depends on forming a viscous glass bond [1].

Important advances in liquid phase sintering arose with qualitative and then quantitative treatments, as summarized elsewhere [2–9].

Liquid phase sintering usually starts with mixed or milled powders. The composition and sintering temperature are selected based on the phase diagram, as indicated in Figure 9.2. Favorable attributes are an increase in solid solubility in the newly formed melt, solid solubility in the liquid, but low liquid solubility in the solid, and decreasing liquidus and solidus temperatures with increased additive. Since diffusion rates scale with melting temperature, the melting temperature decrease gives a dramatic improvement in sintering. Higher temperatures play a double role in increasing the liquid quantity and diffusion rate, so temperature is a critical control parameter. More liquid aids densification, but excessive liquid causes distortion; most common is less than 15 vol.% liquid to retain shape during sintering.

The sequence of steps is outlined in Figure 9.3. Starting with mixed powders, the compact is heated to form a liquid, which spreads to penetrate between the solid grains, dissolving any sinter bonds formed during heating. Dissolution of the solid

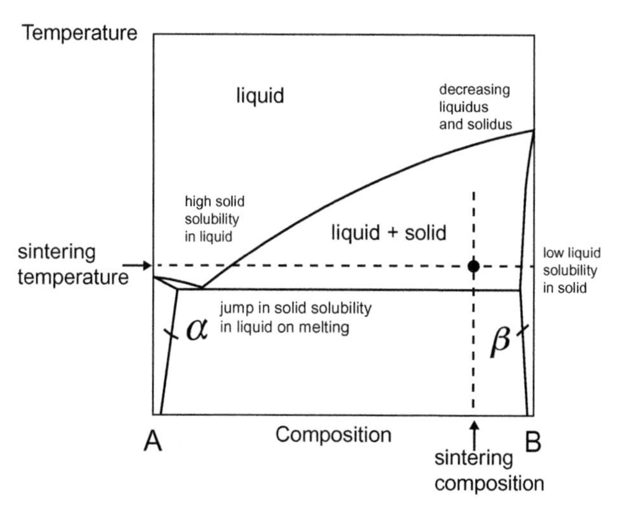

Figure 9.2 A hypothetical binary phase diagram showing the characteristics favorable for liquid phase sintering the B component by the addition of A.

bonds results in a semisolid, viscous mixture of solid grains and wetting liquid. The softened structure undergoes rearrangement due to lubrication of the grains and capillary attraction from the wetting liquid. Pendular bonds induce a capillary force to pull the grains together, resulting in touching grains with liquid bonds as illustrated in Figure 9.4. Grain rearrangement induces higher packing density, releasing liquid to fill any remaining pores. Small particles result in a high capillary stress; reaching upwards of 100 MPa for 1 μm particles. This large local stress makes the structure increasingly dense [10].

Usually the solid is soluble in the liquid, so it moves rapidly within it. Fast diffusion leads to rapid sinter densification, grain growth, and grain reshaping. As sintering progresses, densification results in isolated pores, as sketched in Figure 9.5. Now the solid grains are bonded together, with liquid filling much of the space between the grains.

The densification trajectory consists of overlapping stages as illustrated in Figure 9.6. Early densification occurs by solid-state sintering during heating. Rapid densification follows with melt formation. As the liquid spreads, solution-reprecipitation becomes active. Finally as full density is reached solid skeleton sintering is dominant.

Pores have a high surface energy and are preferentially filled during sintering shrinkage. Solid diffusion though the liquid is fast if the solid is soluble in the liquid. The change in grain size by solution-reprecipitation is most notable during liquid phase sintering, which is also responsible for grain reshaping. Once the solid bonds, the solid skeleton continues to densify while the liquid remains in the gaps between the solid grains. Eventually the solid skeleton is sufficiently dense that all pores are filled by liquid, which solidifies on cooling.

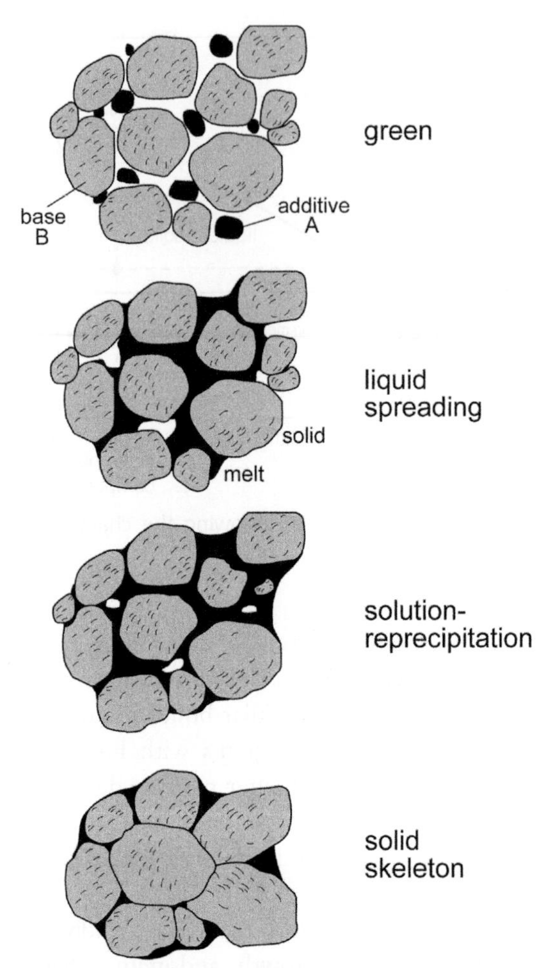

Figure 9.3 A schematic of the microstructure changes during liquid phase sintering (LPS), starting with mixed powders and pores between the particles. During heating the particles sinter, but when a melt forms and spreads the solid grains rearrange. Subsequent densification is accompanied by coarsening. For many products there is pore annihilation as diffusion in the liquid accelerates grain shape changes and facilitates pore removal.

The actual microstructure pathway depends on the solid-liquid solubility. The most common situation has these characteristics:

1. Newly formed liquid wets the solid grains due to solid solubility in the liquid.
2. Solid grains partly dissolve into the liquid to reach the solubility limit.
3. Liquid is low in solubility in the solid.
4. Solid alloying with the additive causes a decrease in solidus and liquidus.

These characteristics are evident in phase diagrams, such as that given earlier in Figure 9.2. Characteristic 1 corresponds to the solubility jump from solid solution to

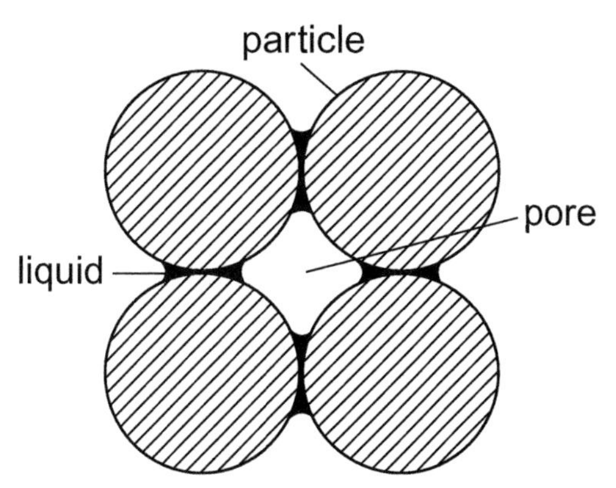

Figure 9.4 Pendular liquid bonds form at the particle contacts, and the resulting capillary force induces grain rearrangement while providing strength prior to sinter neck growth.

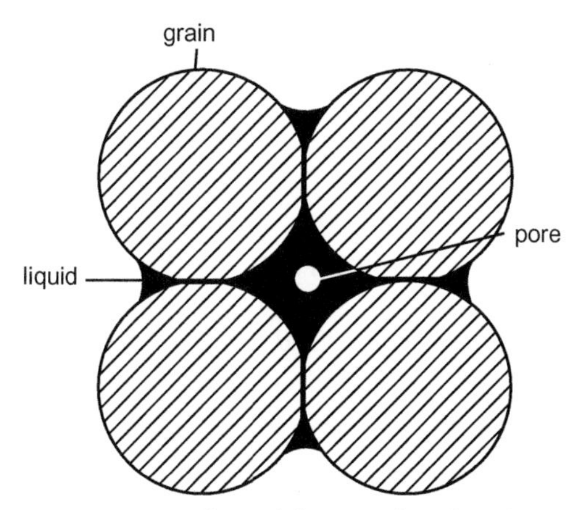

Figure 9.5 As densification progresses, the solid grains bond and shape to accommodate the release of liquid to fill the pores. The funicular state occurs as the isolated spherical pores collapse.

liquid; characteristic 2 is evident from the solid's solubility at the sintering temperature; characteristic 3 corresponds to the low solubility for the liquid species in the solid at the peak temperature; and characteristic 4 corresponds to liquid segregation to the interparticle boundaries.

The sequence of events during liquid phase sintering depends on the liquid volume. If there is no liquid, then only solid-state sintering occurs. With a high liquid content the system densifies as soon as liquid forms and flows to fill pores. In between these extremes, as plotted in Figure 9.7, a sequence of steps occurs. The details of these steps depend on individual system characteristics, as discussed in this chapter.

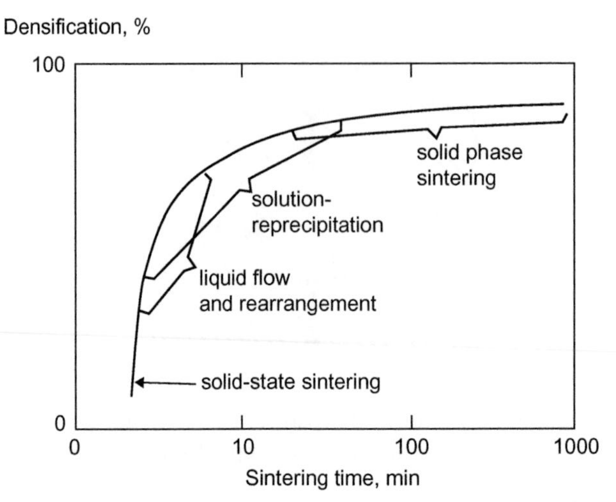

Figure 9.6 A schematic of the overlapping events in liquid phase sintering; densification is rapid at short times since chemical gradients induce early solid diffusion. Rearrangement occurs when liquid forms and solution-reprecipitation quickly follows. Final sintering of the solid skeleton is a slower process. As the particle size, liquid content, and other factors are adjusted the shape and placement of this curve changes.

Because of rapid densification, liquid phase sintering is used in the fabrication of hard materials that are difficult to form by other approaches. The WC-Co system is an example; the eutectic near 1310°C (1583 K) forms a liquid in which the solid is soluble. Accordingly, via rearrangement, solution-reprecipitation, and solid skeletal sintering the WC grains are sintered into a dense component with exceptional hardness and wear resistance.

As well as mixed powders, liquid phase sintering is also possible using alloy powders. Again a semisolid structure is formed at the peak temperature. This approach is used for tool steels, stainless steels, cobalt-chromium alloys, and nickel superalloys. Another variant involves a transient liquid, where the liquid forms during heating to then dissolve into the solid. Tin-copper bearings are formed in this way, as are silver-mercury amalgams (fillings). Some systems are insoluble, such as Mo-Cu, Co-Cu, WC-Ag, WC-Cu, and W-Cu. Here the solid skeleton forms with liquid filling the pores, but the liquid has little role in determining the densification rate.

MICROSTRUCTURE DEVELOPMENT

The composite microstructure resulting from liquid phase sintering offers many opportunities to adjust properties. This section surveys some options.

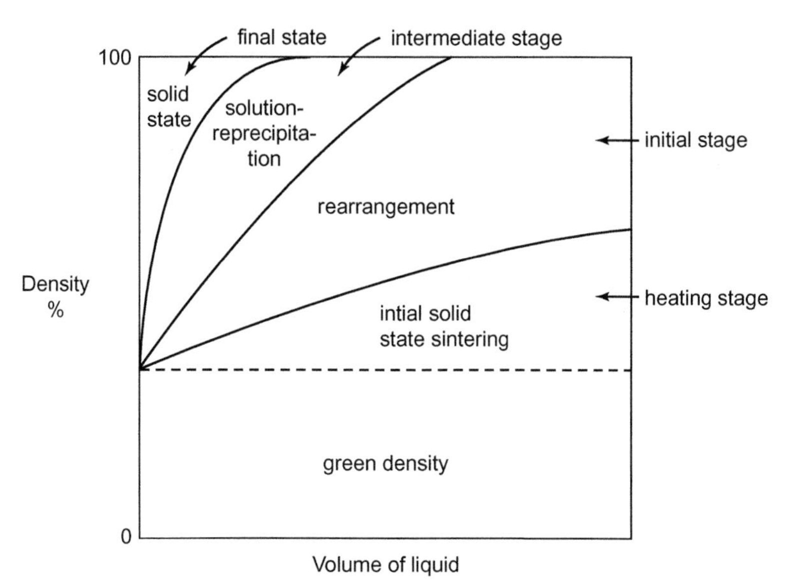

Figure 9.7 A map of the density and liquid content interplay to determine the densification events and how the liquid content plays an important role.

Typical Microstructures

Microstructure is a signature of the processing. It continuously evolves due to bonding and coarsening. Indeed, a "final" microstructure is not seen in liquid phase sintering. What is observed is a glimpse of the evolving structure. With prolonged sintering, the terminal condition would consist of a single grain with an associated liquid cap, such as is illustrated in Figure 9.8 for 20 vol.% liquid and a 20° contact angle [11]. Prior to this condition, the microstructure is characterized by its porosity, pore size, and grain size. The typical sintering for 1 h or less captures the body with a mixture of small grains and a network of liquid surrounding the grains. The liquid network is discernible by selective leaching, giving the structure shown in Figure 9.9. This microstructure undergoes significant changes during the sintering process. Solid interfaces move with velocities in the μm/s range, liquid penetration of grain boundaries is in the μm/s range, and solid coarsening is in the $\mu m^3/s$ range. However, years are required to transform micrometer sized particles into millimeter sized grains. During a typical sintering cycle, thousands of particles coalesce to form each grain.

The sintered microstructure determines the properties. Models for liquid phase sintering link composition, processing, and properties. A homogeneous green structure improves sintering [12]. The amount and placement of the liquid phase is also significant. Most effective is placement of the liquid phase on the interface between the solid grains, possibly using coated powders [13–16]. Further alloying additions are used to modify wetting, diffusion, or hardening.

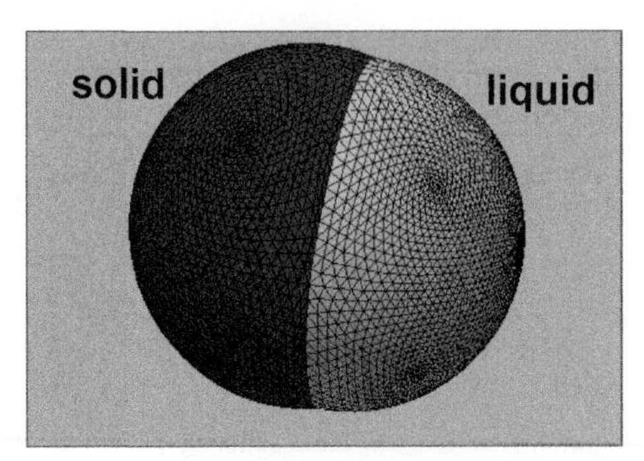

Figure 9.8 An example of the terminal microstructure for liquid phase sintering based on a minimum energy configuration [11]. The calculation is the lowest energy in a system of 80 vol.% solid −20 vol.% liquid at a contact angle of 20°. The liquid placement results in the minimum energy. Another low energy terminal configuration consists of a liquid sphere located inside the solid grains, and this is evident in several of the microstructures shown later in this chapter.

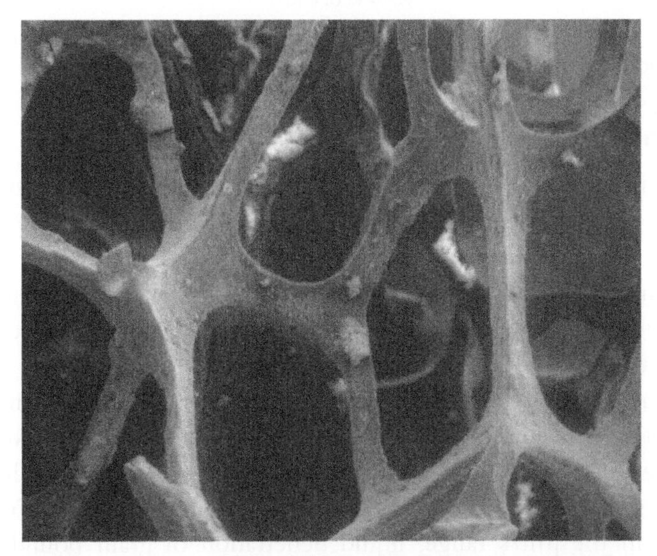

Figure 9.9 Scanning electron micrographs of the solidified liquid network after dissolution of the solid grains.

Contact Angle and Dihedral Angle

Both the contact angle and dihedral angle influence liquid phase sintering. The contact angle depends on the solid-liquid-vapor balance. Liquid spreading on the solid replaces solid-vapor interfaces with liquid-solid and liquid-vapor interfaces. Figure 9.10 contrasts good and poor wetting based on the contact angle θ, also known

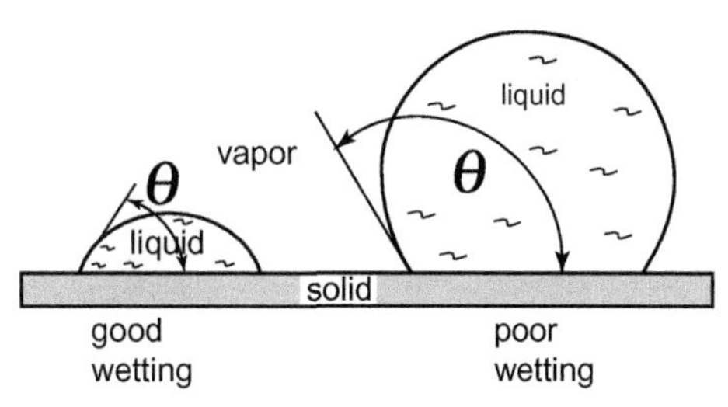

Figure 9.10 Contrast of wetting behavior for a liquid on a horizontal plane showing how a low contact angle supports wetting while a high contact angle resists wetting. Densification in liquid phase sintering requires a low contact angle to ensure that the grains are pulled together.

as the wetting angle. It is defined by the balance of three interfacial energies, γ_{SV}, γ_{SL}, and γ_{LV} as follows:

$$\gamma_{SV} = \gamma_{SL} + \gamma_{LV} \cos(\theta) \tag{9.1}$$

where the subscripts *S*, *L*, and *V* represent solid, liquid, and vapor, respectively. The contact angle depends on solubility or solid surface chemistry [17–19]. For example, the addition of Mo to the TiC-Ni system decreases the contact angle from 30° to 0°. A low contact angle induces the liquid to spread over the solid grains, providing a capillary force to induce densification. For small grains, contact stress can rival that obtained in pressure-assisted sintering [20].

A range of capillary conditions exist since the microstructure consists of various grain sizes, grain shapes, pore sizes, pore shapes, and liquid contents. A wetting liquid occupies the lowest energy configuration, so it preferentially flows into smaller pores and attacks gain boundaries. This gives nearly instantaneous rearrangement densification once the liquid forms [21,22]. The slow step is heat flow and melting in the powder compact.

A high contact angle indicates poor wetting, and the liquid exudes from the pores as pictured in Figure 9.11. This is a picture of liquid droplets on the surface after liquid phase sintering for a poor wetting liquid composition. Thus, depending on the contact angle, liquid formation causes densification or swelling.

Similar to the contact angle, the dihedral angle corresponds to a surface energy balance. In this case it is observed where a grain boundary intersects the liquid. As illustrated in Figure 4.10, the vertical components of the solid-liquid surface energy γ_{SL} balance the grain boundary energy γ_{SS} based on the dihedral angle ϕ as follows:

$$2\gamma_{SL} \cos\left(\frac{\phi}{2}\right) = \gamma_{SS} \tag{9.2}$$

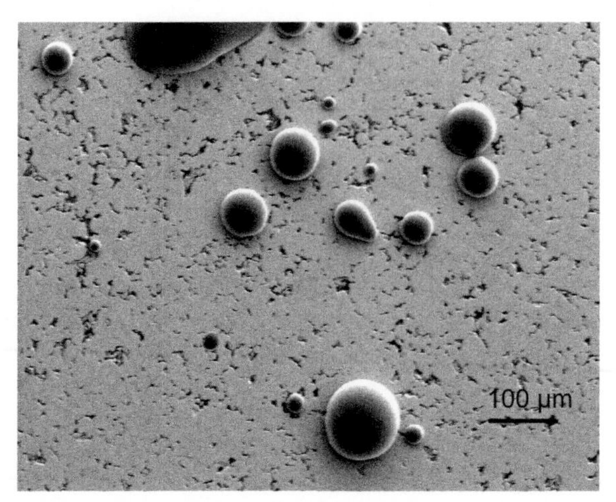

Figure 9.11 A scanning electron micrograph of the surface when the liquid is not wetting, showing exuded beads of liquid on the compact surface in the form of small spheres.

If the ratio of the solid-solid to solid-liquid surface energy is relatively high (>1.8), then the dihedral angle approaches 0° and liquid penetrates the solid grain boundaries.

Some solid-solid contacts form with low crystal misorientation, giving a low grain boundary energy and high dihedral angle. These contacts rotate to give grain growth by coalescence. Since grain boundary energy varies with crystallographic misorientation and chemical segregation, there is a distribution to the dihedral angles. Figure 9.12 plots the dihedral angle distribution for two Fe-Cu systems with 0.5 and 1.0% carbon [23]. Carbon is soluble in iron at the liquid phase sintering temperature. As is evident, the dihedral angle is distributed within the microstructure and varies with carbon content. Depending on which grain contact is examined, it is possible to find extremes of liquid on the grain boundaries (zero dihedral angle or very large dihedral angles).

Low dihedral angles and low contact angles promote liquid phase sintering, typically because the solid is soluble in the liquid.

Volume Fraction

The liquid content in sintering usually varies from 5 to 15 vol.%. Solid bonding increases with high solid contents. Solid grain bonds provide skeletal rigidity to retain component shape. Figure 9.13 is a photograph of right circular cylinders after liquid phase sintering using three solid-liquid ratios. All three compacts reached full density, but slumping increases with the lower solid content.

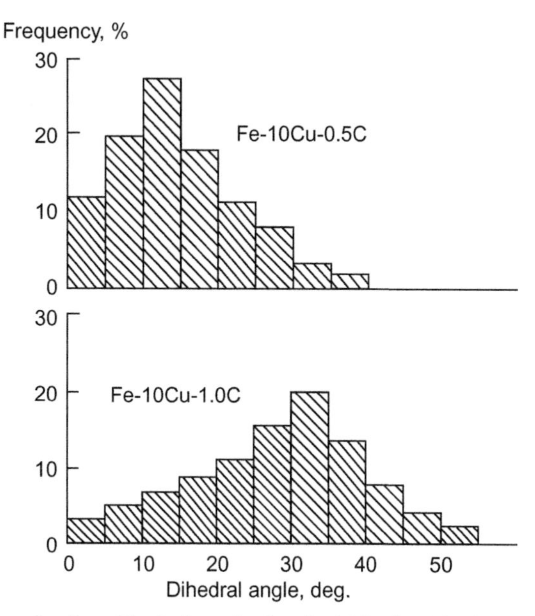

Figure 9.12 Distributions in the dihedral angle for Fe-10Cu liquid phase sintered compositions without and with carbon [23]. Carbon is soluble in solid iron at the sintering temperature and it increases the dihedral angle. The dihedral angle is a distributed parameter since it depends on the crystal misorientation at the grain contact, as evident in these two plots.

Generally the higher the liquid content, the faster full density is attained. A good example is seen in the time-dependent density data for alumina-glass mixtures in Figure 9.14 [24]. At the sintering temperature of 1600°C (1873 K) the sintered density progressively increases with the liquid content.

Porosity, Pore Size, and Pore Location

With proper processing, the pores are smaller than the grains, as is evident in Figure 9.15. Predictions exist for the ideal pore geometry but usually inhomogeneities exist. The dominant feature is the grains. Because of wetting, smaller pores fill first, delaying the filling of the larger pores [25,26]. Thus, the mean pore size increases while the porosity and number of pores decrease. Because of pore buoyancy, there is progressive migration of the pores to the top of the component.

In cases where the particles forming the melt are large, then large pores form. Figure 9.16 is a micrograph of a quenched example where a large particle melted and spread outward. Swelling results from this situation, with densification at long sintering times [27,28]. Swelling is favored when the liquid has solubility in the solid. Large pores fill late in sintering if there is no gas trapped in the pores [29]. During liquid phase sintering, a critical grain size and pore size combination arises due to grain

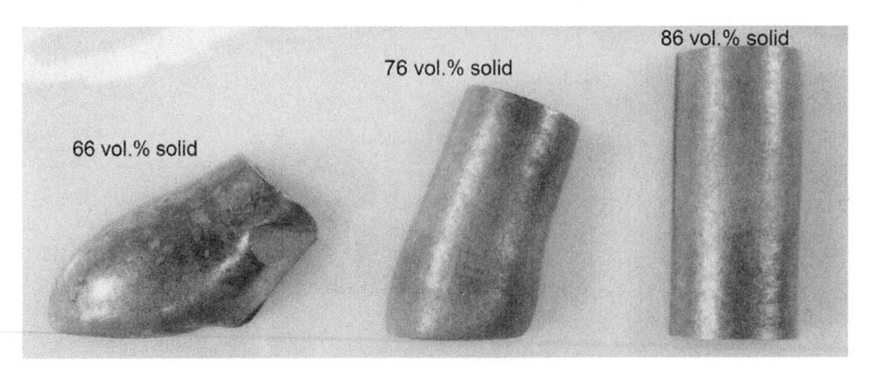

Figure 9.13 Liquid phase sintered tungsten alloys cylinders after sintering. These are 66, 76, and 86 vol.% solid and all have been sintered to full density. Shape distortion is resisted by a high solid content as evident in the 86 vol.% composition.

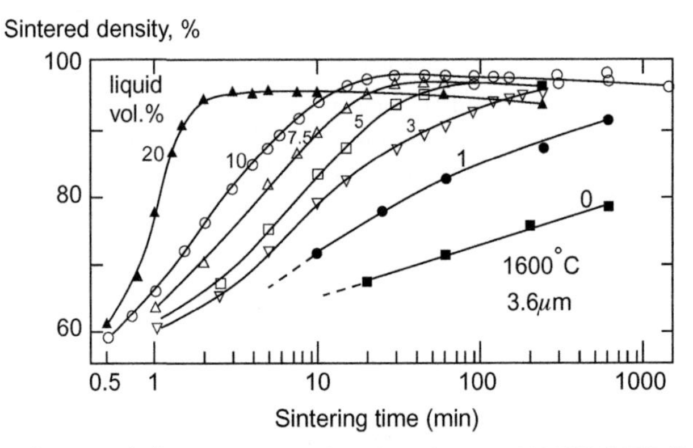

Figure 9.14 Fractional sintered density versus sintering time at 1600°C (1873 K) for alumina-glass mixtures based on a 3.6 μm alumina particle size. Here the volume fraction of liquid phase ranged from 20 vol.% glass to 0 vol.%, showing a progressive detriment in sintering densification (from left to right the curves are 20, 10, 7.5, 5, 3, 1, and 0 vol.% glass) [24].

growth, after which liquid flows to fill the large pores. This occurs when the ratio of the grain size G to pore size d is favorable:

$$\frac{G}{d} = \frac{\gamma_{SS}}{2\gamma_{SV}} = \cos\left(\frac{\phi}{2}\right) \tag{9.3}$$

where ϕ is the dihedral angle. Since grain size increases with sintering time, liquid filling of larger pores is delayed by grain growth. After refilling the pore appears as a liquid lake, such as that seen in Figure 9.17. These are defects, since the liquid is typically weak.

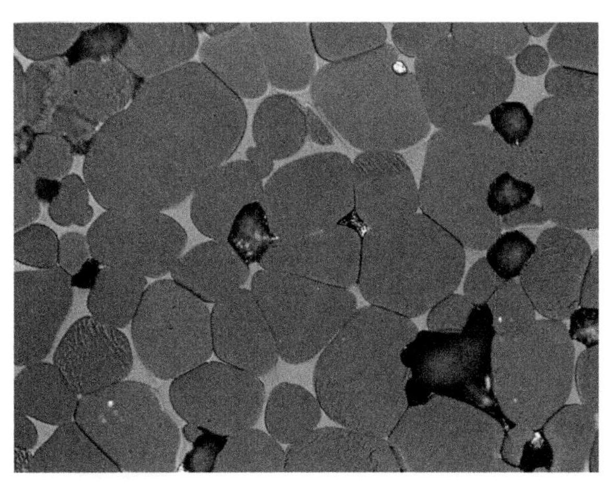

Figure 9.15 A quenched microstructure taken during liquid phase sintering. Note the pores are located on the solid-liquid interface during densification.

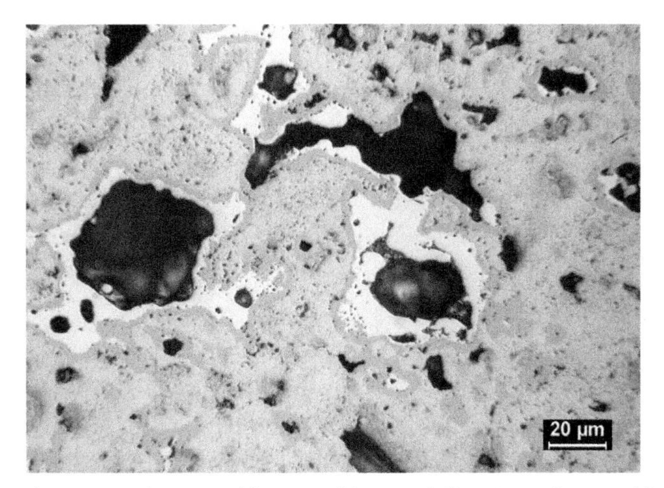

Figure 9.16 Pore formation due to additive melting and flow into the neighboring capillaries. This micrograph of Cu – 10% Sn shows dark pores where the tin grains were prior to melting, surrounded by molten tin, with a reaction layer at the copper interface.

Trapped gas in the pores acts to inhibit final densification. Gas-filled pores are spherical, reflecting the surface energy balance against the pressure of the gas in the pore P_G:

$$P_G = \frac{4\gamma_{LV}}{d_P} \tag{9.4}$$

It is common to seal the pores using an external pressure or by sintering in vacuum.

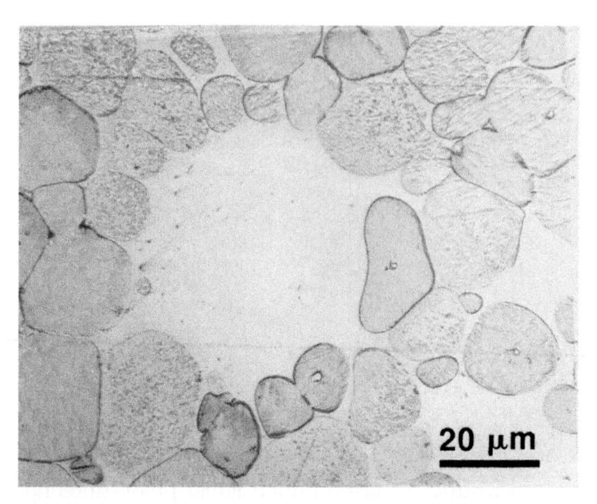

Figure 9.17 A liquid lake formed when a pore was filled during liquid phase sintering. This image is from 90W-7%Ni-3%Fe composition sintered at 1470°C (1743 K) for 30 minutes. Grain growth led to a condition that enabled liquid flow into the pore.

Thus, pore changes during liquid phase sintering start with irregular pores, that transform into tubular pores, and at about 5 to 8% porosity the pores spheroidize. Final pore closure depends on liquid release during grain growth. Unfortunately, some systems exhibit delayed pore generation due to reactions that produce an insoluble gas. An example is shown in Figure 9.18 for mullite ($3Al_2O_3 \cdot 2SiO_2$) sintering with an oxide liquid [30]. The peak density corresponds to open pore closure at 1300°C (1573 K), followed by a swelling of the gas-filled closed pores at higher temperatures.

Grain Shape

Grain shape in a liquid phase sintered microstructure depends on the liquid fraction, dihedral angle, and the anisotropy of the surface energy. Contacts between grains produce a flat face. A high solid content induces a high grain coordination number with more flat faces, while the shape and connectivity also depends on the dihedral angle [31,32]. Effectively, grain size depends on sintering time, while grain shape depends on liquid content and dihedral angle.

As the solid content increases, the grains take on shapes that release liquid to fill pores. This fitting together is grain shape accommodation. During sintering, the grains undergo both size and shape changes in the solution-reprecipitation process. The large grains coarsen and the small grains disappear. For isotropic solid-liquid surface energy and liquid contents over about 30 vol.%, the grains are spherical except for the contact faces. At lower liquid contents, the grains distort to better release

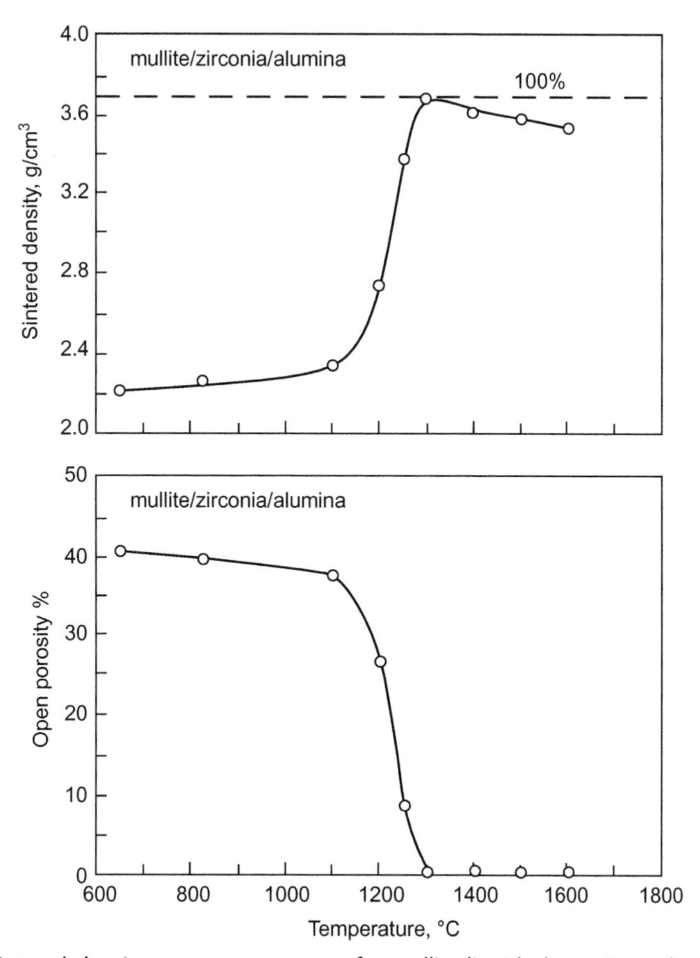

Figure 9.18 Sintered density versus temperature for mullite liquid phase sintered with a zirconia-alumina additive [30]. The composition reaches the highest sintered density when the open pores disappear at about 1300°C (1573 K), but swells at higher temperatures due to thermal expansion of gas trapped in the closed pores.

liquid to fill pores. At very low liquid contents, densification requires significant distortion to eliminate porosity. An example of grain shape accommodation is evident in Figure 9.19. A small amount of residual porosity is evident but most of the grains are not spherical.

Another cause of non-spherical grains is anisotropic solid-liquid surface energy. During liquid phase sintering, the grains converge toward low energy configurations. If the surface energy varies with crystal orientation, then lower energy orientations are preferred for growth [33]. Figure 9.20 is a two-dimensional illustration of this behavior, plotting the calculated grain shape as a function of the relative surface

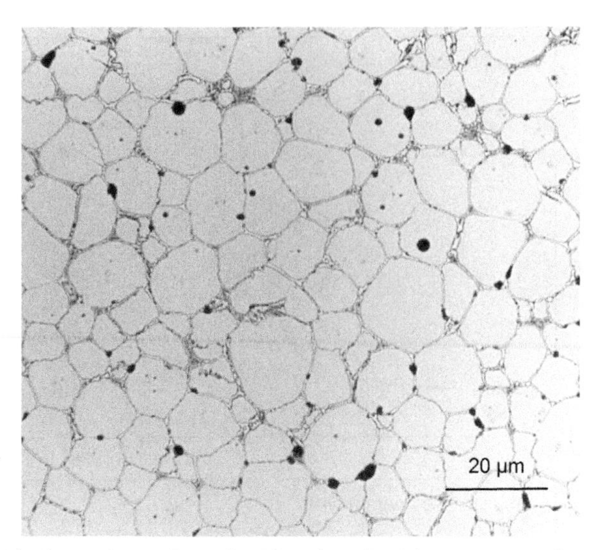

Figure 9.19 A liquid phase sintered steel with a low liquid content so the grains deformed to rounded polygonal shapes to enable fitting together with minimum porosity.

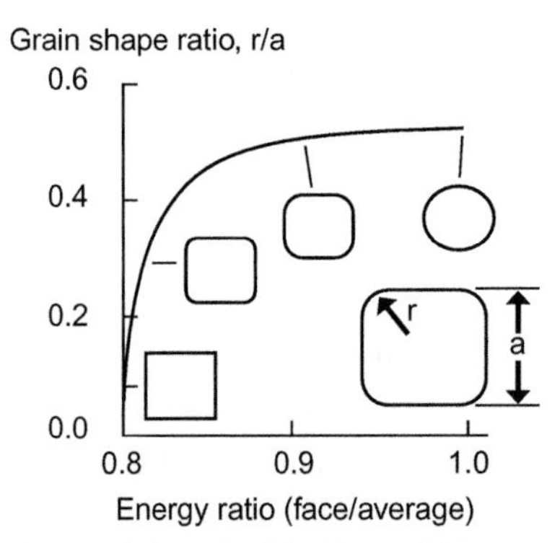

Figure 9.20 Grain shape variation with surface energy anisotropy [33]. These two-dimensional drawings illustrated the grain shape for the lowest energy configuration and how that depends on the interfacial energy of the right-face as a ratio to the mean interfacial energy. The sketches illustrate how sharp edges emerge in terms of the corner radius *r* with respect to the flat face separation distance *a*.

energy. The grain shape is more angular with an anisotropic surface energy. Such grain faceting is an indication of anisotropic surface energy, in which the low energy crystallographic planes are favored.

Flat-faced grains are common in liquid phase sintering cemented carbides. The crystal structure of WC gives a prism shape, as illustrated in Figure 9.21. As a

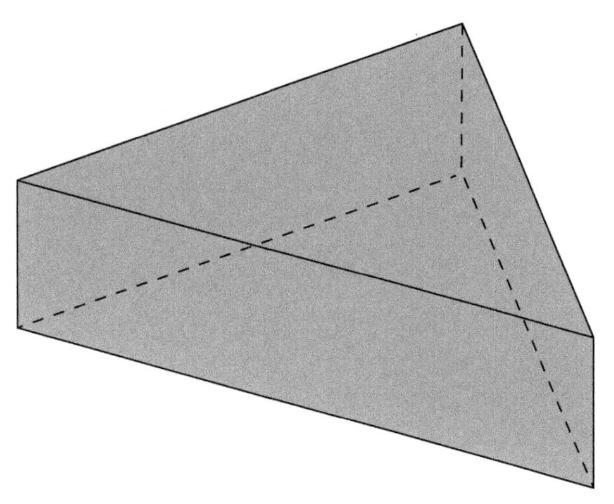

Figure 9.21 Hexagonal crystals tend to favor prismatic grain shapes, such as the one illustrated here. Thus, liquid phase sintered WC-Co favors flat-faced grains such as this, organized in a random array of size.

consequence the microstructure for WC-Co after liquid phase sintering consists of randomly sliced prisms, as evident in Figure 9.22. A variety of grain shapes appear in the section.

Grain Size Distribution

The grain size distribution after liquid phase sintering tracks to a self-similar form. The early efforts found that the distribution was broader than anticipated from simple mean field coarsening concepts. As outlined in Table 9.1, the grain structure is often far from model assumptions. Attempts to isolate a grain growth mechanism from random intercepts are flawed. While early models assumed isolated spherical grains, the actual microstructures consist of connected non-spherical grains [34].

Neck growth between grains, leading to grain agglomeration, is inherent to liquid phase sintering, even in dilute systems. Accordingly, coalescence is part of the grain structure evolution [35,36]. Another problem relates to the assumed diffusion field around each grain. Each grain exhibits a growth or shrinkage trajectory that depends on its local environment, not on the mean field [37].

The grain size distribution in liquid phase sintering becomes self-similar, independent of the starting particle size distribution [38,39]. The sintered grain size distribution converges to the same characteristic shape. The median size determines the

Figure 9.22 Cross-sectional micrograph of sintered WC-Co, illustrating how the prismatic carbide grains appear in a random section plane.

Table 9.1 Contrast of Assumptions and Reality in Microstructure Models

Parameter	Assumption	Reality in LPS
Grain shape	Spherical	Rounded or prismatic
Skeletal structure	Isolated	Highly connected
Grain separation	Uniform	Distributed
Coalescence	Ignored	Fairly common
Size measured	Grain diameter	Random intercept
Grain Distribution	Normal, log-normal	Weibull

normalization or scaling parameter, giving the cumulative distribution $F(L)$, with L being the intercept size, given as:

$$F(L) = 1 - exp\left[\ln\left(\frac{1}{2}\right)\left(\frac{L}{L_{50}}\right)^{N}\right] \tag{9.5}$$

where L_{50} is the median size, where half the grains are smaller. The exponent N is near 2 for intercept size measurements (but near 3 if three-dimensional grain size is measured).

Grain Separation, Population, and Interface Area

Grain separation is important for mechanical behavior, since often the matrix phase provides toughness. The mean separation λ between grains at full density is estimated using microscopy:

$$\lambda = \frac{V_L}{N_L} \quad (9.6)$$

where V_L is liquid volume fraction and N_L is the number of grains per unit length of a sampling scan line. If the solid-solid grain contacts that correspond to no grain separation are ignored, then the separation is skewed to a higher value. Since the grain separation only depends on the number of grains per unit measurement length, it should include the zero separation instances. The mean grain intercept size L (proportional to the true 3-D grain size) is related to the mean separation for zero porosity as below:

$$L = \frac{1}{N_L} - \lambda \quad (9.7)$$

When measured versus sintering time, the grain separation scales with the grain size, usually tracking the cube-root of the sintering time. Thus, the grain separation increases with as illustrated by data in Figure 9.23 for NbC-Fe sintered at 1800°C (2073 K) [40]. More liquid further increases grain separation.

Similarly, the number of grains decreases over time. In liquid phase sintering, the grain size usually increases with the cube root of time, so the number of grains per unit volume declines with inverse time as shown in Figure 9.24 for NbC-Fe at three temperatures. At the same time the solid-liquid interface area per unit volume decreases with the cube root of the sintering time, as plotted using Fe-30Cu data in Figure 9.25 [41].

Neck Size and Shape

Grain contacts grow in liquid phase sintering in a similar way to neck growth in solid-state sintering. Neck growth continues until it reaches a stable size as determined by the grain size and dihedral angle. Consequently the ratio of bond size to grain size is stable in spite of grain growth. A variation in contact sizes arises because of the differences in crystal orientations for the grain boundaries. Simultaneous measurements of the neck size and grain size in two dimensions provide a measure of the dihedral angle.

The distance between contacting grain centers, given by δ, depends on the grain size G and dihedral angle ϕ as follows:

$$\delta = G \cos\left(\frac{\phi}{2}\right) \quad (9.8)$$

If the dihedral angle is zero, $\delta = G$. Many solid-solid contacts involve grains of differing sizes. In such cases the grain boundary curves to favor grain coalescence, with

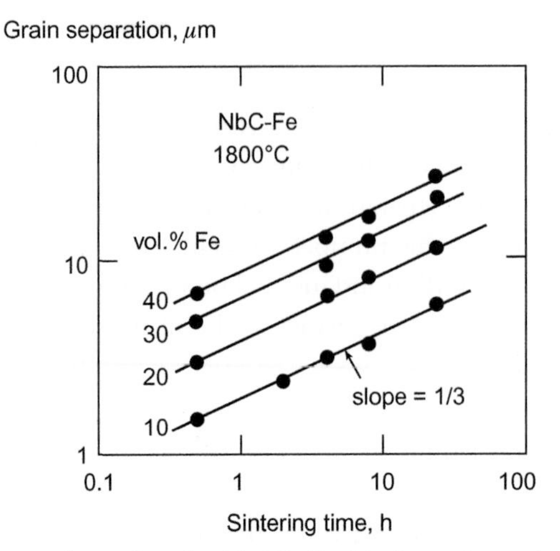

Figure 9.23 Grain separation data taken for NbC-Fe liquid phase sintered at 1800°C (2073 K) [40]. Since grain growth follows time to the one-third powder, likewise the grain separation tracks a similar response. Also the grain separation increases with the liquid content.

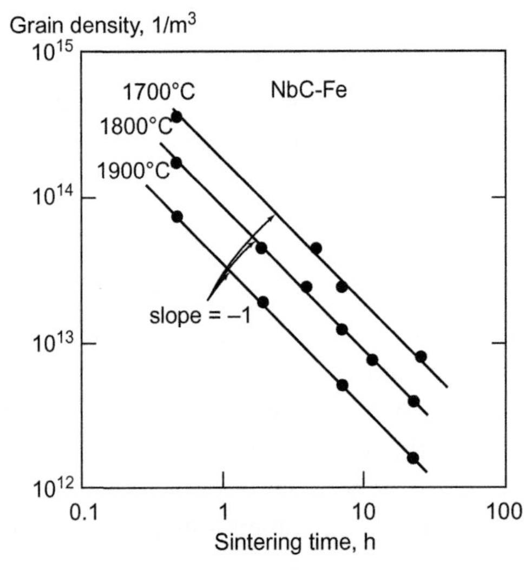

Figure 9.24 Grain density (number per unit volume) versus sintering time for NbC-Fe liquid phase sintered at 1700 to 1900°C (1973 to 2173 K) [40]. Since the mean grain volume increases with time, solid mass conservation dictates the grain population density decline as an inverse function of time.

the large grains absorbing the smaller grain. This process continues throughout sintering, but is most evident soon after liquid forms.

The neck size ratio reaches a constant value during liquid phase sintering. Thus, neck growth over long times is paced by the rate of grain growth. Typically the mean

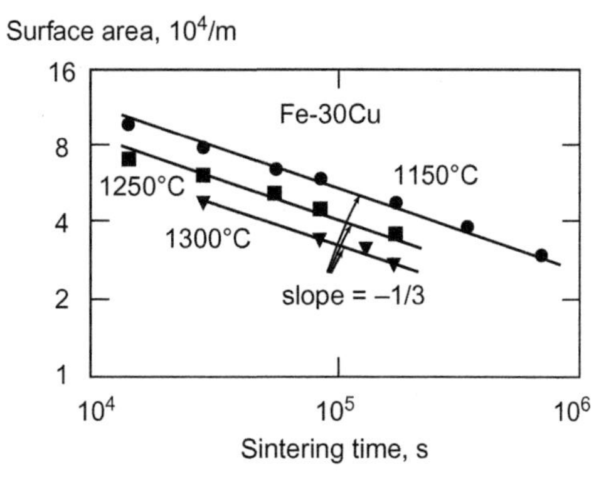

Figure 9.25 Surface area reduction versus time during liquid phase sintering of Fe-30Cu at three temperatures [41]. Each temperature gives a similar slope.

grain size increases with the cube root of time, so the neck size shows a similar dependence ($X \sim t^{1/3}$), which is the same as seen for neck growth in solid-state sintering.

Solid Skeletal Measures

The skeletal solid formed in liquid phase sintering has an interwoven liquid network. The resulting microstructure depends on the solid-liquid ratio. High solid contents increase the grain coordination number. At contents near 50 vol.% solid, the coordination number ranges near 5 and reaches a high of 14 at 100 vol.% solid (no liquid). The three-dimensional coordination number N_C depends on the solid volume fraction V_S and the dihedral angle ϕ as follows:

$$V_S = -0.83 + 0.81\,N_C - 0.056\,N_C^2 + 0.0018\,N_C^3 - 0.36\,A + 0.008\,A^2 \quad (9.9)$$

where $A = N_C \cos(\phi/2)$.

This grain coordination is hard to measure, so the convention is to use two-dimensional measures of contiguity or connectivity [42]. Contiguity C_{SS} is the relative solid-solid interface area in the microstructure, defined by the microstructure interfacial areas:

$$C_{SS} = \frac{S_{SS}}{S_{SS} + S_{SL}} \quad (9.10)$$

where S_{SS} is the solid-solid surface area per grain and S_{SL} is the solid-liquid (matrix) surface area per grain [43]. Contiguity is measured by quantitative microscopy based on the number of intercepts per unit length of test line N:

$$C_{SS} = \frac{2\,N_{SS}}{2\,N_{SS} + N_{SL}} \quad (9.11)$$

The subscript SS denotes the solid-solid intercepts and SL denotes the solid-matrix (solidified liquid) intercepts. The factor of 2 is necessary because the solid-solid grain boundaries are only counted once by this technique, but are shared by two grains.

Contiguity varies during liquid phase sintering, but tends to stabilize to a value reflective of the solid volume fraction and dihedral angle, independent of the grain size. Figure 9.26 shows the relation between volume fraction of solid, dihedral angle, and contiguity for monosized spheres, and includes data from three carbide systems for comparison [44]. The VC-Co system has a low dihedral angle and lower contiguity trace. At full density the relation between contiguity C_{SS}, volume fraction of solid V_S, and dihedral angle ϕ is as follows:

$$C_{SS} = V_S^2[0.43\, sin(\phi) + 0.35\, sin^2(\phi)] \tag{9.12}$$

This relation is not accurate when grain shape accommodation is evident. Figure 9.27 plots the contiguity variation with a dihedral angle for 80 vol.% solid. Contiguity proves useful in linking microstructure to mechanical properties.

Connectivity is a related parameter. It is the average number of grain-grain connections per grain observed in a two-dimensional section. Grain connectivity C_G depends on the coordination number N_C and dihedral angle ϕ as:

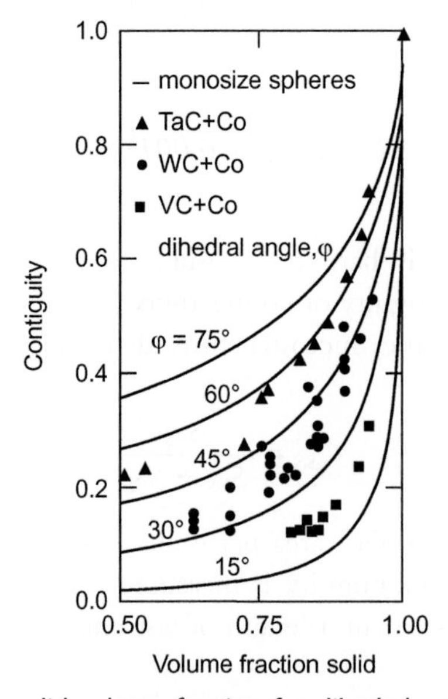

Figure 9.26 Contiguity versus solid volume fraction for dihedral angles ranging from 15° to 75° [44]. For comparison the experimental results for three carbide systems are included, but these do not have monosize grains.

$$C_G = \frac{2}{3} N_C sin\left(\frac{\phi}{2}\right) \qquad (9.13)$$

A typical grain coordination of 6 with a solid content of 60 vol.% at a dihedral angle of 60° has two contacts per grain in cross section. The microstructure will be distributed about this average since the section plane makes a radon slice through the solid skeleton. Figure 9.28 plots the two-dimensional grain connectivity versus the solid volume for three liquid phase sintered materials, illustrating the behavior.

Connectivity is effective in explaining resistance to distortion during liquid phase sintering [45]. Early in sintering, the bonds between solid grains grow, so contiguity increases over time. Any change in interfacial energies changes the dihedral angle and contiguity; thus, contiguity drops on first melt formation, with a subsequent time-dependent recovery to the equilibrium behavior.

The growth of bonds between grains results in a rigid skeleton. Associated with the formation of the solid skeleton is an underlying percolation concept. This refers to the formation of a continuous chain of solid bonds. Without sufficient connections the microstructure is unstable, so it settles to a condition of stability. At this percolation limit, the grain coordination number is 1.5, but sufficient rigidity to resist distortion during liquid phase sintering requires at least three contacts per grain, corresponding to about 30 vol.% solid [46].

Sensitivities in Analysis

Microstructure measures describe the amount of each phase, its distribution, and its composition using descriptors of size, shape, and relations between the phases. Liquid phase sintering is a normalization process. Although the starting point depends on

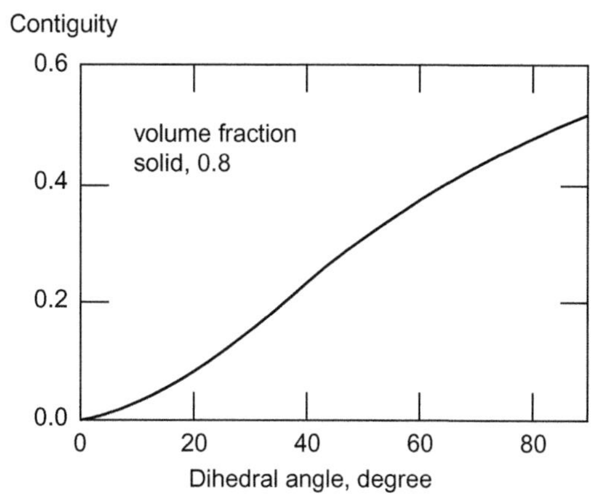

Figure 9.27 Contiguity variation with dihedral angle at 0.8 volume fraction solid, with two curves corresponding to the typical grain size distribution found in liquid phase sintering.

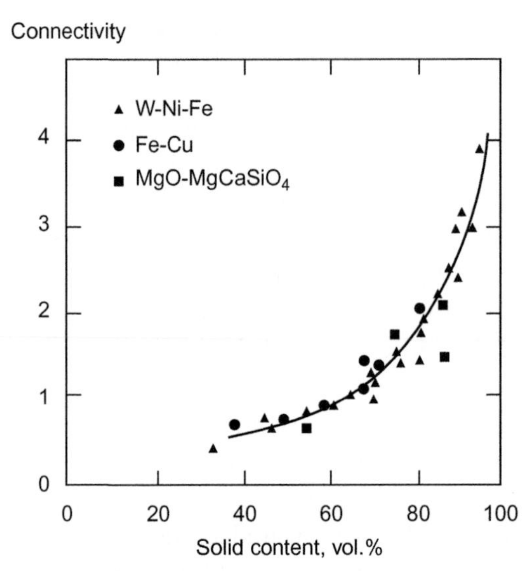

Figure 9.28 Two-dimensional grain connectivity as a function of solid content for three liquid phase sintered materials.

the green body, the sintered microstructure converges to common characteristics during sintering. For most systems, porosity declines, smaller pores are annihilated first, and the mean pore size increases while grain shape accommodation and solid skeletal densification release liquid to fill the smaller pores first. In spite of large differences in starting conditions the sintered material often is the same. Table 9.2 compares features from liquid phase sintered microstructures to illustrate some combinations.

Interfacial energies guide microstructure evolution. These change when the first liquid forms and are sensitive to segregation and temperature, so they vary during sintering [47]. Thus, the microstructure depends on processing conditions and will be different with location within the sintered body. Curiously, controversies arise with regard to these factors without proper vetting of the processing conditions. Proper understanding comes from freezing the microstructure to avoid temperature-dependent shifts in solubility or interfacial energy. Otherwise, reports on the microstructure are only valid with respect to the "sintered" condition and do not represent the conditions existing during "sintering." This is frequently seen in disagreements over grain boundary films. Slow cooling induces decreasing solubility and segregation to heterogeneous sites, resulting in precipitation of liquid-like compositions segregated on grain boundaries. Slow cooling results in a grain boundary layer while quenching finds the grain boundaries free of matrix (liquid). Without proper processing there is poor rationalization of the conditions during liquid phase sintering.

Table 9.2 Example Mean Microstructure Parameters in Liquid Phase Sintering

System (wt.%)	W-8Mo-7Ni-3Fe	WC-8Co	Fe-50Cu	W-7Ni	Mo-46Cu
Sintering Cycle	1480°C 2 h	1400°C 1 h	1200°C 1 h	1540°C 1 h	1400°C 1 h
Liquid, vol.%	14	12	40	30	50
Porosity, vol.%	0.4	0	10	2	12
Grain size, μm	17	3	38	35	10
Dihedral angle, deg.	15	–	22	27	100
Contiguity	0.52	0.39	–	–	–
Connectivity	–	–	0.9	0.2	3.2

PRELIQUID STAGE

Microstructure changes occur rapidly when the melt first forms. Microstructure evolution prior to liquid formation is similar to solid-state sintering. Factors favorable for densification prior to liquid formation, such as a small particle size, are favorable for densification during liquid phase sintering. A complication arises from the solubility characteristics and chemical gradients of mixed powders.

Chemical Interactions

Mixed powders with different compositions represent a non-equilibrium condition. The microstructure is out of equilibrium in the pre-liquid stage of sintering. Phase diagrams provide a first estimate of the potential for densification for mixed powders. As a conceptual outline, Table 9.3 summarizes dimensional change observations for several metallic binary systems. Swelling is observed if the liquid forming species dissolves into the solid.

Systems such as W-Cu with no intersolubility (less than 10^{-3} or 0.001 at %) are termed non-interacting. In such systems, particle size is the dominant factor with respect to densification. Chemical gradients play a role in systems where the solubility exceeds about 0.1 at 0.001%. Systems with high solubility (>5 at 0.001%) of solid in the additive phase, but little reverse solubility, are ideal for liquid phase sintering. This is the situation with systems exhibiting densification, such as WC-Co or W-Ni-Fe. On the other hand, a high solubility of liquid in the solid leads to swelling, as is the case with Cu-Sn.

The chemical gradients from mixed powders add to the solid-state sintering during heating. Activated sintering is employed where the additive induces significant solid-state sintering. It is especially effective when densification of the solid skeleton proves difficult. An example is in the W-Cu system. Densification usually requires sintering at high temperatures where the copper evaporates. The addition of 0.3 wt.% cobalt promotes densification in the form of activated sintering.

Table 9.3 Observations on Solubility and Dimensional Change for Binary Powder Mixtures

Solid Phase	Added Phase	Temperature, °C	Dimensional Change
Al	Zn	500	swell
Cu	Al	600	swell
Cu	Sn	760	swell
Cu	Ti	950	shrink
Fe	Al	1300	swell
Fe	B	1200	shrink
Fe	Sn	800	swell
Fe	Ti	1300	shrink
Mo	Ni	1400	shrink
Ti	Al	700	swell
W	Fe	1100	shrink
W	Ni	1100	shrink

Microstructure Changes

Substantial changes in density, neck size, and grain size occur before the liquid forms. Particle size sets several of the scale parameters—curvature gradients, capillary stress, and diffusion distance. Usually particles in the micrometer size range are used to ensure rapid densification. Prior to liquid formation, solid state sintering gives the neck size ratio (X/D) and the sintering shrinkage ($\Delta L/L_0$) according to the following generic relations:

$$\left(\frac{X}{D}\right)^n = K_1 \frac{t}{D^m} \tag{9.14}$$

$$\left(\frac{\Delta L}{L_O}\right)^n = K_2 \left[\frac{t}{D^m}\right]^2 \tag{9.15}$$

where t is the isothermal sintering time and the constants m and n are mechanism dependent exponents. The most common values are $m = 4$ and $n = 6$. Because of the inverse dependence on particle size, smaller particles induce significant shrinkage prior to liquid formation.

Newly formed liquid dissolves the solid, giving a grain size decrease. Figure 9.29 plots grain size data taken during heating of W-5Ni-2Fe, quenched at various points during heating at 10°C/min (150 min) followed by isothermal conditions. Solid state grain growth prior to liquid formation near 150 min is slow. The grain size undergoes a small decrease on liquid formation, followed by rapid grain growth once the liquid spreads.

Phenomenological relations link the densification rate df/dt to the fractional density f, the grain size G, and the grain growth rate dG/dt during liquid phase sintering:

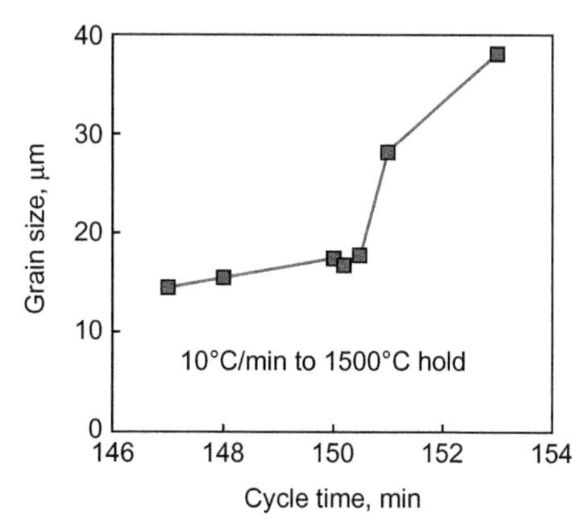

Figure 9.29 Grain size versus cycle time obtained by quenching samples during heating to 1500°C (1773 K) at 10°C/min (first 150 min). At 150 min or 1500°C, isothermal time starts. Grain growth prior to liquid formation is slow, but on liquid formation the grain size undergoes a small decrease followed by rapid grain growth after the liquid forms. These data are from W-5Ni-2Fe.

$$\frac{df}{dt} = K_3 \frac{(1-f)^k}{G^m} \tag{9.16}$$

$$\frac{dG}{dt} = K_4 \frac{1}{G^n(1-f)^l} \tag{9.17}$$

where the exponents k, l, and m are characteristic of different mechanisms, while K_3 and K_4 are collections of geometric and material constants. Densification is faster at lower densities and smaller grain sizes, while grain growth is faster at higher densities and smaller grain sizes.

LIQUID FORMATION

Melt formed liquid dissolves solid and this preferentially is by the attack of grain boundaries. The newly formed liquid solvation of solid dissolves grain boundaries. Solid wetting traces to solid solubility in the liquid. Wetting induces liquid spreading to fill pores and penetration of grain boundaries to enable grain rearrangement.

Fragmentation

Solid bonds formed during heating are dissolved by newly formed liquid [48–50]. This is captured in Figure 9.30. The micrograph on the left is a cross-section of a

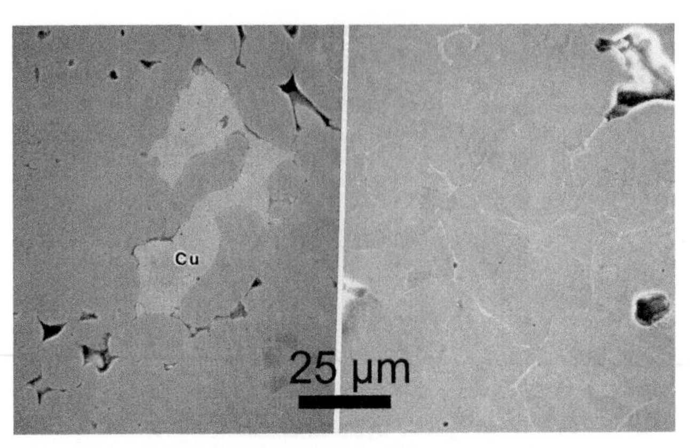

Figure 9.30 Grain boundary penetration by newly formed liquid as seen in Fe-8Cu compacts heated to (left) 1075°C (1348 K) just prior to melt formation and (right) 1110°C (1383 K) after melt formation. The molten copper rapidly penetrates along the iron grain boundaries.

Fe-Cu powder mixture heated to just below the copper melting point, exhibiting copper as discrete grains. The micrograph on the right is the same composition, but heated slightly over the melting point of copper. The liquid copper has spread and penetrated the iron sinter bonds. The initial sinter bonds fragment by liquid penetration at a velocity from 0.1 to 2 μm/s.

Liquid penetration occurs because of the surface energy changes that take place on solid dissolution. This is summarized by examining the first derivative of Equation (9.2):

$$\frac{d\gamma_{SL}}{\gamma_{SL}} = \frac{d\phi}{\phi}\frac{\phi}{2}\tan\left(\frac{\phi}{2}\right) \tag{9.18}$$

The relative dihedral angle change is proportional to the solid-liquid surface energy change associated with solid dissolving into the liquid. Penetration of a grain boundary requires a dihedral angle change of $d\phi = -\phi$, giving:

$$\frac{d\gamma_{SL}}{\gamma_{SL}} = -\frac{\phi}{2}\tan\left(\frac{\phi}{2}\right) \tag{9.19}$$

Small changes in the solid-liquid surface energy induce liquid penetration of grain boundaries. For example, a dihedral angle of 30° requires a 7% decrease in the solid-liquid surface energy for grain boundary penetration.

Newly formed liquid spreads to fill small pores and preferentially penetrate grain boundaries. Dissolution reactions during spreading decrease the solid-liquid interfacial energy below equilibrium. This causes a momentary reduction in dihedral angle, leading to liquid penetration of the grain boundaries. An example is evident with the

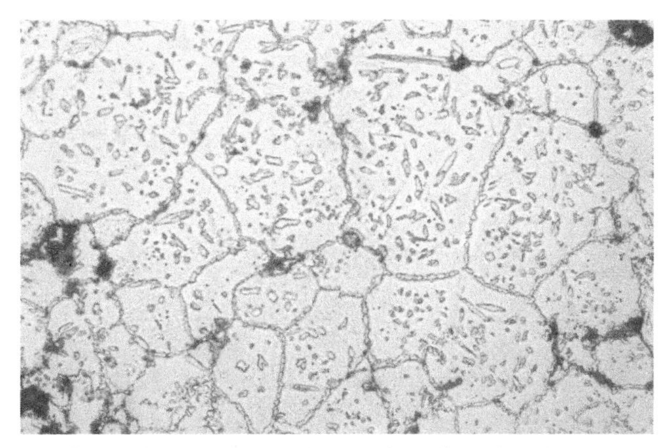

Figure 9.31 An example necklace microstructure resulting from liquid penetration along the solid grain boundaries on first melt formation, followed by a pinching off of the liquid film into discrete islands on the grain boundary. This micrograph is from a Fe – 7% Ti product after sintering.

residue of liquid decorating the grain boundaries, giving a necklace microstructure, as shown in Figure 9.31. After liquid formation and spreading, the solid-liquid system approaches equilibrium, leaving islands on the grain boundary.

Non-interacting systems such as W-Cu and Al_2O_3-Ni have no solubility of liquid in solid or solid in the liquid. The lack of solubility reduces chemically driven sintering. Likewise, on liquid formation the lack of interaction gives poor wetting, and little rearrangement.

Grain Rearrangement

Liquid penetration between grains converts the structure into a semisolid body. In most cases, the liquid wets the solid, and densification occurs on liquid formation. A capillary force arises from a wetting liquid, leading to a contact force approximated as:

$$F_C = 5\gamma_{LV} D \tag{9.20}$$

where F_C is the contact force (distributed over the contact region), γ_{LV} is the liquid surface energy, and D is the particle diameter. The capillary force induces grain rearrangement by pulling the grains together.

Densification by grain rearrangement depends on contact angle, as illustrated in Figure 9.32. Three situations are sketched in this figure. In case (a) the compressive capillary force pulls the grains together, giving shrinkage. In case (b) the contact angle and liquid content is neutral, giving no dimensional change on melt formation. In case (c) the high contact angle of the melt induces grain separation. Since powders often will pack to 60% density, about 40 vol.% liquid is sufficient to give complete

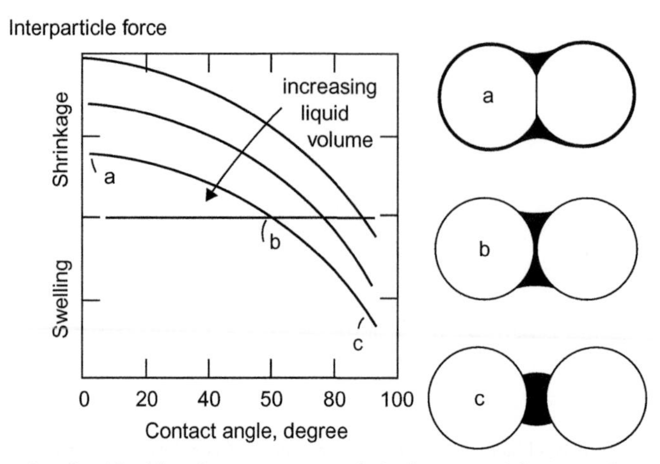

Figure 9.32 A wetting liquid with a low contact angle induces particle attraction and formation of a flat face, leading to sintering shrinkage, while a high contact angle forces swelling and particle separation.

densification on liquid flow into the pores. Lower liquid contents require grain rearrangement, giving full density at about 30 vol.% liquid.

Non-spherical grains provide an additional inducement to rearrangement, since a wetting liquid generates a rearrangement torque that brings flat surfaces into contact.

SOLUTION-REPRECIPITATION

Touching grains bind together after melt spreading. The capillary force pulls the grains into contact, and diffusion initiates neck growth. Both solid state diffusion and diffusion in the liquid are active, assuming the solid is soluble in the liquid. The solid skeletal microstructure emerges, providing component rigidity. Otherwise, liquid phase sintering would not be able to fabricate components while preserving their shape. As with solid state sintering, the initial neck growth and densification rates slow with neck growth. If there is insufficient liquid to fill all the pores, then continued densification relies on diffusion through the liquid. If the solid is not soluble in the liquid, then densification occurs by slower solid state skeletal densification. Because transport rates in liquids are orders of magnitude faster than those in solids, solution-reprecipitation is dominant as long as the solid is soluble in the liquid.

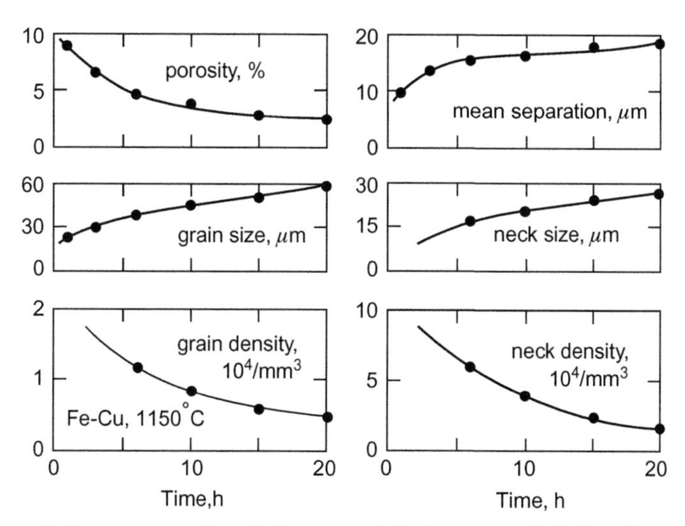

Figure 9.33 Several changes occur in solution-reprecipitation controlled densification. These results are from Fe-20Cu sintered at 1150°C (1423 K) [54]. The plots show the changes in porosity, grain size, number of grains per unit volume, pore separation, neck size, and neck density.

Solution-reprecipitation is a three stage mass transport process [3,4,8,9]:

1. The solid dissolves from the grain surface into the liquid, preferentially from higher energy asperities, convex points, or areas under compression; the dissolution of small grains is preferred [51].
2. Dissolved solid diffuses in the liquid [52].
3. In concave regions or larger grains, the dissolved solid precipitates out of solution to preferentially deposit on areas under tension, resulting in grain growth [53].

Neck growth, pore filling, and microstructure coarsening are the result of solution-reprecipitation. They depend on the same diffusion steps as grain shape accommodation. Figure 9.33 are microstructure data taken during Fe-20Cu liquid phase sintering at 1150°C (1423 K) [54]. Grain size and neck size increase as porosity is eliminated, and the number of grains and number of necks declines due to coarsening. Solution-reprecipitation produces simultaneous changes in several features.

Usually diffusion through the liquid is the controlling step, but for angular grains the limited availability of surface atoms or precipitation sites leads to interfacial control. Usually the initial process is also diffusion controlled, but as flat-faced grains form, then interfacial control takes over. Rounded grains are characteristic of diffusion control and angular grains are characteristic of interfacial control. A curved surface has a high density of defects while flat-faced grains have very few. Most liquid phase sintering systems are diffusion controlled at least for a portion of the solution-reprecipitation process.

Grain Shape Accommodation

A sketch of densification by solution-reprecipitation is given in Figure 9.34. Key changes are a decreasing number of grains, increasing grain size, and grain shape accommodation. The latter reshapes the grains for improved density, releasing liquid to fill pores. Grain shape accommodation reduces overall interfacial energy, since the pore surface energy is greater than the penalty from the increased solid-liquid interface [55]. The process is analogous to the way a stone mason fits flagstones to minimize mortar. The result is a rounded, but polygonal grain shape at full density. A low liquid content causes more grain shape accommodation. If a compact is immersed in a melt of the liquid phase, the compact absorbs more liquid and relaxes the grain shape to become more spherical.

Densification

For most liquid phase sintering compositions, there is insufficient liquid to reach full density without neck growth and grain shape accommodation. Solution-reprecipitation is the means to densify the structure and three mechanisms are active—contact flattening, small grain dissolution, and migration along the contact interface [4–9,56,57]. These options are sketched schematically in Figure 9.35.

Contact flattening is the first mechanism. The wetting liquid pulls the grains together, generating a compressive stress at the contact. This stress causes preferential

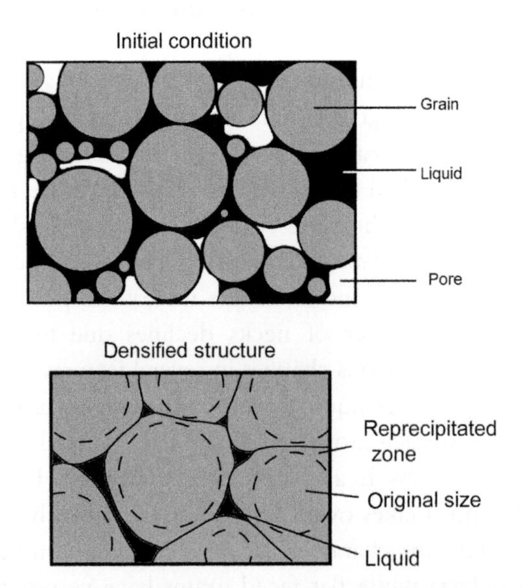

Figure 9.34 A conceptual outline of the microstructure changes associated with solution-reprecipitation densification where both grain growth and grain shape accommodation act to release liquid to fill residual pores.

dissolution of the solid at the contact point with reprecipitation at regions away from this pont. Densification results from grain attraction [58]. Diffusion in the liquid is the controlling step. For small grains, the contact stresses are large, so contact flattening is dominant. Contact flattening does not explain grain growth and the decrease in the number of grains.

The second mechanism involves dissolution of small grains with reprecipitation from large grains. Small grains disappear and large grains grow while undergoing shape accommodation. Again diffusion in the liquid is the controlling transport mechanism. Grain growth occurs, but no shrinkage is evident. This mechanism does not explain densification beyond that which is associated with solid grain packing gains.

The third mechanism involves growth of the intergrain contact by diffusion along liquid films [59]. The contact zone enlarges, resulting in a grain shape change with simultaneous shrinkage. Other than coalescence, this does not involve grain coarsening, but it does require a cooperative redistribution of the mass deposited where the grain boundary intersects the liquid.

These three mechanisms differ in the solid source, transport path, grain growth, and densification. Together they explain solution-reprecipitation controlled grain shape accommodation, grain growth, and densification. Indeed, grain size and density tend to follow a common trajectory during liquid phase sintering over a range of heating rates, peak temperatures, and hold times [60–62]. In diffusion controlled growth, the grains remain rounded with an abundance of atomic steps, so there is no limitation from the population of interfacial sites available for dissolution or precipitation [4,63]. Interfacial control is observed in complex cemented carbides of WC, VC, TiC, or TaC with a cobalt-based liquid [64]. In these cases interfacial events are slowed because of a limited population of reaction sites, giving flat-faced grains.

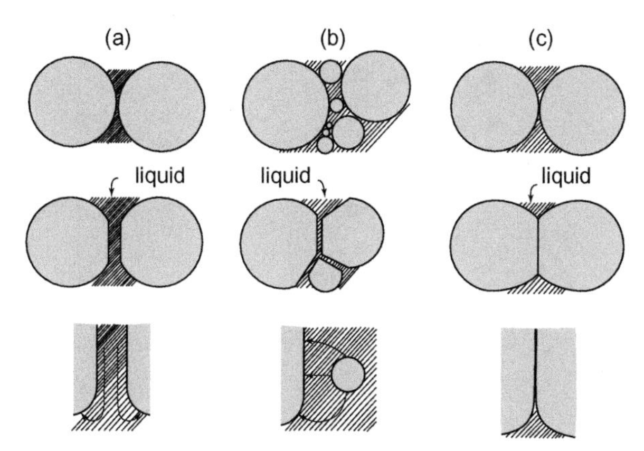

Figure 9.35 The three mechanisms of grain shape accommodation and neck growth during solution-reprecipitation controlled liquid phase sintering densification; (a) contact flattening, (b) dissolution of small grains, and (c) solid state bonding.

Early sintering shrinkage $\Delta L/L_0$ by diffusion controlled solution-reprecipitation is treated as follows [4,65]:

$$\left(\frac{\Delta L}{L_O}\right)^3 = \frac{g_1 \delta_L \Omega \gamma_{LV} D_S t\ C}{R\ T\ G^4} \tag{9.21}$$

In this equation δ_L is the liquid layer thickness between the grains, γ_{LV} is the liquid-vapor surface energy, Ω is the atomic volume of the solid, D_S is the diffusivity of the solid in the liquid, C is the solid concentration in the liquid, t is time, R is the gas constant, T is the absolute temperature, G is the solid grain size which changes with sintering time (typically $G^3 \sim t$), and g_1 is a numerical constant near 192. Example shrinkage data during solution-reprecipitation are plotted in Figure 9.36 on a log-log basis [5,66]. The one-third power slope is expected for diffusion controlled solution-reprecipitation.

For interface reaction control, a modified form results:

$$\left(\frac{\Delta L}{L_O}\right)^2 = \frac{g_2 \kappa \Omega \gamma_{LV} t\ C}{R\ T\ G^2} \tag{9.22}$$

In this equation κ is the reaction rate constant and g_2 is a numerical constant of 16. There is no effect from the diffusion rate in the liquid, since the reaction site availability determines the shrinkage rate, not the arrival rate.

In both cases, densification is faster with higher temperatures, smaller grains, and more solid solubility in the liquid. Small particles are beneficial, as demonstrated by

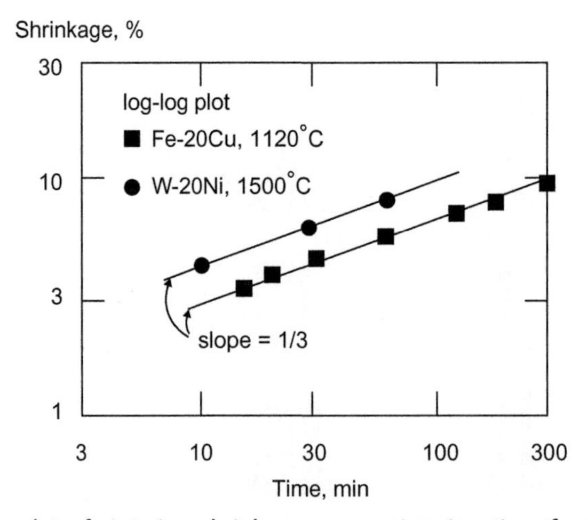

Figure 9.36 A log-log plot of sintering shrinkage versus sintering time for Fe-20Cu and W-20Ni undergoing diffusion controlled solution-reprecipitation densification with the expected one-third slope [5,66].

the densification data for liquid phase sintered alumina-glass in Figure 9.37 [24]. One problem with the models is their inability to predict how the liquid quantity influences densification, although experiments show an effect.

Grain boundary diffusion along the grain contact is a possible densification mechanism. The predicted neck growth rate is the same as that given by solid state grain boundary diffusion models. Since solid state diffusivities are low when compared to liquid diffusivities, solid state sintering is only significant in those cases where there is no solid solubility in the liquid; for example, in electrical contacts and heat sinks made from Mo-Ag, Mo-Cu, W-Cu, SiC-Al, and WC-Ag.

Neck Growth and Shrinkage

Neck growth by solution-reprecipitation occurs after new liquid forms. The rate of neck growth is initially rapid, but then slows and eventually stops. For isothermal conditions, neck growth is similar to grain boundary diffusion controlled solid state sintering with a provision for solid solubility in the liquid:

$$\left(\frac{X}{G}\right)^6 = \frac{g_3 D_S C \gamma_{SL} \Omega t}{R\ T\ G^3} \tag{9.23}$$

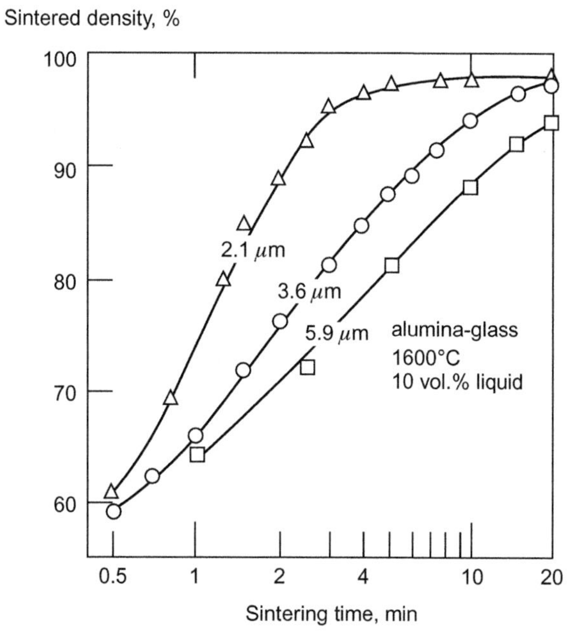

Figure 9.37 Fractional density for liquid phase sintered alumina-glass with 10 vol.% glass, sintered for various times to demonstrate how smaller particles improve densification [24].

where X is the neck diameter, G is the grain diameter, g_3 is a numerical constant with a value of about 160, D_S is the diffusivity of the solid in the liquid, C is the solid concentration in the liquid, γ_{SL} is the solid-liquid surface energy, Ω is the atomic volume, t is the sintering time, R is the gas constant, and T is the absolute temperature. Many parameters are temperature sensitive, with a dominant role from the diffusivity changes with temperature.

If the neck is covered, then the amount of liquid is not a factor. Initially, there is no sensitivity to the dihedral angle, so this model is only useful for initial bonding. Eventually, the neck size reaches a stable size, as dictated by the dihedral angle. For grains of size G with a bond of size X, the equilibrium neck size ratio is plotted in Figure 9.38. A distribution in neck sizes is expected due to crystal orientation and grain size variations.

After neck growth, large-small grain combinations cause curvature in the grain boundary. This is illustrated in Figure 9.39 for a 60° dihedral angle and grain size ratios of 1.0, 0.7,0 .5, and 0.3. The neck size divided by the larger grain size decreases as the grains differ in size, while the neck size divided by the smaller grain size increases. A curved grain boundary provides a driving force for grain coalescence [36].

Shrinkage occurs with neck growth, so neck size equations are often transformed into shrinkage equations, predicting shrinkage increases with the cube root of sintering time. There is a limit to neck growth in liquid phase sintering, determined by the dihedral angle. For a dihedral angle of 60° the neck size ratio grows to a limit of $X/G = 0.5$, corresponding to a shrinkage of 6.25%. At a dihedral angle of 23° the corresponding shrinkage termination from neck growth is 1%. Once the limiting

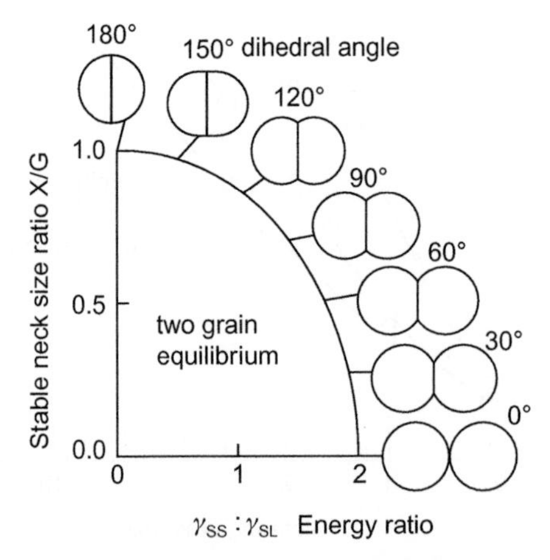

Figure 9.38 Plot of the stable neck size ratio (X/G) versus the solid-solid to solid-liquid surface energy ratio, with sketches of the expected two particle junctions.

neck size ratio is attained, X/G remains constant and further neck growth depends on grain growth.

Coalescence

A wetting liquid induces particle contact. Amorphous particles coalesce since there is no grain boundary. For crystalline solids, there is a 5 to 10% probability of grain contact with a low-angle grain boundary, so grain rotation produces coalescence.

The driving force of coalescence is the grain boundary curvature. Denoting the boundary radius of curvature as r gives a dependence on the dihedral angle ϕ, and grain sizes G_1 and G_2 (G_1 is larger than G_2) as:

$$r = cos\left(\frac{\phi}{2}\right)\left[\frac{G_1 G_2}{G_1 - G_2}\right] \tag{9.24}$$

Because of the curvature, large-small grain combinations favor coalescence through rotation of the small grain and its absorption into the large grain. Several transport paths are possible, including grain boundary migration, solution-reprecipitation of a liquid film on the boundary, solution-reprecipitation between small and large grains, and grain rotation into a coincidence. Grain rotation is favored by high liquid contents and solid diffusion by high solid contents.

Pores undergo coalescence as well. The pore's buoyancy drives migration toward the component top. Buoyancy-driven pore migration leads to larger pores occurring near the top of the component, and in some instances the formation of surface blisters. Simultaneously, pore coarsening leads to a decrease in the number of pores. Gas-filled pores change in volume as they grow, since the internal pressure depends

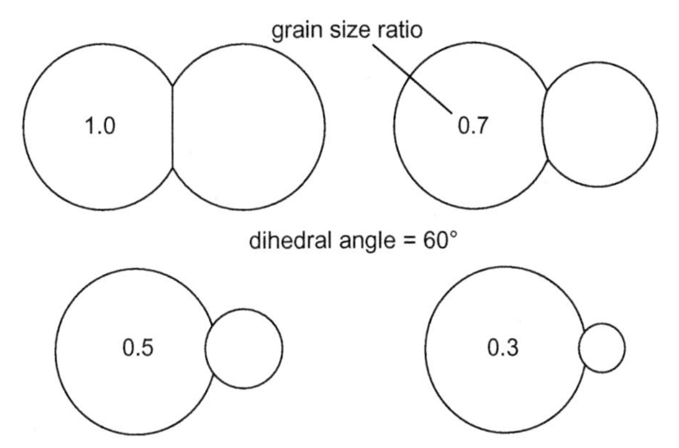

Figure 9.39 Equilibrium grain boundary configurations for a 60° dihedral angle and grain size ratios of 1.0, 0.7,0 .5, and 0.3. The neck size divided by the larger grain size decreases as the grains differ in size and the boundary curvature increases the driving force for grain coalescence [36].

on the inverse of the pore size. Thus, pore volume increases due to coalescence, resulting in long-term swelling as plotted in Figure 9.40 for the liquid phase sintering of MgO-$CaMgSiO_4$ [67]. Gas diffusion in the liquid is the prime means of pore growth [68].

Evidence for pore coarsening has been gathered from microgravity liquid phase sintering experiments. The absence of gravity allows the pores to coalesce into massive pores. Figure 9.41 is a picture taken from a W-15Ni-7Fe composition sintered at 1507°C for 180 min in microgravity, showing two large coarsened pores undergoing coalescence.

Grain Growth

Grain coarsening models for liquid phase sintering are complicated by the several interfaces and different transport rates as those interfaces grow or shrink. Early models assumed an average behavior, so the growth or shrinkage rate for a grain would be a function of its size compared to the average. However, the local environment neighboring a grain is a dominant factor. Some larger grains shrink and some of the smaller grains grow.

The rate of grain growth shifts as the grain shape changes. Thus, grain growth is rapid until the grains become faceted. Further shifts in grain growth are possible by

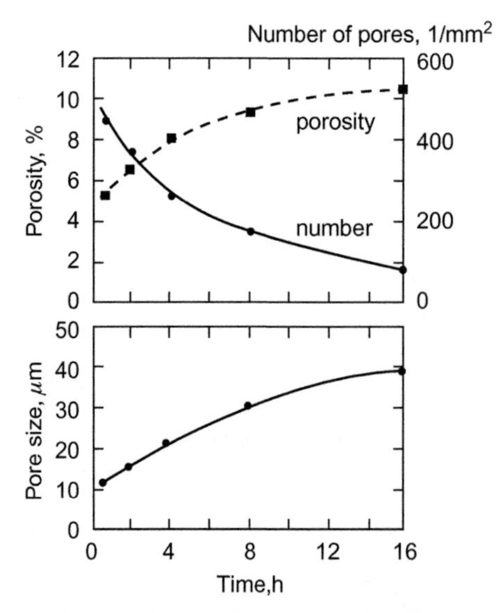

Figure 9.40 Swelling during solution-reprecipitation for MgO-$CaMgSiO_4$ at 1600°C (1873 K) in nitrogen [67]. Porosity increases with pore coarsening by gas diffusion through the liquid, decreasing the number of pores, increasing the size of the pores, and since pore pressure decreases as the pores enlarge to give swelling.

the addition of species that segregate to the solid-liquid interface to lower the active site population. This is evident when VC is added to the WC-Co system prior to sintering. Likewise, systems consisting of two solid phases show inhibited grain growth at intermediate compositions, as demonstrated in Figure 9.42 for the MgO-CaO-Fe_2O_3 system [69].

Pore Filling

Pore filling preferentially follows capillarity—the smaller pores fill first. High green density regions correspond to smaller pores and fill early. Larger pores fill only when the grain size is enlarged. The process of liquid flow into large pores is depicted in Figure 9.43. Pore filling depends on the capillary forces, and for small grains the liquid is retained between the grains. Grain growth eventually reaches a favorable condition for liquid to flow into the pore [70]. Mathematically the onset of pore filling depends on the liquid meniscus radius r_m at the pore-liquid-grain contact given as:

$$r_m = \frac{G}{2}\left[\frac{1 - \cos\alpha}{\cos\alpha}\right] \tag{9.25}$$

G is the grain diameter and α is the angle from the grain center to the solid-liquid-vapor contact point shown in the drawing. In practice, pore filling occurs when the pore size and meniscus radius are about the same and is favored by a low contact angle [26]. The result is a lake of liquid in the microstructure. If grain growth has a cube-root dependence on time, then the time to fill large pores is estimated from the pore size.

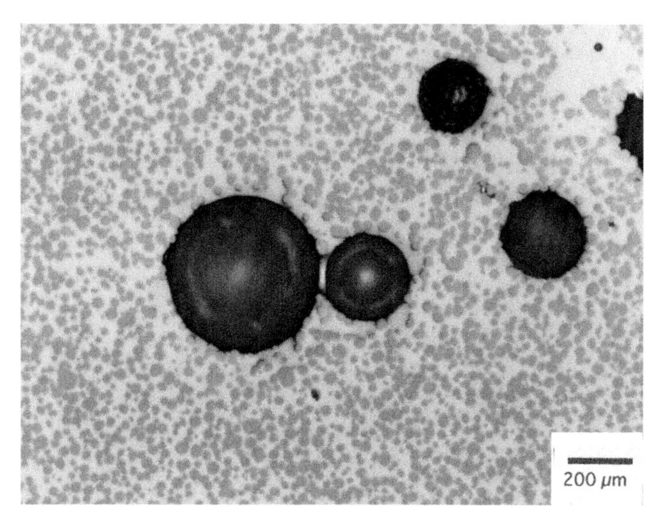

Figure 9.41 Optical micrograph from a sectioned heavy alloy liquid phase sintered in microgravity, capturing coalescence of large pores.

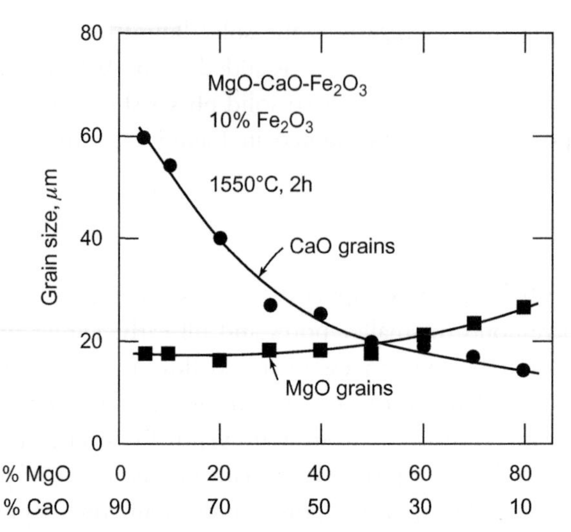

Figure 9.42 Inhibited grain growth during liquid phase sintering for a mixture of MgO and CaO grain with Fe_2O_3, showing how the mixture of two solids reduces coarsening during sintering, resulting in a smaller sintered grain size [69].

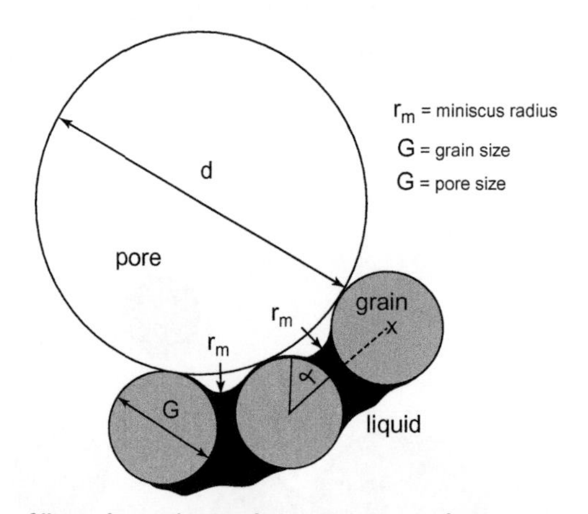

Figure 9.43 Large pore filling depends on the pore size and grain size, and filling is expected when grain growth allows the meniscus radius to reach a condition favoring liquid flow into the large pore.

Large pores represent processing defects, so are generally avoided. Likewise, refilled large pores, termed liquid lakes, are considered defects. Sources of these large pores are polymer inclusions that burn out to leave residual pores, packing or forming defects, pockets of reaction gases, or agglomerated powders.

FINAL STAGE SOLID SKELETAL SINTERING

The final stage of liquid phase sintering involves slow densification of the solid skeleton with liquid occupying the cusps between the grains. Grain growth continues, even after full density is attained. Any gas trapped in the pores slows densification and often induces component swelling. For low solubility systems, such as Mo-Cu, densification is paced by the solid-phase sintering rate with liquid filling the voids. For systems with solid solubility in the liquid, solid skeleton densification couples with solution-reprecipitation events to give densification.

In the final stage, the pores are isolated spheres totaling less than 8% of the volume of the sintered body. For this case, the densification rate is given as follows:

$$\frac{df}{dt} = \frac{12\, D_S C \Omega}{R\, T\, G^2} \beta \left\{ \frac{4\gamma_{LV}}{d} - P_G \right\} \tag{9.26}$$

where f is the fractional density, t is the isothermal time, D_S is the diffusion rate of the solid in the liquid, C is the solubility of the solid in the liquid, Ω is the atomic volume of the solid, R is the gas constant, T is the absolute temperature, G is the grain size, d is the pore size, γ_{LV} is the liquid-vapor surface energy, and P_G is the gas pressure in the pore. The parameter β is a pore density factor given as follows:

$$\beta = \frac{\pi\, N_V\, G^2 d}{6 + \pi\, N_V\, G^2 d} \tag{9.27}$$

where N_V is the number of pores per unit volume. It is best to avoid trapped atmosphere if full density is desired. An insoluble atmosphere such as argon stabilizes pores, and over time the pores coarsen to form blisters on the compact surface.

TRANSIENT LIQUIDS

If the liquid is soluble in the solid it disappears over time, so the process is termed transient liquid phase sintering. For example, a mixture of copper and nickel powders heated over the melting temperature of copper forms a solid solution [71,72]. The final composition is single phase solid, as shown in Figure 9.44. Several applications rely on transient liquids during sintering—Al_2O_3-MgO-SiO_2, $BaTiO_3$-TiO_2, Cu-Al, Cu-Sn, Fe-Al, Fe-Mo-C, Fe-P, Fe-Si, Fe-Ti, Ni-Cu, Ni-Ti, Si_3N_4-AlN-SiO_2, and other mixed powder compositions.

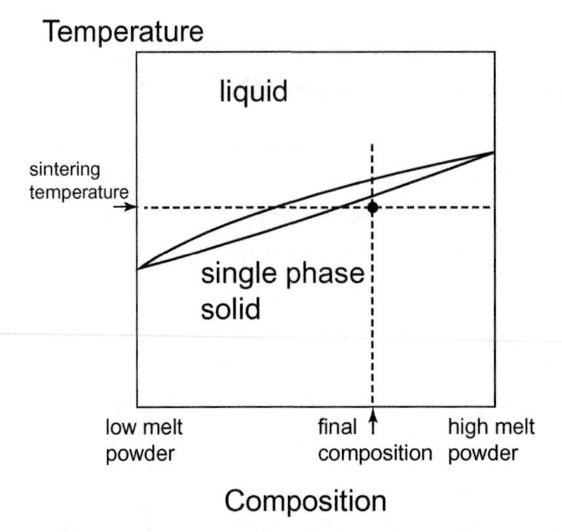

Figure 9.44 An example of the phase diagram characteristics associated with transient liquid phase sintering, showing the composition and temperature in a solid region, but during heating the low melt powder would have melted.

A common example is dental amalgam, based on silver and mercury. Silver-based alloy powder is mixed with liquid mercury, and the slurry is compressed into a dental cavity prior to disappearance of the liquid in a few minutes. If there is no pressure and large particles are used, then large pores remain. Thus, for dental fillings the mixture is compressed to ensure pore closure. Bronze bearings are another application. These are formed by mixing low melting tin with copper, heating over the tin melting temperature, dissolving the molten tin to form bronze, with a pore at the former tin particle site.

The events during transient liquid phase sintering are:

- swelling by interdiffusion prior to melt formation,
- melt formation,
- spreading of the melt to generate pores at additive particle sites
- melt penetration along solid-solid contacts,
- rearrangement of the solid grains,
- solution-reprecipitation induced densification,
- diffusional homogenization (solidification) with a loss of melt,
- formation of a rigid solid structure with continued solid state sintering.

The steps depend on variables such as particle sizes, composition, heating rate, and maximum temperature.

Pores form due to spreading of the melt. If the initial density is high, then liquid flow is inhibited. The greater the initial homogeneity the more swelling in the powder structure, with less liquid duration [73]. Pore refilling is inhibited by the short liquid duration, resulting in large pores in the final microstructure.

In transient liquid phase sintering, the heating rate is a dominant factor; more so than with other processes. Swelling occurs during heating, and more swelling occurs at slower heating rates. This is due to diffusional homogenization with less liquid forming at the peak temperature. The liquid quantity and duration determine shrinkage. The relation between liquid volume fraction V_L, additive concentration C, and heating rate dT/dt is:

$$1 - \left(\frac{V_L}{\zeta C}\right)^{1/3} = \kappa\left[T_L \frac{dT}{dt}\right]^{1/2} \tag{9.28}$$

where ξ and κ are constants, and T_L is the liquid formation temperature. Fast heating suppresses solid interdiffusion prior to liquid formation, leading to better densification.

The amount of liquid is reduced by use of smaller particles. Depending on the phase diagram, solid-state interdiffusion can form an intermediate compound around the additive particles. This envelope inhibits subsequent interdiffusion. The thickness of the intermediate compound increases with the square root of time, giving progressive swelling. But for strongly exothermic reactions, the compact is self-heating. In such cases programmed heating rates are meaningless once the reaction initiates.

Faster heating rates, smaller particle sizes, and lower additive contents produce the best mechanical properties. However, pore formation during heating is a major difficulty. If the pores are large, they remain stable during the balance of the sintering cycle. Of course that is the intent when forming porous bearings using transient liquid phase sintering.

SUPERSOLIDUS SINTERING

Liquids provide for rapid densification. In supersolidus liquid phase sintering, the initial powder is an alloy, and when heated the liquid forms within each particle. Densification in supersolidus liquid phase sintering is a viscous flow process. This is because liquid forming inside the particles fragments the particles into a mixture of solid grains and liquefied grain boundaries. Because of the internal melting, even large particles exhibit rapid densification. The process is most fruitful when applied to alloyed materials, including stainless steels, nickel-base superalloys, cobalt-chromium alloys, and tool steels [74].

Micrographs show that the liquid disintegrates the particles. Figure 9.45 is a micrograph of large bronze particles quenched from the liquid formation temperature of

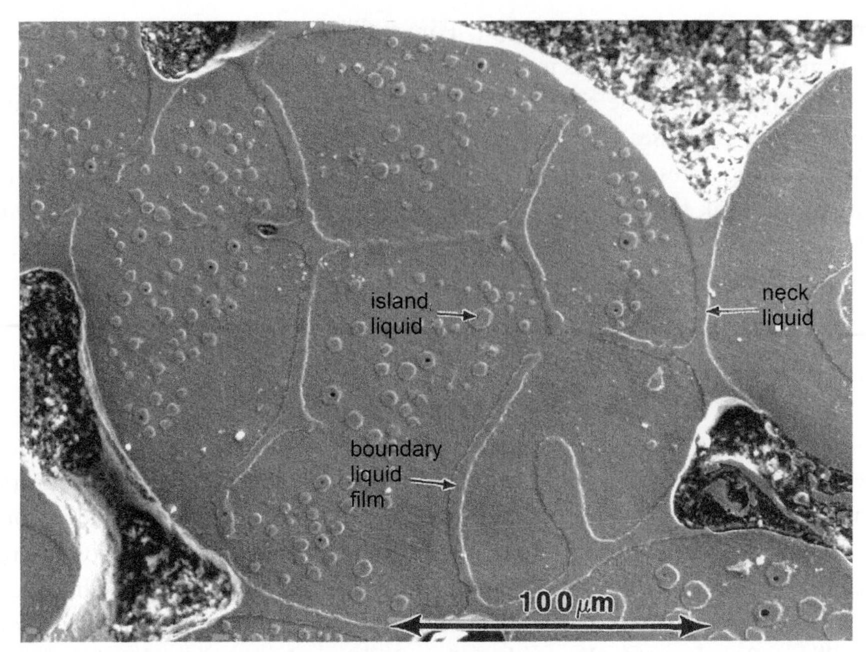

Figure 9.45 Supersolidus liquid phase sintering is captured in this micrograph quenched from just over the solidus temperature for bronze, illustrating the three locations for liquid at the interparticle neck, along grain boundaries, and as discrete islands inside the grains.

863°C (1136 K). A liquid grain boundary film is evident as well as liquid pockets inside the grains. Liquid bridges have formed between particles, providing capillary pressure to densify the soft structure. The wetted grain boundaries soften the structure to allow rapid viscous flow densification in a nearly instantaneous manner, unlike traditional liquid phase sintering.

Densification Mechanism

Densification during supersolidus liquid phase sintering is analogous to viscous flow sintering. As depicted in Figure 9.46, prealloyed particles nucleate liquid at the particle neck, inside the grains, and along grain boundaries. This combination turns the solid particle mushy, and once sufficient liquid forms along the grain boundaries the capillary stress induces densification. The higher the temperature the more liquid forms, and in turn the viscosity of the solid-liquid mixture decreases. This makes for rapid sintering. With too much liquid the system distorts, so a temperature balance is required to induce densification and avoid distortion. The important role of temperature is illustrated by the nickel alloy data in Figure 9.47. The alloy was doped with boron to nucleate liquid on the grain boundaries at 980°C (1253 K), resulting in rapid densification.

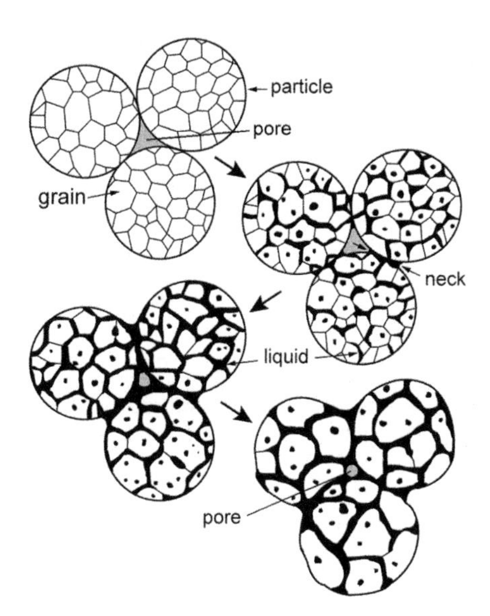

Figure 9.46 Schematic of the supersolidus liquid phase sintering process, where polycrystalline particles heated to the solidus temperature nucleate liquid and soften the semisolid structure to enable rapid viscous flow densification of the liquid-solid mixture.

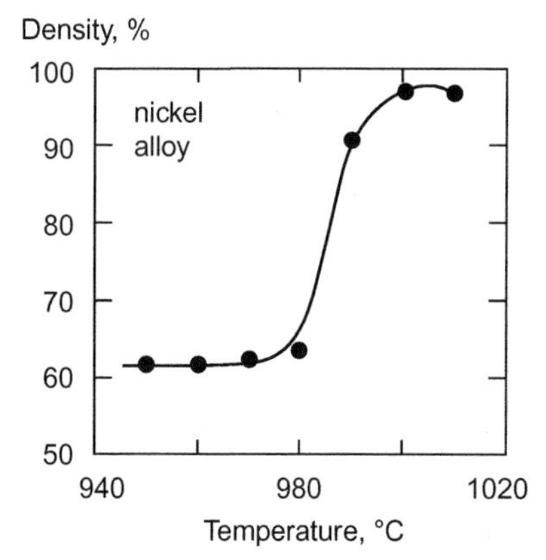

Figure 9.47 Density versus hold temperature (15 min) for a 70 μm nickel alloy powder densified by supersolidus liquid phase sintering.

Densification occurs soon after passing a critical temperature. Prolonged high temperature holds give relatively minor densification gain and cause property decrements. For example with a nickel-base superalloy, 97% density is achieved in 30 minutes and 99% in 60 minutes at temperature; but heating for longer than 60 minutes results in a density decrease. Figures 9.48 illustrates the role of hold time. These data are for tool steel sintered for 3 min or 10 min holds 1220 to 1260°C (1493 to 1533 K) [75].

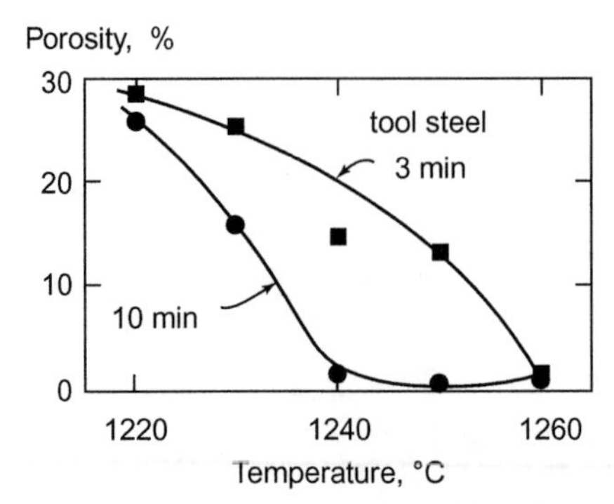

Figure 9.48 Porosity for supersolidus liquid phase sintered tool steel held for 3 or 10 min at temperatures from 1220 to 1260°C (1493 to 1533 K) [75].

For the higher sintering temperature (approximately 1260°C or 1533 K), hold time is not a factor

Such data are characteristic of powders that melt over a temperature range. Rapid densification occurs when heated above the solidus temperature. Of course the same liquid film that softens and induces rapid densification also contributes to rapid grain growth. Such behavior is illustrated in Figure 9.49 [76]. These data are for a tool steel sintered between 1230 and 1240°C (1503 to 1513 K) for 1 to 100 min. Within the optimal temperature range the quantity of liquid is 20 to 40 vol.%. An excess of liquid, from too high a temperature, lowers the viscosity to the condition where the component distorts.

Particle Attributes

Supersolidus liquid phase sintering successfully densifies large particles. This is an advantage in comparison with classic liquid phase sintering, which relies on 1 μm or smaller particles. Full density is attained with powders as large as 500 μm. A particle size near 80 μm or smaller works well for many systems. One important aspect is the solidification rate in forming the powder. Smaller particles cool quickly in atomization, and that favorably influences the grain boundary liquid formation needed for densification.

Densification Model

Densification requires a liquid film on the grain boundaries. For a particle diameter *D*, the grain size and film thickness between grains dictate the volume of liquid needed to coat the grain boundaries. Viscous flow exhibits a transition point from solid to semisolid behavior as the amount of liquid increases to coat the grain boundaries. Thus, sintering shrinkage is rapid once sufficient liquid is formed.

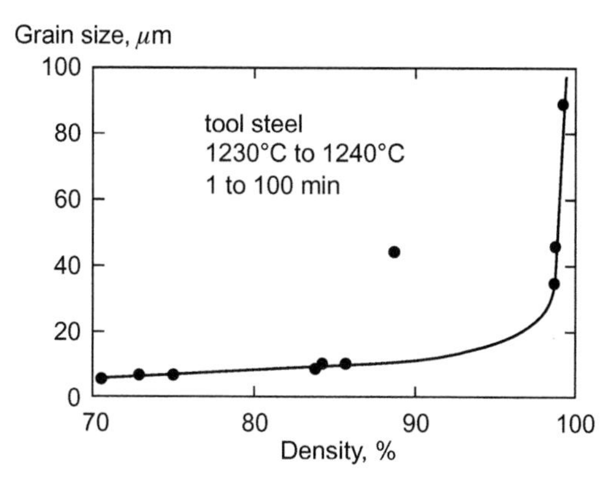

Figure 9.49 Concomitant with densification, the liquid formed on the grain boundaries in supersolidus liquid phase sintering contributes to grain growth. This plot shows grain size versus density for a tool steel [76]. The behavior is similar to that seen in other sintering cycles, where the grain size is proportional to the inverse square root of the fractional porosity.

Densification depends on two parameters, α and φ_C, where φ_C represents the critical solid fraction for viscous flow and α is an energy dissipation term:

$$\alpha = \frac{\gamma_{LV} t}{D\,\eta_O} \tag{9.29}$$

and:

$$\varphi_C = 1 - \frac{g_S \delta\, F_C}{2\,G(1 - F_I) g_V} \tag{9.30}$$

where F_C is the fractional coverage of grain boundaries by liquid, F_I is the fraction of liquid captured inside the grains, G is the grain size, g_V and g_S are geometric constants that depend on the grain shape, t is the isothermal sintering time, D is the starting particle diameter, and η_O is the viscosity of the liquid phase. The link between the semisolid viscosity η and solid fraction φ is:

$$\eta = \frac{\eta_O}{(1-\varphi/\varphi_C)^2} \tag{9.31}$$

The sintered fractional density f is given as follows:

$$f = \frac{f_O}{\left[1 - \frac{3}{4}\alpha(1-\varphi/\varphi_C)^2\right]^3} \tag{9.32}$$

where the solid fraction φ is a function of composition and temperature. Higher temperatures increase the liquid fraction and induce rapid densification. A large value for

α gives a sharp increase in density once sufficient liquid forms. The temperature of the onset of densification is dictated by the critical volume fraction of solid φ_C. Low values of φ_C let densification occur soon after first liquid formation. The density versus temperature curve is dictated by α, while the temperature position where densification begins is dictated by φ_C, which is usually about 20 to 40 vol.% liquid.

REACTIVE LIQUIDS

Some mixed powders react on heating to give reactive sintering, also known as self-propagating high temperature synthesis. Usually the reaction is difficult to control and produces a porous compound, which is the case for borides, carbides, nitrides, sulfides, oxides, hydrides, aluminides, or silicides. Heat is required to initiate the reaction, but like a forest fire, once ignited the reaction propagates by itself. In diffusion-controlled reactions, a kinetic model links the fraction reacted β to the isothermal reaction time t as follows:

$$1-(1-\beta)^{\frac{1}{3}} = \Gamma t^{1/2} \tag{9.33}$$

where the factor Γ is a temperature-dependent diffusion parameter.

Sintering the shape memory compound NiTi from mixed powders that react during heating is one system of great interest [77]. Densification requires an idealized green microstructure. If a reactant is isolated, then it forms pockets of liquid that flow into the surrounding solid leaving behind a pore. On the other hand, a homogeneous green microstructure consisting of interlaced networks of both constituents leads to densification.

Due to the exothermic reaction energy, component self-heating can reach 1200°C or more above the initiation temperature. The initial reaction occurs asymmetrically where the first melt forms, generating a wave that propagates through the compact similar to a combustion wave. The porous powder in front of the combustion wave is heated to initiate the reaction, so the reaction velocity depends on heat transfer. Velocities of up to 150 mm/s have been measured, but slower rates are more typical. Figure 9.50 plots the combustion velocity versus composition for Ni + Al powders. The peak velocity corresponds to the NiAl intermetallic compound.

Control is attained via the green microstructure and heating conditions. Preheating the compact before initiation boosts the peak temperature to improve densification, but the addition of inert particles, including previously reacted products, dilutes the reaction and reduces the peak temperature. This helps maintain compact shape.

Reactive liquid phase sintering can densify and react simultaneously [78–81]. The heat needed for sintering is delivered by the synthesis reaction. The most desirable situations provide long duration thermal pulses to ensure sufficient time and

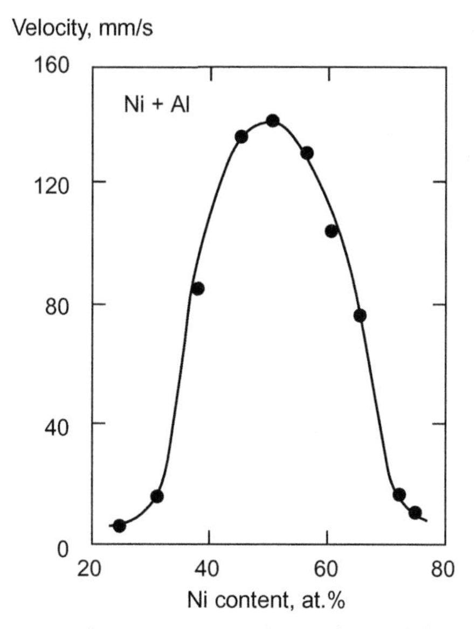

Figure 9.50 Reactive sintering combustion wave velocity for nickel and aluminum powders, illustrating how composition influences the rate of heating. The peak velocity corresponds to the largest exotherm during sintering at the NiAl intermetallic compound.

temperature for sintering. Thus, smaller components with high surface areas over which heat is lost are difficult to fabricate.

So far only a few materials are produced using reactive liquid phase sintering—WC, Ni_3Al, NiAl, $MoSi_2$, Fe-Al, and TiB_2 are examples. A reactive sintered microstructure is shown in Figure 9.51 for Ni_3Al sintered without pressure. In this instance, the particle size ratio was selected to give a green body with both the nickel and aluminum being interconnected. Unfortunately, reactive liquid phase sintering is sensitive to component size. As the component surface area to volume ratio varies, heat loss changes the optimal heating rate.

INFILTRATION SINTERING

A variant of liquid phase sintering is used when composite properties are sought from a system that will not densify. Full density is achieved by infiltrating a liquid during sintering, a process historically termed sinter-casting. There are two variants:

- Two step approach—first the solid is sintered to form a skeleton and in a second heating cycle the liquid is formed to infiltrate the solid skeleton.

Figure 9.51 Reactive sintered Ni_3Al produced from mixed elemental powders heated to about 600°C (873 K). Both reaction and sintering occur simultaneously once the liquid forms and spreads, leading to the reactive sintering event.

- One step approach—a laminated green body is formed where the solid sinters during heating and the melt infiltrates the just-sintered skeleton.

The composite microstructure essentially has full density, as evident from Figure 9.52. The solid structure is filled with liquid. Wetting is important to induce capillary filling of the pores. In the case of poor wetting, an external pressure is used to force infiltration such as boron nitride and aluminum [82,83].

Applications are primarily for systems such as SiC-Al, Cr-Cu, CdO-Ag, Fe-Ag, TiC-Ni, W-Cu, Mo-Ag, W-Ag, TiC-Fe, and WC-Cu. It is possible to adjust particle size to form the sintered solid skeleton prior to melt formation. Otherwise, shape loss occurs when the liquid forms.

Liquid infiltration requires interconnected pores with at least 10% porosity. The melt flows through the pores from an external surface. After the melt forms, the depth of infiltration h varies with the square-root of time t as:

$$h = \left(\frac{d\ t\ \gamma_{LV} cos(\theta)}{4\,\eta} \right)^{1/2} \tag{9.34}$$

where d is the pore diameter, γ_{LV} is the liquid-vapor surface energy, ϑ is the solid-liquid contact angle, and η is the liquid viscosity.

A variant is reaction infiltration, where the infiltrating liquid reacts with the solid, similarly to transient liquid phase sintering. One example is a diamond skeleton infiltrated with liquid silicon. The product is a composite of diamond particles in a silicon

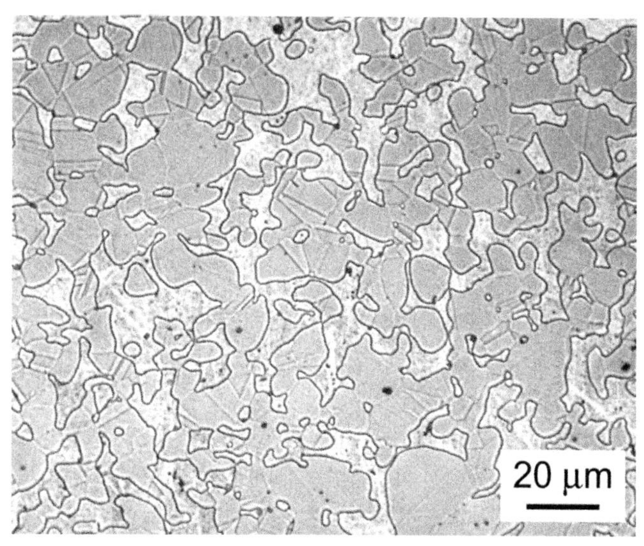

Figure 9.52 Infiltration sintering involves sintering a solid skeleton coupled with filling the pores with a liquid phase as illustrated here for invar (Fe-36Ni) as the solid and silver as the liquid.

carbide matrix, the SiC being formed by reacting silicon with the diamond. This material exhibits excellent wear properties, but the process is difficult to control.

ACTIVATED LIQUID PHASE SINTERING

In cases where sintering is slow, an activator might be used to promote diffusion [84]. For example, in the W-Cu system, copper and tungsten have little solubility, so sintering over the copper melting temperature is slow. Use of a high temperature to induce tungsten sintering results in copper evaporation. Adding about 0.3% Co provides activated sintering of the tungsten without the reacting with the copper liquid. The solid sinters with liquid in the voids between the solid. The primary transport process is via activated sintering. It adds a rapid interface transport path, namely cobalt, to induce solid sintering in the presence of the copper liquid.

In order for activated liquid phase sintering to work, the activator must be insoluble in the liquid. A similar process relies on infiltration sintering with an activator. The solid skeleton is densified prior to filling the pores with infiltrant. This is especially effective when densification of the solid skeleton proves difficult.

Examples where activated liquid phase sintering or activated infiltration sintering are used are W-Cu and W-Ag [85]. The normally high temperature skeleton sintering is from the added activator. One application is in making high density birdshot using tungsten-bronze-iron mixtures, with iron as the activator for the tungsten solid. These systems are effective since the activator is not dissolved by the liquid, so it remains on the solid interfaces.

PRACTICAL ASPECTS

Besides the typical time-temperature aspects of sintering, liquid phase sintering is especially sensitive to green density, particle size, and liquid homogeneity. The liquid flows on melting. Coated powders are one means of ensuring homogeneous liquid formation [86,87]. It is possible to design systems for zero dimensional change where the swelling and shrinkage compensate, which is desirable for the fabrication of complex components. Here, the green body homogeneity is critical, since tool pressure gradients prior to sintering induce dimensional distortion in sintering.

Particle size is important to liquid phase sintering. Smaller particles are more suitable. The exceptions are those cases where controlled porosity is desired, and also supersolidus liquid phase sintering. For persistent liquid phase sintering, shrinkage is inversely dependent on particle size. Thus, as plotted in Figure 9.53, densification is enhanced by smaller particles as has been long recognized [88,89]. However, grain coarsening is rapid, so the benefit of a small starting particle size is lost during sintering. As one example, starting with 11 to 32 nm WC-12Co particles, the grain size after sintering to nearly full density (99.52%) at 1400°C (1673 K) was 1.25 μm.

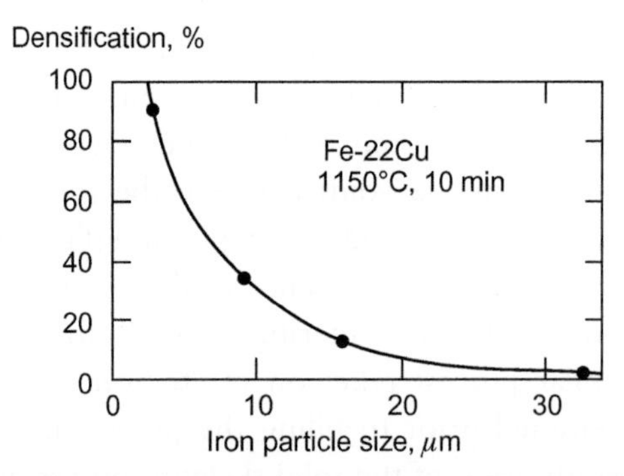

Figure 9.53 Sintering densification versus iron particle size for the liquid phase sintering of Fe-22Cu at 1150°C (1423 K) with a 10 min hold [88,89]. Smaller particles have more capillary force, shorter transport distances, and generally accelerates liquid phase sintering.

SUMMARY

There are two trajectories possible when mixed particles are heated above the melt formation temperature. Swelling occurs when the liquid is soluble in the solid and is useful in forming porous structures. The more typical situation is where the solid is soluble in the liquid. This is associated with densification.

For most liquid phase sintering cycles, densification is the goal. Depending on the amount of liquid, densification will require some or all of the stages—rearrangement, solution-reprecipitation, and solid skeleton sintering. Usually a short sintering time is required once the liquid forms. Other factors are important in terms of the detailed trajectory, including particle size, green density, heating rate, amount of liquid, hold time, and process atmosphere.

The models for liquid phase sintering treat the process as a sequence of steps, but actually the steps overlap. With no liquid, only slower solid-phase sintering occurs. Once liquid forms, it flows to fill the gaps between the solid grains. The capillary force combines with dissolution of solid-solid bonds to give rapid grain rearrangement. Next solution-reprecipitation and solid skeleton sintering act. Densification by solution-reprecipitation occurs with microstructure coarsening, wherein pores are annihilated as the grain size and grain separation increase. Solid diffuses through the liquid and deposits on convex surfaces, allowing grain shape changes, releasing liquid to fill pores.

The initial chemical gradients are important to the initial sintering trajectory, as is the green body density homogeneity. Integral work concepts have successfully lumped parameters for densification, grain growth, distortion, and strength evolution [60,61,90–94]. These ideas are integrated into process simulations to enable predictions of size, shape, density, grain size, distortion, and strength.

REFERENCES

[1] S. Kemethmuller, M. Hagymasi, A. Stiegelschmitt, A. Roosen, Viscous flow as the driving force for the densification of low-temperature Co-Fired ceramics, J. Amer. Ceram. Soc. 90 (2007) 64–70.
[2] F.V. Lenel, Sintering in the presence of a liquid phase, Trans. TMS-AIME 175 (1948) 878–896.
[3] H.S. Cannon, F.V. Lenel, Some observations on the mechanism of liquid phase sintering, in: F. Benesovsky (Ed.), Plansee Proceedings, Metallwerk Plansee, Reutte, Austria, 1953, pp. 106–121.
[4] W.D. Kingery, Sintering in the presence of a liquid phase, in: W.D. Kingery (Ed.), Ceramic Fabrication Processes, John Wiley, NY, 1958, pp. 131–143.
[5] W.D. Kingery, Densification during sintering in the presence of a liquid phase 1. Theory, J. Appl. Phys. 30 (1959) 301–306.
[6] D. Kingery, M.D. Narasimhan, Densification during sintering in the presence of a liquid phase 2. Experimental, J. Appl. Phys. 30 (1959) 307–310.
[7] V.N. Eremenko, Y.V. Naidich, I.A. Lavrinenko, Liquid Phase Sintering, Consultants Bureau, New York, NY, 1970.
[8] R.M. German, Liquid Phase Sintering, Plenum Press, New York, NY, 1985.

[9] R.M. German, P. Suri, S.J. Park, Review liquid phase sintering, J. Mater. Sci. 44 (2009) 1–39.
[10] K.S. Hwang, R.M. German, F.V. Lenel, Capillary forces in the early stage of liquid phase sintering, Rev. Powder Metall. Phys. Ceram. 3 (1986) 113–164.
[11] J. Fikes, S.J. Park, R.M. German, Equilibrium states of liquid, solid, and vapor and the configurations for copper, tungsten, and pores in liquid phase sintering, Metall. Mater. Trans. 42B (2011) 202–209.
[12] E. Liden, E. Carlstrom., L. Eklund, B. Nyberg, R. Carlsson, Homogeneous distribution of sintering additives in liquid-phase sintered silicon carbide, J. Amer. Ceram. Soc. 78 (1995) 1761–1768.
[13] A. Nakajima, G.L. Messing, Liquid phase sintering of alumina coated with magnesium aluminosilicate glass, J. Amer. Ceram. Soc. 81 (1998) 1163–1172.
[14] B. Ozkal, A. Upadhyaya, M.L. Ovecoglu, R.M. German, Comparative properties of 85W-15Cu powder prepared using mixing, milling, and coating techniques, Powder Metall. 53 (2010) 236–243.
[15] K. Flemming, K.P. Wieters, B. Kieback, The sintering behavior of coated particles, in: D. Bouvard (Ed.), Proceedings of the 4th International Conference on Science, Technology and Applications of Sintering, Institut National Polytechnique de Grenoble, Grenoble, France, 2005, pp. 21–24.
[16] J. Konstanty, Powder metallurgy diamond tools, Elsevier, Amsterdam, Netherlands, 2005.
[17] R. Oro, M. Campos, C. Gierl, H. Danninger, J.M. Torralba, Atmosphere effects on liquid phase sintering of PM steels modified with master alloy additions, Proceedings PM 2010 World Congress, Florence Italy, European Powder Metallurgy Association, Shrewsbury, UK, 2010.
[18] G.B. Schaffer, J.Y. Yao, S.J. Bonner, E. Crossin, S.J. Pas, A.J. Hill, The effect of tin and nitrogen on liquid phase sintering of Al-Cu-Mg-Si alloys, Acta Mater. vol. 56 (2008) 2615–2624.
[19] M. Humenik, N.M. Parikh, Cermets: I, fundamental concepts related to microstructure and physical properties of cermet systems, J. Amer. Ceram. Soc. 39 (1956) 60–63.
[20] K.S. Hwang, R.M. German, F.V. Lenel, Capillary forces between spheres during agglomeration and liquid phase sintering, Metall. Trans. 18A (1987) 11–17.
[21] L. Froschauer, R.M. Fulrath, Direct observation of liquid-phase sintering in the system iron-copper, J. Mater. Sci. 10 (1975) 2146–2155.
[22] A. Belhadjhamida, R.M. German, A model calculation of the shrinkage dependence on rearrangement during liquid phase sintering, Advances in Powder Metallurgy and Particulate Materials – 1993, vol. 3, Metal Powder Industries Federation, Princeton, NJ, 1993, pp. 85–98.
[23] S.J. Jamil, G.A. Chadwick, Investigation and analysis of liquid phase sintering of Fe-Cu and Fe-Cu-C compacts, Powder Metall. 28 (1985) 65–71.
[24] O.H. Kwon, G.L. Messing, Kinetic analysis of solution-reprecipitation during liquid phase sintering of alumina, J. Amer. Ceram. Soc. 73 (2) (1990) 275–281.
[25] T.M. Shaw, Model for the effect of powder packing on the driving force for liquid phase sintering, J. Amer. Ceram. Soc. 76 (1993) 664–670.
[26] D.Y. Yang, D.Y. Yoon, S.J.L. Kang, Abnormal grain growth enhanced densification of liquid phase sintered WC-Co in support of the pore filling theory, J. Mater. Sci. 47 (2012) 7056–7063.
[27] P. Lu, X. Xu, W. Yi, R.M. German, Porosity effect on densification and shape distortion in liquid phase sintering, Mater. Sci. Eng. A318 (2001) 111–121.
[28] N.K. Xydas, L.A. Salam, Transient liquid phase sintering of high density Fe_3Al using Fe and Fe_2Al_5-$FeAl_2$ powders, Part 2 – densification mechanism, Powder Metall. 49 (2006) 146–152.
[29] S.M. Lee, S.J.L. Kang, Theoretical analysis of liquid phase sintering pore filling theory, Acta Mater. 46 (1998) 3191–3202.
[30] M.D. Sacks, N. Bozkurt, G.W. Scheiffele, Fabrication of mullite and mullite-matrix composites by transient viscous sintering of composite powders, J. Amer. Ceram. Soc. 74 (1991) 2428–2437.
[31] P.J. Wray, The geometry of two-phase aggregates in which the shape of the second phase is determined by its dihedral angle, Acta Metall. 24 (1976) 125–135.
[32] W. Beere, A unifying theory of the stability of penetrating liquid phases and sintering pores, Acta Metall. 23 (1975) 131–138.
[33] R. Warren, Microstructural development during the Liquid-Phase sintering of two-phase alloys with special reference to the NbC/Co system, J. Mater. Sci. 3 (1968) 471–485.

[34] A. Tewari, A.M. Gokhale, R.M. German, Effect of gravity on three-dimensional coordination number distribution in liquid phase sintered microstructures, Acta Mater. 47 (1999) 3721–3734.
[35] R.M. German, Grain agglomeration in solid-liquid mixtures under microgravity conditions, Metall. Mater. Trans. 26B (1995) 649–651.
[36] W.A. Kaysser, S. Takajo, G. Petzow, Particle growth by coalescence during liquid phase sintering of Fe-Cu, Acta Metall. 32 (1984) 115–122.
[37] W.J. Boettinger, P.W. Voorhees, R.C. Dobbyn, H.E. Burdette, A study of the coarsening of liquid-solid mixtures using synchrotron radiation microradiography, Metall. Trans. 18A (1987) 487–490.
[38] Z. Fang, B.R. Patterson, M.E. Turner, Influence of particle size distribution on coarsening, Acta Metall. Mater. 40 (1992) 713–722.
[39] R.M. German, E.A. Olevsky, Modeling grain growth dependence on the liquid content in liquid-phase sintered materials, Metall. Mater. Trans. 29A (1998) 3057–3067.
[40] S. Sarin, H.W. Weart, Kinetics of coarsening of spherical particles in a liquid matrix, J. Appl. Phys. 37 (1966) 1675–1681.
[41] A.N. Niemi, T.H. Courtney, Microstructural development and evolution in liquid-phase sintered Fe-Cu alloys, J. Mater. Sci. 16 (1981) 226–236.
[42] N. Limodin, L. Salvo, M. Suery, F. Delannay, Assessment by microtomography of 2D image analysis method for the measurement of average grain coordination and size in an aggregate, Scripta Mater. 60 (2009) 325–328.
[43] J. Gurland, The measurement of grain contiguity in two-phase alloys, Trans. TMS-AIME 212 (1958) 452–455.
[44] R. Warren, M.B. Waldron, Microstructural development during the liquid-phase sintering of cemented carbides I. Wettability and grain contact, Powder Metall. 15 (1972) 166–180.
[45] A. Upadhyaya, R.M. German, Shape distortion in liquid-phase-sintered tungsten heavy alloys, Metall. Mater. Trans. 29A (1998) 2631–2638.
[46] A.V. Shatov, S.S. Ponomarev, S.A. Firstov, R. Warren, The contiguity of carbide crystals of different shapes in cemented carbides, Inter. J. Refract. Met. Hard Mater. 24 (2006) 61–74.
[47] J.L. Cahn, J.R. Alcock, D.J. Stephenson, Supersolidus liquid phase sintering of moulded metal components, J. Mater. Sci. 33 (1998) 5131–5136.
[48] A. Eliasson, L. Ekbom, H. Fredriksson, Tungsten grain separation during initial stage of liquid phase sintering, Powder Metall. 51 (2008) 343–349.
[49] S. Farooq, A. Bose, R.M. German, Theory of liquid phase sintering: model experiments on W-Ni-Fe heavy alloy system, Prog. Powder Metall. 43 (1987) 65–77.
[50] S. Pejovnik, D. Kolar, W.J. Huppmann, G. Petzow, Sintering of Al_2O_3 in presence of liquid phase, in: M.M. Ristic (Ed.), Sintering – New Developments, Elsevier Scientific, New York, NY, 1979, pp. 285–292.
[51] G.W. Greenwood, The growth of dispersed precipitates in solutions, Acta Metall. 4 (1956) 243–248.
[52] N.M. Parikh, M. Humenik, Cermets: II, wettability and microstructure studies in liquid-phase sintering, J. Amer. Ceram. Soc. 40 (1957) 315–320.
[53] L.P. Skolnick, Grain growth of titanium carbide in nickel, Trans. TMS-AIME 209 (1957) 438–442.
[54] R. Watanabe, Y. Masuda, The growth of solid particles in Fe-20 wt.% Cu alloy during sintering in the presence of a liquid phase, Trans. Japan Inst. Metals 14 (1973) 320–326.
[55] W.J. Huppmann, The elementary mechanisms of liquid phase sintering 2. Solution – reprecipitation, Z. Metall. 70 (1979) 792–797.
[56] M.N. Rahaman, Kinetics and mechanisms of densification, in: Z.Z. Fang (Ed.), Sintering of Advanced Materials, Woodhead Publishing, Oxford, UK, 2010, pp. 33–64.
[57] D.J. Srolovitz, M.G. Goldiner, The thermodynamics and kinetics of film agglomeration, J. Met. 47 (3) (1995) 31–36.
[58] J.E. Marion, C.H. Hsueh, A.G. Evans, Liquid phase sintering of ceramics, J. Amer. Ceram. Soc. 70 (1987) 708–713.
[59] G.H. Gessinger, H.F. Fischmeister, H.L. Lukas, A model for second-stage liquid-phase sintering with a partially wetting liquid, Acta Metall. 21 (1973) 715–724.

[60] S.J. Park, J.M. Martin, J.F. Guo, J.L. Johnson, R.M. German, Grain growth behavior of tungsten heavy alloys based on the master sintering curve concept, Metall. Mater. Trans. 37A (2006) 3337–3346.

[61] S.J. Park, J.M. Martin, J.F. Guo, J.L. Johnson, R.M. German, Densification behavior of tungsten heavy alloy based on master sintering curve concept, Metall. Mater. Trans. 37A (2006) 2837–2848.

[62] S.J. Park, S.H. Chung, J.M. Martin, J.L. Johnson, R.M. German, Master sintering curve for densification derived from a constitutive equation with consideration of grain growth – application to tungsten heavy alloys, Metall. Mater. Trans. 39A (2008) 2941–2948.

[63] S. Sarin, H.W. Weart, Factors affecting the morphology of an array of solid particles in a liquid matrix, Trans. TMS-AIME 233 (1965) 1990–1994.

[64] H.E. Exner, Ostwald-reifung von ubergangsmetallkarbiden in flussingem nickel und kobalt, Z. Metall. 64 (1973) 273–279.

[65] O.H. Kwon, G.L. Messing, A. Theoretical, Analysis of solution-reprecipitation controlled densification during liquid phase sintering, Acta Metall. Mater. 39 (1991) 2059–2068.

[66] W.J. Huppmann, H. Riegger, Liquid phase sintering of the model system W-Ni, Inter. J. Powder Metall. Powder Tech. 13 (1977) 243–247.

[67] U.C. Oh, Y.S. Chung, D.Y. Kim, D.N. Yoon, Effect of grain growth on pore coalescence during the liquid phase sintering of MgO-$CaMgSiO_4$ systems, J. Amer. Ceram. Soc. 71 (1988) 854–857.

[68] R.M. German, K.S. Churn, Sintering atmosphere effects on the ductility of W-Ni-Fe heavy metals, Metall. Trans. 15A (1984) 747–754.

[69] J. White, Sintering of oxides and sulfides, in: G.C. Kuczynski, N.A. Hooton, C.F. Gibbon (Eds.), Sintering and Related Phenomena, Gordon and Breach, New York, NY, 1967, pp. 245–269.

[70] S.J.L. Kang, K.H. Kim, D.N. Yoon, Densification and shrinkage during liquid phase sintering, J. Amer. Ceram. Soc. 74 (1991) 425–427.

[71] D.M. Turriff, S.F. Corbin, L.M.D. Cranswick, M.J. Watson, Transient liquid phase sintering of copper-nickel powders in situ neutron diffraction, Inter. J. Powder Metall. 44 (6) (2008) 49–59.

[72] D.M. Turriff, S.F. Corbin, Quantitative thermal analysis of transient liquid phase sintered Cu-Ni powders, Metall. Mater. Trans. 39A (2008) 28–38.

[73] C.M. Kipphut, R.M. German, Alloy phase stability in liquid phase sintering, Sci. Sintering 20 (1988) 31–40.

[74] R.M. German, Supersolidus liquid phase sintering part I, process review, Inter. J. Powder Metall. 26 (1990) 23–34.

[75] S. Takajo, M. Nitta, M. Kawano, Behavior of liquid phase sintering of high speed steel powder compacts, J. Japan Soc. Powder Metall. 36 (1986) 398–401.

[76] S. Takajo, M. Kawano, M. Nitta, W.A. Kaysser, G. Petzow, Mechanism of liquid phase sintering in Fe-Cu, Cu-Ag and high speed steel, in: S. Somiya, M. Shimada, M. Yoshimura, R. Watanabe (Eds.), Sintering '87, 1, Elsevier Applied Science, London, UK, 1988, pp. 465–470.

[77] M. Whitney, S.F. Corbin, R.B. Gorbet, Investigation of the mechanisms of reactive sintering and combustion synthesis of NiTi using differential scanning calorimetry and microstructural analysis, Acta Mater 56 (2008) 559–570.

[78] Z.A. Munir, Synthesis of high temperature materials by self-propagating combustion methods, Ceram. Bull. 67 (1988) 342–349.

[79] A. Bose, B.H. Rabin, R.M. German, Reactive sintering nickel-aluminide to near full density, Powder Metall. Inter. 20 (3) (1988) 25–30.

[80] K. Taguchi, M. Ayada, K.N. Ishihara, P.H. Shingu, Near-net shape processing of TiAl intermetallic compounds Via Pseudo HIP-SHS route, Intermetallics 3 (1995) 91–98.

[81] L.C. Pathak, D. Bandyopadhyay, S. Srikanth, S.K. Das, P. Ramachandrarao, Effect of heating rates on the synthesis of Al_2O_3-SiC composites by the self-propagating high temperature synthesis (SHS) technique, J. Amer. Ceram. Soc. 84 (2001) 915–920.

[82] J.M. Molina, R. Prieto, M. Duarte, J. Narciso, E. Louis, On the estimation of threshold pressures in infiltration of liquid metals into particle preforms, Scripta Mater. 59 (2008) 243–246.

[83] H.S.L. Sithebe, D. Mclachlan, I. Sigalas, M. Hermann, Pressure infiltration of boron nitride preforms with molten aluminum, Ceram. Inter. 34 (2008) 1367–1371.

[84] J.L. Johnson, R.M. German, A theory of activated liquid phase sintering and its application to the W-Cu system, Advances in Powder Metallurgy and Particulate Materials, vol. 3, Metal Powder Industries Federation, Princeton, NJ, 1992, pp. 35–46.
[85] N.C. Kothari, Factors affecting tungsten-copper and tungsten-silver electrical contact materials, Powder Metall. Inter. 14 (1982) 139–159.
[86] S.C. Colbeck, Sintering and compaction of snow containing liquid water, Phil. Mag. A 39 (1979) 13–32.
[87] V. Yaroshenko, D.S. Wilkinson, Phase evolution during sintering of mullite/zirconia composite using silica coated alumina powders, J. Mater. Res. 15 (2000) 1358–1366.
[88] W.D. Kingery, Sintering in the presence of a liquid phase, in: W.D. Kingery (Ed.), Kinetics of High Temperature Processes, John Wiley, New York, NY, 1959, pp. 187–194.
[89] F.V. Lenel, Sintering with a liquid phase, in: W.E. Kingston (Ed.), The Physics of Powder Metallurgy, McGraw-Hill, New York, NY, 1951, pp. 238–253.
[90] R. Bollina, S.J. Park, R.M. German, Master sintering curve concepts applied to full density supersolidus liquid phase sintering of 316L stainless steel powder, Powder Metall. 53 (2010) 20–26.
[91] G. Sethi, S.J. Park, R.M. Johnson, German: Linking homogenization and densification in W-Ni-Cu alloys through Master Sintering Curve (MSC) Concepts, Inter. J. Refract. Met. Hard Mater 27 (2009) 688–695.
[92] G.A. Schoales, R.M. German, Combined effects of time and temprature on strength evolution using integral work-of-sintering concepts, Metall. Mater. Trans. 30A (1999) 465–470.
[93] S.J. Park, R.M. German, Master curves based on time integration of thermal work in particulate materials, Inter. J. Mater. Struct. Integ 7 (2007) 128–147.
[94] D.C. Blaine, S.J. Park, P. Suri, R.M. German, Application of work of sintering concepts in powder metallurgy, Metall. Mater. Trans. 37A (2006) 2827–2835.

CHAPTER TEN

Sintering With External Pressure

ROLE OF EXTERNAL PRESSURE

Sintering has difficulty densifying materials that naturally resist high temperatures. Normally the sintering driving force comes from the surface energy associated with small particles. An external pressure increases the driving force, giving faster densification. Thus, pressure-assisted sintering is a means to densify materials which are normally resistant to sintering.

The inherent sintering stress σ_S is proportion to the surface energy γ_{SV} divided by microstructure scale G (typically the grain size):

$$\sigma_S = \frac{g\gamma_{SL}}{G} \tag{10.1}$$

where g is a geometric parameter near 4. For a 40 μm particle with a surface energy of 2 J/m^2, the sintering stress is about 0.2 M Pa (two atmospheres pressure). This is the pressure pulling the particles together. An external pressure applied to the outside of the body supplements the sintering stress. In the early stage of sintering, an external stress is amplified at the particle contacts, and greatly enhances the internal sintering stress. The temperature required for sintering decreases as the available pressure increases.

Sintering techniques relying on external pressures give improved densification in materials otherwise resistant to sintering [1]. Because of the densification, and by implication improved performance, sintering with an external pressure focuses on demanding applications in aerospace, biomedical, and electronic materials. Many of the products are composites which are designed for high temperature use. This same attribute inherently makes sintering consolidation difficult. Consequently, sintering with an external pressure, also known as pressure-assisted sintering, is associated with demanding applications and higher performance materials.

Major property gains are associated with pressure-assisted sintering because of the elimination of residual porosity. Further, the homogeneity of the microstructure proves valuable in tool steels. Inclusions, such as sulfides associated with traditional metalworking, produce highly anisotropic mechanical properties. Figure 10.1 shows wrought steel with inclusions present as stringers along the flow direction. In this case, the tensile strength is dramatically different depending on orientation; varying by 150%. Such defects

Sintering: From Empirical Observations to Scientific Principles
DOI: http://dx.doi.org/10.1016/B978-0-12-401682-8.00010-0

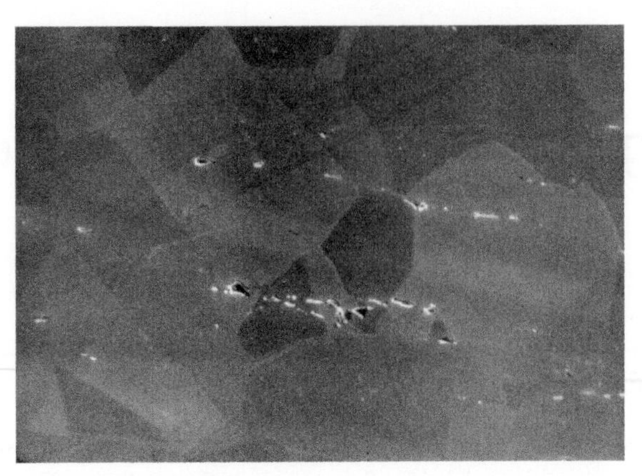

Figure 10.1 Micrograph of wrought steel showing MnS stringers in the microstructure aligned in the rolling direction. Such inclusions lead to anisotropic properties. Consolidation by pressure-assisted sintering delivers homogeneous properties by avoiding these defects.

are avoided by the pressure consolidation of small particles, an approach widely used in applications ranging from artificial knees to oil well drilling tools.

Several materials rely on pressure-assisted sintering. These include composites. Particular attention is directed to thermally unstable ceramics and instances where microstructure coarsening is a concern. For example, pure WC is difficult to form, but is consolidated in pressure-assisted sintering cycles. Likewise, Si_3N_4 and other nitrogen ceramics are stabilized with a nitrogen overpressure during sintering. Indeed, since the application of pressure during sintering adds to the expense and slows consolidation, most of the applications are in high value materials and devices.

To introduce pressure-assisted sintering, let the densification rate be expressed as:

$$\frac{df}{dt} = \frac{(1-f)}{(1-f_O)} B\left[\frac{g\,\gamma_{SV}}{G} + P_E - P_G\right] \quad (10.2)$$

where f is the fractional density, t is time, f_O is the starting fractional density, B is a thermally activated mass transport parameter that depends on the mass transport mechanism, P_E is the effective external pressure, P_G is the gas pressure in the pores, g is a geometric parameter, γ_{SV} is the surface energy, and the microstructure size is given by the grain size G. The densification rate goes to zero as full density is attained. The parameter B depends on material properties with an exponential temperature dependence.

Early in sintering the external pressure is amplified in the microstructure, since contact zones are small. This effective pressure P_E is several times the applied pressure. As a consequence, the densification rate is dominated by the applied pressure. To illustrate this behavior, Figure 10.2 plots data for copper and magnesia, showing an increase in sintered density with applied pressure for a 30 min hold [2]. The density at

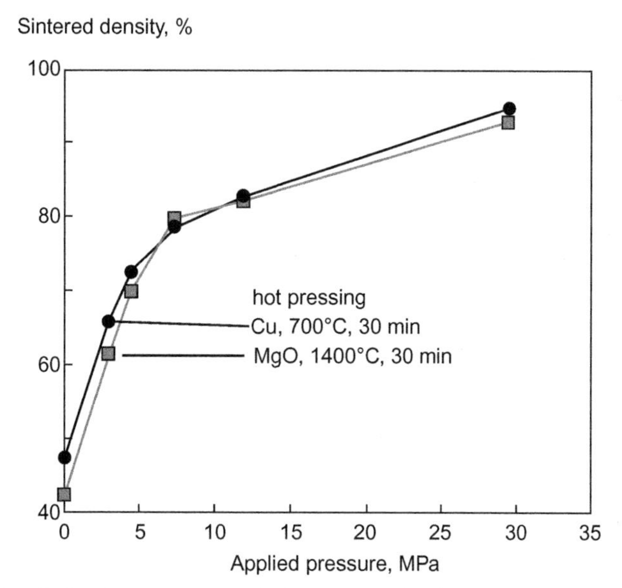

Figure 10.2 Fractional density versus applied pressure for hot pressed copper and magnesia (copper at 700°C and magnesia at 1400°C), showing the significant benefit from high pressures during the sintering cycle [2].

zero time increases with applied pressure, reflecting the initial powder compaction. After initial densification, slower diffusion events take over. Thus, plots of density versus time reflect a declining rate of densification, as is evident in Figure 10.3 [2].

Thus, sintering with an external pressure shows some important features:

- External pressure is amplified inside the body by the initially small particle contacts.
- Initial densification increases with pressure due to plastic flow.
- The rate of densification declines over time.
- The rate of densification declines as density increases.
- Temperature is important since diffusion events control late stage densification.
- Densification depends on the microstructure scale, particle size, and grain size.
- The sintering stress from surface energy is a minor term.
- Gas trapped in the pores retards densification.

THERMAL SOFTENING

Materials lose strength when heated, meaning that plastic flow occurs at lower stresses at elevated temperatures. As an example, Figure 10.4 plots the yield strength

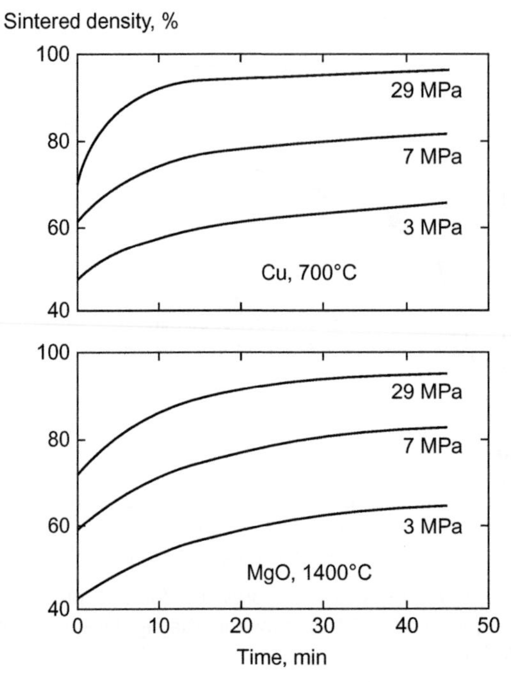

Figure 10.3 Another example of the hot pressing behavior for copper and magnesia [2] in terms of fractional density versus time for three applied pressures and one temperature.

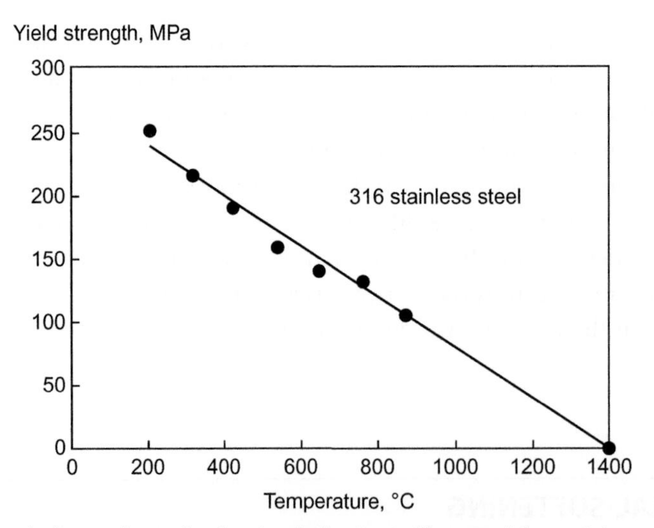

Figure 10.4 A nearly linear thermal softening behavior is illustrated for stainless steel in terms of the yield strength versus test temperature. Such thermal softening is important to pressure-assisted sintering.

of a stainless steel versus test temperature. This alloy melts at about 1400°C (1673 K) and has no strength once molten. This is relevant to pressure-assisted sintering, since even very low stresses are effective in deforming (densifying) a material if the temperature is sufficiently high. Almost all materials exhibit similar strength declinations, except for a few intermetallics, graphite, and nonoxide ceramics.

A model for strength (full density) as a function of processing parameters follows [3]:

$$\sigma = (A + B\,\varepsilon^n)\left[1 + C\ln\left(\frac{d\varepsilon^*}{dt}\right)\right]\{1 - T_H^m\} \tag{10.3}$$

where A, B, C, n, and m are material specific constants, ε is the plastic strain and $d\varepsilon^*/\mathrm{dt}$ is the strain rate in dimensionless form (effectively normalized to 1/s strain rate). Most importantly, T_H is the homologous temperature equal to the absolute temperature divided by the absolute melting temperature. The temperature sensitivity m is generally unity, as seen in Figure 10.4. The importance of thermal softening is that even low pressures will be sufficient to densify a powder by plastic flow if the temperature is sufficiently high.

Thermal softening means that powders otherwise difficult to densify will soften with enough heat, and this is the basis for pressure-assisted sintering. Glass is a good example, since at room temperature it is hard, brittle, and resists deformation. At high temperatures glass is easily formed at modest pressures. In the same manner, elevated temperatures aid sintering because of the material's softness and increased workability. A demonstration of the important role of temperature and pressure is captured in Figure 10.5. This plots the temperature-pressure combinations required to reach 99% density for iron powder by hot pressing [4]. Thermal softening lowers the required pressure.

A related plot, shown in Figure 10.6, shows the density attained in Fe-2Ni with a constant pressure using various temperatures [5]. Plastic flow at high temperatures gives fast consolidation. For pressure-assisted sintering there is substantial plastic flow that is often absent in traditional sintering.

PRESSURE EFFECTS

When pressure is applied, the particles flatten at the contact points. Even with low applied pressures, small contacts significantly amplify the pressure. The corresponding density change is dramatic. The contacts enlarge until the local pressure falls below the yield strength at the processing temperature.

Stress Concentration

When pressure is applied to a powder compact, stress concentrates at particle contacts. Plastic flow gives rapid densification. Likewise, during creep the contact stress promotes

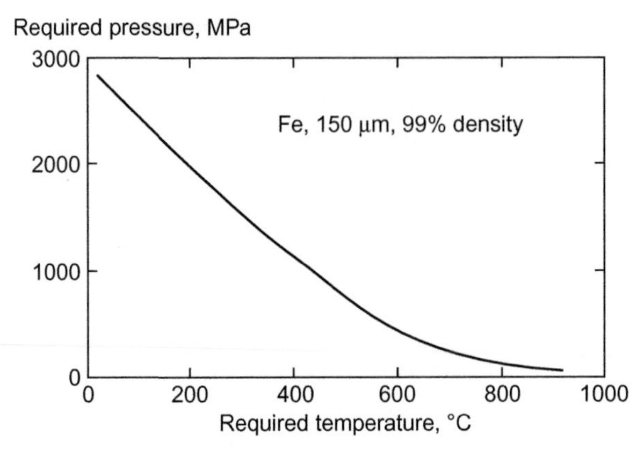

Figure 10.5 Temperature and pressure combination required to hot press 150 μm iron powder to 99% density [4]. A higher temperature softens the material such that a lower pressure is required to densify the powder.

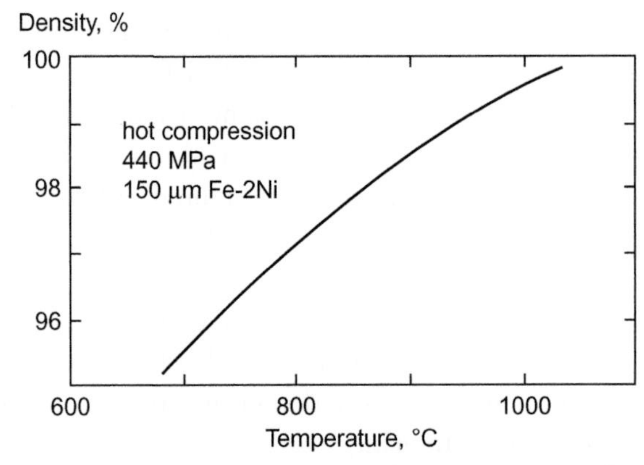

Figure 10.6 For the situation of constant pressure, in this case 440 MPa, the effect of temperature is to increase the consolidated density. These data are for a low alloy steel [5].

rapid flow. Initially the contact stress is much larger than the applied pressure. Calculation of the contact or effective stress compared to the applied pressure depends on the density. This ratio of effective contact pressure divided by the applied pressure P_E/P_A varies with the contact size. Figure 10.7 is a drawing of the grain, where pressure is transmitted at several grain contacts. The relation between contact size X, grain diameter D, and fractional density f when shear is present to induce particle sliding (such as hot pressing) is given by:

$$\left(\frac{X}{D}\right)^2 = 1 - \left[\frac{f_O}{f}\right]^{2/3} \tag{10.4}$$

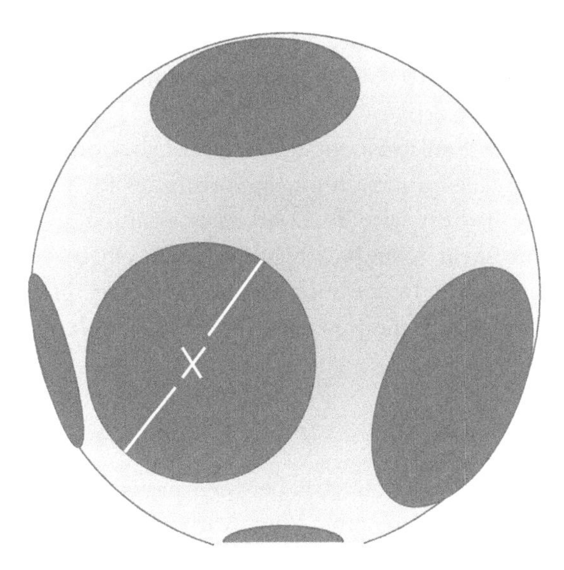

Figure 10.7 Sketch of a spherical grain during pressure-assisted sintering, where stress concentration at the contact points forms flat bonds on the grain surface. The diameter of the contact is designated *X*. The number of contacts per grain depends on the fractional density, reaching about 13 to 14 at 100% density.

For situations of pure hydrostatic compaction without shear, such as in hot isostatic pressing, the following applies:

$$\left(\frac{X}{D}\right)^2 = \frac{1}{3}\left[\frac{f - f_O}{1 - f_O}\right] \tag{10.5}$$

Experimental measurements indicate that these two relations slightly overestimate the neck size ratios [6]. An approach by Kumar [7] relies on the relative solid-solid contact S_R as a function of fractional density f:

$$S_R = -\log\left[\frac{(1 - f)f_O}{(1 - f_O)f}\right] \tag{10.6}$$

giving the pressure ratio as:

$$\frac{P_E}{P_A} = \frac{1}{S_R} \tag{10.7}$$

where f_O is the green density. A different relation is used by Ashby [8,9]:

$$\frac{P_E}{P_A} = \frac{1 - f_O}{f^2(f - f_O)} \tag{10.8}$$

which gives a higher effective pressure. The two relations are compared as functions of fractional density in Figure 10.8. At low densities the pressure amplification at the grain contacts is at least ten-fold.

Concomitant with neck enlargement under pressurization, the particle coordination increases. More contacts provide more opportunities for particle bonding during sintering. A tetrakaidecahedron with 14 contacts is assumed as the full density grain shape, but if there is no shear to induce rearrangement, then the terminal coordination number only reaches 12. In pressure-assisted sintering, pressure combines with temperature and time to densify the powder. The pressure role is pronounced at short times and higher temperatures.

Plastic Flow

On initial pressurization, the small contacts between particles amplify the applied stress to exceed the material's yield strength. Thus, early densification is by plastic flow. Powders densify by grain rearrangement and plastic flow until the contact stress falls below the yield strength. At low temperatures the yield strength is high, so pressures in the GPa range are required for full densification. For example, Figure 10.9 plots density versus applied pressure for a tool steel powder compacted at room temperature. A pressure of 3 GPa gives a density of near 99%, with no heating. In this case the grain size is unchanged. Thus, one benefit of high pressure consolidation is the avoidance of

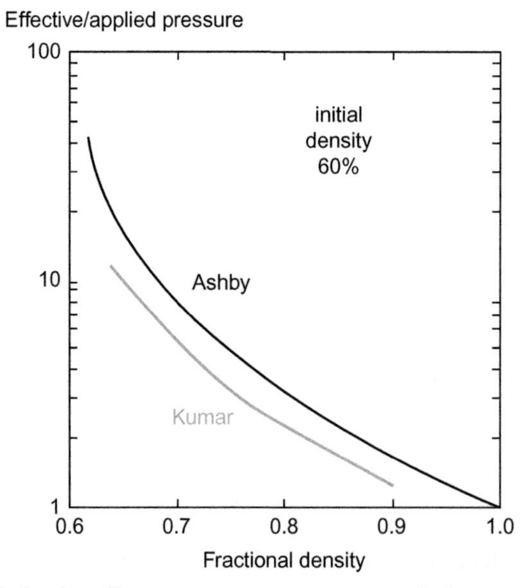

Figure 10.8 Two models for the effective contact pressure to applied pressure ratio versus the fractional density for a starting condition of 60% dense, showing the models from Kumar (Equation 10.7) and Ashby (Equation 10.8).

coarsening and in those cases the grain size versus density trajectory is flat. Heating reduces the pressure required for densification. To reach full density by plastic flow requires an applied pressure (P_A) of about three times the yield strength σ_Y at that temperature:

$$P_A \geq 3\,\sigma_Y \tag{10.9}$$

Likewise, the final fractional density f attainable by plastic flow depends on the applied pressure P_A as follows [8]:

$$f = \left[\frac{(1-f_O)P_A}{1.3\,\sigma_Y} + f_O^3\right]^{1/3} \tag{10.10}$$

where f_O is the starting powder density. This is valid for fractional densities less than 0.9.

At higher final densities with closed pores the limiting fractional density by plastic flow is estimated as follows:

$$f = 1 - \mathrm{epx}\left[-\frac{3\,P_A}{2\,\sigma_Y}\right] \tag{10.11}$$

The yield strength σ_Y is corrected for thermal softening. Example data for steel (Fe-2Ni-0.5Mo-0.5C) and stainless steel (Fe-18Cr-8Ni-2Mo) are plotted in Figure 10.10 [10].

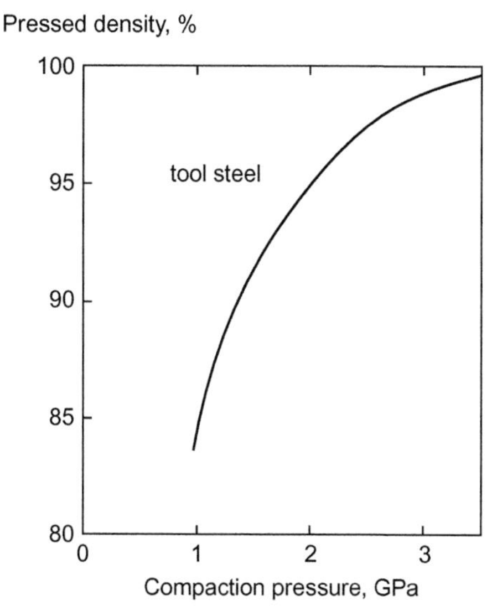

Figure 10.9 Densification by plastic flow is possible if high pressures are applied, as evident in this plot of fractional density versus compaction pressure for a tool steel powder.

Full density by plastic flow requires an applied pressure of about three times the yield strength at the consolidation temperature [11]. However, there are few tool materials able to withstand the consolidation pressure-temperature combinations. Thus, thermal softening is employed to lower the necessary pressures. For example, a powder billet is first heated and then pressurized while hot, which is the common practice for tool steels.

Particle Work Hardening

For crystalline materials, plastic flow results in work hardening, a progressive increase in hardness and strength due to dislocation entanglements. At the same time, diffusion at elevated temperatures relaxes these dislocation tangles, resulting in recovery and decreased resistance to deformation. Usually, deformation-induced work hardening starts to recover or relax at about 30% of the absolute melting temperature. Usually work hardening is not a concern because of the high consolidation temperatures. Further, relaxation is a time-dependent phenomenon, so the time to relax work hardening τ at the grain contacts decreases as the temperature T increases:

$$\tau = \tau_O \exp\left[-\frac{Q}{R\,T}\right] \tag{10.12}$$

where τ_O is a material relaxation time constant, Q is an activation energy similar to that for volume diffusion, and R is the gas constant 8.31 J/(mol K). Since sintering

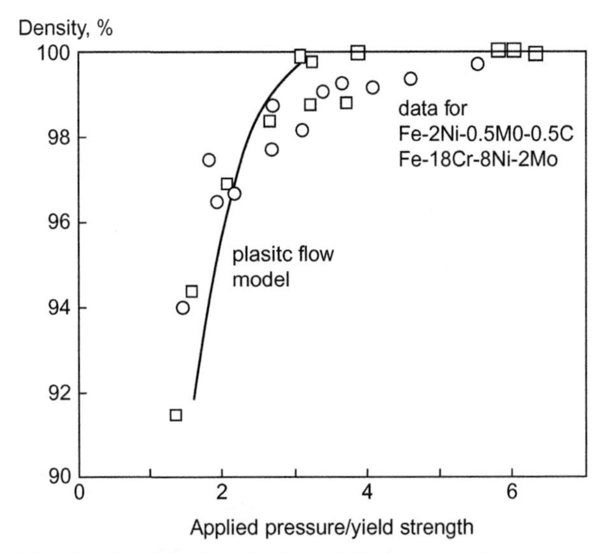

Figure 10.10 Plot of fractional density versus the applied compaction pressure normalized to the yield strength for a low alloy steel (4650) and stainless steel (316 L) [10]. The model from Equation 10.11 is included for comparison.

tends to be undertaken at temperatures much higher than the recovery temperature and is associated with long cycles, particle work hardening is generally ignored. Only the peak pressure is a concern with respect to plastic flow induced densification.

DIFFUSION AND CREEP

Diffusion and stress combine to give creep, a time-dependent deformation of a material at a stress below its yield strength. Besides green density and grain size, the main parameters in creep are temperature-pressure-time. Creep occurs by atomic motion from regions in the grain structure under compression (grain contacts) to regions under tensile stress (pores). Mechanisms of transport include diffusion in the lattice, along grain boundaries, and diffusion coupled to dislocation motion. Creep is the dominant process for densification in most pressure-assisted sintering techniques [12].

The generalized rate df/dt represents the instantaneous change in fractional density f. The powder compact starts at a fractional density of f_O. Under isothermal conditions the densification rate depends on the porosity $(1-f)$, as well as the temperature and microstructure:

$$\frac{df}{dt} = \frac{(1-f)}{(1-f_O)} B \left[\frac{4\,\gamma_{SV}}{d} + P_E \right] \tag{10.13}$$

where P_E is the effective pressure, γ_{SV} is the solid-vapor surface energy, d is the pore size, and B is a collection of terms that depend on the microstructure and transport mechanisms, as will be detailed below. At high porosities $(1-f)$, the densification rate is high, but as porosity is depleted the rate declines, thus the need for the ratio $(1-f)/(1-f_O)$. Inherently the densification rate reflects the rate of atomic motion.

At a fixed temperature, an increase in pressure results in faster densification. Figure 10.11 plots densification rate data at 1630°C (1903 K) for 0.3 μm alumina doped with 200 ppm magnesia [13]. The densification rate is essentially linear with pressure. At zero applied stress there is a non-zero densification rate resulting from the surface energy effect. However, the external pressure role is significant.

According to the above equation (Equation 10.13), the densification rate decreases as porosity is eliminated. This is the combined effect of larger diffusional transport distances, from microstructure coarsening, and reduced contact stress as the sinter bonds enlarge. Such behavior is illustrated in Figure 10.12 for hot pressed tantalum carbide [14]. At shorter times the densification rate is high, indicated by a steeper slope, but at longer times it slows. Note also the temperature effect, where a higher temperature increases the transport rate and improves sintering.

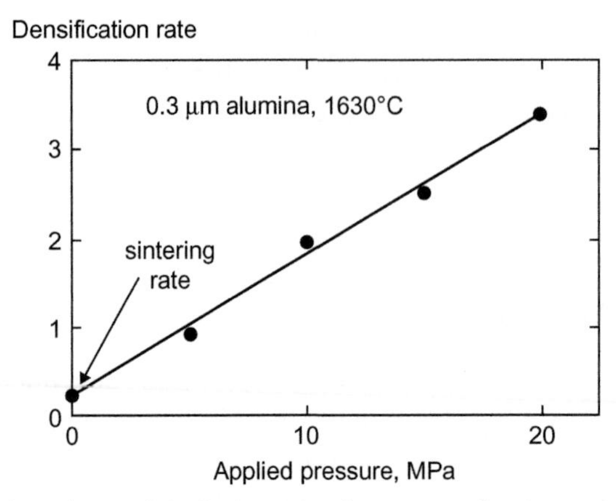

Figure 10.11 Data for the relative densification rate for 0.3 μm alumina at 1630°C (1903 K) to illustrate the enhancement possible from an externally applied pressure [13].

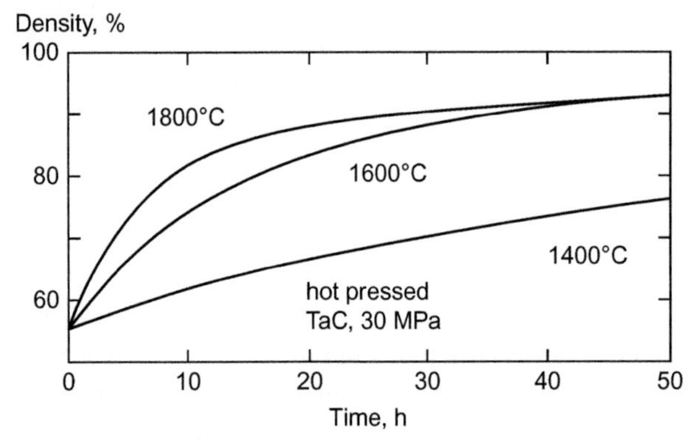

Figure 10.12 Data of fractional density versus hold time for tantalum carbide at 30 MPa pressure and three different hold temperatures [14]. The initial densification rate increases with temperature.

Three diffusion processes contribute to creep in pressure-assisted sintering. The atomistic events are similar to those identified for solid-state sintering. But now a pressure is applied that generates a diffusion gradient from compressive to tensile surfaces. In final stage sintering, even with an applied pressure, gas filled pores retard densification. If the applied pressure is high compared to the gas pressure in the pores, the pores collapse, but do not disappear. If the material is reheated without pressure, the pores reappear [15]. This nuisance is known as thermally-induced porosity and is a detrimental attribute of materials sintered with an external pressure.

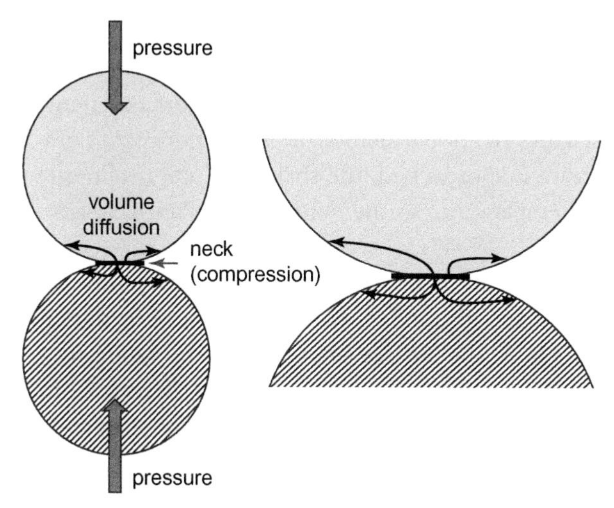

Figure 10.13 A schematic of two particle contact under compressive pressure, showing a grain boundary at the neck and volume diffusion from the region of compression to give volume diffusion creep as a densification mechanism.

Volume Diffusion Creep

For densification controlled by volume diffusion, creep moves mass from the contact under compression to deposition sites on pore surfaces under tension. This occurs by the motion of vacancies and atoms. The population of vacancies depends on temperature, as does atomic motion. Thus, faster densification occurs at higher temperatures and pressures. Early models lumped the pressure and temperature terms in pressure-assisted sintering to assume the powder was a viscous body [4,16–19].

In reality, the behavior is more complex. Initially, grain contacts grow by plastic flow, but as the effective pressure falls due to contact enlargement, diffusion becomes dominant. Diffusional creep controls final pore elimination. Volume diffusion controlled creep is also known as Nabarro-Herring creep, and it occurs when the vacancy flow is directed by the stress gradient from interfaces in tension to those in compression. The mass flow is illustrated in Figure 10.13, moving mass from the contact face, through the lattice, to deposit at the pore surface away from the neck. Vacancy flow is in the opposite direction, with vacancy annihilation at the grain boundary.

The volume diffusion-controlled shrinkage rate $d(\Delta L/L_o)/dt$ is modeled as follows [20]:

$$\frac{d\left(\frac{\Delta L}{L_O}\right)}{dt} = \frac{13.3\,\Omega\, D_{Vo} \exp\left[-\frac{Q_V}{R\,T}\right]}{R\,T\,G^2} \frac{(1-f)}{(1-f_O)} \left(\frac{4\,\gamma_{SV}}{d} + P_E\right) \qquad (10.14)$$

where T is the absolute temperature, R is the gas constant, Ω is the atomic volume, D_{Vo} is the volume diffusivity frequency factor, Q_V is the activation energy for volume diffusion, G is the grain size, γ_{SV} is the solid-vapor surface energy, f is the fractional density, f_O is the starting fractional density, d is the pore size, and P_E is the effective pressure. As full density is approached, the shrinkage rate declines to zero. Volume diffusion is sensitive to temperature, so the densification rate increases rapidly with higher temperatures.

Grain boundary annihilation of vacancies is key to volume diffusion creep. The flow of vacancies to the grain boundary results in rotation or migration of that boundary. For a typical situation, the surface energy is in the range from 1 to $2\,J/m^2$, and the microstructure scale is often in the order of 0.1 to 20 μm. Consequently, the magnitude of the internal sintering stress term associated with curved pore surfaces is from 1 to 20 MPa. To significantly impact, the inherent densification rate requires a higher effective stress, as previously illustrated in by the alumina data in Figure 10.11.

Grain Boundary Diffusion Creep

Most crystalline powders sinter by grain boundary diffusion, and this is also true in pressure-assisted sintering. The grain boundary forms at the grain contact and is a fast path for mass transport, especially when enhanced by a stress gradient [21,22]. For pressure-assisted sintering, the compact shrinkage rate depends on the mobility along the boundary; a layer just a few atoms thick. A schematic of the flow along the grain boundary is shown in Figure 10.14. The boundary thickness depends on segregated films and impurities, but for a pure substance it is about 5 to 10 atoms thick. The sintering shrinkage rate for grain boundary diffusion control $d(\Delta L/L_o)/dt$ is as follows [23–26]:

$$\frac{d\left(\frac{\Delta L}{L_O}\right)}{dt} = \frac{47.5\delta\,\Omega\,D_{Bo}\exp\left[-\frac{Q_B}{RT}\right]}{R\,TG^3}\frac{(1-f)}{(1-f_O)}\left(\frac{4\gamma_{SV}}{d}+P_E\right) \tag{10.15}$$

where f is the fractional density, f_O is the starting fractional density, T is the absolute temperature, R is the gas constant, δ is the grain boundary width, Ω is the atomic volume, D_{Bo} is the grain boundary diffusivity frequency factor, Q_B is the activation energy for grain boundary diffusion, G is the grain size, γ_{SV} is the solid-vapor surface energy, d is the pore size or a similar microstructure scale feature, and P_E is the effective pressure. The effective pressure is calculated from the applied pressure and relative density. Note that the rate of densification declines as densification progresses. Grain boundary diffusion is sensitive to temperature.

Small grain structures with a high grain boundary area are very responsive to pressure-assisted sintering due to the large number of active grain boundaries and small transport distances. Grain boundary diffusion is important in such situations. This is evident in Figure 10.15 for alumina consolidated from different particle sizes

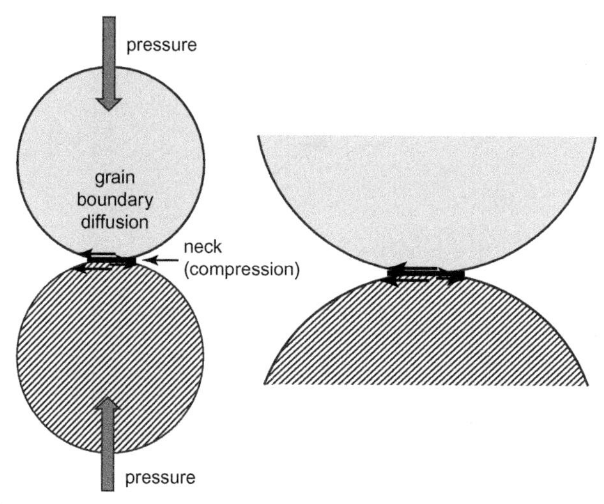

Figure 10.14 Similar to Figure 10.13, grain boundary diffusion creep transports mass from the contact region to the pore surface, in this case by grain boundary diffusion.

using a cycle at 1200°C (1473 K) for 1 h and 50 MPa. Smaller particles promote densification, since the diffusion distance scales with the particle size.

Liquid films or glassy phases on grain boundaries provide fast diffusion paths. For many ceramics, such films prove important to densification [27,28]. Figure 10.16 plots porosity data after pressure-assisted sintering for silicon nitride. The consolidation cycle is at 1800°C (2073 K), 90 MPa, and 180 min with variations in the added yttria [29]. Segregation of yttria to the grain boundaries provides faster diffusion and produces a lower final porosity.

Dislocation Climb and Power Law Creep

Dislocation climb is a high temperature creep process that involves a combination of volume diffusion and dislocation motion. It is active when both the pressure and temperature are high [9,30,31]. The deformation rate is initially high since the dislocation population increases due to early plastic flow. Atom motion from the grain boundary to the dislocations is a shorter diffusion distances compared to the distance for traveling to a free surface. Accordingly, the rate of densification depends on the dislocation climb rate described by a power law creep model for shrinkage rate $d(\Delta L/L_o)/dt$ as[32,33]:

$$\frac{d\left(\frac{\Delta L}{L_O}\right)}{dt} = \frac{C\, b\, U\, D_{Vo} \exp\left[-\frac{Q_V}{R\,T}\right]}{R\,T} \frac{(1-f)}{(1-f_O)} \left(\frac{P_E}{U}\right)^n \tag{10.16}$$

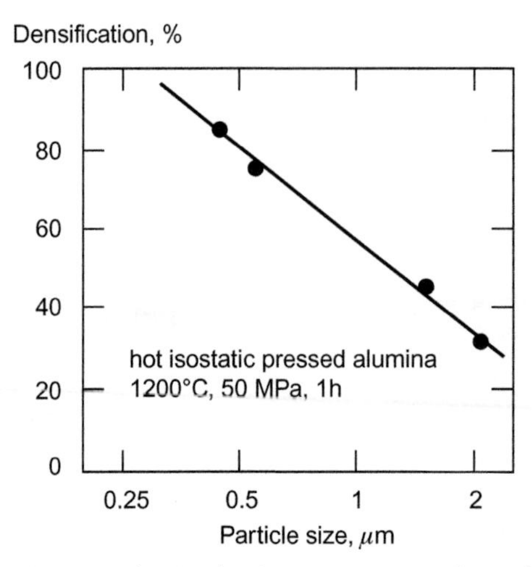

Figure 10.15 A smaller starting particle size leads to more grain boundary area during densification, and significant changes in densification during hot isostatic pressing as illustrated in this plot for alumina at 1200°C (1473 K), 50 MPa, with 1 hour hold. The smaller particle size gives shorter transport distances, sharper curvature gradients, and more grain boundary area for diffusion.

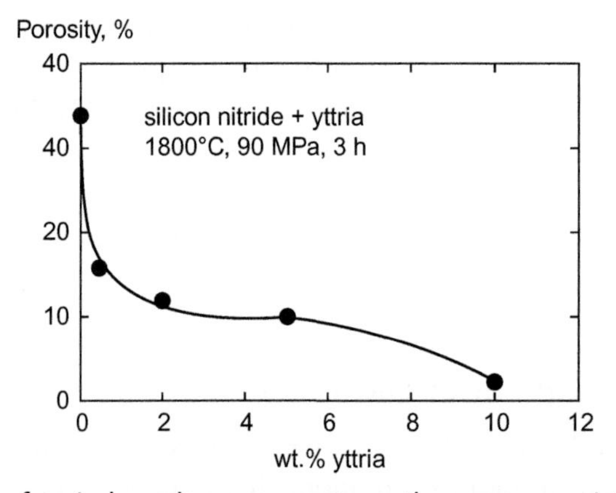

Figure 10.16 The role of grain boundary segregants on the pressure-assisted sintering of silicon nitride treated with various yttria concentrations to form a grain boundary viscous phase that assists in sintering densification [29]. The consolidation cycle is at 1800°C (2073 K), 90 MPa, and for 3 hours.

In this equation C is a material constant, T is the absolute temperature, R is the gas constant, b is the Burgers vector length (about the same as atomic spacing), D_{Vo} is the volume diffusion frequency factor, Q_V is the activation energy for volume diffusion, U is the shear modulus, f is the fractional density, and f_O is the starting fractional density. In this form there is no dependence on grain size and the pressures are typically so high that solid-vapor surface energy is ignored. The exponent n is related to the strain rate sensitivity.

This is an effective model for explaining experimental data obtained under many conditions. For example, superplastic flow is accomplished in pressure-assisted sintering when the stress and creep strain rate are related by $n = 2$. This is a special condition that occurs in a two-phase microstructure with a stable, small (<1 μm) grain size, such as is found in high carbon steels and ceramic-ceramic composites.

A linear combination of the densification rates by the individual processes gives the total densification rate, lumped into a viscosity approximation for implementation in finite element codes [34–36]. The instantaneous rates are integrated over time to predict final density, component size, or microstructure. Many refinements have been offered for the treatment outlined here, but the essential features are evident and provide a good representation of experimental data.

Liquid and Viscous Phases

Fast diffusion is associated with liquids, so a liquid or amorphous phase on the interfaces between grains accelerates pressure-assisted densification. To illustrate the advantage, Figure 10.17 plots sintering shrinkage versus applied pressure for a nickel-base superalloy consolidated with 20 vol.% liquid [37]. Shrinkage increases nearly linearly with the applied pressure. Relatively small external pressures, up to 0.4 MPa (approximately four atmospheres), significantly change the sintering behavior.

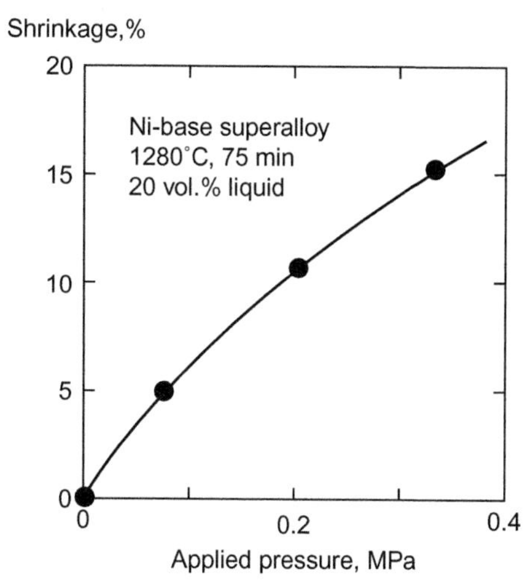

Figure 10.17 A liquid phase promotes faster transport during pressure-assisted sintering. This plots the sintering shrinkage for a nickel-base superalloy powder heated to 1280°C (1553 K) for 75 min, where the liquid content was approximately 20 vol. % [37].

One approach to liquid phase hot consolidation involves sintering to a closed pore condition in a vacuum. Subsequently, pressure is applied while the compact is hot with a thermal pulse used to form liquid at the solidus temperature. Liquid forms on the grain boundaries, resulting in rapid densification. This is essentially pressure-assisted supersolidus liquid phase sintering.

For mixtures of solid grains in a liquid or glassy phase, the solid-liquid-pore viscosity η varies with the composition as follows:

$$\eta = \frac{\eta_O \exp(-a(1-f))}{\left[1-\frac{\varphi}{\varphi_C}\right]^2} \tag{10.17}$$

In this expression, η_O is the viscosity of the amorphous phase, φ is the fraction of the system that is solid, and φ_C is the critical fraction required to initiate viscous flow (about 0.2). There is no deformation if φ is greater than the critical value. For mixtures of solid grains and viscous liquids, the viscous phase lubricates particle sliding and promotes viscous flow. A viscous phase is very useful if solid grain decomposition occurs, for example in systems containing diamond. Viscosity decreases at higher temperatures, so this aids densification. Also, the system viscosity depends on the solid-liquid ratio and porosity, so as pores are removed the viscosity increases. A higher pressure gives faster densification.

In the instances where the densification involves crystalline solids with liquid grain boundary phases or purely amorphous powders, densification takes place by viscous flow. A liquid cannot sustain a shear stress; thus, a viscous material will densify in proportion to the effective pressure P_E:

$$\frac{df}{dt} = \frac{3 P_E (1-f)}{4 \eta (1-f_O)} \tag{10.18}$$

where f is the fractional density, f_O is the initial fractional density, and η is the system viscosity. Viscosity and diffusivity are inversely related—a high diffusivity corresponds to a low viscosity.

The combination of a viscous phase and shear stress induces grain rearrangement. The external pressure supplements the capillary force. For the case of solid diffusion through a liquid, shrinkage $\Delta L/L_O$ follows [38]:

$$\frac{\Delta L}{L_O} = \left(\frac{g\,\delta\, D_L\, C\,\Omega\, t}{G^3\, R\, T}\right)\left(\frac{4\gamma_{LV}}{d} + P_E - P_P\right) \tag{10.19}$$

where g is a numerical constant, δ is the liquid film thickness between grains, D_L is the diffusion rate in the liquid which varies with temperature, C is the solid solubility in the liquid, Ω is the atomic volume for the solid, t is the sintering time, G is the solid grain size, R is the gas constant, T is the absolute temperature, γ_{LV} is the liquid-vapor surface energy, d is the pore diameter, P_E is the effective pressure, and P_P is the gas pressure in the pores. This model is valid while there is ample porosity, since shrinkage ends when full density is reached. Shrinkage and shrinkage rates are linked to density and densification rates using geometric models introduced in Chapter 7.

Reactions and Exothermic Processes

Reactions occur in some mixed powders to form compounds. Exothermic reactions generate heat useful in sintering the compound. When coupled to simultaneously applied external pressure the variants are known by such names as reactive hot isostatic pressing, exothermic hot pressing, reactive quasi-static pressing, reactive extrusion, pulse discharge pressure combustion sintering, flash sintering, and pressure-assisted self-propagating high temperature synthesis—they are favorite routes for intermetallics, compounds, and composites [39—42].

As an example, a mixture of titanium and boron powders is encapsulated and heated to 700°C (973 K) under 100 MPa pressure to produce dense TiB_2. During heating an exothermic reaction releases 293 kJ/mol to self-heat the material, so the peak temperature exceeds the furnace set temperature. Products with grain sizes of 5 μm and ultrahigh hardness are fabricated by reactive pressure-assisted sintering. If there is excess titanium, then a composite of titanium diboride in titanium results.

Explosive options also are employed, but shock waves often damage the compact, resulting in fragments. This approach is applied to ceramics, composites, and intermetallics [43]. It may have the greatest benefit in the simultaneous synthesis and consolidation of high temperature composites, such as Al_2O_3-TiC. In this case, alumina, titanium, and graphite are mixed and react to form the carbide during pressing at 30 MPa and 1750°C (2023 K). The basic concept remains the same—use exothermic heat from a synthesis reaction to supplement heating with an external pressure to induce densification of the reaction product. The exothermic heat, component mass, and heat capacity determine the peak temperature, while heat loss depends on the component surface area to volume. Unfortunately, the process is difficult to scale and for that reason remains a curiosity.

One option is to infuse gas into the material during sintering. It is useful for forming aluminum nitride during consolidation. Another example is the use of nitrogen strengthening steels by pressure-assisted sintering in a nitrogen atmosphere. Experiments with metals sintered in high pressure hydrogen resulted in improved densification.

Electric Fields

Pressure accelerates sintering densification. An approach dating back to the late 1890s is to use electric current discharge through a pressurized powder to induce densification.

The concept was initially termed spark sintering [44], and when taken up by the elctro-discharge machining industry, the name was rephrased spark plasma sintering [45]. Subsequent efforts show there is no plasma, so spark sintering is most representative, but a variety of descriptor terms are in use [46]:

- current activated pressure-assisted sintering
- current activated tip-based sintering
- electric current assisted sintering
- electric pulse assisted consolidation
- electrical discharge compaction
- electro-consolidation
- field assisted sintering technique
- field effect activated sintering
- pressure plasma consolidation
- pulse discharge sintering
- pulsed electric current sintering
- spark plasma sintering
- spark sintering

Process control is complicated by substantial shifts in material conductivity during heating and densification. At the start of the consolidation cycle even metallic powders are poor conductors, but as densification progresses they become conductive. In some cases the compact densifies to become so conductive that it fails to properly heat. On the other hand, nonconductor materials rely on die heating. For materials that have oxide coatings, such as aluminum, the situation is more complex, but spark sintering delivers 99.7% density using a cycle of 0.3°C/s to 600°C (873 K) with a 30 min 40 MPa hold. Hot pressing gives the same result, and when the strength and elongation are compared, the spark sintered material is 93 MPa and 48% versus the hot pressed material at 90 MPa and 50%. Spark sintering resulted in higher properties at short times, probably due to erosion of the surface films, still both approaches plateaued at similar properties when fully densified.

The technological merits of spark sintering come from rapid heating, short hold times, and possible electric field induced diffusion [45–51]. The first two are possible in a variety of hot consolidation approaches. For example, rapid heating in hot pressing is possible using exothermic reactions, induction heating, and capacitive discharge. Likewise, quick hot isostatic pressing is performed using cycles of just one minute. Exothermic hot pressing and hot isostatic pressing do not require electric current, are fast, and deliver small grain sizes in full dense materials [51–57]. By heating rapidly, lower temperature surface diffusion is avoided, carrying more of the surface energy to high temperatures where grain boundary diffusion is active.

Speculation arises on new phenomena in spark sintering and the unanswered influences includes the following [58]:

- Grain boundary structure changes, such as improved ledge formation and migration.
- Altered grain boundary migration and segregation.
- Electromigration enhanced diffusion.

In spark sintering it is possible to reach current densities where electromigration supplements diffusion, usually at current densities over 1000 A/cm^2. Electromigration was first postulated as a means to accelerate the atom migration [59]. The supplemental diffusion effect scales with the additional atomic drift velocity V due to the electric field E,

$$V = \frac{D\,Z\,e\,E}{k\,T} \tag{10.20}$$

where D is the inherent temperature-dependent diffusivity, Z is a material dependent experimental factor measured to be near 0.1, e is the electric charge (1.6 $10-^{19}$ C), k is Boltzmann's constant, and T is the absolute temperature.

For a material such as copper at 800°C (1073 K) the volume diffusivity is 2.5 10^{-15} m^2/s and in 1 s the mean atomic displacement is 0.123 μm. With 5 V applied over a typical 3 mm compact thickness, the electric field is 1.67 10^3 V/m, so the additional displacement due to electromigration is 4.5 10^{-6} μm. This is a trivial supplemental contribution (about 0.004%). However, at the onset of spark sintering the small particle contacts concentrate the current to much higher levels. Such concentrated current could give an early contribution to particle bonding. As the necks enlarge, the current density declines and the beneficial effect is lost. However, the electron wind also works to accumulate vacancies, resulting in large pore growth. Normally, diffusion fluxes during sintering are based on concentration gradients, but electromigration adds a supplemental term that includes charge, field, and mobility [60]. The main electromigration gains come with DC pulses of 1 s or more, but many efforts rely on 2 to 3 ms pulses that do not induce electromigration.

Smaller grain sizes result when large voltage gradients are used, even without the application of pressure [61]. Grain boundary films preferentially interact with the electric field. For metastable materials, such as tungsten carbide or diamond, decomposition occurs during consolidation; the decomposition behavior varies with the time-temperature-pressure combination as follows [62]:

$$\frac{dx}{dt} = \beta \exp\left[-\frac{E + P\,\Delta V}{R\,T}\right] \tag{10.21}$$

where x is the mass fraction decomposed, t is the time, T is the temperature, E is the decomposition activation energy, P is the pressure, ΔV is the activation volume, and

R is the gas constant. The parameter β is a rate constant that depends on the atmosphere composition (argon, oxygen, air, or such). For example, the decomposition activation energy for diamond is in the 728 to 1159 kJ/mol range depending on crystal orientation and the activation volume is 10 cm^3/mol. For unstable materials, such as diamond, this model shows the benefits of sintering at lower temperatures using shorter times as possible by spark sintering. However, as full density is approached, the current concentration decreases and the benefits versus hot pressing or other consolidation routes are lost. The key gain is fast heating, an advantage well known in sintering and pressure-enhanced sintering without the need for electric fields.

MULTIPLE MECHANISM DENSIFICATION RATES

The features of pressure-assisted sintering are combined in densification maps, where multiple mechanisms act simultaneously to determine the overall response [1,8,63]. These are graphical combinations of plastic flow, creep, and diffusion processes based on input parameters of green density, particle size, temperature, pressure, and time.

To construct a densification map representative of pressure-assisted sintering requires definition of the starting structure and determination of the mass flow from each of the mechanisms [9,34,64,65]. The rates are summed to provide net changes in neck size, shrinkage, pore size, grain size, and related parameters. In a simple sense, fractional density f is calculated as follows:

$$f = f_O + \left[\frac{(1-f_O)\,P_A}{1.3\,\sigma_Y} + f_O^3\right]^{1/3} + \int_0^t \sum \frac{d\left(\frac{\Delta L}{L_O}\right)}{dt}\,dt \qquad (10.22)$$

Consolidation starts at the initial density f_O, progresses in a time independent mode via plastic flow, and further progresses in a time-dependent manner based on the summed shrinkage rates from the creep mechanisms, volume diffusion, grain boundary diffusion, and power law creep. The relative contributions change during densification, for example due to grain growth and densification. Instantaneous rates are integrated over time.

Implementation of these models is via computer simulations. Some of the early simulations were for neck size, but calculations shifted to predictions of density, component size and shape, and finally defects. The defects arise from nonuniform heating or buckling of the powder container [66,67].

Integral work laws for pressure-assisted sintering allow for densification predictions in a single curve that includes the key processing variables [68]. An example curve for the hot pressing of alumina is shown in Figure 10.18. The parameter Θ is based on the integral time-temperature attributes,

$$\Theta = \int_0^t \frac{1}{T} \exp\left[-\frac{Q}{R\,T}\right] dt \tag{10.23}$$

where T is the absolute temperature, t is the time, R is the gas constant, and Q is the apparent activation energy. This integral reflects then the combined effects of density changes, thermal softening, diffusion and creep, grain growth, and other factors that impact densification, all lumped into a single factor. For the alumina data, the heating rates varied from 35 to 150°C/min, giving 290 kJ/mol as the apparent activation energy.

DENSITY AND MECHANISM MAPS

Densification maps are graphical presentations of the processing features involved in pressure-assisted sintering [9,69–71]. The instantaneous mass flow is calculated for each contribution and numerically integrated over time. Several input parameters are required describing the material, powder, and processing conditions. The result is a density map versus processing parameters. Two such maps are given in

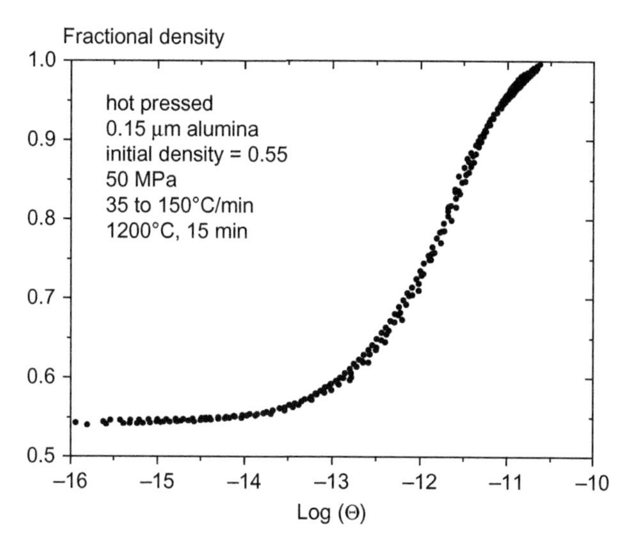

Figure 10.18 Hot pressing data for 0.15 μm alumina sintered from a 55% starting density using 50 MPa applied pressure and various heating rates to 1200°C (1473 K) with a 15 min hold. The fractional density is plotted versus the integral time-temperature work of sintering.

Figure 10.19 for 50 μm tool steel consolidated at either a constant temperature or constant pressure. The dominant region for each sintering, creep, or plastic flow contribution is outlined. The left plot is for 1200°C (1473 K) with sintered density plotted versus applied pressure for isothermal hold times of 0.25, 0.5, 1, 2, or 4 hours. Densification at that temperature is rapid once pressure increases. Typical consolidation conditions would be 200 MPa, giving rapid densification. Boundary lines indicate where the transport mechanisms are making equal contributions (but not necessarily 50% contributions, because other processes are active). The right plot corresponds to a variable temperature at 100 MPa applied pressure; for example the plot predicts full density at 1100°C (1373 K) in 1 h. At shorter times densification is dominated by creep, especially at lower temperatures.

It is possible to assess interplay between process variables to optimize pressure-assisted sintering. For example, the pressure and temperature combinations necessary to attain 99% density are shown in Figure 10.20 for 50 and 100 μm tool steel powders. There is particle size role, with a slight edge for the smaller powder. However, sintering temperature provides the greatest leverage.

A similar densification map is plotted in Figure 10.21 for the consolidation of a titanium aluminide (Ti_3Al) intermetallic using 10 μm powder. This map is for the pressure and temperature conditions that give 100% density using a 10 μm powder. It shows how longer times reduce the temperature or pressure. Grain boundary diffusion is dominant, especially at the lower temperatures.

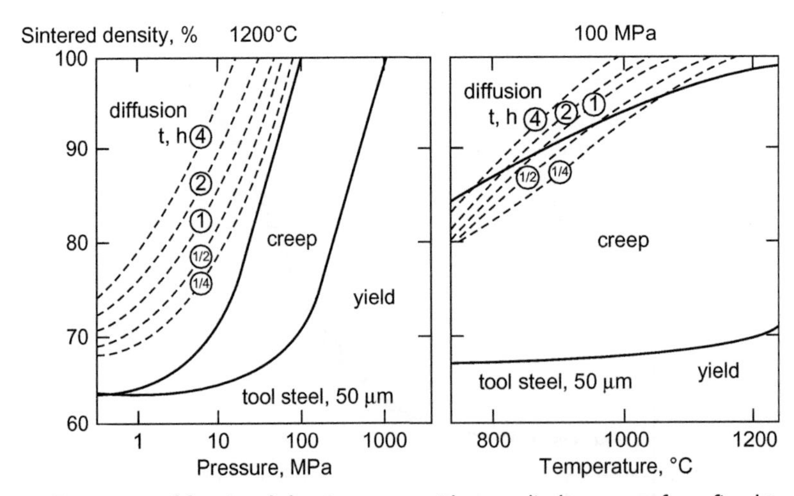

Figure 10.19 Two maps of fractional density versus either applied pressure for a fixed temperature of 1200°C (1473 K) for times of 0.25, 0.5, 1, 2, or 4 hours on the left, or fractional density versus hold temperature for a fixed 100 MPa pressure again for 0.25, 0.5, 1, 2, or 4 hours on the right. Both maps are for 50 μm tool steel powder.

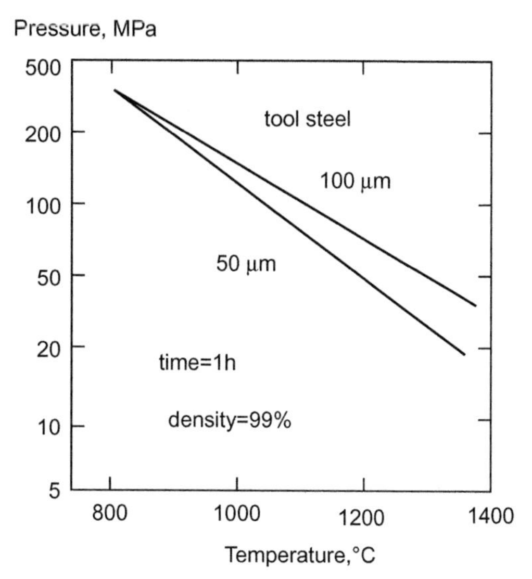

Figure 10.20 Similar maps of hot isostatic pressing for two particles sizes of tool steel, showing the temperature and pressure combinations required to attain 99% final density in 1 hour.

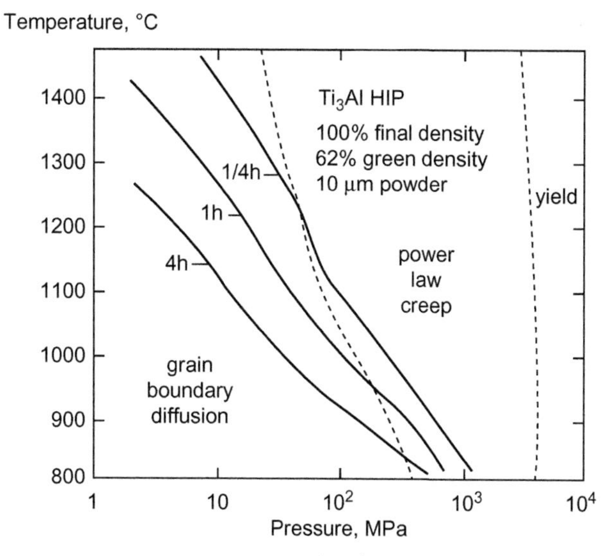

Figure 10.21 A pressure-temperature pressure-assisted sintering map for densification of 10 μm titanium aluminide powder starting at 62% density. This map plots time contours to show lower temperatures or pressures result from longer cycles.

Beyond these calculations of density, related models are used to predict grain size, final shape, component size, and defects [72–74].

MICROSTRUCTURE EVOLUTION

Microstructure controls the performance of a sintered material. It is second only to density in determining properties. Even at full density, mechanical properties depend on grain size, defects, and interfacial segregation. Pressure-assisted sintering provides a means to eliminate porosity. However, the dimensional control is poor and problems arise with microstructure coarsening. Pressure-enhanced sintering is often performed in a die or container, which contaminates the material. The surface contamination might be evident as locally large grains or reactions products [47].

Pressure induces grain contact to accelerate grain growth during pressure-assisted sintering. An example of grain growth is shown in Figure 10.22 for alumina consolidated at 1300°C (1573 K) for 1 h [75]. Grain size enlarges because pressure induces early and large grain contacts over which grain growth occurs.

Faster densification cycles reduce the time for microstructure coarsening, refining the microstructure scale while preventing reactions or decomposition of phases. This is a major benefit from pressure-assisted sintering. On the other hand, several contamination sources must be controlled. Surface films are problems with small particles, sometimes arising from exposure to air. If the surface is enriched with an impurity,

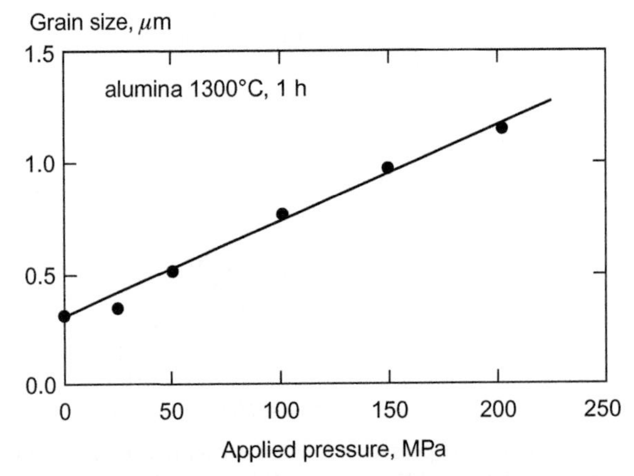

Figure 10.22 Grain size versus applied pressure for alumina consolidated at 1300°C (1573 K) for one hour [75]. The higher pressure increases the grain-grain contact, resulting in larger grain sizes as well as higher densities.

undesirable phases form in what is termed prior particle boundary precipitation [76]. Figure 10.23 shows this feature in a powder consolidated by hot isostatic pressing. There are no pores in the structure, but oxides outline the particle boundaries, and provided an easy fracture path with low properties. A demonstration of preferential failure at the prior particle boundaries is given in Figure 10.24 for a dense ferrous alloy. Normally this alloy would exhibit extensive ductility and a dimpled fracture surface. Instead failure was along the weak prior particle boundaries arising from impurity contamination. Full density by pressure-assisted sintering is not a guarantee of success. Usually a shear stress during consolidation is helpful in disrupting these surface films.

Contamination from tool materials is a concern. Figure 10.25 compares microstructures at the same magnification from the inside and surface of a steel consolidated using pressure-assisted sintering. The outside surface was contaminated by the container and performance differences arise from such gradients. A related problem in compounds arises with grain boundary films. The films precipitate on cooling and can result in continuous layers of weak phase, greatly degrading fracture resistance, in spite of full density [77,78].

PRESSURE APPLICATION TECHNIQUES

Sintering is accelerated by the application of an external pressure. The stress state depends on the means used to apply the pressure. Options range from isostatic

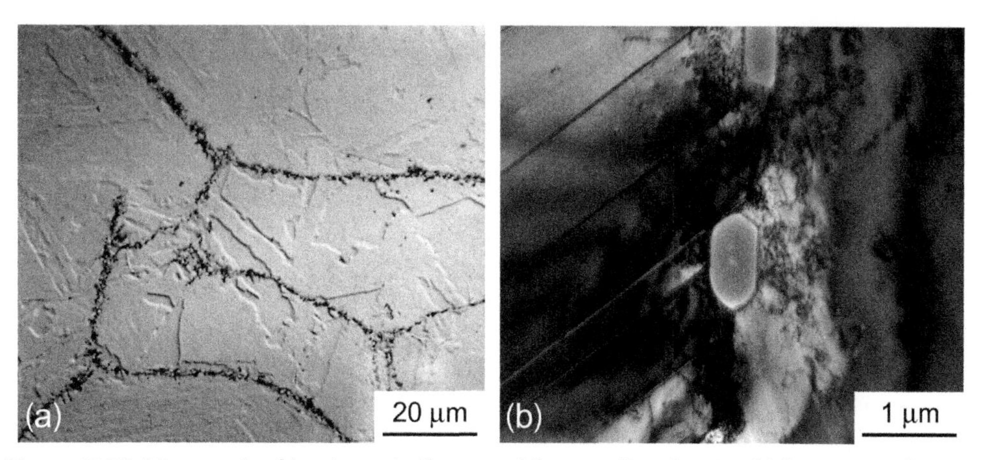

Figure 10.23 Micrograph of hot isostatically pressed ferrous alloy showing (a) lower magnification evidence of decorations on the prior particle boundaries, and (b) transmission electron microscopy image of an individual TiC precipitate.

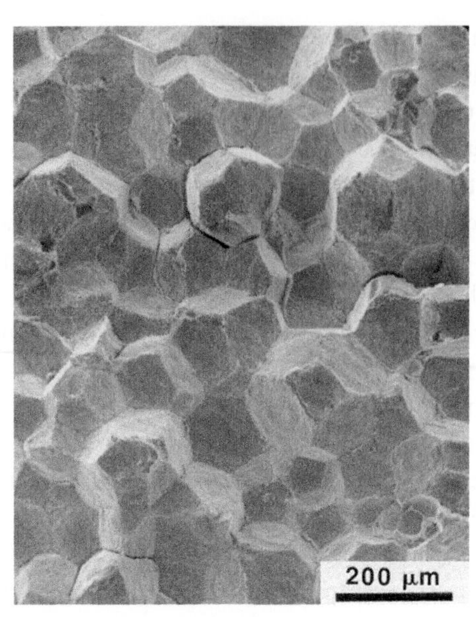

Figure 10.24 Scanning electron micrograph of the fracture surface of a hot isostatically pressed powder that reached full density, but the prior particle boundary precipitation allowed easy fracture without ductility.

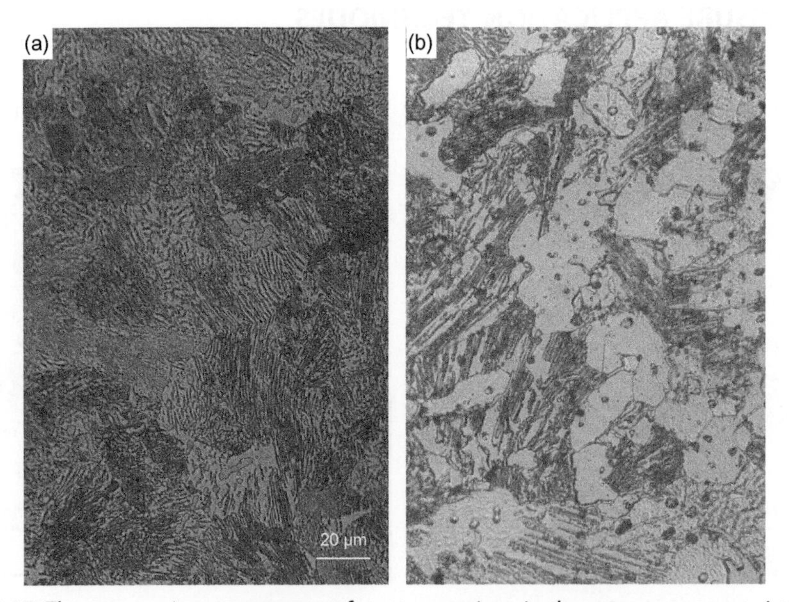

Figure 10.25 These two microstructures are from two regions in the same pressure-assisted sintering billet, with (a) corresponding to the interior and (b) corresponding to the near surface region which showed contamination effects.

compaction with no shear to approaches such as hot extrusion with significant shear. The most common approaches are:

- Hot isostatic pressing—heating and pressurization by hot gas; powder is sealed in a flexible container or sintered to closed pore condition prior to hot gas pressurization.
- Sinter-hot isostatic pressing—two stages with vacuum sintering to closed pore condition in the first stage followed by pressurization of the furnace to collapse remaining pores in the second stage.
- Uniaxial hot pressing—high pressure, low strain rate forming of a powder in a die using simultaneous heating with pressurization from punches aligned along one axis.
- Upset forging—uniaxial pressing of a powder compact with intentional lateral flow at high strain rates, usually relies on presintered materials to have sufficient strength to resist cracking.
- Sinter forging—uniaxial pressing at low pressure and low strain rate to provide shear, deformation, and densification during sintering.
- Granular forging—uniaxial compaction of heated granules (graphite or ceramic) in a die to provide quasi-static pressure on a heated powder compact embedded in the granules.
- Hot extrusion—hot powder is forced through a narrow die to form a straight wall geometry, usually relies on presintered porous powder billet.
- Spark sintering—similar to hot pressing but heating is by electric current passage through the die and powder with rapid heating and uniaxial compression.

Pressure-assisted sintering techniques employ combinations of temperature and pressure to eliminate pores. Net-shaping is used to imply fabrication to the final size and shape to eliminate machining. A schematic map of the processes is given in Figure 10.26. On the axes are temperature relative to the melting temperature and pressure relative to the yield strength (at that temperature). The high temperature processes rely on less pressure, while the high pressure processes rely on lower temperatures. In terms of process variables, smaller particle sizes, high pressures, and high temperatures contribute to rapid densification. There are many combinations that produce full density [1,49,79–84].

The desire is to understand densification, especially for composites. Composites inherently resist sintering, so pressure-assisted sintering is attractive. In systems involving a reaction, it seems the appropriate strategy is to initiate the reaction and then pressurize to avoid inhomogeneities. Consequently, several of the processing techniques apply pressure after heating.

Stress States and Strain Rates

The pressure-assisted sintering approaches differ in stress state during densification. This influences particle sliding and dimensional change. The mean pressure, or hydrostatic

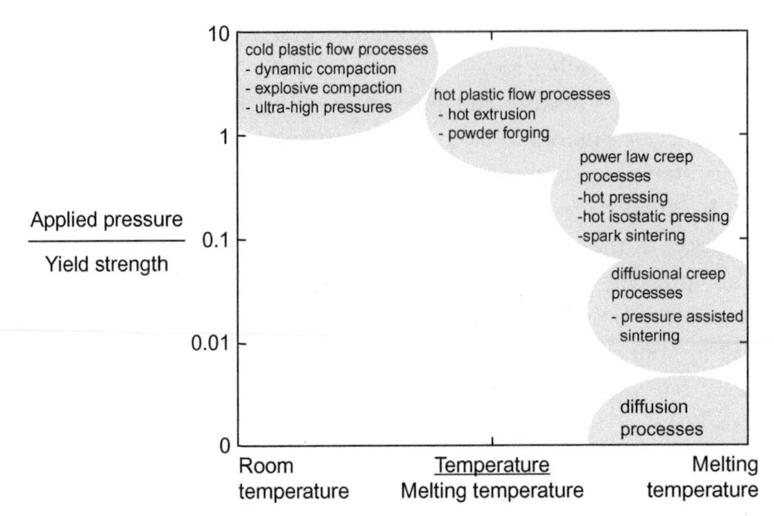

Figure 10.26 A schematic plot of the various pressure-assisted sintering techniques versus the relative temperature (compared to the melting temperature) and relative pressure (applied pressure normalized to the yield strength). The third axis would be time, with diffusion processes being slow and requiring much time and rapid consolidation processes taking place at lower temperatures.

stress, determines pore shrinkage and controls the densification rate. Shear induces particle rearrangement by sliding, collapsing large pores while disrupting particle surface films. Figure 10.27 contrasts the pore collapse from shear and hydrostatic conditions. Nonuniform dimensional change is a disadvantage from shear, meaning less precision in the final component shape. Even so powder forging is widely used because of the high mechanical properties resulting from the particle shear.

Table 10.1 summarizes the main consolidation routes to provide a contrast and comparison. The sketches in Figure 10.28 illustrate the pressure states for these approaches. Note triaxial and hot pressing (and other quasi-static approaches such as granular forging) are essentially the same stress states, giving equivalent densification.

In hot isostatic pressing (HIP), a hot gas provides uniform stress in all directions. This is similar to the hydrostatic state encountered in pressureless sintering as generated by surface energy. Because the pressure is uniform, there is no shear and all points in the compact experience identical stress states. An advantage is uniform densification, but surface films are not disrupted, so the powders are handled under conditions that avoid surface contamination. The sinter-HIP variant is the same, but usually the pressure vessel is rated for a lower pressure.

Uniaxial hot pressing occurs in an externally heated die. During densification, the radial stress is proportional to the applied axial stress. Shear exists in uniaxial hot pressing.

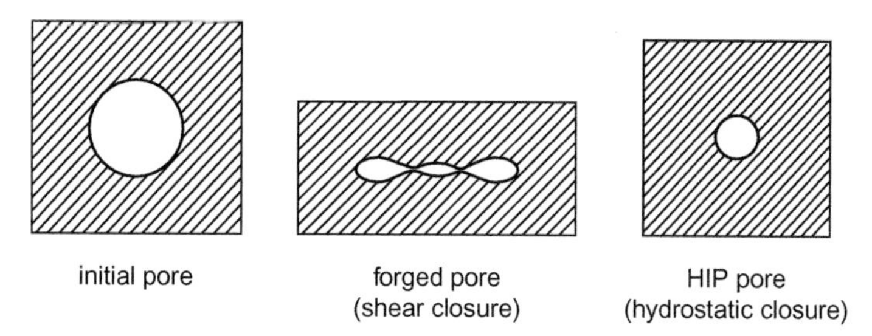

Figure 10.27 A spherical pore densifies differently depending on the stress state during pressure-assisted sintering. Generally a shear stress, such as in forging, leads to more smearing of the pore but anisotropic densification. On the other hand, hydrostatic pore closure shrinks the pore with no new sinter bonds.

Table 10.1 Contrast of Pressure-Assisted Sintering Routes

Process	Peak Temperature °C	Peak Pressure MPa	Stress State	Hold Time Min	Strain Rates
uniaxial hot pressing	to 2200	100	compressive, some shear	15 to 60	low
hot isostatic pressing	to 2200	200	hydrostatic	60 to 120	very low
sinter–HIP	to 1500	20	hydrostatic	60	very low
spark sintering	2200	100	compressive, some shear	10	intermediate
powder forging	1000	500	compressive, shear, some tensile	0.1	high
sinter forging	800	10	compressive, shear, minor tensile	60	low
triaxial	1500	100	compressive, some shear	15 to 60	intermediate
hot extrusion	1300	50	compressive, extensive shear	<1	very high
gas forging	1200	500	hydrostatic	<1	very high
granular forging	1300	200	quasi-hydrostatic, some shear	10	intermediate
dynamic compaction	no heating	1000	compressive, extensive shear	1 μs	very high

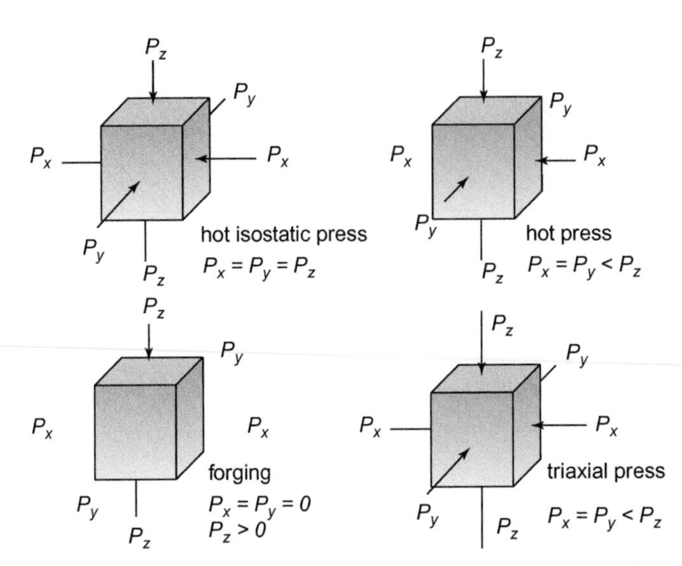

Figure 10.28 Sketches of the different pressure states associated with some of the popular pressure-assisted sintering processes. Hot isostatic pressing has equal pressure in all directions, while hot pressing has pressure applied along the vertical axis and die wall constrain pressure. Forging is also uniaxial, but initially without constraint. Triaxial pressing is a hybrid of hot pressing and hot isostatic pressing.

The most common tooling is graphite, and this limits the temperature, pressure, and atmosphere combination.

Powder forging (PF) starts with a presintered, unconstrained porous billet that is forced to pancake and flow against the die sides. There is a large shear component since the material deforms to encounter radial constraint from a die. Variants on powder forging include granular forging and gas forging. In granular forging the powder compact is embedded in hot deformable graphite pellets that provide some constraint during uniaxial pressing. Gas forging is more like hot isostatic pressing, but the pressure pulse is generated by a squirt of liquid argon or liquid nitrogen into the hot chamber. The quick evaporation of the liquid creates a high strain rate forging pressure pulse.

Triaxial compaction is a hot isostatic pressing approach with supplemental stress on one axis. Effectively it is a combination of hot pressing and hydrostatic pressing. As opposed to hot pressing, triaxial pressing allows independent control of the radial stress. Accordingly, the stress state is intermediate between the two technologies. The stress condition is the same as found in granular forging. Note the shear stress is highest for forging operations and lowest for hot isostatic pressing.

Gas-based pressing techniques, such as hot isostatic pressing, provide hydrostatic pressures without shear. The axial and radial stresses are the same and the absence of

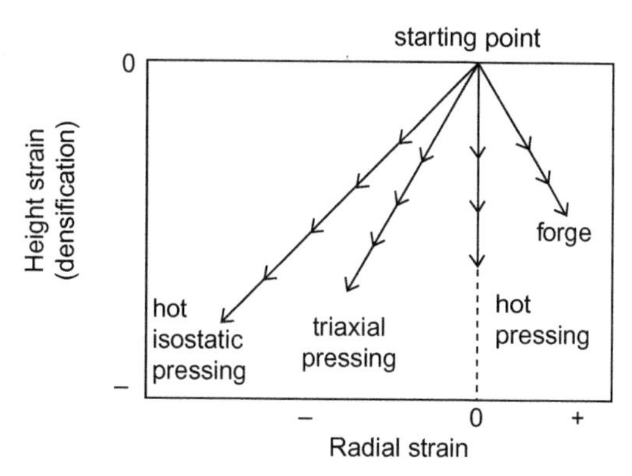

Figure 10.29 Strain trajectories for the same four processes illustrated in Figure 10.28. The pressure-assisted processes give densification, but the radial strain differs.

shear makes the collapse of large pores ineffective at low stresses. A benefit is that dimensional change is uniform and somewhat predictable. Uniaxial hot pressing involves radial constraint from the die wall, with the radial stress generated at the die wall being proportional to the applied pressure. The uniaxial pressure is controlled during processing, creating shear that improves bonding and the collapse of large pores. Sinter forging also produces shear. Depending on the applied pressure and the sintering rate, sinter forging can result in either radial swelling or shrinking. At low stresses, the sintering stress creates a higher densification shrinkage than the radial flow. Alternatively, at high applied stresses lateral flow dominates.

Figure 10.29 captures the strain trajectories for these different densification approaches. Powder compression gives the negative strains. During sintering, hot isostatic pressing, and other hydrostatic densification processes, the axial and radial strains are equal. Accordingly, the trajectory is along a 45 degree incline for these processes. In the ideal case the component size is uniform after densification. Uniaxial hot pressing has no radial strain because of die wall constraint, so all densification is associated with a change in the compact height. The benefit with respect to dimensional change is that the radial dimensions are determined by the tooling and only the height strain or starting mass needs to be controlled. Forging is usually performed at a high stress to exceed the yield strength. Initially the height decreases while the radius increases. The outer surface cracks if the forging stroke is too aggressive. If the applied stress is low, then incomplete densification results. Finally, with triaxial loading the radial strain is less than the height strain, leading to a path between hot pressing and sintering.

In all cases, densification increases with the applied pressure. Hydrostatic situations give uniform dimensional change. However, other forces, including gravity and container friction, distort larger structures. A shear stress improves densification, but usually the effect is not large. For example, in hot isostatically pressed titanium alloy

99.5% density is attained using hot isostatic pressing at 850°C (1123 K) and 117 MPa for 30 min. The same powder under triaxial conditions at the same temperature and time required an axial pressure of 169 MPa and radial pressure of 90 MPa. This corresponds to an equivalent hydrostatic pressure of 116 MPa, essentially the same as required in hot isostatic pressing.

Uniaxial Hot Pressing

Hot pressing is performed in a rigid die using loading along the vertical axis as sketched in Figure 10.30. To avoid die damage, the graphite die is enclosed in a protective atmosphere or vacuum chamber. Loading is along the vertical axis on punches pressurized from an external hydraulic system. Although the pressure is applied along the vertical axis, there is a radial pressure against the die wall. The differential stress between the axial and radial directions generates shear that is effective in particle bonding. This shear stress is proportional to the applied stress. Initial densification includes particle rearrangement and plastic flow. As densification progresses, creep by grain boundary diffusion and volume diffusion becomes controlling.

For induction heating, the die is made from graphite, but resistance heating is also used. Other die materials include refractory metals and their alloys, and sometimes ceramics such as alumina or silicon carbide at low stresses. Thermal expansion compatibility between the powder compact and tooling is required to avoid cracking of

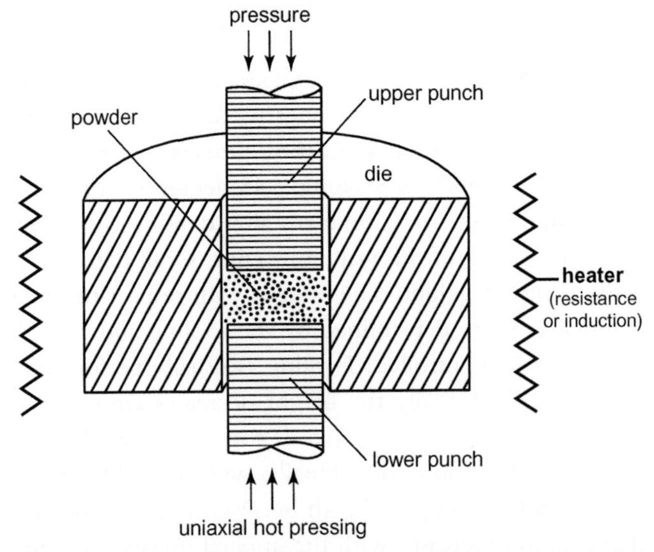

Figure 10.30 A schematic of uniaxial hot pressing, where pressure is applied along the vertical axis by graphite punches and the powder is contained in a heated die. Heating is from resistance or more rapidly using induction, and very rapidly using exothermic reactions. Other punch and die materials besides graphite are possible, and usually the process is performed under vacuum.

one or the other during cooling. An option is to eject the compact at the pressing temperature.

Uniaxial hot pressing is usually slow because of the thermal mass associated with the tooling. Typical maximum temperatures are 2200°C (2473 K) and maximum pressures are 50 MPa. Overall hot pressing is expensive and restricted to diameters smaller than 400 mm. Compact contamination from the die is a perpetual problem. Even so, hot consolidation is widely used to fabricate unique materials; especially brittle materials.

A most important use for uniaxial hot pressing is in the consolidation of diamond-metal cutting tools. A typical composite consists of 1.5 μm particles of cobalt that form a matrix containing about 10 vol.% diamond. This composite is hot pressed in graphite tooling at 35 MPa with a peak temperature of 900°C (1173 K) using a hold time of 2 min to avoid diamond decomposition.

It is possible to induce superplastic flow using uniaxial hot pressing [85]. Multiple phase materials with a small grain size are candidates; the strain rate is controlled to allow diffusion to keep pace densification. The formation of a liquid phase also allows for rapid densification during hot pressing [86]. Steels with 1.5 to 2% carbon densify at temperatures under 1200°C (1473 K) in 10 min using 100 MPa pressure [87].

A variant used since the 1950s relies on first hot pressing a porous ceramic structure. This structure is infiltrated with a high temperature alloy to form a composite. This sinter-casting approach was developed for Fe-TiC, a material known as ferritic, producing the composite microstructure shown in Figure 10.31. In nonwetting situations, the uniaxial pressure forces the liquid metal into the pores, and is useful in Al-SiC and Mg-SiC.

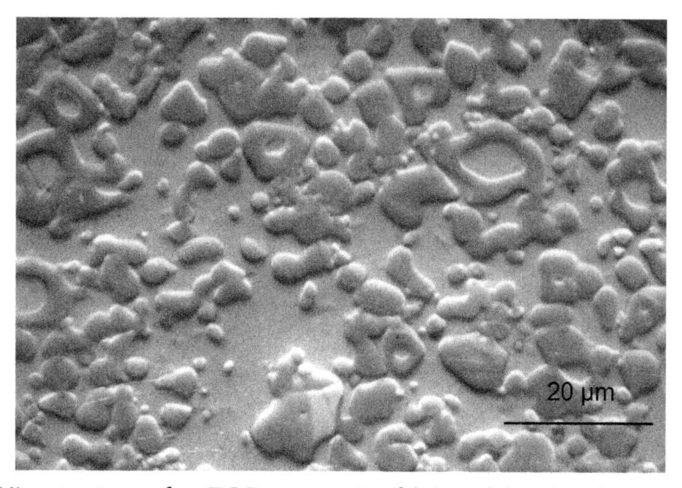

Figure 10.31 Microstructure of a TiC-Fe composite fabricated by the sinter-casting approach, where the TiC is first hot pressed to form a skeleton and molten tool steel is infiltrated into the structure. The TiC is the rounded phase.

Hot Isostatic Pressing

HIP relies on high pressure vessels and hydrostatic gas for pressurization. The pressure chambers are machined from forged steel, backed by wire windings, and water cooled to prevent failure during pressurization. Four options exist in hot isostatic pressing:

- place powder in container (soft metal or glass), transfer filled container to pressure vessel, heat and pressurize over several hours
- place powder in container, heat powder filled container, insert hot canister into pressure vessel where it is pressurized quickly, without additional heating
- vacuum sinter a powder compact to closed pore condition, about 95% dense, and pressurize at temperature in the same cycle, termed sinter-HIP
- vacuum sinter a powder compact to closed pore condition, cool and subsequently heat and pressurize in a separate cycle to remove pores.

For the first option, Figure 10.32 is a schematic of the sequence where powder is filled into a can and then the can is subjected to simultaneous heating and pressurization. The primary control parameters are pressure, temperature, and time, and Figure 10.33 is an example of the cycle applied to a large cemented carbide body, such as used in oil well drilling. Prior to HIP consolidation, the powder-filled container is heated and vacuum degassed to remove volatile contaminants, and then sealed under vacuum. Failure to adequately degas the powder leads to thermally induced porosity, where gas filled pores reform in the full density compact when reheated after HIP. Since pressures are relatively low, final densification is by creep which requires an extended hold for diffusion.

One variant is to use previously sintered compacts with densities over 95% with closed pores. The shaped component does not need a container. Figure 10.34 compares the microstructure before and after such densification. The pores remaining after sintering are removed, but the grain size is enlarged. This approach is used to

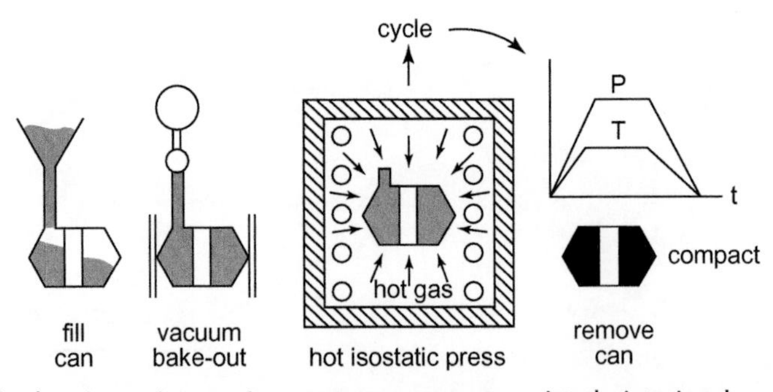

Figure 10.32 The hot isostatic pressing route to pressure-assisted sintering has several variants. Shown here is the situation where a container is filled with powder, evacuated, and subjected to simultaneous heating and pressurization. The container is removed to leave a shrunken version of the starting shape.

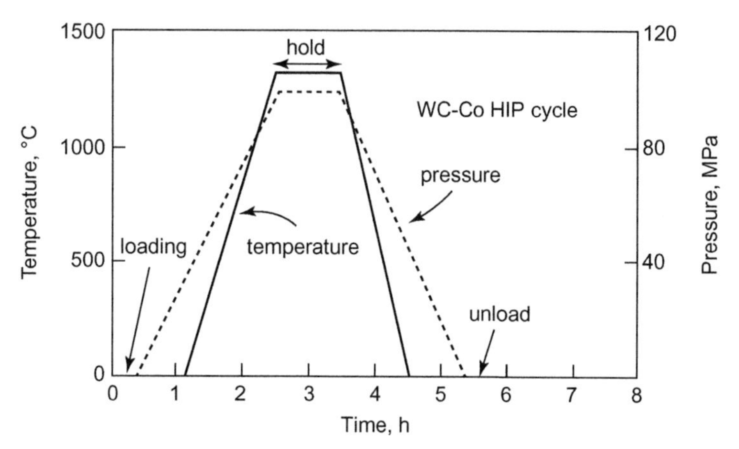

Figure 10.33 A plot of temperature and pressure versus time to consolidate a typical cemented carbide by hot isostatic pressing, with a peak temperature of near 1350 to 1400°C (1623 to 1673 K) and 100 MPa.

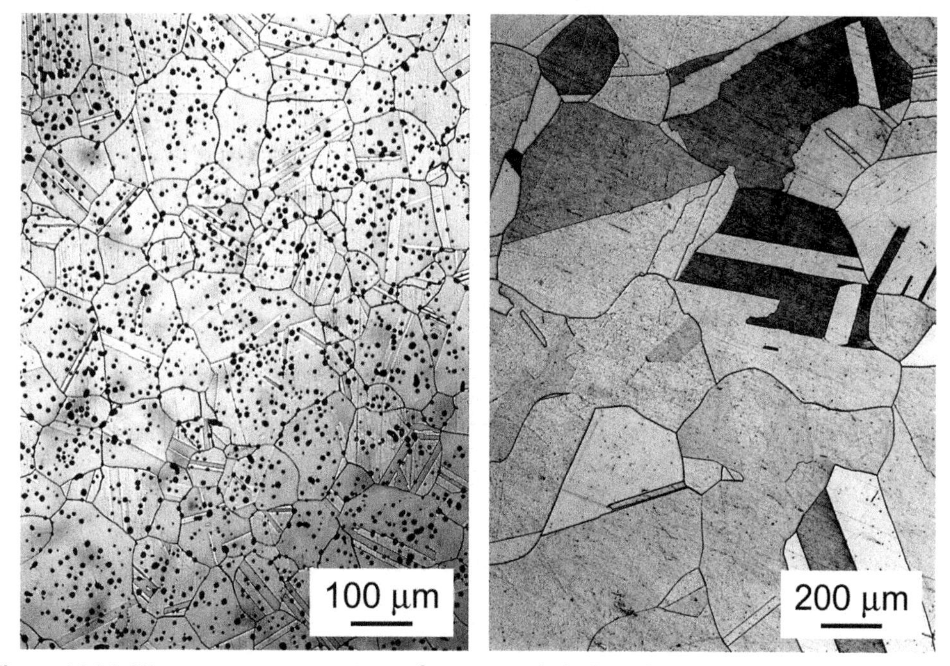

Figure 10.34 Microstructure comparison of an iron-nickel alloy after sintering to the closed pore condition (on the left) and after pressure-assisted sintering via hot isostatic pressing. Note the significant loss of pores (black dots) and grain size enlargement with containerless HIP *(courtesy of Gerald Camus).*

consolidate cemented carbides, wear materials, covalent ceramics, injection molded components, and titanium implants.

Production machines where there is no internal heating cost less to fabricate. For large billets, internal heating is not needed. Instead the hot billet is transferred into the pressure vessel and consolidated before significant cooling occurs. Small components cool too quickly, so this is largely used for large tool steel billets. Pressurization is controlled by the amount of gas introduced in the chamber; the gas pressure is driven mostly by pumps and gas thermal expansion. There is no intentional shear stress in hot isostatic pressing, so prior particle boundary films are not ruptured. Numerous reports tell of reaching 100% density in HIP, but with low mechanical properties.

Containerless HIP applies to the case where the powder is sintered to the closed pore condition and no container was required. For powder, a gas-tight container is used to shape the powder. Otherwise, the HIP container is fabricated from a material that is deformable at the consolidation temperature, and glass, steel, stainless steel, titanium, and tantalum are common, depending on the maximum temperature. High pressure argon or nitrogen are used to transfer heat and pressure to the compact. Temperatures up to 2200°C (2473 K) and pressures up to 200 MPa are common, but a few designs reach higher. Such methods are applied to a wide range of materials.

The sinter-HIP cycle combines vacuum sintering to a closed pore condition followed by hot isostatic pressing in the same cycle. At a strategic point in the cycle the vacuum sintering furnace is pressurized with gas sufficient to collapse lingering pores. Figure 10.35 sketches such as cycle. Initial vacuum sintering densifies the component and avoids trapped gas. After pore closure the furnace is pressurized to eliminate final pores. A less productive variant is to separate the sintering and hot isostatic pressing into two separate runs in separate furnaces.

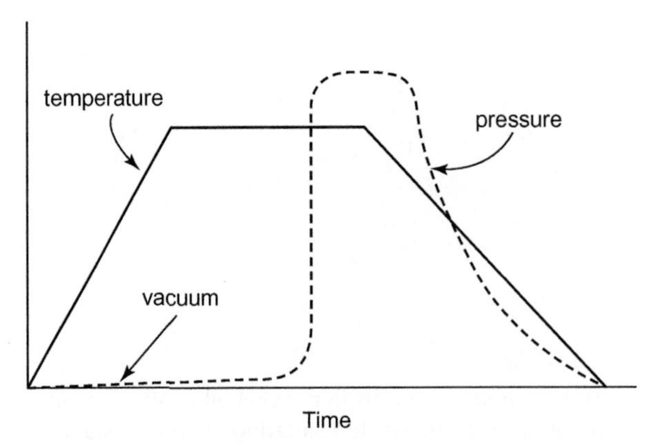

Figure 10.35 Sinter-HIP relies on one heating cycle but a shift from vacuum early in the cycle to pressurization late in the cycle. Proper pressure application results in collapse of the residual pores.

Triaxial Pressurization

Hot isostatic pressing generates a hydrostatic pressure. Triaxial techniques supplement the HIP pressure with additional uniaxial pressure. One means to achieve this is via a granular or semifluid medium that holds the powder preform. The concepts are termed rapid omnidirectional compaction, ceramic granule consolidation, and steel container powder forging. Force transmission is through a low melting point, soft phase, such as copper, or graphite. The soft phase converts the uniaxial stress into a pseudo-hydrostatic stress; the hydrostatic stress is about one-third of the applied stress. The resulting stress state is a mixture of hydrostatic and uniaxial stress; a hybrid between hot isostatic pressing and forging. The resulting shear proves beneficial in disrupting packing irregularities and particle surface films, but with anisotropic dimensional change.

Spark Sintering

The initial idea of an electric discharge pressing and heating dates from early efforts on WC, W, and refractory metals. An early success in 1910 was the sintering of tungsten billets later drawn into lamp filaments. Subsequently, considerable insight has arisen on the role of electric and magnetic fields on sintering, especially for the consolidation of diamonds and diamond composites.

Figure 10.36 sketches one of the concepts. Spark sintering occurs in a vacuum hot press using graphite electrodes to pass current directly through the die and powder compact. The die is also formed from graphite so it acts as a resistor, heating with current passage. If the powder compact is conductive, some of the current passes through the sintered powder [50]. Resistive heating contrasts with other pressure-assisted

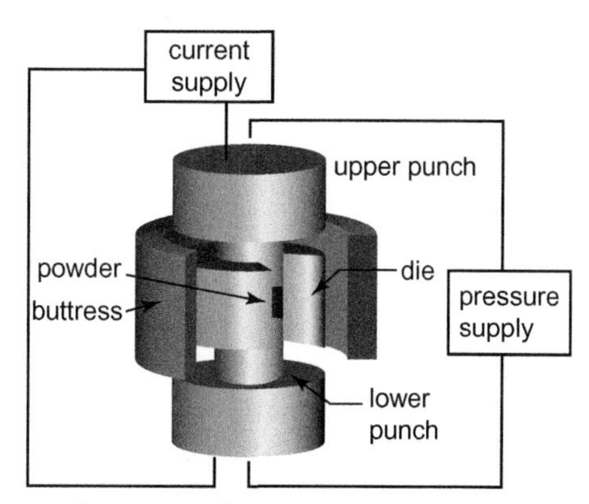

Figure 10.36 Spark sintering relies on an electric current passing through the die and the powder compact while simultaneously undergoing pressurization from the punches. Typically the die and punches are formed from graphite, which limits the peak pressure but provides rapid heating.

sintering cycles where heat is generated outside the compact by induction, radiation, microwave, combustion, or thermal conduction.

For spark sintering, typical operating conditions depend on the material and component size, but might be in the range of 5 to 15 V, up to 20,000 A, pulsed DC ranging from 2 ms up to 15 s duration pulses, and pressure up to 200 MPa. For diamond consolidation the pressures reach to 6 GPa. Most efforts rely on DC cycles, but AC cycles are used for some metals. Current densities reach up to 10,000 A/cm^2, mandating small devices because of available current limitations. The consolidation time ranges up to a few hours, but often is just minutes with rapid heating. To avoid concentrated current in corners it is best to rely on thin cylinder shapes, similar to poker chips; the typical thickness is 3 mm. Diameters depend on power supplies and hydraulic systems; ranging from 8 mm to 200 mm.

Densification is rapid once temperature and pressure are attained, but temperature distribution in the powder compact is not uniform [88,89]. Graphite as the die material requires a protective environment. Thus, most units operate in vacuum or argon. Because the die is closed, gas pressure around the powder compact is high, often consisting of carbon monoxide. Graphite is relatively weak and this limits applied pressures, although temperatures up to 2400°C (2673 K) are possible if the power supply is large. Buttressed dies enable higher pressures. For diamond sintering at 5 to 6 GPa, special dies are employed based on refractory metals such as tantalum. Expansion to shapes beyond the typical poker chip geometry is possible using granular graphite to fill the cavity between the punches and component.

Much disagreement exists on the direct merits of these technologies. Temperature errors of 150°C are common, and density, as extracted from displacement records, tends to be in error by 5%. Thus, it is difficult to make direct comparisons to standard hot pressing; however, in systems where cycles and equipment designs have been matched (time, temperature, heating rate, pressure, tool layout), rapid heating without the electric current via hot pressing shows similar merits [90–92]. The microstructure trajectories are the same and follow the classic grain size dependence on density given earlier. However, the data in Figure 10.37 show both hot pressing and spark sintering give smaller grain sizes when compared to sintering [91].

Hot Forging

Forging is a high strain rate deformation process, conducted at elevated temperatures where the material has a low strength and high malleability. Under certain conditions a porous material will deform by superplastic flow, allowing considerable reshaping. With respect to sintering, there are two hot forging variants:

- high strain rate forging at high stresses for rapid densification by exceeding the material yield strength, usually applied to metallic powders

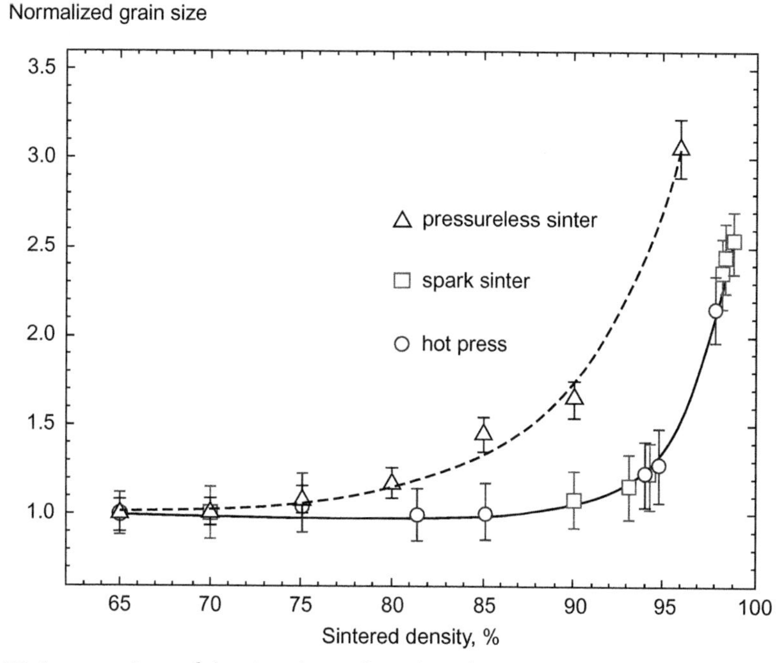

Figure 10.37 A comparison of the sintering trajectories of grain size versus sintered density for alumina [91]. The pressureless sintering data show larger grains, but the spark sintering and hot pressing are essentially identical traces.

- low strain rate forging at low stresses to deform a fine grained compact by diffusional creep, usually applied to ceramic powders.

A powder preform will densify in a single high strain rate forging strike. This is the blacksmith's approach to sintering. If the preform weight is closely monitored, then forging can produce the final size at full density. As illustrated in Figure 10.38 sintering is used to fabricate the preform. The sintered compact is densified by plastic flow. Some of the axial flow results in bulging to cause the compact to reach the die walls. Although mass remains constant, volume decreases with pore collapse. Die wall constraint determines the peak stress condition and too high a stress results in fracture. Friction with the die and punches results in tensile stresses and cracking, as illustrated by the circumferential tensile stresses in Figure 10.39. Lubrication of the tool surface reduces cracking. The initial preform density is not significant with respect to fracture conditions, although the preform mass must be sufficient to reach full density within the forging stroke.

Temperature determines the stress necessary to achieve densification. Typically ferrous forging operations do not exceed 1200°C (1473 K), and many preforms are forged at temperatures closer to 800°C (1073 K). The preform size and density are

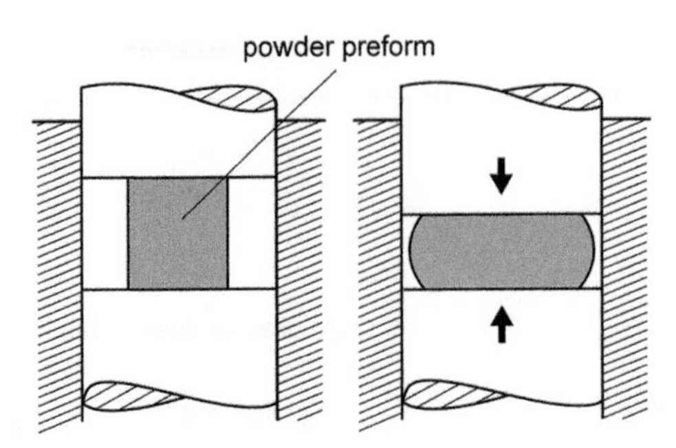

Figure 10.38 Powder forging or sinter forging applies a pressure along the vertical axis. In one variant the loading rate is very high and radial constraint is required to avoid cracking. In another variant the loading rate is very low and the compact densifies at the inherent sintering rate without need for radial constraint.

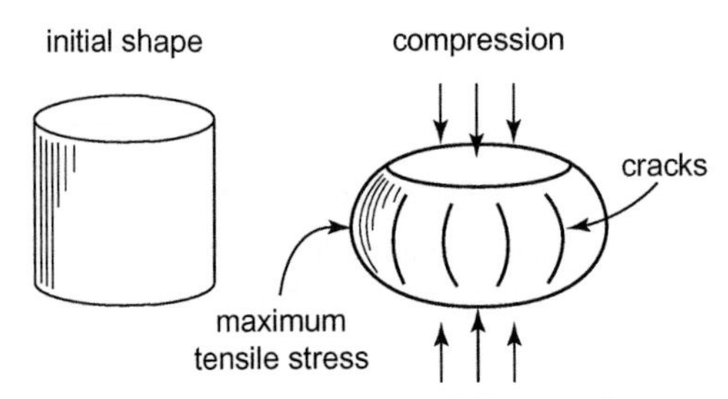

Figure 10.39 One of the difficulties with any form of forging during sintering is that the large tensile stresses on the outside of the component cause cracking.

determined by the lateral die constraints, fracture conditions, and the need to obtain full density in a single strike.

Sinter forging is applicable to high temperature ceramics that fracture under high strain rate conditions. It involves compression at stresses and strain rates compatible with the inherent sintering rate. The pressures are low (0.1 to 20 MPa), on the same order of magnitude as the sintering stress. Consequently, the time to full density is not greatly different from that required for sintering. Figure 10.40 illustrates the applied pressure effect alumina densification at 1500°C (1773 K) [93]. At low stresses, both the axial and radial dimensions shrink. Alternatively, at high stresses, axial shrinkage occurs with radial expansion. An advantage to sinter forging is that uniaxial pressurization generates a shear stress to remove defects in the compact.

Gas Forging

Gas forging is similar to hot isostatic pressing, but the peak pressures are higher, more like those encountered in powder forging. Densification is fast since the strain rate and

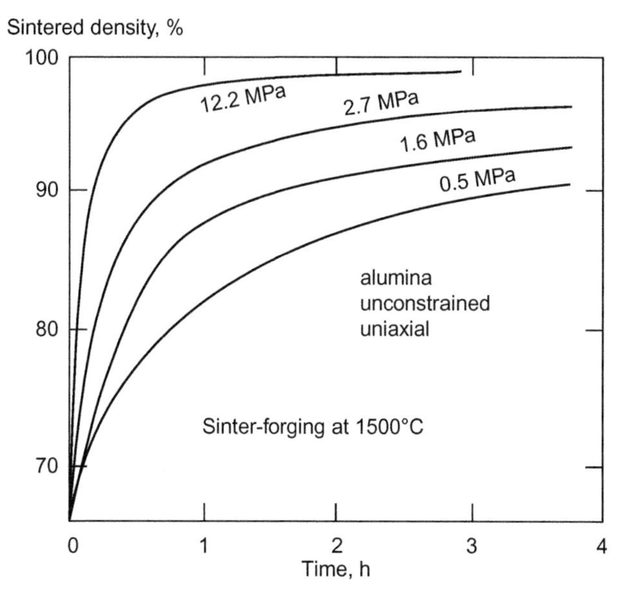

Figure 10.40 Sinter forging data for alumina during uniaxial compression at low stresses [93]. The plot shows density grains from higher pressures and longer times.

effective pressure are both high. To attain these conditions, a powder is sealed and heated in a furnace capable of substantial pressures. Once the compact is at temperature, liquid argon or liquid nitrogen is introduced into the pressure chamber. This liquid immediately converts to gas to generate a rapid pressure surge. The final pressure is determined by the working temperature, available chamber volume, and amount of liquid introduced into the chamber. Such an approach eliminates the time required to pressurize a HIP vessel. The lower pressure and longer time required with conventional HIP cycles result in lower final properties because of greater microstructural coarsening. One variant is to use previously sintered components, so a canister is not required.

Dramatic gains in mechanical properties are possible from the forging strain rate during densification. For example, with iron:

- Conventional hydrogen sintering at 1300°C (1573 K) for 1 h gives 249 MPa strength.
- Conventional HIP after sintering bumps the tensile strength to 385 MPa.
- Gas forging after sintering produces 732 MPa strength.

Powder Extrusion

Hot extrusion involves a high strain rate to plastically deform a powder with little involvement of diffusion. Long shapes with a constant cross section are the main products of extrusion. By canning the powder in a vacuum tight container, it is degassed prior to hot extrusion. A high extrusion deformation is required to obtain full densification.

The concept of powder extrusion is sketched in Figure 10.41. A small penetrator in the extrusion ram directly presses on the powder to avoid buckling the can. The extrusion constant C measures the deformation work and related to the extrusion force F and extrusion area reduction as follows:

$$F = C\,A\ln(R) \tag{10.24}$$

where A is the cross-sectional area of the feed material and R is the reduction ratio, equal to the ratio of the stating and final cross-section areas. Although intrinsic material properties influence extrusion, temperature is a main control variable. Too high a temperature damages the product and shortens tool life. Alternatively, too low a temperature makes extrusion difficult because of higher pressures and tool wear. Typically, extrusion is performed at temperatures over two-thirds of the absolute melting temperature for the material. The force to initiate extrusion flow increases with smaller particle sizes, and is higher than the force to maintain flow.

Hot extrusion is used to fabricate tubes or constant cross section high temperature components including dispersion strengthened copper, aluminum alloys, stainless steel, and beryllium. It is also used for materials with limited room temperature plasticity such as metal-matrix composites. Mixed powders are induced to react during extrusion, for example to form NiTi rod from mixed nickel and titanium powders [94].

Shock Wave Consolidation

Shock waves rapidly densify a powder with modest heating and high stresses [95]. If the material is brittle, then preheating is required prior to compaction. Densification is rapid by plastic flow while the material thermally softens from frictional self-heating.

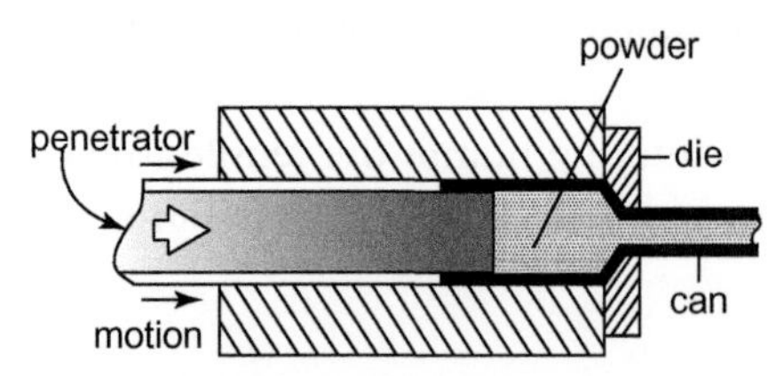

Figure 10.41 Hot extrusion relies on a preheated powder pressurized and forced though a die to produce a long product. It is similar to how toothpaste is extruded. As with HIP, the powder is usually encapsulated in a canister and evacuated prior to extrusion.

An example layout for explosive consolidation is given in Figure 10.42. The detonator initiates an explosive wave that pressurizes the powder and heating at the particle contacts causes sinter bonding. Related techniques rely on impacting a high velocity mass against the powder compact, launched by a gas gun, collapsing magnetic field, or capacitive discharger.

Applications for shock consolidation are brittle materials otherwise difficult to sinter. For Si_3N_4 and SiC, densification to 96% of theoretical density is possible using peak pressures in the 20 to 60 GPa range. The densification event occurs in approximately 4 μs. The consolidated density increases as the relative input energy per unit mass increases [96]. Excessive energy damages the compact while too little energy

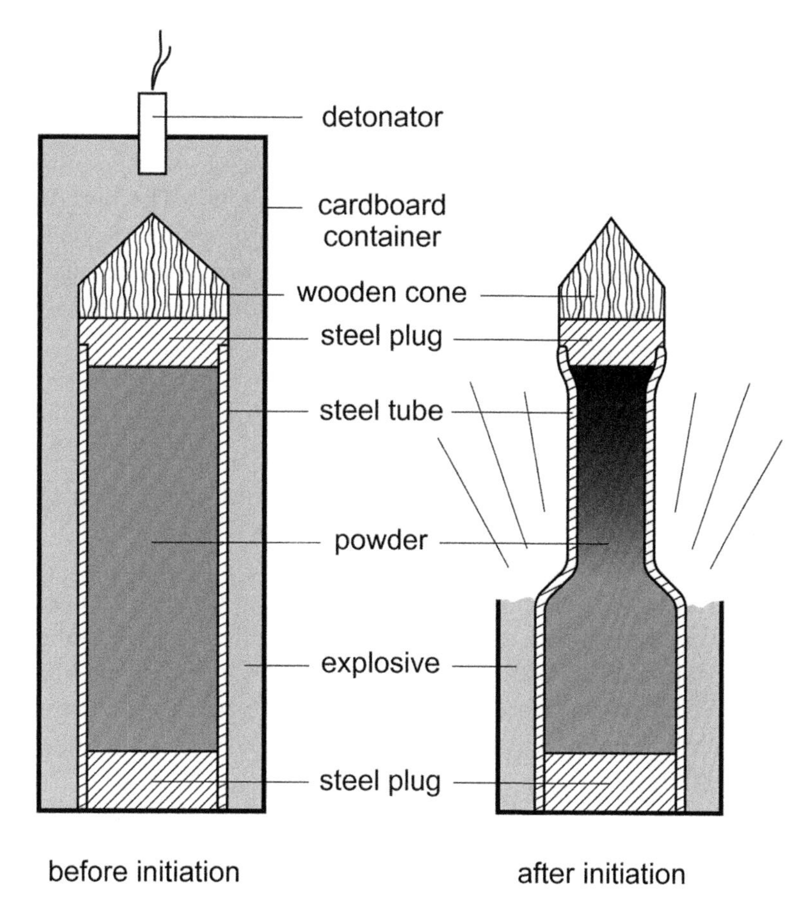

Figure 10.42 The extreme in pressure-assisted sintering is seen in explosive compaction, where an explosive charge propagates a shock wave though the powder. The rapid consolidation induces particle heating. The combination of pressure and temperature induces near instantaneous densification.

results in porosity. Shock wave consolidation is not precise and the compacts have considerable residual strain.

A related idea is dynamic compaction. It is used for combinations of synthesis and densification. Generally a delay of a few seconds between reaction initiation and densification is beneficial. This allows self-heating, and thermal softening, prior to initiating plastic flow.

DENSIFICATION LIMITATIONS

Pressure offsets inherent densification limitations encountered in sintering. For example, large pores resist sintering densification, but can be collapsed by an external pressure.

A problem in pressure-assisted sintering results from gas trapped in the pores. In processes such as hot isostatic pressing the pressurization gas must be insoluble in the compact. But high applied pressures induce some solubility. The dissolved gas then forms pores on reheating, such as in heat treatment. It is important to minimize gas in closed pores by vacuum degassing the compact and removing species that can evaporate or react. The final fractional density after pressurization is estimated as,

$$f = \frac{\beta}{1+\beta} \tag{10.25}$$

where

$$\beta = 11\frac{P_A}{P_O} \tag{10.26}$$

assuming pore closure at about 8.25% porosity with P_O as the gas pressure in the pores at pore closure. For a compact sintered in one atmosphere gas ($P_O = 0.1$ MPa) to a closed pore condition ($f = 0.92$) and subsequently pressurized to 20 MPa, the final density is 99.96%. If pressurization is 10 atmospheres (1 MPa), the limiting density is about 99.1%. This constraint is evident when internal gas pressures are high or processing pressures are low.

REFERENCES

[1] H.V. Atkinson, S. Davies, Fundamental aspects of hot isostatic pressing: an overview, Metall. Mater. Trans. 31A (2000) 2981–3000.
[2] R. Ge, Development of a new hot pressing equation in powder metallurgy, Powder Metall. Inter. 24 (1992) 229–232.

[3] S. Brown, An internal variable constitutive model for hot working of metals, Inter. J. Plasticity 5 (2) (1989) 96–130.
[4] C.G. Goetzel, Principles and present status of hot pressing, in: W.E. Kingston (Ed.), The Physics of Powder Metallurgy, McGraw-Hill, New York, NY, 1951, pp. 256–277.
[5] W.B. James, Powder forging, Rev. Particulate Mater. 2 (1994) 173–213.
[6] H.F. Fischmeister, E. Arzt, Densification of powders by particle deformation, Powder Metall. 26 (1983) 82–88.
[7] J.V. Kumar, Physics of pressure transmission in fluids and powders, in: J.V. Kumar (Ed.), Frontiers of Metallurgy and Materials Technology, BS Publications, Hyderabad, India, 2011, pp. 285–306.
[8] A.S. Helle, K.E. Easterling, M.F. Ashby, Hot isostatic pressing diagrams new developments, Acta Metall. 33 (1985) 2163–2174.
[9] M.F. Ashby, Background Reading HIP 6.0, Engineering Department Cambridge University, Cambridge, UK, 1990.
[10] A.A. Hendrickson, P.M. Machmeier, D.W. Smith, Impact forging of sintered steel preforms, Powder Metall. 43 (2000) 327–344.
[11] E.Y. Gutmanas, High-pressure compaction and cold sintering of stainless steel powders, Powder Metall. Inter. 12 (1980) 178–182.
[12] R.L. Coble, Mechanisms of densification during hot pressing, in: G.C. Kuczynski, N.A. Hooton, C.F. Gibbon (Eds.), Sintering and Related Phenomena, Gordon and Breach, New York, NY, 1967, pp. 329–347.
[13] M.P. Harmer, R.J. Brook, The Effect of MgO additions on the kinetics of hot pressing in Al_2O_3, J. Mater. Sci. 15 (1980) 3017–3024.
[14] L. Ramqvist, Theories of hot pressing, Powder Metall. 9 (1966) 1–25.
[15] G. Wegmann, R. Gerling, F.P. Schimansky, Temperature induced porosity in hot isostatically pressed gamma titanium aluminide alloy powders, Acta Mater. 51 (2003) 741–772.
[16] C.G. Goetzel, Hot pressed and sintered copper powder compacts, in: J. Wulff (Ed.), Powder Metallurgy, American Society for Metals, Cleveland, OH, 1942, pp. 340–351.
[17] C.G. Goetzel, Treatise on Powder Metallurgy, vol. 1, Interscience Publishers, New York, NY, 1949, pp. 259–312.
[18] M.Y. Chu, L.C. De Jonghe, Effect of temperature on the densification/creep viscosity during sintering, Acta Metall. 37 (1989) 1415–1420.
[19] L.L. Seigle, Atom movements during solid state sintering, Prog. Powder Metall. 20 (1964) 221–238.
[20] M.R. Notis, R.H. Smoak, V. Krishnamachari, Interpretation of hot pressing kinetics by densification mapping techniques, in: G.C. Kuczynski (Ed.), Sintering and Catalysis, Plenum Press, New York, NY, 1975, pp. 493–507.
[21] R.L. Coble, A model for boundary diffusion controlled creep in polycrystalline materials, J. Appl. Phys. 34 (1963) 1679–1682.
[22] R.L. Coble, Diffusion models for hot pressing with surface energy and pressure effects as driving forces, J. Appl. Phys. 41 (1970) 4798–4807.
[23] A.C.F. Cocks, N.D. Aparicio, Diffusional creep and sintering—the application of bounding theorems, Acta Metall. Mater. 43 (1995) 731–741.
[24] H. Riedel, V. Kozak, J. Svoboda, Densification and Creep in the Final Stage of Sintering, Acta Metall. Mater. 42 (1994) 3093–3103.
[25] R.M. German, Sintering Theory and Practice, Wiley-Interscience, New York, NY, 1996.
[26] H.C. Yang, K.T. Kim, Creep densification behavior of micro and nano metal powders, grain size dependent model, Acta Mater. 54 (2006) 3779–3790.
[27] X.F. Zhange, Q. Yang, L.C. De Johghe, Microstructure development in hot pressed silicon carbide: effects of aluminum, boron, and carbon additives, Acta Mater. 51 (2003) 3849–3860.
[28] G. Pezzotti, K. Ota, H.J. Kleebe, Grain boundary relaxation in high purity silicon nitride, J. Amer. Ceram. Soc. 79 (1996) 2237–2246.
[29] O. Yeheskel, Y. Gefen, M. Talianker, HIP of Si_3N_4 processing – properties – microstructure relationships, Proceeding of Third International Conference on Isostatic Pressing, vol. 1, Metal Powder Report, Shrewsbury, UK, 1986, pp. 20.1–20.12.

[30] W.R. Cannon, T.G. Langdon, Review creep of ceramics, Part 2, an examination of flow mechanisms, J. Mater. Sci. 23 (1988) 1–20.

[31] Y.S. Kwon, K.T. Kim, Densification forming of alumina powder – effects of power law creep and friction, J. Eng. Mater. Tech. 118 (1996) 471–477.

[32] D.S. Wilkinson, M.F. Ashby, Pressure sintering by power law creep, Acta Metall. 23 (1975) 1277–1285.

[33] Y.M. Liu, H.N.G. Wadley, J.M. Duva, Densification of porous materials by power law creep, Acta Metall. Mater. 42 (1994) 2247–2260.

[34] L. Sanchez, E. Ouedraogo, L. Federzoni, P. Stutz, New viscoplastic model to simulate hot isostatic pressing, Powder Metall. 45 (2002) 329–334.

[35] M.K. Spencer, R.B. Alley, T.T. Creyts, Preliminary firn-densification model with 38-site dataset, J. Glaciology 47 (2001) 671–676.

[36] A.M. Laptev, H.P. Buchkremer, R. Vassen, Investigation of input data for compaction modelling of hot isostatic pressing, in: A. Zavaliangos, A. Laptev (Eds.), Recent Developments in Computer Modeling of Powder Metallurgy Processes, ISO Press, Ohmsha, Sweden, 2001, pp. 151–159.

[37] M. Jeandin, J.L. Koutny, Y. Bienvenus, Rheology of solid-liquid p/m Astroloy – application to supersolidus hot pressing of P/M superalloys, Inter. J. Powder Metall. Powder Tech. 18 (1982) 217–223.

[38] W.D. Kingery, J.M. Woulbroun, F.R. Charvat, Effects of applied pressure on densification during sintering in the presence of a liquid phase, J. Amer. Ceram. Soc. 46 (1963) 391–395.

[39] K. Morsi, The diversity of combustion synthesis processing, a review, J. Mater. Sci. 47 (2012) 68–92.

[40] J.C. Murray, R.M. German, Reactive sintering and reactive hot isostatic compaction of niobium aluminide $NbAl_3$, Metall. Trans. 23A (1992) 2357–2364.

[41] W. Misiolek, R.M. German, Reactive sintering and reactive hot isostatic compaction of aluminide matrix composites, Mater. Sci. Eng. A144 (1991) 1–10.

[42] R.M. German, R.G. Iacocca, Powder metallurgy processing, in: N.S. Stoloff, V.K. Sikka (Eds.), Physical Metallurgy and Processing of Intermetallic Compounds, Chapman and Hall, New York, NY, 1996, pp. 605–654.

[43] T. Aizawa, S. Kamenosono, J. Kihara, T. Kato, K. Tanaka, Y. Nakayama, Shock reactive synthesis of TiAl, Intermetallics 3 (1995) 369–379.

[44] C.G. Goetzel, V.S. De Marchi, Electrically activated pressure sintering (Spark Sintering) of titanium powders, Powder Metall. Inter. 3 (1971) 80–87 and 134–136.

[45] Z.A. Munir, U. Anselmi-Tamburini, M. Ohyanagi, The effect of electric field and pressure on the synthesis and consolidation of materials: a review of the spark plasma sintering method, J. Mater. Sci. 41 (2006) 763–777.

[46] S. Grasso, Y. Sakka, G. Maizza, Electric Current Activated/Assisted Sintering (ECAS): a Review of Patents 1906–2008; Sci. Tech. Adv. Mater. 10 (2009) article 053001.

[47] M. Suganuma, Y. Kitagawa, S. Wada, N. Murayama, Pulsed electric current sintering of silicon nitride, J. Amer. Ceram. Soc. 86 (2003) 387–394.

[48] J. Zhang, A. Zavaliangos, J. Groza, Field activated sintering techniques: a comparison and contrast, P/M Sci. Tech. Briefs 5 (4) (2003) 5–8.

[49] J.F. Garay, Current activated, pressure assisted densification of materials, Ann. Rev. Mater. Res. 40 (2010) 445–468.

[50] Z.A. Munir, D.V. Quach, M. Ohyanagi, Electric current activation of sintering: a review of the pulsed electric current sintering process, J. Amer. Ceram. Soc. 94 (2011) 1–19.

[51] A. Accary, R. Caillat, Study of mechanism of reaction hot pressing, J. Amer. Ceram. Soc. 45 (1962) 347–351.

[52] L. Rangaraj, S.J. Suresha, C. Divakar, V. Jayaram, Low temperature processing ZrB_2-ZrC composites by reactive hot Pressing, Metall. Mater. Trans. 39A (2008) 1496–1505.

[53] A.L. Chamberlain, W.G. Fahrenholtz, G.E. Hilmas, Low temperature densification of zirconium diboride ceramics by reactive hot pressing, J. Amer. Ceram. Soc. 89 (2006) 3638–3645.

[54] A. Sewchurran, L.A. Cornish, Microstructure-hardness relationships in reactively HIPPed Ruthenium-Aluminum Alloys, in: R.M. German, G.L. Messing, R.G. Cornwall (Eds.), Sintering Science and Technology, The Pennsylvania State University, State College, PA, 2000, pp. 63–68.

[55] L. Chen, E. Kny, Reaction hot pressed submicron Al_2O_3 + TiC ceramic composite, Inter. J. Refract. Met. Hard Mater. 18 (2000) 163–167.
[56] E. Paransky, E.Y. Gutmanas, I. Gotman, M. Koczak, Pressure-assisted reactive synthesis of titanium aluminides from dense 50Al-50Ti elemental powder blends, Metall. Mater. Trans. 27A (1996) 2130–2139.
[57] L. Rangaraj, C. Divakar, V. Jayaram, Reactive hot pressing of titanium nitride – titanium diboride composites at moderate pressure and temperature, J. Amer. Ceram. Soc. 87 (2004) 1872–1878.
[58] Y. Aman, V. Garnier, E. Djurado, Pressureless spark plasma sintering effect on nonconventional necking process during the initial stage of sintering of copper and alumina, J. Mater. Sci. 47 (2012) 5766–5773.
[59] D.R. Campbell, H.B. Huntington, Thermomigration and electromigration in zirconium, Phys. Rev. 179 (1969) 609–612.
[60] C.M. Hsu, D.S.H. Wong, S.W. Chen, Generalized phenomenological model for the effect of electromigration on interfacial reaction, J. Appl. Phys. 102 (2007), article 023715.
[61] M. Cologna, B. Rashkova, R. Raj, Flash sintering of nanograin zirconia in less than 5 s at 850 C, J. Amer. Ceram. Soc. 93 (2010) 3556–3559.
[62] G. Davies, T. Evans, Graphitization of diamond at zero pressure and at a high pressure, Proc. Royal Soc. London A 328 (1972) 413–427.
[63] G.C. Davies, D.R.H. Jones, Creep of Metal-type organic compounds – IV. Application to hot isostatic pressing, Acta Mater. 45 (1997) 775–789.
[64] L. Mahler, M. Ekh, K. Runesson, A class of thermo-hyperelastic-viscoplastic models for porous materials: theory and numerics, Inter. J. Plasticity 17 (2001) 943–969.
[65] A. Svoboda, H.A. Haggblad, L. Karlsson, Simulation of hot isostatic pressing of a powder metal component with an internal core, Comp. Meth. Appl. Mech. Eng. 148 (1997) 299–314.
[66] H. Yoshimura, T. Nomoto, Sintering deformation behavior during hot isostatic pressing, J. Japan Soc. Powder Powder Metall. 40 (1993) 488–496.
[67] S. Shima, A. Inaya, Simulation of Pseudo-isostatic pressing of powder compact, in: M. Koizumi (Ed.), Hot Isostatic Pressing Theory and Application, Elsevier Applied Science, London, UK, 1992, pp. 41–47.
[68] O. Guillon, J. Langer, Master sintering curve applied to the field assisted sintering technique, J. Mater. Sci. 45 (2010) 5191–5195.
[69] B.K. Lograsso, D.A. Koss, Densification of titanium powder during hot isostatic pressing, Metall. Trans. 19A (1988) 1767–1773.
[70] S.V. Nair, J.K. Tien, Densification mechanism maps for hot isostatic pressing (HIP) of unequal sized particles, Metall. Trans. 18A (1987) 97–107.
[71] E. Arzt, M.F. Ashby, K.E. Easterling, Practical applications of hot-isostatic pressing diagrams: four case studies, Metall. Trans. 14A (1983) 211–221.
[72] A. Nissen, L.L. Jaktlind, R. Tegman, T. Garvare, Rapid computerized modelling of the final shape of HIPed axisymmetric containers, in: R.J. Schaefer, M. Linzer (Eds.), Hot Isostatic Pressing Theory and Applications, ASM International, Materials Park, OH, 1991, pp. 55–61.
[73] L. Sanchez, E. Ouedraogo, C. Dellis, L. Federzoni, Influence of container on numerical simulation of hot isostatic pressing: final shape profile comparison, Powder Metall. 47 (2004) 253–260.
[74] S.H. Chung, H. Park, K.D. Jeon, K.T. Kim, S.M. Hwang, An optimal container design for metal powder under hot isostatic pressing, J. Eng. Mater. Tech. 123 (2001) 234–239.
[75] J. Besson, M. Abouaf, Grain growth enhancement in alumina during hot isostatic pressing, Acta Metall. Mater. 39 (1991) 2225–2234.
[76] Y. Wu, J. Wang, T. Lan, L. Zhou, Y. Wu, Z. Jin, Variation rule of the primary powder boundaries during the densification, J. Adv. Mater. 36 (2004) 18–21.
[77] L.M. Peng, Preparation and properties of ternary Ti_3AlC_2 and its composites from Ti-Al-C powder mixtures with ceramic particulates, J. Amer. Ceram. Soc. 90 (2007) 1312–1314.
[78] S. Ochiai, H. Takeda, Y. Kojima, S. Kikuhara, High temperature deformation properties in TiB_2-TiAl composites produced by HIP method, J. Japan Soc. Powder Powder Metall. 44 (1997) 647–652.

[79] R.M. German, High density powder processing using pressure-assisted sintering, Rev. Particulate Mater. 2 (1994) 117–172.
[80] F.H. Froes, C. Suryanarayana, Powder processing of titanium alloys, Rev. Particulate Mater. 1 (1993) 223–276.
[81] C.A. Kelto, E.E. Timm, A.J. Pyzik, Rapid omnidirectional compaction (ROC) of powder, Ann. Rev. Mater. Sci. 19 (1989) 527–550.
[82] H.F. Fischmeister, Modern techniques for powder metallurgical fabrication of low alloy and tool steels, Ann. Rev. Mater. Sci. 5 (1975) 151–176.
[83] W.B. Eisen, Mathematical modelling of hot isostatic pressing, Rev. Particulate Mater. 4 (1996) 1–42.
[84] A. Bose, Overview of several non-conventional rapid hot consolidation techniques, Rev. Particulate Mater. 3 (1995) 133–170.
[85] K. Isonishi, M. Tokizane, Superplastic deformation of a P/M ultrahigh carbon steel, Inter. J. Powder Metall. 25 (1989) 187–194.
[86] M.A. Eudier, High density sintering of metal powder compacts, in: W. Leszynski (Ed.), Powder Metallurgy., Interscience, New York, NY, 1961, pp. 137–156.
[87] T. Kimura, A. Majima, T. Kameoka, Supersolidus hot pressing of ferrous P/M alloys, Inter. J. Powder Metall. Powder Tech. 12 (1976) 19–23.
[88] X. Wang, S.R. Casolco, G. Xu, J.E. Garay, Finite element modeling of electric current activated sintering: the effect of coupled electrical potential, temperature and stress, Acta Mater. 55 (2007) 3611–3622.
[89] U. Anselmi-Tamburini, J.E. Garay, Z.A. Munir, Fundamental investigations on the spark plasma sintering/synthesis process III. Current effects on reactivity, Mater. Sci. Eng. A407 (2005) 24–30.
[90] N. Tamari, I. Kondoh, T. Tanaka, M. Kawahara, M. Tokita, Y. Makino, et al., Effect of spark plasma sintering on densification, mechanical properties and microstructure of alumina ceramics, J. Japan Soc. Powder Powder Metall. 46 (1999) 816–819.
[91] J. Langer, M.J. Hoffmann, O. Guillon, Direct comparison between hot pressing and electric field assisted sintering of submicron alumina, Acta Mater. 57 (2009) 5454–5465.
[92] J. Langer, M.J. Hoffmann, O. Guillon, Electric field-assisted sintering in comparison with the hot pressing of yttria stabilized zirconia, J. Amer. Ceram. Soc. 94 (2010) 24–31.
[93] K.R. Venkatachari, R. Raj, Shear deformation and densification of powder compacts, J. Amer. Ceram. Soc. 69 (1986) 499–506.
[94] K. Morsi, S.O. Moussa, J.J. Wall, Simultaneous combustion synthesis (Thermal Explosion Mode) and extrusion of nickel aluminides, J. Mater. Sci. 40 (2005) 1027–1030.
[95] B.H. Rabin, G.E. Korth, R.L. Williamson, Fabrication of titanium carbide-alumina composites by combustion synthesis and subsequent dynamic compaction, J. Amer. Ceram. Soc. 73 (1990) 2156–2157.
[96] H. Miura, T. Honda, H. Muraba, Explosive compaction of SiC ceramic powder, J. Japan Soc. Powder Powder Metall. 35 (1988) 655–661.

CHAPTER ELEVEN

Mixed Powders and Composites

IMPORTANCE

Mixed powder sintering is a widely employed means of forming composites. Powder mixtures also induce chemical gradients that can alter the rate of sintering. For example, in sintering polycrystalline diamond the starting composition is a mixture of five different particle sizes, ranging upwards from nanoscale diamonds, included to preferentially form sinter bonds between the larger particles. This case relies on a mixture of powders of the same composition, while other examples use mixed powders of differing compositions to induce chemical gradients to drive sintering.

Sintering mixed powders offers several benefits. It is a means of fabricating composites; one phase modifies the properties of the matrix phase. For example, tungsten has a high melting temperature and low thermal expansion coefficient, whereas copper has high thermal and electrical conductivity. A composite of W-20Cu, consisting of insoluble phases, is sintered to form electrical contacts (tungsten resists arcing and copper provides conductivity), heat sinks (tungsten lowers thermal expansion and copper dissipates heat), electro-discharge machining electrodes (tungsten provides stiffness and arcing resistance and copper provides electrical conduction), and missile heat deflectors (tungsten resists evaporation while copper evaporates to remove heat).

Alloys are formed by mixing powders that homogenize during sintering to form a high strength combination—examples include iron and graphite mixed to produce sintered steel, and copper and tin mixed to produce sintered bronze. Additionally, second phases modify sintering behavior, for example during liquid phase sintering. Other objectives rely on additives to accelerate grain boundary diffusion for faster sintering.

In mixed powders, the dominant factors in sintering are the physical and chemical interactions. The physical interactions depend on the particle size ratio and the quantity of each powder. These factors determine the starting structure. The chemical effects typically dominate events during heating. Although mixed powders are far from equilibrium, phase diagrams help to understand the chemical interactions and the resulting atomic flow trajectories. This is because the chemical driving gradient for mass transport is large compared to the sintering stress, so concentration gradients

Sintering: From Empirical Observations to Scientific Principles
DOI: http://dx.doi.org/10.1016/B978-0-12-401682-8.00011-2

dominate over particle size effects. In this chapter the particle size effects will be separated from the chemical interaction effects.

Five options are covered in this chapter. The first is where the powders are similar in composition so there are no chemical gradients. In this case, the physical interactions are dominant and refer to the amount and size ratio of the two powders. This is the bimodal problem dealing with mixtures of two powders of different sizes. In a particle size distribution, the most populous particle size is the mode, so when two different powder sizes are mixed prior to sintering the mixture is bimodal.

The chemical interaction between the mixed powders induces mass transport during sintering. The options include swelling, enhanced densification, compositional homogenization, and the formation of two phase composites.

PHYSICAL INTERACTIONS

Two parameters help understand the sintering of mixed powders with the same composition. The first is the particle size ratio and the second is the composition. A high green density is usually desirable. When two powders of differing size are mixed, the particle packing density depends on composition and particle size ratio [1,2]. Monosized spherical powders vibrate to a packing density of 64%. Mixtures of spherical particles with differing particle sizes improve on this density; however, at the same time the sintering response might be degraded. Thus, density gains from improved packing are offset by degraded sintering.

To rationalize the physical effects of mixed powders, consider bimodal powders where the only difference is particle size, designated by effective sizes D_L and D_S [3]. The subscripts L stands for large and S for small. Figure 11.1 plots the variation in packing density versus composition. For convenience the composition is measured by X in terms of the relative large particle mass. Five conditions occur:

1. 100% small particles packing to a density of f_S,
2. majority of small particles with some added large particles,
3. maximum packing density mixture $f^\star$ at composition $X^\star$,
4. majority of large particles with some small particles, and
5. 100% large particles packing to a density of f_L.

The packing density improves in the terminal region rich in small particles (low X) where large particles substitute dense regions for porous clusters of small particles. Alternatively, for high concentrations of large particles, the density improves with small particles filling the interstices between large particles. The packing density $f^\star$ peaks at composition $X^\star$, corresponding to a dense packing of large particles with small particles fitting into all of the interstices.

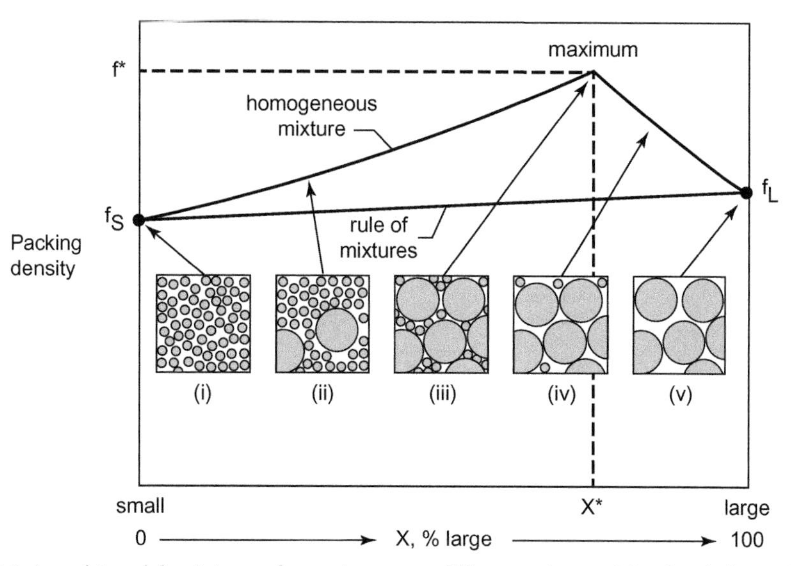

Figure 11.1 In a bimodal mixture of powders, any difference in particle size induces improved fitting together to improve packing density. This schematic plot shows the structure changes with composition of small to large particles. Optimal packing occurs when all of the pores between the large particles are filled by the small particles at a large particle size difference.

The specific density and composition of the maximum depends on the particle size ratio. While the packing density improves in a bimodal mixture, the sintering response is intermediate between that of each individual powder. Sintering shrinkage declines nearly linearly as the large particle content increases, reflecting the effect of an increase in the average particle size. This is evident in Figure 11.2, where the green density, sintered density, and shrinkage are plotted versus composition for a bimodal mixture of 4 and 66 μm iron powders. These mixtures were sintered at 1100°C (1373 K) for 60 min. The green density peaks at about 70% large particles, but declines beyond that point. The sintering shrinkage declines nearly linearly with the large particle content and is essentially zero above the X^* maximum composition. This maximum density composition corresponds to a rigid large particle skeleton. The large particles restrain densification by the small particles. Although the small particles sinter to the large particles, the overall densification is restricted by the large particle skeleton. Cracking can occur in the microstructure due to the sintering restraint. This is a problem with fast heating, high contents of large particles, and large differences in particle sizes.

Bimodal powder mixtures improve the starting density. But large particles degrade sintering, so the sintering response of a mixture is variable with composition, particle size ratio, and sintering response of the two powders. The green density variation

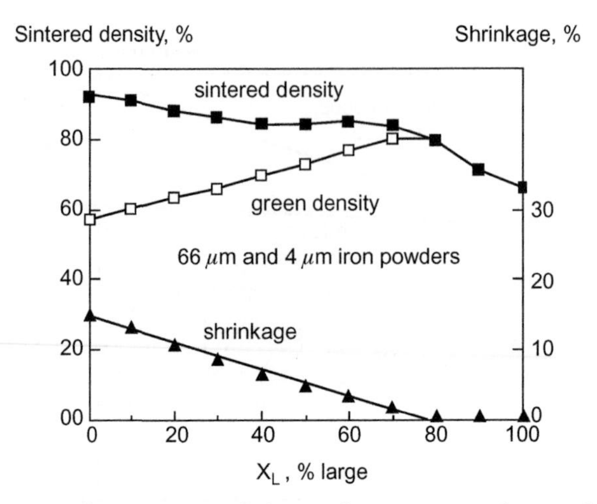

Figure 11.2 Composition effect on sintered density for a mixture of 66 and 4 μm iron particles. The large particles degrade sintering shrinkage, so in spite of improved packing the sintered density is lower for the mixed powder situations.

with composition is predictable if the powder characteristics are known [1]. At the extreme of 100% small powder, the packing density is f_S. At the extreme of 100% large powder the packing density is f_L. For a mixture consisting of very different particle sizes, the maximum packing density f^* occurs when the large particles constitute a close packing with the interstitial spaces filled by small particles:

$$f^* = f_L + (1 - f_L) f_S \tag{11.1}$$

This occurs at a composition of X^* given as follows:

$$X^* = \frac{f_L}{f^*} \tag{11.2}$$

This peak density assumes that the powders are homogeneously mixed. Separation is a problem if the powders are vibrated or otherwise allowed to segregate. As the particle size ratio between the large and small particles decreases, the packing gains prior to sintering are less attractive [4,5]. Usually a ratio of about 70% large gives the highest packing density.

Once the green density is understood for mixed powders, then attention turns to the sintering response. Three variants are possible with mixed particles of the same composition:

1. The sintered density is nearly constant with composition because packing gains are offset by inhibited sintering shrinkage as the large particle content increases.

2. With low sintering shrinkages, the highest sintered density occurs near the composition giving the highest packing density, near 70% large particles.
3. When the small particles exhibit high sintering shrinkage, then the large particles inhibit sintering and the highest sintered density is obtained with 100% small powder.

The third case dominates many sintered composites, especially where a large fiber or particle is added to a small particle matrix. This added phase strengthens the matrix, but inhibits matrix sintering to the point where cracks form [6,7]. If the small particles sinter to a high density, then there is no mixture with large particles as dense after sintering. In principle, fibers, whiskers, or particles are added for strength or wear attributes, yet they inhibit sintering. Thus, bimodal mixtures provide a means to start sintering with higher densities; but in such mixtures densification declines with the addition of large particles. An offset to this difficulty is to add another phase to enhance sintering. For example, in a mixture of large and small iron powders, the addition of a small quantity of phosphorus results in a high sintered density [8].

For low levels of sintering shrinkage, the packing density dominates the sintered density. This is evident in the 750°C (1023 K) zinc oxide sintering data in Figure 11.3 [9]. For sintering for 1 min the sintered density improves with large powder additions due to packing density gains. But over longer times, where more small particle shrinkage occurs, the inhibition from the large particles dominates. Thus, the 60 min data shows declining sintering density with large particle additions. Another example is given in Figure 11.4, using data from mixed 5 and 0.5 μm alumina powders heated to 1600°C (1873 K) for 10 min [10]. The highest green density is near 70% large powder, but the highest sintered density is at 100% small powder.

If the small powder has little sintering shrinkage, then a mixture of large and small improves sintered density. Accordingly, a map helps determine the benefit of large particles as given in Figure 11.5 [3]. The vertical axis is the small particle shrinkage and the horizontal axis is the large particle shrinkage for the same thermal cycle. Three large to small particle size ratios of 10, 30, and 100 are shown. From this plot, the behavior is contrasted between 100% small particles and a bimodal mixture of large and small particles. For the mixture, the maximum sintered density occurs at the composition defined by Equation 11.2. Generally, at high sintering temperatures or long sintering times the maximum sintered density occurs at the 100% small particle composition, even though the maximum packing density might occur with a mixture. This is because large particles inhibit sintering densification. At the terminal compositions (100% small particles or 100% large particles), densification depends only on particle size. The small particles undergo more densification than the large particles, so as the volume of large particles increases the sintering densification degrades.

Other practical factors work against sintering bimodal powder mixtures, including tendencies for size segregation during handling and difficulties in sourcing powders with large size differences.

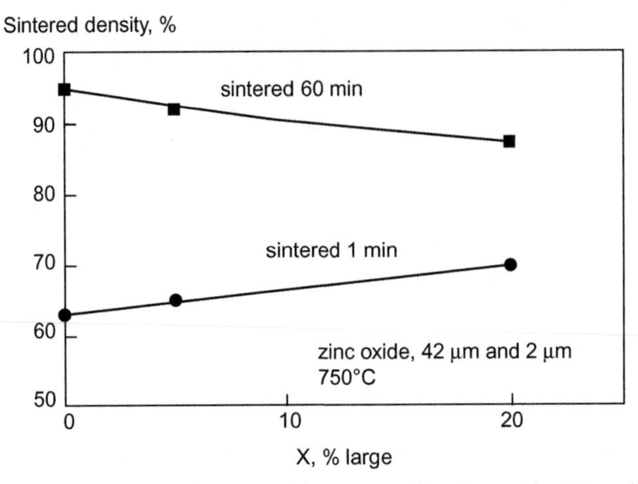

Figure 11.3 Sintered density versus large particle content for zinc oxide (42 and 2 μm) for 1 and 60 min at 750°C (1023 K) [9]. At short sintering time the packing benefit dominates, but at longer sintering time the inhibited shrinkage from the larger particles dominates.

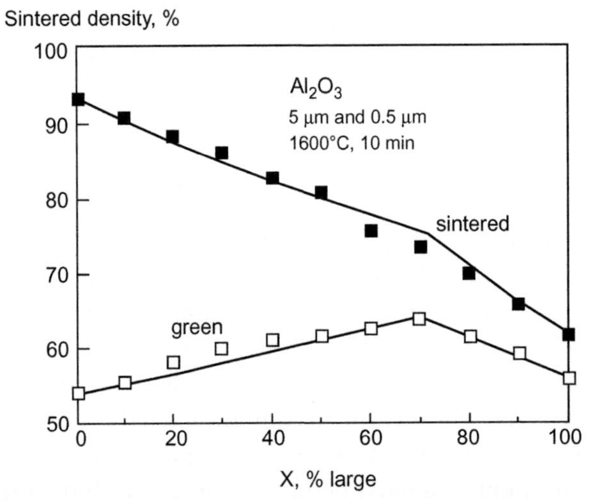

Figure 11.4 Mixed alumina powder sintering at 1600°C (1873 K) for 5 and 0.5 μm particles [10]. In this case the highest packing density is near 70% large powder, but the highest sintered density is with 100% small powder.

Another related case of inhibited densification occurs with agglomerates. The difficulty derives from the larger pores between the agglomerates, as sketched in Figure 11.6. The small pores inside the particle agglomerates sinter and densify early in the heating cycle, but the larger pores between agglomerates with smaller curvatures

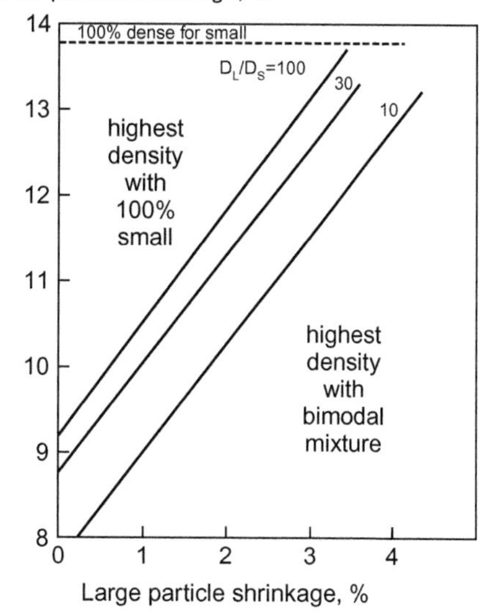

Figure 11.5 Calculation map of the sintering density versus composition for large-small powder mixtures for particle size ratios of 100, 30, or 10 (D_L = large size, D_S = small size). Depending on the small particle shrinkage and the large particle shrinkage, this map shows if the highest sintered density is with the bimodal mixture or simply the small particles.

and longer diffusion distances resist densification and retard overall sintering [11–13]. Milling is a common means of removing such agglomerates [14]. One option is to apply a high compaction pressure to collapse the large pores prior to sintering.

A sintering map helps to select the particle size ratio and composition relation; Figure 11.7 is an illustration of such a map for cordierite—$(Mg,Fe)_2Al_4Si_5O_{18}$—and glass mixtures sintered at 850°C (1123 K) for 60 min [15]. Glass is the active sintering phase at this temperature and cordierite is inert. As the volume fraction of glass increases, densification improves. Making the glass particle size relatively small offsets the sintering decrement from the larger cordierite particles; to compensate for retarded densification, the glass particle size needs to be selectively reduced as the volume fraction of cordierite increases.

One option for offsetting the inhibited sintering densification in mixed powders is to apply pressure during sintering. In compositions consisting of mostly large particles, the small particles have little influence on the mixture sintering densification. Efforts to add nanoscale particles found little benefit and a significantly higher cost. In compositions where the small particles constitute the matrix, large particles, agglomerates, whiskers, and fibers resist densification, especially at shorter times and lower sintering temperatures. In such cases, cracks are often observed in the microstructure.

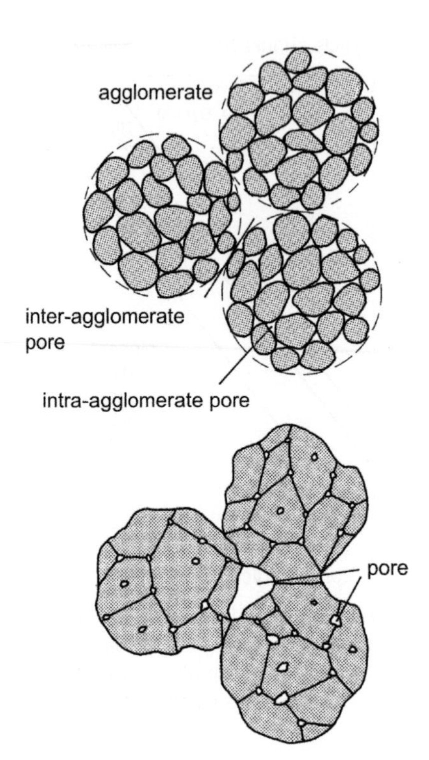

Figure 11.6 Agglomeration causes a mixture of small and large pores, the smaller pores are inside the agglomerates and the large pores are between the agglomerates. A high compaction pressure will eliminate the pore size difference, but without care the structure will sinter to form stable large pores.

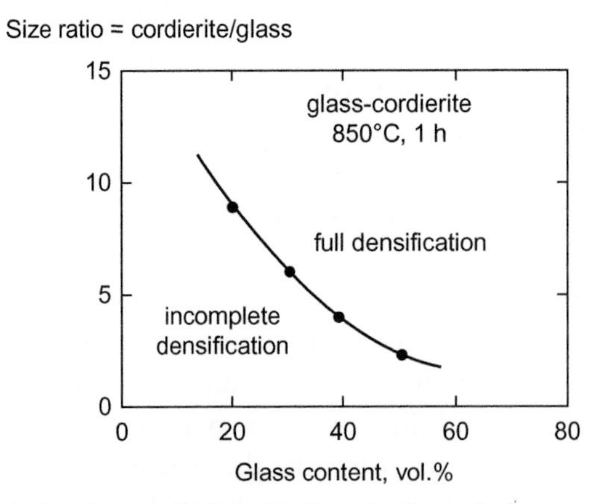

Figure 11.7 Experimental sintering results for mixed cordierite and glass powders [15]. Densification requires a large size ratio or high glass content.

CHEMICAL INTERACTIONS

Mixed powder sintering is a common means of forming a new chemistry from two or more elements or compounds. Several options arise; because the sintering stress, due to surface curvature, is overwhelmed by the chemical gradients. To demonstrate, assume 1 μm particles and $2\,J/m^2$ surface energy at $10\,kg/m^3$ density, corresponding to molybdenum. This powder has stored energy in the form of surface area of 1.2 J/g. If that powder is mixed with silicon at an atomic stoichiometric ratio of 1 Mo to 2 Si, the reactive sintering to produce $MoSi_2$ has an energy release is 861 J/g. This is 700-fold more than the energy from the chemical interaction. Thus, unlike single phase sintering where curvature gradients dominate sintering, for mixed powders the chemical gradients are dominant.

SOLUBILITY ROLE

The most common case of mixed powder sintering involves a liquid phase, covered in Chapter 9. Mixed powders enable the formation of compounds, alloys, composites, and enhanced sintering compositions. As such, mixed powders are widely employed in sintering. There are four options when the powders differ in composition, as outlined in Figure 11.8. This figure shows solubility combinations involving a majority powder **M** and an additive powder **A**. The majority powder is over 50 vol.% of the total and the additive is at a lower volume fraction, often in the 2 to 30 vol.%

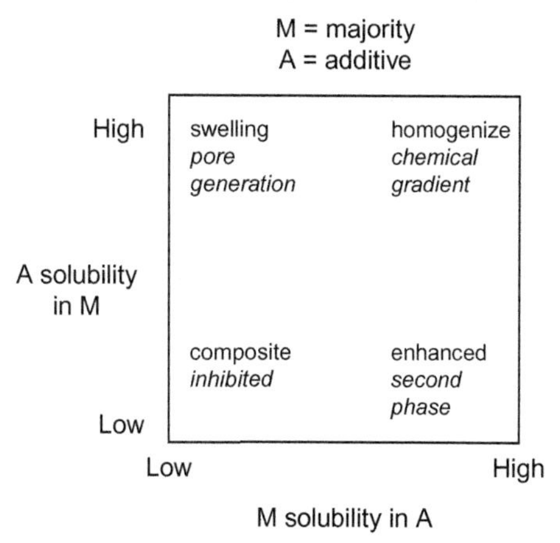

Figure 11.8 A grid showing the four possible solubility relations between the majority and additive phases, where depending on the relative solubility various different effects will dominate sintering.

range. What happens during sintering depends on the chemical interactions as evident from solubility effects [16,17]:

Homogenization

M is soluble in **A**, **A** is soluble in **M**—the two powders dissolve one another and form a homogeneous product, as is possible using alumina and chromia to produce artificial ruby.

Swelling

M is insoluble in **A**, **A** is soluble in **M**—the biased additive solvation and diffusion into the majority powder leads to pore generation during alloy formation, as is used to sinter porous bronze bearings from mixed tin and copper powders.

Enhanced

M is soluble in **A**, **A** is insoluble in **M**—this is employed to induce rapid sintering where the diffusion path in the additive short circuits the otherwise slow sintering process, for example small contents of nickel are added to tungsten to reduce the sintering temperature.

Composite

M is insoluble in **A**, **A** is insoluble in **M**—this leads to a composite of reinforcing phase distributed in a sintered matrix, the composite sintering depends on the majority phase constrained by the additive, an example is aluminum reinforced with silicon carbide.

Fundamentally, when the molar volume of the reaction product is smaller than the molar volume of the starting powders, swelling occurs. Heat liberation is a factor, since intense energy release occurs during some of the chemical interactions. Due to these intense reaction enthalpies, the systems self-heat and self-sinter by reactive sintering [18–22]. Another consideration, for those systems where there is solubility, comes from the diffusion rates. Significant changes in sintering are evident when chemical solvation enhances diffusion, especially in activated sintering. This is seen where refractory metals (W, Mo, and Ta) are treated with transition metal additives (Ni, Fe, Co, Pd, Pt).

Homogenization

Homogenization is seen in mixed powder sintering for systems of mutual solubility. The isomorphous Cu-Ni system is one example. Homogenization during sintering is an alternative to forming compacts from pre-compounded or pre-alloyed powders. In metals, the mixtures are soft and easier to compact than harder pre-alloyed powders.

The Cu-Ni phase diagram is given in Figure 11.9. A copper-nickel powder mixture has mutual intersolubility, meaning that chemical gradients at the mixed particle interfaces drive mass flow during sintering. Homogenization is measured by quantitative metallography, X-ray diffraction, magnetization, electron microprobe, differential

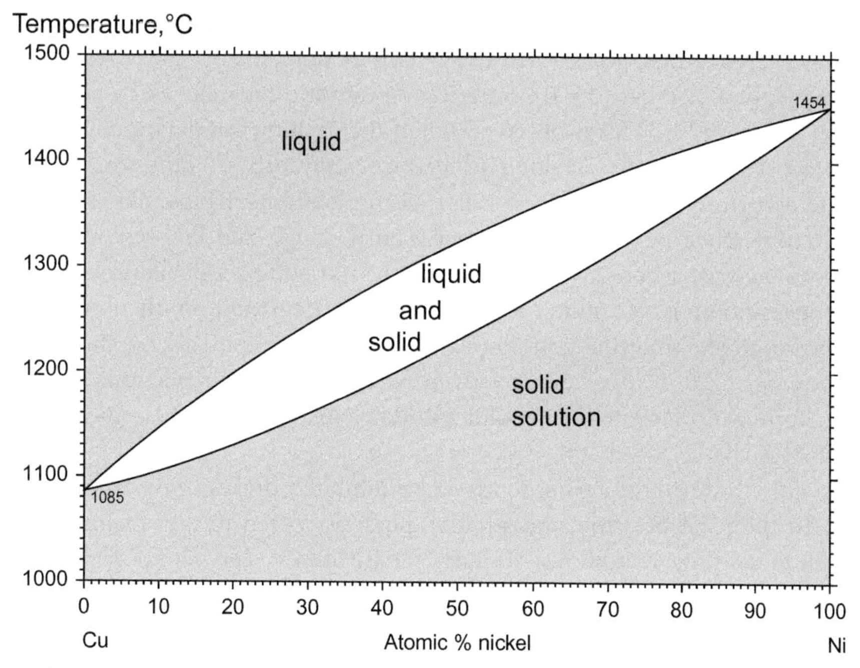

Figure 11.9 Homogenization in sintering is illustrated by a mixture of copper and nickel powders, which have intersolubility at all compositions. If sintering starts with a mixture of the powders, over time the two materials will chemically homogenize to remove chemical gradients. Chemical gradients are strong driving forces for mass flow and tend to dominate early sintering.

scanning calorimetry, or related techniques [22,23]. The degree of homogenization H is defined as the point-to-point chemistry variation, and it depends on the diffusion rate and particle size as follows [24]:

$$H \approx \frac{D_V t}{\lambda^2} \tag{11.3}$$

where λ is the scale of the microstructural segregation, D_V is the volume diffusivity, and t is the isothermal hold time. Diffusivity increases rapidly with temperature, since it follows an Arrhenius temperature dependence. The segregation scale measured by λ depends on the particle size, additive or minor phase concentration, and initial powder mixing. Thorough mixing and small particle sizes aid homogenization by reducing λ. Smaller particles reduce the diffusion distances and increase the diffusion interface area.

If there is minimal chemical reaction during sintering, solvation leads to a homogeneous final product. Homogenization occurs by each species diffusing into the other. In cases where strong chemical reactions occur, the chemical gradients bias diffusion to induce pore formation and swelling. If diffusion rates for the two components are different, then pores form [25]. This is observed in sintering the Ni-Cu system. As plotted in Figure 11.10, sintering densification is reduced by an increase in the nickel content [26]. This is observed when the melting temperatures of the mixed powders are very different. Melting points are indicators of diffusion rates (lower melting temperature materials have weaker atomic bonding which allows faster diffusion). The difference in diffusion rates between the high and low temperature powders leads to vacancy accumulation, pore growth, and compact swelling.

Homogenization is quantified by the standard deviation in the point-to-point chemistry. So, unlike sintering composites where there is essentially no chemical interaction, homogenization involves a relatively weak chemical interaction. Examples of sintering homogenization to form solid solutions are Cr_2O_3-Al_2O_3, W-Mo, Ni-Cu, and NiO-MgO.

Grain growth during sintering tends to be inhibited during mixed powder sintering [27]. In the ideal situation, the additive phase is coated on the majority phase to ensure ideal starting conditions. Lower temperatures slow homogenization and densification.

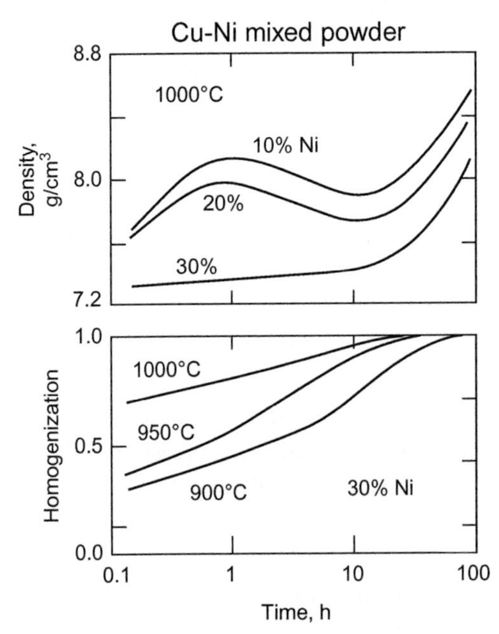

Figure 11.10 Density and homogenization versus log time for sintering mixed copper and nickel powders, showing the effects of composition and temperature [26]. The loss of density at intermediate times and low nickel contents is from unbalanced diffusion rates of copper into nickel versus nickel into copper.

Homogenization depends on a combination of particle size, diffusivity (which changes with temperature), and sintering time. A near linear dependence is assumed:

$$H = H_O + \beta \frac{D_O\, t \exp\left[-\frac{Q}{RT}\right]}{\lambda^2} \quad (11.4)$$

where H is the homogeneity which peaks at 1.0, H_O is the starting homogeneity and depends on the initial powder mixing, λ is the typical separation distance of particles and is approximated by the particle size, β is an adjustable parameter near 10, D_O is the diffusion rate frequency factor, t is the sintering time, T is the sintering temperature, R is the gas constant, and Q is the activation energy for atomic diffusion. Homogeneity is measured by chemical analysis at several random spots, given by one minus the standard deviation σ in those random composition measurements (where $H = 1 - \sigma$). Likewise, H_O is the starting homogeneity measured in the green structure. In this formulation, the homogeneity increases rapidly with time and temperature and as the scale of the system decreases. Smaller particles decrease the segregation scale, since at a given composition the number of additive particles increases with the inverse cube of the additive particle size. Since temperature enters the diffusion term as an exponential factor, it has a dominant effect. If the diffusion rates of the two powders are very different, then swelling occurs, since the more rapidly diffusing species moves into the other phases without offsetting diffusion.

Swelling

Many mixed powder systems sinter with energy release. When a reaction occurs there is a tendency to swell [16]. Swelling is typified by a low solubility of the majority phase in the additive, but a high solubility of the additive in the majority phase. A simple phase diagram, as shown in Figure 11.11, provides an illustration. The additive **A** has little solubility for the majority phase **M**, but the opposite solubility of **A** in **M** is large. This leads to one direction flow of **A** into **M** during sintering. In some cases the reactions are exothermic with extreme swelling [25]. Alternatively, the reaction forms a new intermediate composition compound, and the use of mixed nickel and titanium powders to form Ni-Ti or nickel and aluminum powders to form Ni_3Al are examples [28,29].

Unbalanced diffusion between the additive and matrix is a prime cause of swelling. Since melting temperatures are first indications of diffusion rates, usually the lower temperature additive diffuses faster. Additive particle dissolution into the surrounding majority phase powder results in biased diffusion away from the additive. This outward radial diffusion pattern leaves a pore behind. Compound layers might form, depending on the phase diagram, and these usually inhibit diffusion.

The resulting large pore microstructure is evident as captured in Figure 11.12. This is a cross-sectional micrograph of an aluminum particle in an iron matrix,

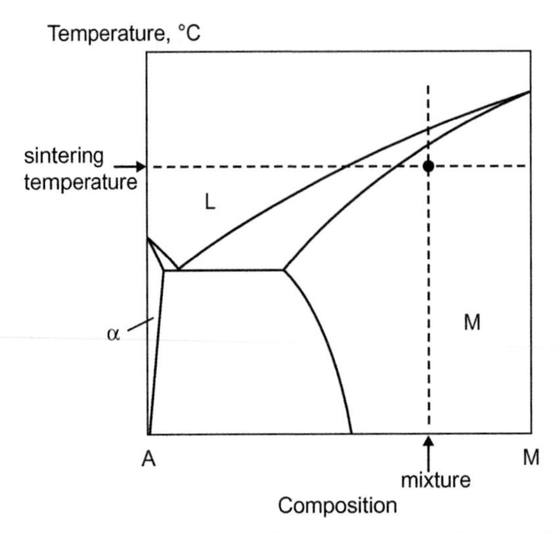

Figure 11.11 Schematic phase diagram for a binary mixed powder situation that would tend to induce swelling during sintering. The lower melting additive phase is soluble in the majority, so it will diffuse from its initial particle sites to leave pores.

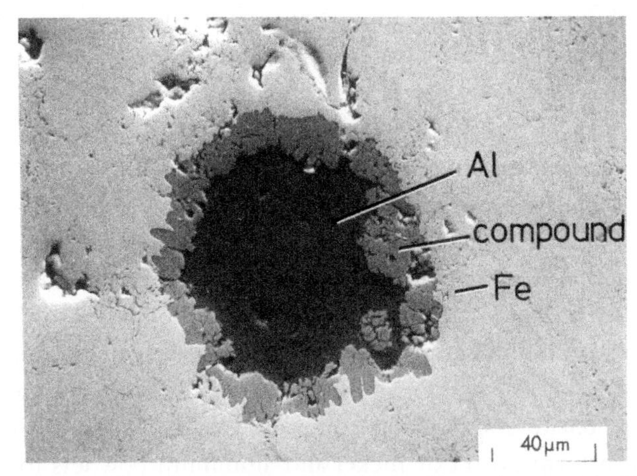

Figure 11.12 Cross-section of an aluminum particle sintered in an iron matrix. The aluminum particle exhibits outward diffusion to form an iron aluminide ring after heating to 1100°C (1373 K).

showing outward diffusion and formation of an iron aluminide ring after heating to 1100°C (1373 K). During prolonged sintering, the intermediate layers are dissolved to eventually achieve an equilibrium composition. Because of the outward flow from the additive particle site, the surrounding matrix expands while a large pore is left behind. Swelling increases with the volume fraction of the additive, but final density depends on green density [17,30].

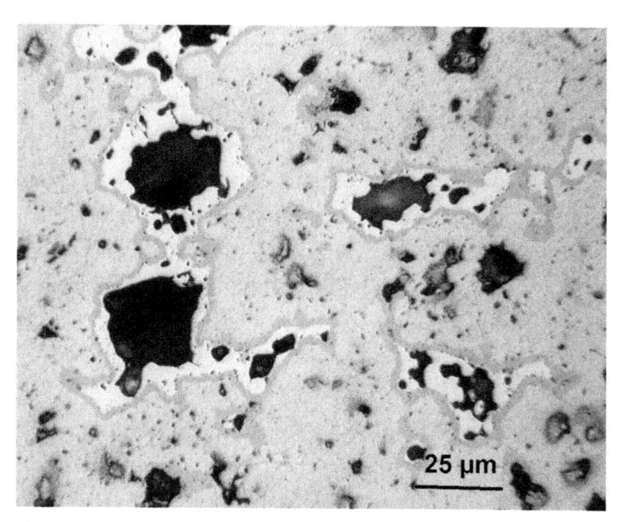

Figure 11.13 Optical micrograph of a mixture of tin and copper particles after low temperature sintering. The tin is diffusing and reacting to form an intermetallic compound, but leaves pores behind in the structure due to the difference in solubility and diffusivity.

An example of solubility driven swelling occurs in the mixed copper-tin system, used to form porous self-lubricating bearings. Mixed Cu and Sn powders are heated and in sintering form a porous bronze structure. Figure 11.13 is a micrograph of a Cu-10Sn mixture after sintering, showing black pores and tin-rich phases formed around the sites that were initially tin particles. At higher temperatures the structure sinters. So the sequential steps are reactions to form intermetallics, and eventually homogenization to a single phase solid [31,32]. The structure shown here at an intermediate level of sintering homogenization is used in sintered frangible ammunition.

The greater the chemical activity, the more the sintering events are dominated by chemical interactions [33,34]. In the extreme, reactive sintering is a means to synthesize and densify compounds in one thermal cycle [29]. A variety of systems are produced this way, including aluminides (Ni_3Al, $NbAl_3$, Ti-Al), composites (TiB_2-TiC, $MoSi_2$-$TiSi_2$), and compounds (TiB_2, Ni-Ti, CrB_2). The reactions are intense, lasting just seconds. Unfortunately, most of the reactions produce unbalanced diffusion and prove hard to control.

Swelling can be offset by pressure-assisted sintering. Large additive particles generate large pores that resist densification, so smaller particles are beneficial. Gas release from the reaction considerably hinders densification. For example in the reaction of TiN and B_4C to produce a composite of TiB_2 and TiC there is a release of nitrogen. A two-stage heating cycle, with a first hold at 1560°C (1833 K) for nitrogen release followed by 60 min sintering at 2000°C (2273 K) delivered densities in the 96% range.

Swelling during heating is evident in many mixed powder systems. A few are listed here to illustrate how frequently sintering fails to densify mixed powders: Ag-Pt, Al-Cu, Al-Mg, Al-Zn, Co-Al, Co-Be, Cu-Sn, Cu-Al, Cu-Ti, Cu-Zn, Fe-Cu, Fe-Ti, Fe-Al, Fe-Cu, Fe-Ti, Nb-Al, Ni-Al, Ni-Be, Ni-Co, Ni-Re, Ni-Ti, Ti-Al, and U-Be. Many of the systems exhibit intermediate compound phase diagrams, where two mixed powders react to form a new compound—such as nickel and titanium reacting to give NiTi. The higher the relative melting temperature of the intermediate compound, the greater the propensity for swelling during sintering.

Enhanced

Most interesting are the enhanced sintering responses which occur with certain mixed powder systems, on which the rate of sintering is accelerated by favorable chemical or microstructural features [35–41]. Sintering enhancements arise from a reduction in the activation energy for diffusion, or adjustments of the microstructure to increase mass flow during sintering. Most common are techniques that lower the activation energy for mass flow, resulting in a shorter sintering time, lower sintering temperature, or an improved degree of sintering [42,43]. The reduced activation energy results from the majority phase dissolving into an additive which is lower in melting temperature. In this way the additives sweeps atoms along at a higher rate. Inherently improved diffusion rates are associated with lower melting temperature. Thus, if a segregated grain boundary phase exists with faster diffusion and solubility for the majority phase, enhanced sintering results. Thus, activated and liquid phase sintering occur in similar situations.

With respect to enhanced sintering, another variant involves polymorphic materials. Sintering treatments that cycle through the polymorphic phase transformation provide a microstructure strain that drives faster sintering, especially in cases where diffusion is faster in one crystal form. Iron diffusion in the body-centered cubic crystal structure at 910°C (1183 K) is about 100 times faster than in the face-centered cubic crystal structure at the same temperature. By oscillating the temperature, the strain on phase transformation couples to the diffusion rate change to induce faster sintering [44,45]. Sintering shrinkage increases with the number of cycles. Similarly, sintering rates are higher near the order-disorder temperature for compounds. Such internal strain effects are analogous to external stresses in hot pressing.

In some cases the additive increases the population of rate controlling crystal defects. For example, in sintering SnO the addition of CuO causes a shift in the vacancy population to improve diffusion. Similar results occur in BeO with MgO additions. In ferrous systems, boron is one of the most potent additives for improving sintering. For covalent ceramics (SiC, AlN, and Si_3N_4), beneficial additives such as yttria, boron, and alumina segregate to the grain boundaries to enhance diffusion.

The role of MgO in promoting sintering densification of Al_2O_3 was observed in the 1950s and eventually traced to preferential grain boundary segregation that removed calcia segregation [46]. Another example is copper sintering, where the addition of iron or graphite leads to oxide reduction and improved sintering. For the refractory metals, an important effect occurs when the added powder has a beneficial solubility and melting temperature, opening up a low temperature, rapid transport process termed activated sintering.

Activated sintering is associated with specific phase diagram characteristics, as illustrated schematically in Figure 11.14. The dashed composition and temperature lines indicate candidate conditions for activated sintering. The melting temperature for the additive **A** is lower than the majority phase **M**. This promotes faster diffusion of the majority phase in the additive. At the sintering temperature, the additive has substantial solubility for the majority species, but the reverse solubility hinders the additive from disappearing. The rapidly decreasing liquidus and solidus curves on the **M**-rich side indicate a propensity for segregation to grain boundaries. Thus, the additive segregates to the grain boundaries, where it acts to sweep the dissolved majority atoms to new sites at a sintering temperature much lower than is typical.

Activated sintering works because diffusion rates scale with melting temperatures [47]. The segregated additive phase provides a more rapid diffusion environment. This is diagramed in Figure 11.15, where the activator-rich phase is present as a thin grain boundary layer [48]. Activated sintering occurs when the high temperature material is soluble in the low melting temperature phase, resulting in a short-circuit sintering path at the interparticle grain boundary. Since diffusion is important to sintering, relative sintering changes depend on solubility and melting temperature in the phase

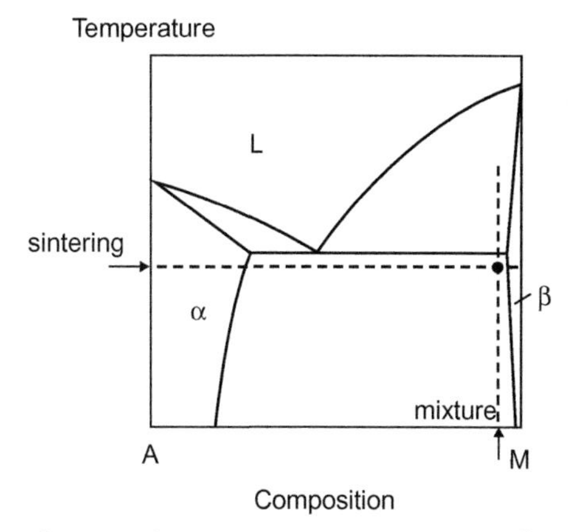

Figure 11.14 Several phase diagram characteristics are indications of enhanced sintering potential, where the target sintering temperature and composition are indicated by the intersection in the two phase solid region.

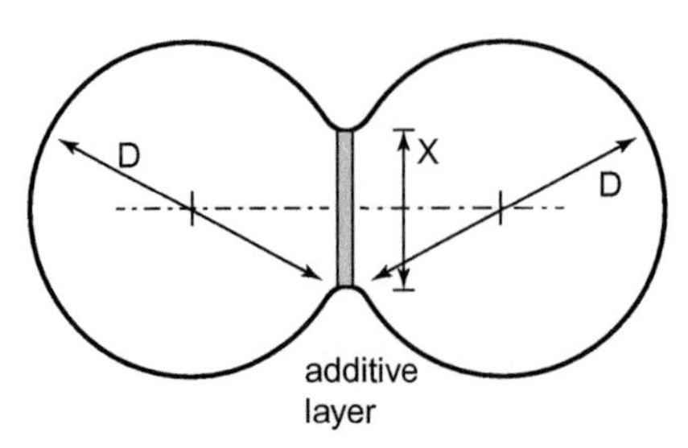

Figure 11.15 Sintering with an additive layer segregated to the interparticle grain boundary provides an analog to grain boundary diffusion, but the additive typically has a lower melting temperature (faster diffusion) and provides solubility for the material being sintered.

diagram. A high melting temperature material requires a high sintering temperature. However, a lower melting temperature phase is inherently faster at the same temperature. Of course, no sintering benefit occurs if there is no solubility.

Examples of activated sintering are found in the refractory metals—molybdenum, tungsten, chromium, rhenium, and tantalum. For example, when 1 μm chromium powder is sintered at 1400°C (1673 K) for 1 h, the sintered density is 78%. The addition of 1% Pd to the same powder prior to the same sintering cycle produces a sintered density of 96%.

The amount of additive is important in activated sintering. Sufficient additive is required to coat the grain boundaries. Less than that amount means interrupted transport on the grain boundaries, and the resulting slow step at those gaps reduces the benefit. An excess of additive simply thickens the grain boundary layer. Thus, as plotted in Figure 11.16, sufficient additive is required to induce activated sintering. This plot gives the sintering shrinkage in 1 h at 1300°C (1573 K) for molybdenum treated with various palladium contents. About 0.5 wt.% is sufficient to activate sintering, giving 10% shrinkage versus 2% shrinkage for molybdenum with no additive.

The activated sintering effect varies between additives. This is evident in Figure 11.17, where the sintering shrinkage is plotted versus the 1 h hold temperature for tungsten with small quantities of several additives. A sintering shrinkage of 22% corresponds to 100% final density. Note that the pure tungsten exhibits minimal sintering over this temperature range and copper produces no benefit. Sintering shrinkage is high with palladium additions. The variation results from a change in diffusion activation energy at the grain boundaries. In turn, sintered strength, hardness, and other properties change dramatically. The activation energy for densification of the W-Fe and W-Co is 380 kJ/mol, while for W-Ni and W-Pd it is about 300 kJ/mol. This difference in activators rationalizes the differences in diffusion rates for tungsten in the segregated second phase.

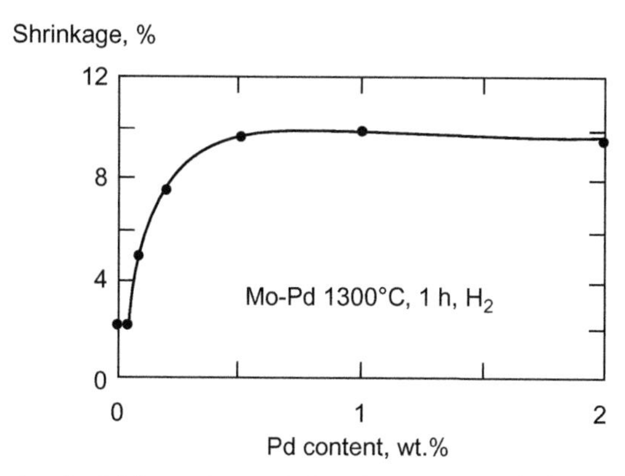

Figure 11.16 An example of additive-enhanced sintering, commonly known as activated sintering, for molybdenum treated with palladium heated to 1300°C (1573 K) for one hour in hydrogen. The enhanced sintering response is dependent on sufficient additive to coat the grain boundaries, often observed with a small additive quantity.

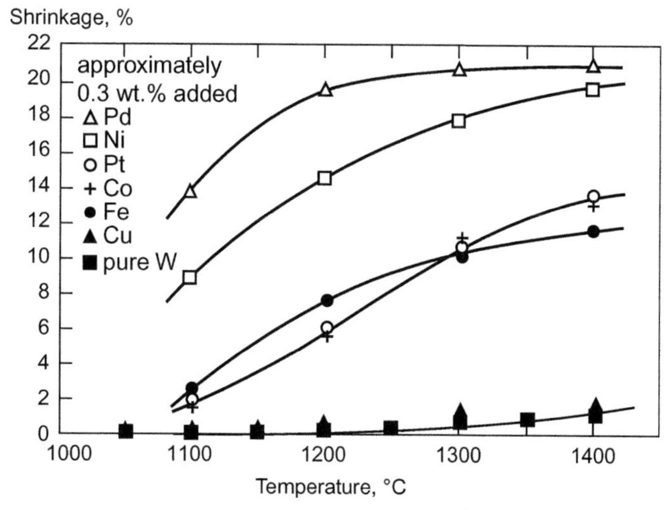

Figure 11.17 Sintering shrinkage versus sintering temperature for one hour holds using tungsten and tungsten treated with 0.3 wt.% of several transition metals. The pure tungsten shows minimal sintering shrinkage over this temperature range, but palladium gives dramatic gains as predicted from the tungsten-palladium phase diagram.

In activated sintering, the effective activator is well distributed in the microstructure. Consequently, smaller additive particles are most effective in promoting sintering. The transport kinetics in the activator layer depend on solubility and diffusivity. The sintering shrinkage $\Delta L/L_O$ tracks to the sintering parameters as follows:

$$\left(\frac{\Delta L}{L_O}\right)^3 = \frac{g\Omega\delta\, C\gamma_{SV}\, t\, D_A\, exp\left[-\frac{Q_A}{RT}\right]}{RTG^4} \tag{11.5}$$

where Ω is the atomic volume, δ is the width of the second phase activator layer on the grain boundaries, g is a geometric constant, C is the solubility of the material being sintered in the additive, γ_{SV} is the solid-vapor surface energy, D_A is the frequency factor for the diffusion of the dissolved phase in the additive, Q_A is the activation energy for diffusion in the additive, t is the sintering time, G is the grain size which initially equals the particle size, R is the gas constant, and T is the absolute temperature. During sintering the grain size increases with approximately the cube root of time.

Inherently, activated sintering is similar to grain boundary diffusion controlled sintering. The difference is that the grain boundary layer provides much faster transport than pure grain boundary diffusion. More sintering densification benefit results as the solubility and diffusion rate in the additive layer increase. The controlling step is the diffusivity in the activator. Because activated sintering is associated with reduced activation energy, temperature is the dominant parameter.

A high final density requires good coupling of the grain boundaries to the pores, especially during periods of grain growth. Mixed second phases are one means of retarding grain growth. Composition adjustments are helpful in stabilizing two phase microstructures that are effective in this manner. A second powder that retards grain growth is useful, as evident when MgO is added to Al_2O_3, or Fe is added to MgO, Ca is added to ThO_2, CaO is added to ZrO_2, or Th is added to Y_2O_3.

Another means of enhancing sintering is via control of the crystal structure in a polymorphic material. Structures with lower atomic packing densities have higher diffusivities than the close-packed crystal structures. Accordingly, stabilization of the lower packing density crystal structure is beneficial. For iron, the volume diffusivity at 910°C (1183 K) is over a hundred times higher in the body-centered cubic ferrite phase versus the face-centered cubic austenite phase. Additives that stabilize the body-centered cubic phase, such as molybdenum, assist sintering. Densification increases with the amount of ferrite stabilized at high sintering temperatures. A mixed phase microstructure also resists grain growth. Similar ideas are found in the sintering of austenitic stainless steels, where an excess of chromium or silicon assists sintering densification. At the same time, an increase in the nickel or nitrogen content stabilizes the face-centered cubic structure and lowers sintering densification [49–52].

A second phase slows grain growth. Thus, composites tend to be smaller in grain size than the pure sintered analogs, but second phases retard densification. The second phase effect depends on the ratio of densification rate to microstructure coarsening rate. Effective second phases shift the ratio to favor densification. Or to put sintering additives in perspective, each gain comes with a penalty. So those systems that exhibit enhanced densification also tend to exhibit rapid microstructure coarsening.

Composites

When two powder form an insoluble system, the sintered product is a composite. The phases are differentiated by differences in chemistry, crystal structure, and possibly size and shape. An example is the mixing of stainless steel powders with an oxide ceramic to form wear resistant knives. Many ceramic-ceramic composites are known; including crystalline oxides sintered in glass phases to form porcelain and graded glasses for fiber optics.

There are several levels of phase connectivity and this influences sintering. At one extreme are the situations where the dispersed phase is isolated. In this case, the second phase content is limited to approximately 30 vol.% or less to avoid particle-particle contacts. A high second phase content microstructure is shown in Figure 11.18. In this example, alumina particles reinforce tungsten carbide formed by hot pressing at 1200°C (1473 K) for 60 min using 50 MPa pressure. At higher second phase contents, the reinforcing phase is connected to form an interlaced three-dimensional network. A cross-sectional micrograph of the latter is in Figure 11.19. This is a composite of an iron-nickel alloy dark phase and silver light phase sintered to form a low thermal expansion composite with high thermal conductivity.

Many variants exist, some relying on solid-state sintering using small particles and some relying on larger particles and liquid phase sintering. In solid-state systems, the options include systems such as Al-AlN, Al_2O_3-B_4C, Al_2O_3-Mo, Al_2O_3-NbC, Al_2O_3-SiC, Al_2O_3-Si_3N_4, Al_2O_3-ZrO_2, B_4C-TiB_2, bronze-glass, cordierite-glass, Cu-Fe, Cu-TiB_2, Cu-Cr, Cu-SiC, diamond-SiC, diamond-W_2C, Fe-Al_2O_3, Fe-cordierite, glass-SiC, HfB_2-SiC, Ni-Al_2O_3, Ni-ZrO_2, Ni_3Al-SiC, SiC-Si_3N_4, stainless steel-Y_2O_3,Ti-TiC, tool steel-Al_2O_3, tool steel-SiC, WC-Mo_2C, WC-TiC, ZrB_2-ZrC, ZrC-Al_2O_3, and ZnO-SiC. Details on sintering such composites are given in several

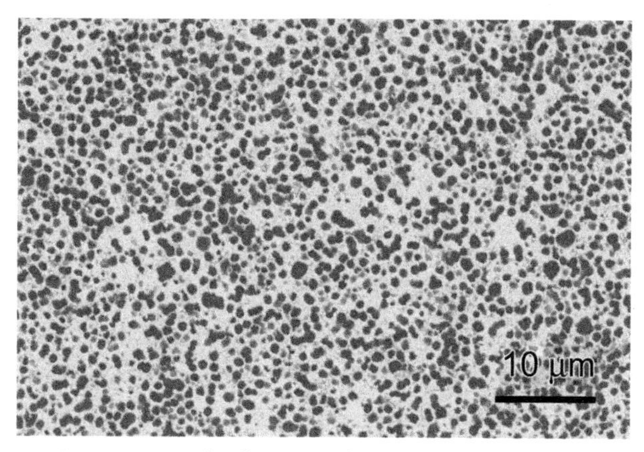

Figure 11.18 Cross-section micrograph of a particulate composite consisting of dispersed alumina particles (dark) in a continuous carbide matrix (white), consolidated by hot pressing at 1200°C (1473 K) for 60 min at 50 MPa applied pressure.

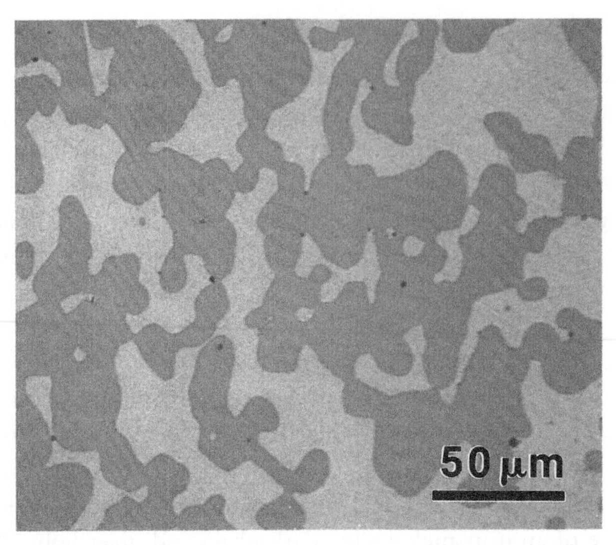

Figure 11.19 A composite microstructure with two interlaced phases, the darker phase is an iron-nickel alloy and the lighter phase is silver. Note in contrast with Figure 11.18, this structure is connected for both phases.

publications [14–21,53–64]. Note the sintering of composites follows similar rules that lend to treatment using the master sintering curve concept [65].

Sintered composites are targeted at novel property combinations—high stiffness and high fracture toughness, high wear resistance and corrosion resistance, or high strength and low thermal expansion coefficient. In these cases, sintering is inhibited by the addition of the inert phase. At short sintering times, packing effects dominate, but at longer sintering times the reinforcing phase inhibits densification. This is evident from the data in Figure 11.20 [9]. Here 0.4 μm ZnO powder has various quantities of 12 μm SiC added, the compacts were initially at 57% density prior to heating for 120 min at 700°C (973 K). The higher the inert silicon carbide concentration added to the zinc oxide, the greater the retardation of sintering densification. Since the silicon carbide sinters at a much higher temperature (versus the zinc oxide), it remains inert and inhibits densification.

Inert particles reduce the volume of the sintering phase and exert a restraint on shrinkage [66–68]. This restraint causes tensile stresses in the matrix which can possibly induce cracking. One means to offset this problem is to coat the inert particles. More typically, pressure-assisted sintering is employed [69–77]. The external stress makes up for the reduced sintering stress and compensates for the retarded densification.

One benefit from an inert second phase is grain boundary pinning during sintering. Grain size for sintered alumina without an additive and with 10 vol.% zirconia is plotted in Figure 11.21 [78]. The benefit of a reduced grain size is often offset by restrained sintering. However, at long sintering times this is not the case, and in these

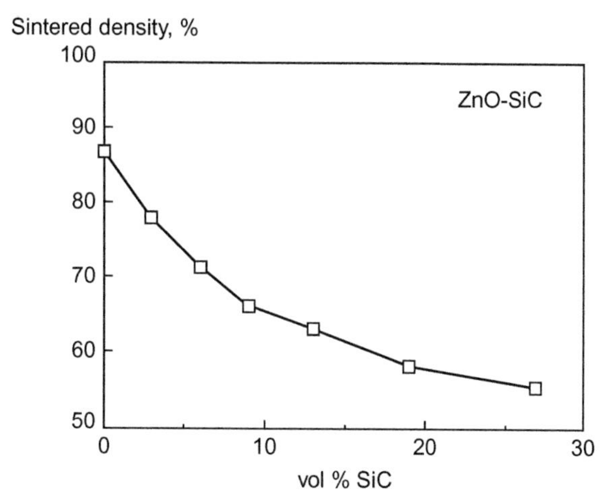

Figure 11.20 Plot of sintered density versus composition for ZnO-SiC particulate composite showing the degradation of sintering density from the added inert silicon carbide [9].

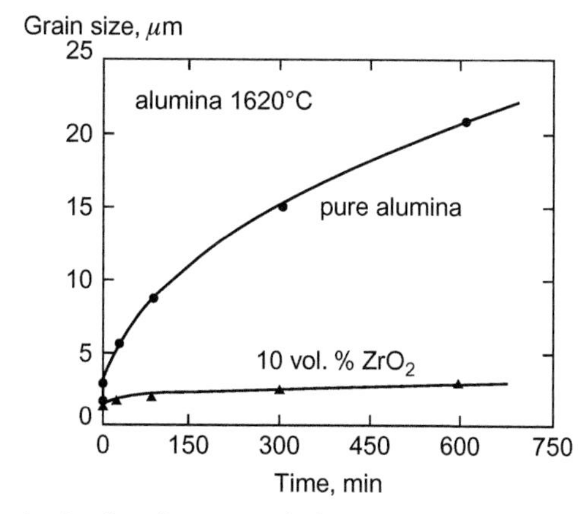

Figure 11.21 Sintered grain size for alumina and alumina with 10 vol.% zirconia, for hold times up to 600 min at 1620°C (1893 K) [78]. The pure alumina exhibited significantly more grain growth, while the composite had less grain growth and reached a slightly higher final density.

alumina experiments the zirconia containing material had a slightly higher density after 600 min (97% versus 96% for pure alumina).

Whisker reinforced composites exhibit differential densification around the inert phase [79]. Shrinkage along the whisker axis is constrained more than perpendicular to the whisker, resulting in tensile stresses and retained pores in the sintering matrix. Because the whisker is anisotropic, the matrix sintering response is orientation sensitive. Each whisker results in residual pores located parallel to its axis.

In sintering a composite, the reinforcing phase retards densification. The retardation increases in direct proportion to the volume fraction of the inert phase. Further, if the matrix is viscous, the retarding stress relaxes over time so the impediment is small. However, during rapid heating, microstructural damage results in cracks. The cracks form early in the sintering cycle while the matrix is poorly sintered. Prolonged low temperature sintering allows matrix strengthening prior to densification to help avoid cracking. This is one reason why rapid heating technologies are avoided in practice.

CO-FIRE, LAMINATED, AND BIMATERIAL SINTERING

One challenging situation is where two powders are combined into distinct regions, such as laminated layers. In one application, concentric cylinders of high hardness on the outside and high toughness on the inside were sintered to form twist drills. The two powders are matched for sintering to avoid defects. Figure 11.22 is a micrograph of a two-material composite, consisting of WC-Co balls (some pores remain in the balls), sintered in a surgical steel matrix, with bronze as the third phase.

A common application for co-firing is multiple-layer structures. This might include magnetic-nonmagnetic, conductive-nonconductive, hard-tough, or expensive-low cost combination structures [80–91]. These are common in capacitors, glass-to-metal seals, wear coatings, magnetic sensors, metal cutting tools, and devices with intentional functional gradients. The initial structure consists of two or more distinct powders, but

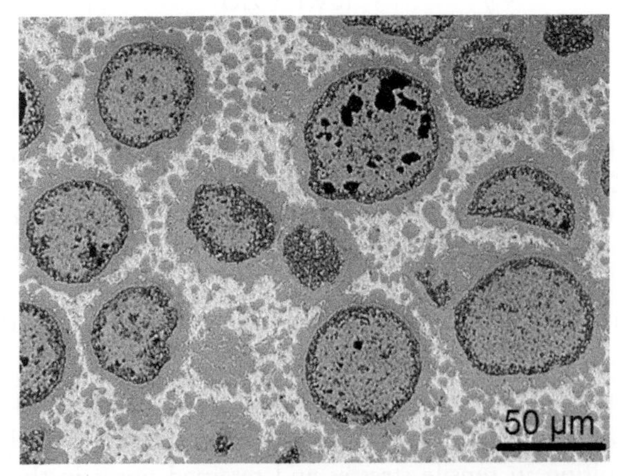

Figure 11.22 Sintered composite consisting of agglomerated WC-Co balls coated with stainless steel (doped with boron) with bronze as the final phase. Residual pores are evident in the carbide phase.

unlike mixed systems, the powders are separated by an interface. The objective is to sinter both materials simultaneously without distortion or defect formation. Co-firing requires that the two powders follow similar shrinkage pathways, even though they differ in properties [92–94]. The balanced firing means the heating rate, particle size, green density, and time-temperature combination are selected to avoid distortion or separation [95].

An early application for co-firing was in computer packaging. An alumina ceramic and molybdenum conduction were combined to provide electrical interconnections in three-dimensional arrays. The green structure was formed by stacking ceramic powder layers with channels filled with metal powder. The powders were selected to provide matched shrinkage profiles during heating.

A wide variety of co-fired structures exist today, including some important biomaterials. The matched sintering determines particle size and firing cycle. The dictates of matched sintering sometimes requires multiple composition steps in layers to give a functional gradient structure. These are materials that gradually transition properties. For example, one end is 100% metallic, and over several steps of increasing ceramic content the final part is 100% ceramic. Each layer has a progressive change, possibly in 10% or 20% increments, to move from the pure metal to the pure ceramic. This gives a gradual property change, thereby minimizing problems from incompatible thermal expansion coefficients or other differences. Examples included electronic circuits, metal cutting tools, biomaterials, and high temperature bodies such as Al_2O_3-ZrO_2. Sintering without defects proves difficult [96,97]. Pressure-assisted sintering is common. Figure 11.23 shows the cross-sectional microstructure of an alumina whisker

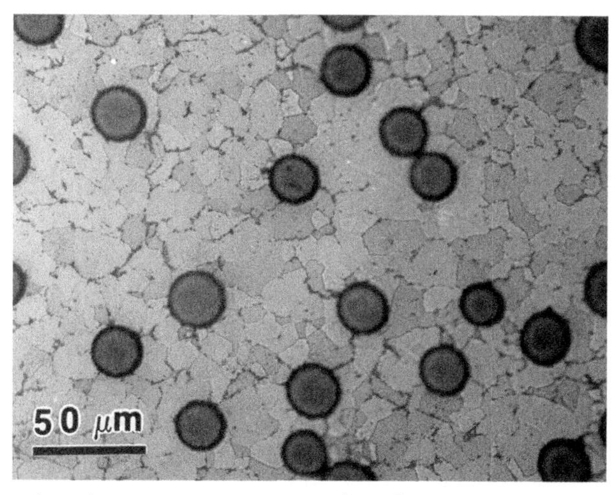

Figure 11.23 Composite microstructure consisting of Ni-Al matrix and Al_2O_3 whiskers consolidated by hot isostatic pressing. In this case the whiskers are aligned perpendicular to the cross-section plane.

reinforced nickel aluminide formed by reactive synthesis under pressure. In this case the fibers are aligned perpendicular to the cross-section plane.

SUMMARY

The sintering of powder mixtures has many variants and is a widespread aspect of sintering practice. It is cheaper to maintain inventories of the powder ingredients and mix the desired composition on demand, than it is to maintain many different compound compositions in inventory. Thus, in several operations the ingredients—such as iron, copper, nickel, graphite—are mixed to tailor the sintered properties. A large number of compositions are possible from a relatively small number of ingredients.

Besides forming alloys by homogenization, mixed powder sintering is a route to the fabrication of composites, functionally gradient structures, and co-fired structures. However, mixed powders add new challenges to the sintering process, largely due to the chemical gradients. The gradients induce diffusion effects which dominate early sintering. Depending on sintering temperature, composition, and solubilities-reactivities-diffusivities, a wide range of responses are possible. Usually the sintering stress is low compared to the chemical gradients and reaction enthalpies. The means it is difficult to broadly characterize the sintering response.

Interactions during heating are dependent on reactions, and most of the process variables play a role. For example, heating rate is more of a factor in mixed powder sintering than in single phase sintering. The usual protocol is to use slow heating to better control the chemical interactions. Examples included in this chapter range from solid solution systems with small interaction energies to reactive sintering systems with exothermic swelling reactions.

One concern with all mixed powder structures is particle sizes, and understanding each powder's role during sintering. If the particles are of similar size and shape, then the packing density is independent of composition. On the other hand, if the particles are different in size, then the larger particles control the packing and sintering behavior. Dimensional changes during sintering are dictated by a hybrid of the sintering for the large and small particle structures. Large, chemically inert particles in a small particle matrix generate constraining stresses that form cracks at the interface. Void growth and compact swelling are possible if the large particles are chemically active

Chemical interactions include four cases treated in this chapter—homogenization (mixed Cu and Ni form an alloy during sintering), swelling (mixed Sn and Cu form porous bronze), enhanced (mixed Ni is added to Mo) and composites (mixed Y_2O_3 is added to stainless steel). A number of variants are encountered in practice [98—106].

REFERENCES

[1] D.J. Cumberland, R.J. Crawford, The Packing of Particles, Elsevier Science, Amsterdam, Netherlands, 1987.

[2] R.M. German, Particle Packing Characteristics, Metal Powder Industries Federation, Princeton, NJ, 1989.

[3] R.M. German, Prediction of sintered density for bimodal powder mixtures, Metall. Trans. 23A (1992) 1455–1465.

[4] M.J. O'Hara, I.B. Cutler, Sintering kinetics of binary mixtures of alumina powders, Proc. Brit. Ceram. Soc. 12 (1969) 145–154.

[5] W.H. Tuan, E. Gilbart, R.J. Brook, Sintering of heterogeneous ceramic compacts Part 1 Al_2O_3–Al_2O_3, J. Mater. Sci. 24 (1989) 1062–1068.

[6] C.P. Ostertag, P.G. Charalambides, A.G. Evans, Observations and analysis of sintering damage, in: C.A. Handwerker, J.E. Blendell, W. Kaysser (Eds.), Sintering of Advanced Ceramics, American Ceramic Society, Westerville, OH, 1990, pp. 710–732.

[7] R. Huang, J. Pan, A. Two, Scale model for sintering damage in powder compact containing inert inclusions, Mech. Mater. 39 (2007) 710–726.

[8] G.J. Shu, K.S. Hwang, High density powder injection molded compacts prepared from a feedstock containing coarse powders, Materials Trans. 45 (2004) 2999–3004.

[9] L.C. De Jonghe, M.N. Rahaman, C.H. Hsueh, Transient stresses in bimodal compacts during sintering, Acta Metall. 34 (1986) 1467–1471.

[10] J.P. Smith, G.L. Messing, Sintering of bimodally distributed alumina powders, J. Amer. Ceram. Soc. 67 (1984) 238–242.

[11] W.H. Rhodes, Agglomerate and particle size effects on sintering yttria-stabilized zirconia, J. Amer. Ceram. Soc. 64 (1981) 19–22.

[12] M.D. Sacks, J.A. Pask, Sintering of mullite containing materials: II, effect of agglomeration, J. Amer. Ceram. Soc. 65 (1982) 70–77.

[13] G.C. Culbertson, J.P. Mathers, The effect of particle size and state of aggregation on the sintering of aluminum nitride, in: R.M. German, K.W. Lay (Eds.), Processing of Metal and Ceramic Powders, Metallurgical Society, Warrendale, PA, 1982, pp. 109–122.

[14] F.F. Lange, B.I. Davis, I.A. Aksay, Processing related fracture origins: III, differential sintering of ZrO_2 agglomerates in Al_2O_3/ZrO_2 composite, J. Amer. Ceram. Soc. 66 (1983) 407–408.

[15] J.H. Jean, T.K. Gupta, Liquid phase sintering in the glass-cordierite system: particle size effect, J. Mater. Sci. 27 (1992) 4967–4973.

[16] Y.S. Kwon, A. Savitskii, Solid-state sintering of metal powder mixtures, J. Mater. Syn. Proc. 9 (2001) 299–317.

[17] A.P. Savitskii, New approaches to the problem of sintering two-component mixtures, Sci. Sintering 30 (1998) 139–147.

[18] E.I. Maksimov, A.G. Merzhanov, V.M. Shkiro, Gasless compositions as a simple model for the combustion of nonvolatile condensed systems, Comb., Expl. Shock Waves 1 (1965) 15–18.

[19] Z.A. Munir, U. Anselmi-Tamburini, Self-propagating exothermic reactions: the synthesis of high temperature materials by combustion, Mater. Sci. Rept. 3 (1989) 277–365.

[20] R.L. Coble, Reactive sintering, in: D. Kolar, S. Pejovnik, M.M. Ristic (Eds.), Sintering–Theory and Practice, Elsevier Scientific, Amsterdam, Netherland, 1982, pp. 145–151.

[21] M.E. Washburn, W.S. Coblenz, Reaction formed ceramics, Ceram. Bull. 67 (1988) 356–363.

[22] D.M. Turriff, S.F. Corbin, Modeling the influences of solid-state interdiffusion and dissolution on transient liquid phase sintering kinetics in a binary isomorphous system, Metall. Mater. Trans. 37A (2006) 1645–1655.

[23] D.M. Turriff, S.F. Corbin, Quantitative thermal analysis of transient liquid phase sintered Cu-Ni powders, Metall. Mater. Trans. 39A (2008) 28–38.

[24] M.S. Masterller, R.W. Heckel, R.F. Sekerka, A mathematical model study of the influence of degree of mixing and powder particle size variation on the homogenization, Metall. Trans. 6A (1975) 869–876.

[25] F. Aldinger, Controlled porosity by an extreme Kirkendall effect, Acta Metall. 22 (1974) 923–928.

[26] R. Watanabe, H. Nagai, Y. Masuda, The Kirkendall effect in the sintering of Cu-Ni Alloys, Sci. Sintering 11 (1979) 31–58.
[27] C. Menapace, P. Costa, A. Molinari, Study of the liquid phase sintering in the Cu-Sn system by thermal analysis, Proceedings PM2004 Powder Metallurgy World Congress, 2, European Powder Metallurgy Association, Shrewsbury, UK, 2004, pp. 172–177.
[28] B. Yuan, X.P. Zhang, C.Y. Chung, M.Q. Zeng, M. Zhu, Comparative study of the porous TiNi shape memory alloys fabricated by three different processes, Metall. Mater. Trans. 371 (2006) 755–761.
[29] A. Bose, B.H. Rabin, R.M. German, Reactive sintering Nickel-Aluminide to near full density, Powder Metall. Inter. 20 (3) (1988) 25–30.
[30] A.P. Savitskii, Liquid Phase Sintering of the Systems with Interacting Components, Russian Academy of Sciences, Tomsk, 1993.
[31] C. Menapace, M. Zadra, A. Molinari, C. Messner, P. Costa, Study of microstructural transformations and dimensional variations during liquid phase sintering of 10% tin bronzes produced with different copper powders, Powder Metall. 45 (2002) 67–74.
[32] N.N. Acharya, P.G. Mukunda, Sintering in the copper-tin system part I: Identification of phases and reactions, Inter. J. Powder Metall. 31 (1995) 63–71.
[33] L.H. Chiu, D.C. Nagle, L.A. Bonney, Thermal analysis of self-propagating high-temperature reactions in titanium, boron, and aluminum powder compacts, Metall. Mater. Trans. 30A (1999) 781–788.
[34] E.H. Sun, T. Kusunose, T. Sekino, K. Niihara, Fabrication and characterization of cordierite/zircon composites by reaction sintering: formation mechanism of zircon, J. Amer. Ceram. Soc. 85 (2002) 1430–1434.
[35] T.M. Puscas, A. Molinari, J. Kazior, T. Piezonka, M. Nykiel, Density and microstructure of duplex stainless steel produced by mixtures of austenitic and ferritic powders, in: K. Kosuge, H. Nagai (Eds.), Proceedings of the 2000 Powder Metallurgy World Congress, Part 2, Japan Society of Powder and Powder Metallurgy, Kyoto, Japan, 2000, pp. 980–983.
[36] I.M. Robertson, G.B. Schaffer, Design of titanium alloy for efficient sintering to low porosity, Powder Metall. 52 (2009) 311–315.
[37] J. Liu, R.M. German, A. Cardamone, T. Potter, F.J. Semel, Boron-enhanced sintering of iron-molybdenum steels, Inter. J. Powder Metall. 37 (5) (2001) 39–46.
[38] T. Osada, H. Miura, Y. Itoh, M. Fujita, N. Arimoto, Optimization of MIM process for Ti-6Al-7Nb alloy powder, J. Japan Soc. Powder Powder Metall. 55 (2008) 726–731.
[39] R.J. Hellmig, J. Ferkel, Using nanoscaled powder as an additive in coarse-grained powder, J. Amer. Ceram. Soc. 84 (2001) 261–266.
[40] T.M. Puscas, A. Molinari, J. Kazior, T. Pieczonka, M. Nykiel, Sintering transformations in mixtures of austenitic and ferritic stainless steel powders, Powder Metall. 44 (2001) 48–52.
[41] P. Marsh, J.V. Wood, J.R. Moon, Diffusion between high speed steel and iron powders, Powder Metall. 44 (2001) 205–210.
[42] R.M. German, Z.A. Munir, Enhanced low-temperature sintering of tungsten, Metall. Trans. 7A (1976) 1873–1877.
[43] J.L. Johnson, R.M. German, Theoretical modeling of densification during activated solid-state sintering, Metall. Mater. Trans. 27A (1996) 441–450.
[44] H.S. Choi, Y.K. Yoon, W.K. Park, Effects of cyclic heating through alpha gamma phase transformations during sintering of Fe-Ni alloy powder compacts, Inter. J. Powder Metall. 9 (1973) 23–37.
[45] W.H. Tuan, G. Matsumura, Effects of cyclic heating on the sintering of iron powder, Powder Metall. Inter. 16 (1984) 16–18.
[46] P. Svancarek, D. Galusek, F. Loughran, A. Brown, R. Brydson, A. Atkinson, et al., Microstructure-stress relationships in liquid phase sintered alumina modified by the addition of 5 wt. % calcia-silica additives, Acta Mater. 54 (2006) 4853–4863.
[47] J.R. Cahoon, O.D. Sherby, The activation energy for lattice self-diffusion and the Engel-Brewer theory, Metall. Trans. 23A (1992) 2491–2500.

[48] Z.A. Munir, R.M. German, A generalized model for the prediction of periodic trends in the activation of sintering of refractory metals, High Temp. Sci. 9 (1977) 275–283.
[49] H.O. Gulsoy, Influence of nickel boride additions on sintering behaviors of injection moulded 17–4 PH stainless steel powder, Scripta Mater. 52 (2005) 187–192.
[50] N. Tosangthum, O. Coovattanachai, R. Tongsri, Sintering of 316L + Ni powder compacts, Advances in Powder Metallurgy and Particulate Materials–2004, Part 5, Metal Powder Industries Federation, Princeton, NJ, 2004, pp. 51–60.
[51] F.A. Corpas Iglesias, J.M. Ruiz Roman, J.M. Ruiz Prieto, L. Garcia Cambronero, F.J. Ingesias Godino, Effect of nitrogen on sintered duplex stainless steel, Powder Metall. 46 (2003) 39–42.
[52] P. Datta, G.S. Upadhyaya, Copper enhances the sintering of duplex PM stainless steels, Met. Powder Rept. 54 (No. 1) (1999) 26–29.
[53] K. Tsukuma, I. Yamashita, T. Kusunose, Transparent 8 mol. % Y_2O_3–ZrO_2 (8Y) ceramics, J. Amer. Ceram. Soc. 91 (2008) 813–818.
[54] M. Khakbiz, A. Simchi, Effect of SiC addition on the compactability and sintering behavior of M2 high speed steel powder, P/M Sci. Tech. Briefs 5 (4) (2003) 23–27.
[55] S.F. Moustafa, Z. Abdel-Hamid, A.M. Abd-Elhay, Copper matrix SiC and Al_2O_3 particulate composites by powder metallurgy techniques, Mater. Lett. 53 (2002) 244–249.
[56] A. Bose, B. Moore, R.M. German, N.S. Stoloff, Elemental powder approaches to nickel aluminide-matrix composites, J. Met. 40 (no.9) (1988) 14–17.
[57] P. Maheshwari, Z.Z. Fang, H.Y. Sohn, Early stage sintering densification and grain growth of nanosized WC-Co powders, Inter. J. Powder Metall. 43 (2) (2007) 41–47.
[58] F.C. Sahin, O.A. Sepin, S.A. Yesilcubuk, O. Addemir, Hot pressing of B_4C/TiB_2 composites, in: D. Bouvard (Ed.), Proceedings of the Fourth International Conference on Science, Technology and Applications of Sintering, Institut National Polytechnique de Grenoble, Grenoble, France, 2005, pp. 410–413.
[59] E.M.J.A. Pallone, J.J. Pierre, V. Trombini, R. Tomasi, Production of Al_2O_3 nanocomposites with inclusions of nanometric ZrO_2, in: D. Bouvard (Ed.), Proceedings of the Fourth International Conference on Science, Technology and Applications of Sintering, Institut National Polytechnique de Grenoble, Grenoble, France, 2005, pp. 504–506.
[60] H. Miura, H. Morikawa, Y. Kawakami, A. Ishibashi, Development of high performance sliding abrasive wear resistant materials through powder injection molding, Advances in Powder Metallurgy and Particulate Materials–1998, Metal Powder Industries Federation, Princeton, NJ, 1998, pp. 5.183–5.191.
[61] T.J. Weaver, J.A. Thomas, S.V. Atre, A. Griffo, R.M. German, Steel rapid tooling via powder metallurgy, Advances in Powder Metallurgy and Particulate Materials–1998, Metal Powder Industries Federation, Princeton, NJ, 1998, pp. 6.15–6.24.
[62] T. Osada, K. Nishiyabu, Y. Karasaki, S. Tanaka, H. Miura, Evaluation of the homogeneity of feedstock for micro sized parts produced by MIM, J. Japan Soc. Powder Powder Metall. 51 (2004) 435–440.
[63] H. Chang, T.P. Tang, K.T. Huang, F.C. Tai, Effects of sintering process and heat treatments on microstructures and mechanical properties of VANADIS 4 tool steel added with TiC powders, Powder Metall. 54 (2011) 507–512.
[64] G. Herranz, G.P. Rodriguez, E. Alonso, G. Matula, Sintering process of M2 HSS feedstock reinforced with carbides, Powder Inj. Mould. Inter. 4 (2) (2010) 60–65.
[65] M.G. Bothara, S.J. Park, R.M. German, A.V. Atre, Spark plasma sintering of ultrahigh temperature ceramics, Advances in Powder Metallurgy and Particulate Materials 2008, Part 9, Metal Powder Industries Federation, Princeton, NJ, 2008, pp. 264–270.
[66] C.H. Hsueh, A.G. Evans, R.M. McMeeking, Influence of multiple heterogeneities on sintering rates, J. Amer. Ceram. Soc. 69 (1986) C64–C66.
[67] A.R. Boccaccini, Sintering of glass powder compacts containing rigid inclusions, Sci. Sintering 23 (1991) 151–161.
[68] F.F. Lange, T. Yamaguchi, B.I. Davis, P.E.D. Morgan, Effect of ZrO_2 inclusions on the sinterability of Al_2O_3, J. Amer. Ceram. Soc. 71 (1988) 446–448.

[69] R. Orru, R. Licheri, A.M. Locci, A. Cincotti, G. Cao, Consolidation/synthesis of materials by electric current activated/assisted sintering, Mater. Sci. Eng. vol. R63 (2009) 127–287.

[70] S.J. Dong, Y. Zhou, Y.W. Shi, B.H. Chang, Formation of a TiB_2 reinforced copper based composite by mechanical alloying and hot pressing, Metall. Mater. Trans. 33A (2002) 1275–1280.

[71] S. Vives, C. Guizard, C. Oberlin, L. Cot, Zirconia-tungsten composites: synthesis and characterization for different metal volume fractions, J. Mater. Sci. 36 (2001) 5271–5280.

[72] W. Acchar, J.L. Fonseca, Sintering Behavior of Alumina Reinforced with (Ti, W) Carbides, Mater. Sci. Eng. vol. A371 (2004) 382–387.

[73] J.T. Neil, D.A. Norris, Whisker orientation measurements in injection molded Si_3N_4- SiC composites, Proceedings ASME Gas Turbine and Aeroengine Congress, American Society of Mechanical Engineers, Amsterdam, Netherlands, 1988, pp. 1–7.

[74] Z.J. Lin, J.Z. Zhang, B.S. Li, L.P. Wang, H.K. Mao, R.J. Hemley, et al., Superhard Diamond/Tungsten Carbide Nanocomposites, Appl. Phys. Lett. 98 (2011) paper 121914.

[75] L. Rangaraj, S.J. Suresha, C. Divakar, V. Jayaram, Low temperature processing ZrB_2-ZrC composites by reactive hot pressing, Metall. Mater. Trans. 39A (2008) 1496–1505.

[76] S. Sugiyama, Y. Kodaira, H. Taimatsu, Synthesis of WC-W_2C composite ceramics by reactive resistance heating hot pressing and their mechanical properties, J. Japan Soc. Powder Powder Metall. 54 (2007) 281–286.

[77] C. Liu, J. Zhang, X. Zhang, J. Sun, Fabrication of Al_2O_3/TiB_2/AlN/TiN and Al_2O_3/TiC/AlN composites, Mater. Sci. Eng. A 465 (2007) 72–77.

[78] J. Zhao, M.P. Harmer, Effect of pore distribution on microstructure development: II, first and second generation pores, J. Amer. Ceram. Soc. 71 (1988) 530–539.

[79] O. Sudre, G. Bao, B. Fan, F.F. Lange, A.G. Evans, Effect of inclusions on densification: II, numerical model, J. Amer. Ceram. Soc. 75 (1992) 525–532.

[80] H. Miura, K. Hasama, T. Baba, T. Honda, Joining for more functional and complicated parts in MIM Process, J. Japan Soc. Powder Powder Metall. 44 (1997) 437–442.

[81] M. Eriksson, M. Radwan, Z. Shen, Spark plasma sintering of WC, cemented carbide, and functional graded materials, Inter. J. Refract. Met. Hard Mater. 36 (2013) 31–37.

[82] Y. Boonyongmaneerat, C.A. Schuh, Contributions to the interfacial adhesion in co-sintered bilayers, Metall. Mater. Trans. 37A (2006) 1435–1442.

[83] O.O. Eso, Z.Z. Fang, A new method for making functionally graded WC-Co composites via liquid phase sintering, Advances in Powder Metallurgy and Particulate Materials – 2004, Part 8, Metal Powder Industries Federation, Princeton, NJ, 2004, pp. 62–72.

[84] C. Pascal, A. Thomazic, A.A. Zdziobek, J.M. Chaix, Co-sintering and microstructural characterization of steel/cobalt base alloy bimaterials, J. Mater. Sci. 47 (2012) 1875–1886.

[85] A. Baumann, M. Brieseck, S. Hohn, T. Moritz, R. Lenk, Development of multi-component powder injection moulding of steel-ceramic compounds using green tapes for inmould label process, Powder Inj. Mould. Inter. 2 (1) (2008) 55–58.

[86] K. Satoh, Y. Itoh, T. Harikoh, H. Miura, Joining of SUS316L and SUS430L stainless steel parts with the paste containing stainless steel powder in MIM process, J. Japan Soc. Powder Powder Metall. 50 (2003) 566–570.

[87] A. Petersson, J. Agren, Numerical simulation of shape changes during cemented carbide sintering, in: F.D.S. Marquis (Ed.), Powder Materials Current Research and Industrial Practices III, Minerals, Metals and Materials Society, Warrendale, PA, 2003, pp. 65–75.

[88] H. Kato, K. Washida, Y. Soda, Microstructure and strength of sinter-bonding of high speed steel and low alloy steel, J. Japan Soc. Powder Powder Metall. 49 (2002) 651–657.

[89] W. Shi, A. Kamiya, J. Zhu, A. Watazu, Properties of titanium biomaterial fabricated by sinter-bonding of titanium/hydroxyapatie composite surface-coated layer to pure bulk titanium, Mater. Sci. Eng. vol. A337 (2002) 104–109.

[90] M. Menon, I.W. Chen, Bimaterial composites via colloidal rolling techniques ii, sintering behavior and thermal stresses, J. Amer. Ceram. Soc. 82 (1999) 3422–3429.

[91] H. He, Y. Li, P. Liu, J. Zhang, Design with a skin-core structure by metal co injection moulding, Powder Inj. Mould. Inter. 4 (1) (2010) 50–54.

[92] J.L. Johnson, L.K. Tan, P. Suri, R.M. German, Design guidelines for processing bi-materials components via powder injection molding, J. Met. 55 (10) (2003) 30–34.
[93] R. Zuo, E. Aulbach, J. Rodel, Shrinkage free sintering of low temperature cofired ceramics by loading dilatometry, J. Amer. Ceram. Soc. 87 (2004) 526–528.
[94] P. Suri, D.F. Heaney, R.M. German, Defect-free sintering of two material powder injection molded components: part II, model, J. Mater. Sci. 38 (2003) 4875–4881.
[95] A.L. Maximenko, O. Van Der Biest, E.A. Olevsky, Prediction of initial shape of functionally graded ceramic preforms for near-net-shape sintering, Sci. Sintering 35 (2003) 5–12.
[96] S.I. Sumi, Y. Mizutani, T. Abe, Fabrication of Ni-PSZ FGM by pulse discharge sintering and observation of its internal defects, J. Japan Soc. Powder Powder Metall. 45 (1998) 1071–1075.
[97] M.L. Pines, H.A. Bruck, Pressureless sintering of particle reinforced metal-ceramic composites for functionally graded materials: part II. Sintering model, Acta Mater. 54 (2006) 1467–1474.
[98] J.F. Rhodes, W.R. Mohn, Ceramic composites – reliable ceramics, in: R.J. Schaefer, M. Linzer (Eds.), Hot Isostatic Pressing Theory and Applications, ASM International, Materials Park, OH, 1991, pp. 179–189.
[99] K. Okazaki, Electro-discharge consolidation of particulate materials, Rev. Particulate Mater. 2 (1994) 215–268.
[100] R. Orru, R. Licheri, A.M. Locci, A. Cincotti, G. Cao, Consolidation/synthesis of materials by electric current activated/assisted sintering, Mater. Sci. Eng. vol. R63 (2009) 127–287.
[101] F. Hussain, M. Hojjati, M. Okamoto, R.E. Gorga, Polymer-matrix nanocomposites, processing, manufacturing, and application: an overview, J. Comp. Mater. 40 (2006) 1511–1575.
[102] H.V. Atkinson, S. Davies, Fundamental aspects of hot isostatic pressing: an overview, Metall. Mater. Trans. 31A (2000) 2981–3000.
[103] A. Bose, W.B. Eisen, Hot Consolidation of Powders and Particulates, Metal Powder Industries Federation, Princeton, NJ, 2003.
[104] W. Dressler, R. Riedel, Progress in silicon-based non-oxide structural ceramics, Inter. J. Refract. Met. Hard Mater. 15 (1997) 13–47.
[105] C.M. Ward-Close, R. Minor, P.J. Doorbar, Intermetallic matrix composites–a review, Intermetallics 4 (1996) 217–229.
[106] R.M. German, R.G. Iacocca, Powder metallurgy processing, in: N.S. Stoloff, V.K. Sikka (Eds.), Physical Metallurgy and Processing of Intermetallic Compounds, Chapman and Hall, New York, NY, 1996, pp. 605–654.

CHAPTER TWELVE

Rapid Heating Approaches

INTRODUCTION

Flash sintering is a term used to describe sintering cycles lasting just a few seconds [1]. There is no formal definition of rapid heating or flash sintering, but a typical target is 10°C/s, or faster. As a benchmark, industrial sintering relies on typical heating rates of 10°C/min (0.167°C/s); about 60-fold slower.

In sintering demonstrations, heating rates have reached 600°C/s. However, such cycles are not realistic. Limitations arise in several practical areas, such as how to deliver large quantities of energy in short times, and how to control the process, as well as how to avoid component cracking and distortion. Several studies report sintering benefits with heating rates in the 10 to 20°C/s range. Thus, the idea of flash sintering is spreading to microelectronics, where printed powders are replacing solders. The powders are present in layers only micrometers thick and are illuminated by intense light sources to sinter the structure in less than a second. This promises to be a large application for sintering and largely breaks away from the traditional slow heating applied for sintering engineering components in the past.

Several factors combine to make rapid heating an interesting twist for sintering [2–4]. These range from shifts in diffusion events as temperature increases to changes in production economics. One contention is that thermal gradients from rapid heating significantly increase diffusion rates in sintering [5,6]. If rapid heating were performed without thermal gradients, little benefit is predicted. This makes for an interesting conjecture, but the two variables of heating rate and temperature gradient are confounded with the sense that rapid heating is the key.

Some important points with respect to rapid heating experiments are as follows:

- **Surface energy**—a powder has energy stored in the form of surface area which equates to surface energy, at lower temperatures surface diffusion is a dominant sintering mechanism due to a high initial surface area and low activation energy, but surface diffusion fails to promote densification while it dissipates surface energy—rapid heating preserves surface energy to higher temperatures where densification processes are active, improving the sintered density

Sintering: From Empirical Observations to Scientific Principles
DOI: http://dx.doi.org/10.1016/B978-0-12-401682-8.00012-4

- **Instability**—materials that are unstable decompose during heating, especially diamond and some oxides used in electronics (lead-containing titanates), rapid heating cycles minimize decomposition to preserve stoichiometry during sintering
- **Microstructure coarsening**—slow heating cycles allows more grain coarsening, while fast cycles reduce grain coarsening by separating densification from grain growth, an approach of great benefit in sintering nanoscale materials
- **Economics**—quick cycles, as associated with rapid heating, reduce furnace time, meaning that more product is sintered without requiring the purchase of more equipment

Much speculation exists on the possible gains from rapid heating. Relatively few measurements directly monitor the effects, so most of this is from conjecture or calculations. The notions include the following ideas on rapid heating and its benefits:

- surface-core temperature gradients induce thermal stresses to cause plastic flow by dislocation motion
- dislocation generation is coupled to lattice diffusion to give dislocation climb
- reduced grain growth occurs versus conventional heating, resulting in substantially enhanced grain boundary diffusion and more densification
- thermal gradients activate additional diffusion as a supplement to that normally contributing to sintering

Thermal stresses induce dislocation motion during rapid heating, giving the benefit of plastic flow and dislocation climb [7]. Further, with small particles the sintering stress is sufficient to induce dislocation motion during heating. The smaller sintered grain size with rapid heating reflects the kinetic differences between grain growth and densification. Fast heating is beneficial when the densification process has a higher activation energy versus the activation energy for grain coarsening. This is the case in alumina, one of the systems responsive to rapid heating. Rapid heating reduces the time during which grain growth occurs, resulting in a smaller grain size at each density level. Figure 12.1 plots this behavior by tracing grain size versus sintered density for alumina doped with 200 ppm magnesia [8]. Rapid heating generates a different trajectory, with a smaller grain size at each density. Heat arrives on the exterior surface of the powder compact, so rapid heating generates thermal gradients that improve bulk transport sintering [9,10].

As the investment community says, "... with all technology the good news comes first." The bad news on rapid heating cycles usually comes down to damage. The thermal gradients responsible for enhanced diffusion also induce stresses on a weak body that result in cracking, warpage, and bloating from trapped gases. In one study applying rapid heating to cutting tools, the maximum thickness possible without cracking was 10 mm. Rapid heating exhibits less advantage as the component mass

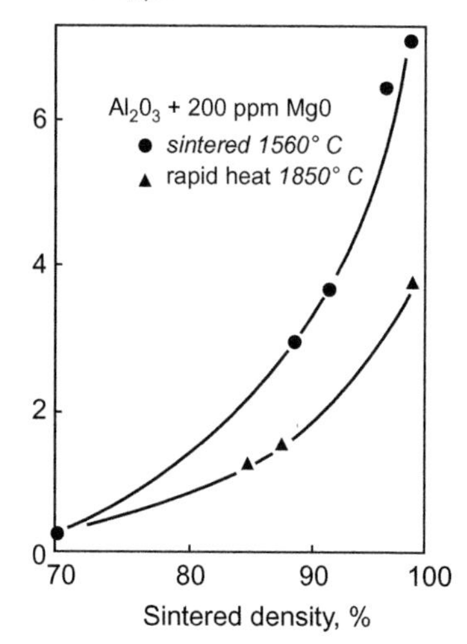

Figure 12.1 Grain size versus sintered density for magnesia doped alumina sintered at 1560°C (1833 K) in a traditional cycle and rapidly heated to 1850°C (2123 K) [8]. Even with the higher temperature, rapid heating resulted in less grain growth for each density level.

increases [11]. Small, thin samples work best. Even without cracking, there is a tendency to warp, giving "potato chip" geometries.

Another difficulty with rapid heating is the "line of sight" requirement for radiant heating of the powder compact, restricting rapid heating to simple geometries. In the electronics field this means sintering one component at a time. Approaches such as microwave heating, claim interior heating, but usually a susceptor powder (such as silicon carbide) is packed around the component and gives only exterior heating. Otherwise a thermal runaway happens as the material heats, making process control difficult [12]. Also, with rapid heating no lubricant or binder is allowed, otherwise the compact ruptures as the polymer decomposes. A separate slow thermal cycle is required to remove volatile species ahead of sintering. In a microwave sintering facility for cemented carbides, sintering occurred in 8 min to 1400°C (1673 K), but dewaxing prior to microwave sintering took 8 h in a separate furnace. The extra handling and steps made microwave sintering slow and costly.

Because of the "one at a time" character of rapid heating technologies, the productivity is low. A production microwave furnace for WC-Co sintering is eight-fold more expensive when compared to similar production rates in a traditional batch furnace. Still the conceptual learning from rapid heating is important in pulling various aspects of sintering into a coherent conceptualization.

EARLY DEMONSTRATIONS

One means to execute rapid heating is to employ a dunk cycle. The concept is sketched in Figure 12.2. A sintering furnace is preheated to the peak temperature and the powder compact is conveyed into the hot furnace [13]. The translation rate into the furnace determines the heating rate, typically about 10°C/s. Early trials produced date showing nearly instant densification. Figure 12.3 plots sintered density for SnO_2 doped with three levels of iron [14]. For these trials, small samples were inserted into a furnace preset to 1200°C (1473 K) to enable densification in the first minute.

Fast firing with dunk cycles followed using other intense energy input routes such as electric discharges, exothermic reactions, as well as heating using microwave, induction, plasma, infrared, or laser sources. In newer approaches, the temperature rise reaches upwards to 200°C/s. A common variant is to discharge an electric current through the powder compact, a process termed spark sintering and later given names such as field affect sintering technology (FAST), spark sintering (SPS), and electric current activated sintering (ECAS) [15–17]. Most of these approaches add supplemental pressure to accelerate densification. The thermal mass of the die and punches slows heating and cooling, giving less differentiation from hot pressing. For example, in sintering zirconia, similar densities were attained using 50 MPa spark sintering at 1250°C (1523 K) for 5 min and conventional sintering at 1500°C (1773 K) for 120 min. The conventional process resulted in over a three-fold larger grain size, but a lower cost.

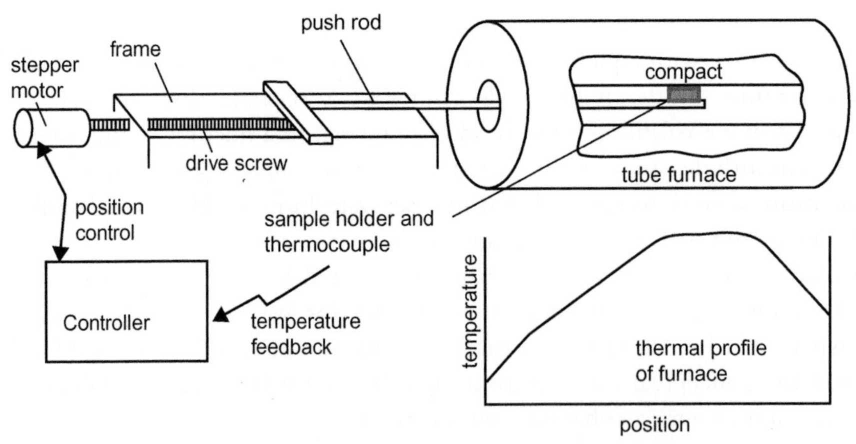

Figure 12.2 Early demonstrations of rapid heating relied on a dunk cycle, where the powder compact is inserted into a preheated furnace; the rate of entry determines the heating rate.

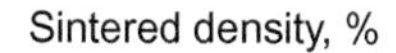

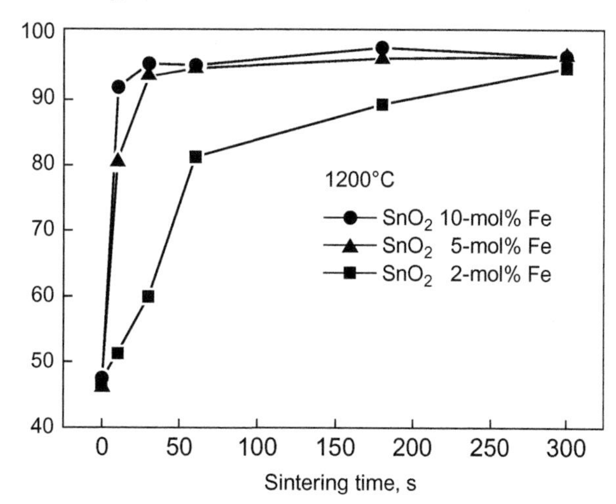

Figure 12.3 Sintered density for tin oxide (SnO_2) doped with various levels of iron (in the form of oxide) showing the remarkably rapid densification possible with dunk heating cycles [14]. The furnace was set to 1200°C (1473 K) for these trials.

Rapid heating results in impressive shrinkage rates, up to 1%/s [2]. Success comes when the temperature sensitivities of sintering and coarsening are separated, using the heating path to drive densification with minimum coarsening [18]. Surface diffusion, with a lower activation energy, dominates low temperature sintering. Grain boundary diffusion tends to dominate high temperature sintering. Surface diffusion consumes the sintering potential without densification, so skipping the lower temperature range provides grain boundary diffusion with a larger driving force, resulting in more densification. With nanoscale powder there is much initial surface area and little grain boundary area. As neck growth progresses the opposite is true, with more grain boundary area and less surface area. Grain growth accelerates as grain boundary area increases. By jumping to the high temperature, the curvature and surface area act to drive densification without the loss of surface area seen with slow heating.

Calculations of rates for grain boundary diffusion, surface diffusion, and grain growth give estimates of the crossover points during different sintering cycles. Initially, at lower temperatures and higher surface areas, surface diffusion dominates. At higher temperatures after grain boundaries form at the particle contacts, a crossover arises where grain boundary diffusion and densification dominate. This is accompanied by grain growth. Rapid heating induces early densification, and if properly managed, the short cycle minimizes grain growth. Such behavior is illustrated in Figure 12.4, where the flux for surface diffusion (nearly horizontal lines) and grain boundary diffusion (upward rising lines) are plotted for silver versus the sintering

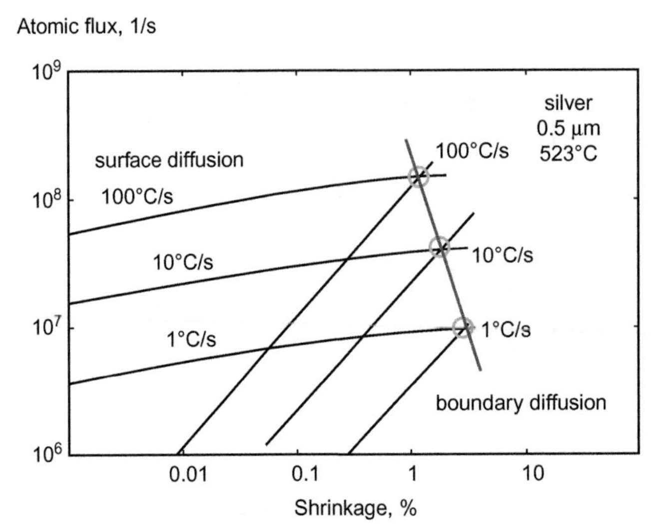

Figure 12.4 Calculations of the surface diffusion and grain boundary diffusion fluxes versus sintering shrinkage for 0.5 μm silver heated to 523°C (800 K) using three heating rates—1, 10, and 100°C/s [2]. The intersection of the two flux traces shows that densification by grain boundary diffusion starts earlier in the sintering process with faster heating rates.

shrinkage for three heating rates—1, 10, and 100°C/s [2]. The calculations are for 0.5 μm powder at 523°C (800 K). The points of intersection indicate where surface and grain boundary diffusion are equal. This locus of transition points at higher heating rates shows that grain boundary diffusion becomes dominant earlier. Fast heating leads to more grain boundary diffusion with concomitant sintering densification. Although dunk cycles are able to generate heating rates of 10°C/s, other approaches are able to better manipulate diffusion mechanisms with higher heating rates. This chapter mentions several of those options in the context of rapid heating trials.

Fast heating rates were part of the early sintering work with electric discharges dating from the early experiments to sinter refractory metal lamp filaments. Traditional furnaces were temperature limited, so direct electric current was a means to reach high temperatures. During these trials, there was no focus on heating rate effects. Indeed, temperature measurement was immature. Plasma and microwave rapid heating was subsequently applied to alumina, tungsten, urania, and zirconia [19–22].

The demonstrations generally show that sintered density improves with fast heating, as demonstrated in Figure 12.5 for titanium consolidated by spark sintering at 50 MPa pressure using three heating rates [23]. Because of the thermal mass from the die, heating rates peaked at 200°C/min or 3.3°C/s. The fastest heating shows an advantage.

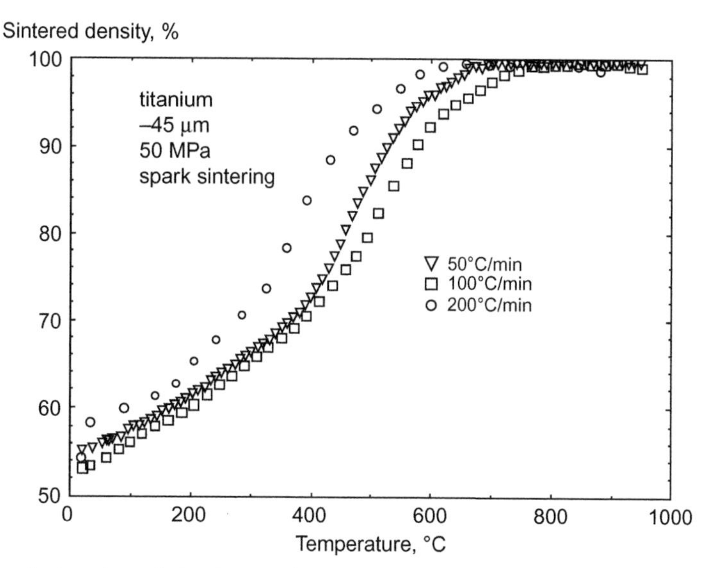

Figure 12.5 Spark sintering of titanium with 50 MPa applied pressure at three heating rates [23]. The faster heating rate shows some densification advantage.

NANOSCALE OPTIONS

Rapid heating experiments during the 1970s demonstrated much success with nanoscale powders [1,24]. The gains with smaller powders came from the high sintering stress. Nanoscale powders heated in dunk cycles produce full density in seconds. For example, 22 nm yttria doped zirconia powder reached 99% density in 60 s after immersion into a furnace at 1300°C (1573 K), giving a final grain size of 100 nm [25].

Rapid heating results in a smaller grain size in the sintered product, improving strength and hardness [26,27]. The effect is more pronounced with smaller particle sizes, as demonstrated with titania [24]. As plotted in Figure 12.6, sintering shrinkage at 700°C (973 K) declined as the particle size increased from 8 to 17 nm. For these trials, 0.2 g samples were subjected to heating rates estimated at 100°C/s. Extensive shrinkage occurs in 20 s for the smaller powder, but decreases with larger particles. Similar results are reported for a host of materials ranging from oxides to pure metals.

Nanoscale particles sinter to a high density with rapid heating. Indeed nanoscale powders show the most dramatic gain. However, at higher peak temperatures density decreases due to impurity evaporation after pore closure. An example is plotted in Figure 12.7 in terms of sintered density data for yttria stabilized zirconia heated at 500°C/min, or slightly more than 8°C/s [27]. The peak temperature was held for 1 min in each case. The sintered density increases up to 1300°C (1573 K), but at

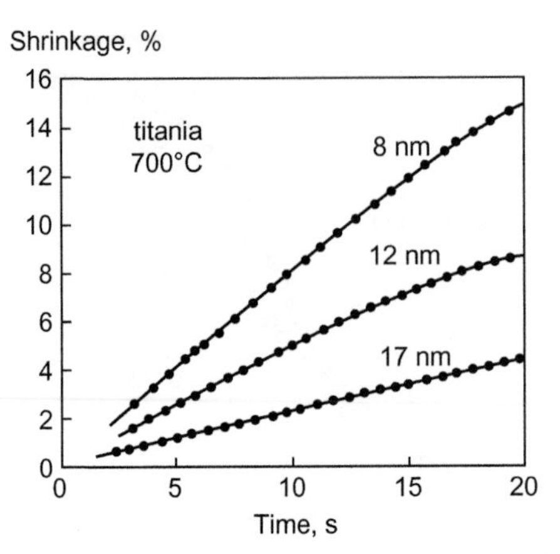

Figure 12.6 Particle size role in rapid heating is evident in this plot of sintering shrinkage at 700°C (973 K) for three particle sizes of titania (TiO_2) [24]. For the 8 nm powder significant shrinkage occurs in the first 20 s of sintering.

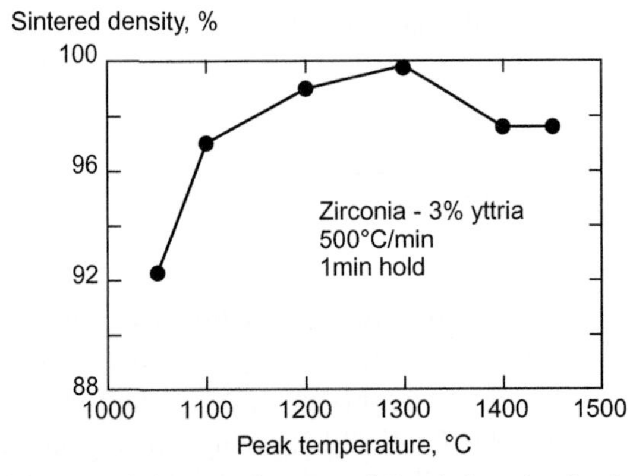

Figure 12.7 Data for the sintered density of yttria stabilized zirconia after heating at 500°C/min to various peak temperatures with a 1 min hold [27]. Higher temperatures decrease the sintered density due to trapped volatile species that expanded residual pores. Other studies find similar de-densification with longer times.

higher temperatures decreased. The vapor pressure of the trapped impurities generated a pore pressure that increases with temperature to offset densification. Trials showed better densification with slower heating (99.3% for 10°C/min versus 90.7% for 500°C/min). A calcination treatment to remove impurities prior to sintering was beneficial [28]. Conventional sintering resulted in a lower density and larger grain size, but did not require the calcination treatment. Small grain sizes and fast densification are evident gains with nanoscale powders, but contamination issues are an offsetting concern.

Thus, the combination of rapid heating and nanoscale particles brings a mixture of benefits and problems:

- High sintering stress to enhance densification.
- Less time at low temperatures where the high surface area promotes surface diffusion sintering without densification.
- More surface energy preserved to higher temperatures where grain boundary diffusion, dislocation climb, plastic flow, and other densification processes are active.
- More contamination with less time to volatilize the contaminants.
- Smaller grain size at density.
- Significant component warping or cracking due to thermal stresses.
- Component mass effects that restrict rapid heating to low mass or thin and flat compacts.

RAPID HEATING TECHNIQUES

Rapid heating in sintering is an area of much invention. The demonstrations are exciting, but the production of consistent components is difficult. Thus, other than for metastable materials (sintered diamonds, thermoelectric compounds, polymer-metal composites), where the time at temperature is short to minimize decomposition, traditional sintering has avoided rapid heating. Even so, inventions abound, ranging from self-heating exothermic processes to closely coupled plasma energy sources. Discussed below are several of the options.

EXOTHERMIC

An exothermic reaction between mixed powders is a means to rapidly heat the compact while simultaneously forming a compound. Portable welding relies on the thermite reaction of aluminum and iron oxide to generate molten iron. Once ignited, the reaction is rapid, so it is a means to rapidly heat a powder compact. One sacrificial reactant Mn is mixed with S. Once ignited to form MnS excess heat is available for heating an embedded powder compact. A variant is exothermic hot pressing where a pressure is applied during the reaction. Process control is always a concern since many of the cycles are not reproducible.

Alternatively, constituent powders are reacted to form a compound that is sintered by the heat of formation. The powders are mixed and compacted. Once ignited, the reaction induces adiabatic heating that gives sintering [29–31]. For example to produce sintered nickel aluminide NiAl, nickel and aluminum powders are mixed and compacted in an equal atomic ratio composition. The reaction initiates at near 600°C (873 K) and is self-heating to 1200°C (1473) or higher with a propagation velocity of a few mm/s. The velocity depends on green density and particle size. Thousands of systems are known, including aluminides (Ni_3Al), silicides ($MoSi_2$), carbides (TiC), nitrides (ZrN), and borides (MgB). Unlike other rapid heating approaches, heat is supplied from within the compact. Often a liquid phase forms to accelerate sintering densification. However, for large powder compacts the exothermic wave creates problems. The reaction propagates by passing heat through the surrounding material, similar to the way a fire advances. The combination of a thermal stress, sintering shrinkage, and weak unreacted structure leads to distortion. Thus, only in a few applications, such as sintering $MoSi_2$ heating elements, have been commercialized.

Electric Current

Electrical discharge heating was employed to lamp filaments. Subsequently, electric current discharge for sintering arose in sintering ferrite magnets with cycles lasting 100 s [32]. Heating rates of 10°C/s arose using electric current passage though graphite tooling. Such ideas spread to become a favorite means for rapid heating with simultaneous pressurization [33].

Flash sintering induced by electric current discharge is applied to ceramics, with energy inputs near 2 MJ/kg, which is lower than with traditional sintering. More recent applications have come with flash sintering of iron powder [34]. Iron powder of 60 μm particle size starting at 73% green density reached full density at 800°C (1073 K) in 6 min using a 600°C/s heating rate. Since iron is conductive, this required a current of nearly 13,000 A/cm^2. Along these lines, subsequent studies applied the idea to lower conductivity powder. One detailed report is on yttria stabilized zirconia using 60 nm particles pressed to 40% green density [3]. When a high voltage gradient (120 V/cm) and rapid heating to 850°C (1123 K) are applied, a small compact (3 mm by 1.58 mm thick) sinters to full density in 5 s. Since the starting material is a mixture of phases, the mechanism was enhanced grain boundary heating from the electric current. Some of the voltage gradient and temperature combinations for flash sintering are as follows:

120 V/cm – 850°C
100 V/cm – 910°C
60 V/cm – 1005°C.

Unfortunately many studies fail to give the whole story. For iron the voltage is not reported and in the study on zirconia the current density is not reported. Technological barriers and capital cost are widespread concerns in direct electric discharge sintering. Even so, intense electric fields accelerate sintering, probably by enhancing grain boundary diffusion, similar to how electromigration operates in electronic circuits [35].

Most devices for electric current sintering add pressure using a punch and die assembly, similar to hot pressing. Loading ensures particle contacts with the electrodes during shrinkage with creep densification as a mechanism. Pulsed direct current causes heating, with a typical pulse duration from 2 ms to 15 s. Longer cycles are more used in production and shorter cycles are employed in research. The current density and pulse duration depend on the material conductivity. If the powder is not conductive, the current passes through the tooling to give rapid hot pressing without an electric current benefit [36,37].

Metallic materials exhibit intense sintering when subjected to electrical discharge activation. As an example of the process, full density titanium results from a cycle of 230 A/cm^2 for 500 s with 14 MPa pressure, giving a peak temperature of 1150°C (1423 K). However, the graphite tooling contaminates some materials.

Plasma Discharge

Plasma discharges provide a novel environment for rapid heating [38]. Plasma discharge heating is applied to diamond composites and several oxide ceramics. Densification is rapid, showing 99% density for 50 nm alumina within about a minute after initiation of the cycle, as plotted in Figure 12.8 [39].

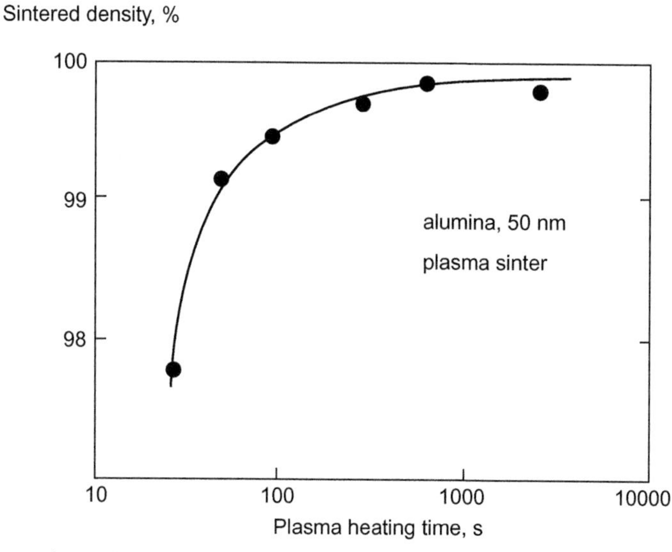

Figure 12.8 Plasma heated 50 nm alumina showing 99% density attained in about 60 s [39]. This study indicates a slight loss of sintered density at the longest time.

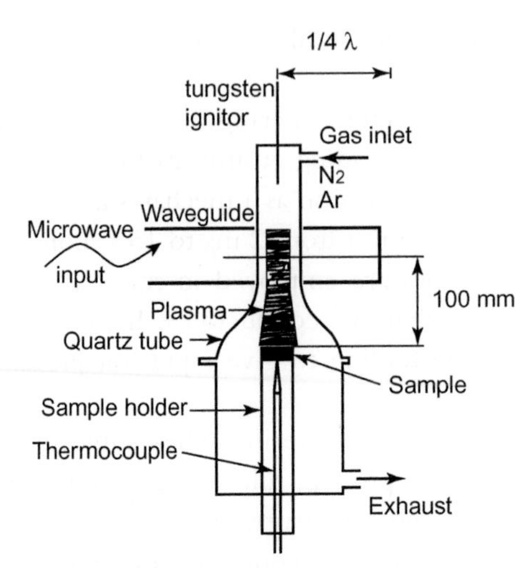

Figure 12.9 A schematic of how microwave energy is used to generate a plasma for sintering. Rapid heating generates thermal gradients in the powder to help promote sintering.

In plasma heating, the current-voltage situation is the opposite from that seen in electric current heating, since plasma discharge relies on a high voltage and low current. In many cases the plasma is generated by a microwave or induction field acting on a low pressure gas. The residual gas is stripped of electrons and the gas nucleus is accelerated through a voltage gradient to impact the plasma on the powder compact. Polyatomic gases such as nitrogen or hydrogen are common. A schematic set-up is sketched in Figure 12.9. The approach gives rapid heating of a single compact with low power loss. The current thinking is that the plasma assists densification via a steep temperature gradient from surface heating.

Plasma sintering delivers a higher sintered density at intermediate temperatures compared to conventional sintering, but at typical peak temperatures there is less advantage. Figure 12.10 compares density versus temperature for plasma and conventional sintering of alumina to illustrate the relative density performance [40]. For this plot, the comparison uses 40 to 50 nm alumina powder treated for 10 min by plasma sintering versus 2 h in conventional sintering.

Plasmas provide possible new mechanisms to drive sintering. The rapid densification is impressive, but other than for rod or tube sintering most of the demonstrations are for a single small compacts, one at a time. These are usually under 1 cm^3 in volume. Although the heating rate is impressive, still the throughput is under 1 kg/h.

Microwave

The advent of low cost home microwave ovens at 2.45 GHz frequency, corresponding to a 122 mm wavelength (designed to couple with water), generated interest in

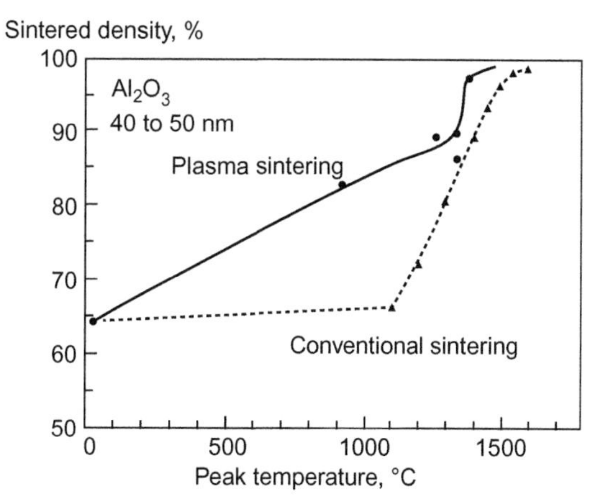

Figure 12.10 Sintered density of 40 to 50 nm alumina powder heated to various temperatures [40]. Plasma heating was for 10 min while conventional heating was for 120 min.

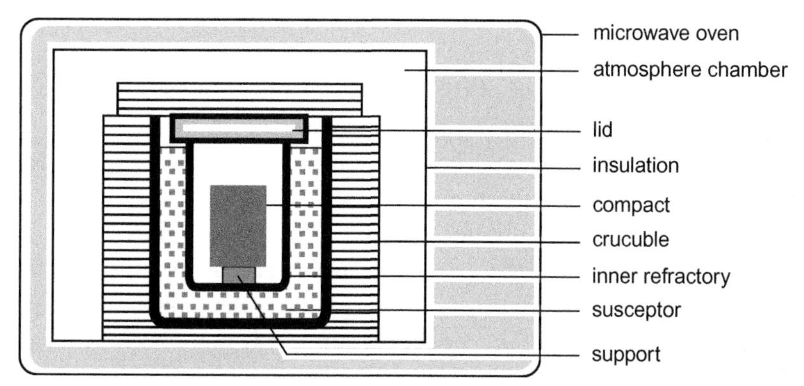

Figure 12.11 Microwave sintering is possible if the heat is properly managed. Sketched here is a profile of the microwave oven containing a coffin of insulation board to contain the radiant energy emission from the compact. A susceptor powder (usually SiC) is packed inside to absorb the microwave energy and the powder compact is either embedded in the susceptor or packaged in a crucible inside the susceptor.

microwave sintering. Reports emerged in the 1980s and included materials such as B_4C heated to 95% density in 12 min [41]. Microwave heating is also used to generate glow discharge plasmas. First efforts relied on microwave heating by direct coupling to the green compact, with subsequent demonstrations on many systems [42–46].

The low cost is the major benefit. For sintering, the home microwave is modified as illustrated in Figure 12.11. The powder to be sintered is positioned in the cavity and surrounded by protective and susceptor ceramics (usually silicon carbide). Microwave energy heats the susceptor and in turn it emits radiant heat to the compact. Small

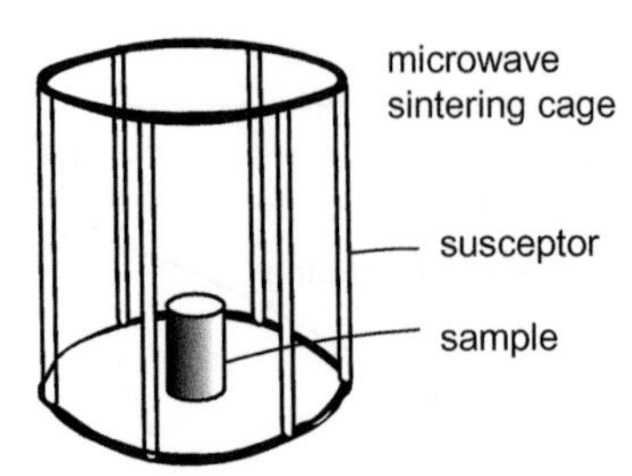

Figure 12.12 A susceptor cage is another way of ensuring uniform heating during microwave sintering. The microwave couples to the susceptor which heats the powder compact via radiant heating.

sample sintering is possible in just 2 min. For metals, the chamber is flooded with inert gas. To avoid damage to the oven, an outer layer of microwave-transparent ceramic prevents heat escape since the home microwave is not designed for high temperatures. The green body needs to be porous to allow impurity evaporation during heating.

Heating by direct coupling of the powder compact is possible, but coupling depends on the dielectric behavior of the material. A measure of effective polarization in an oscillating field is the loss tangent; a low loss tangent means the microwave passes through the material without heating. For many materials at low temperatures the loss tangent is low, but increases at high temperatures. Penetration depths for ceramics are often large, so ceramics are initially not effective in absorbing microwave energy. As temperature increases the penetration depth decreases and a critical condition occurs where absorption becomes rapid and temperature increases quickly. Susceptors help smooth the nonlinear coupling. The susceptor surrounds the component, and might be configured as a cage as sketched in Figure 12.12.

The wave pattern in the microwave cavity determines the uniformity of heating. Multiple-mode cavities avoid localized hot spots, meaning the cavity size is large compared to the wavelength of the electromagnetic field. For uniform heating the sample size and cavity size are matched and often the sample is translated inside the cavity to improve uniformity. Large components exhibit uneven heating, restricting microwave sintering to smaller components to avoid distortion. Since most ceramics exhibit surface heating with temperature gradients up to 20°C/mm, the protocol is to sinter one compact at a time.

A production microwave furnace is illustrated in Figure 12.13. Here microwave energy is delivered to one component at a time with the crucibles acting as susceptors when held in front of the microwave. As a demonstration of the laboratory microwave sintering, Figure 12.14 plots sintered density versus peak temperature for alumina [45]. Small disk samples were processed by both microwave and conventional sintering, where the microwave had zero hold time at temperature and the conventional

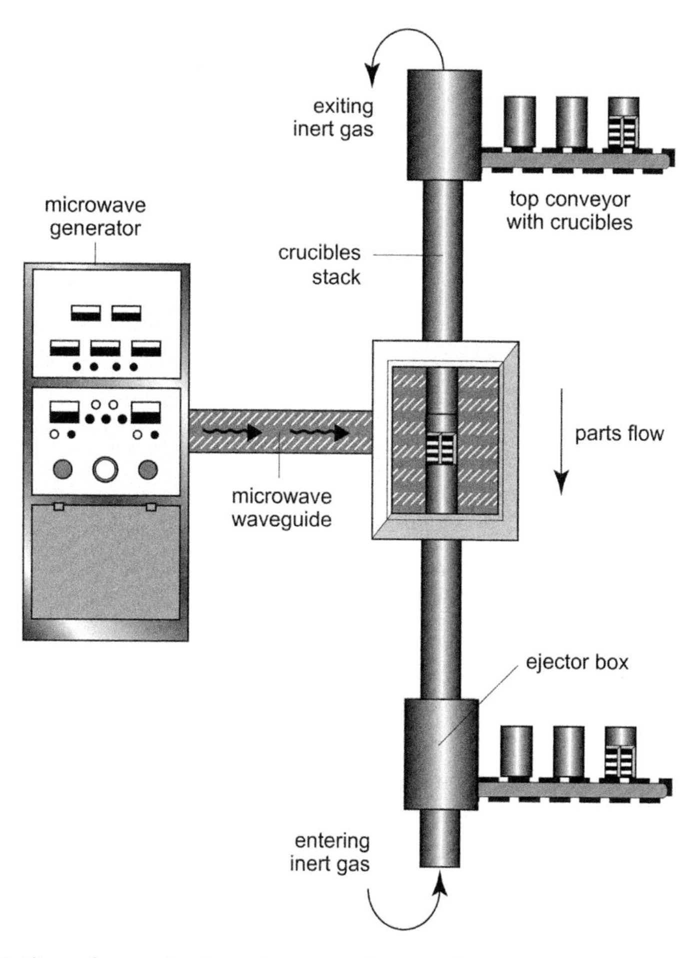

Figure 12.13 Outline of a production microwave sintering furnace, where a compact passes into the microwave zone. An elevator system conveys the susceptor crucibles into the microwave. In this unit, reaching 1400°C (1673 K), the cycle time was 8 min per compact.

cycle relied on 2 h hold at temperature. Substantial density gains are evident with microwave sintering.

The benefit from microwave heating begs the question on the sintering mechanism [47]. There is a general sense grain boundary impurities preferentially interact with the microwave energy, possibly giving higher temperature grain boundaries, thereby promoting faster grain boundary transport. Another thought is that thermal gradients induced by the microwave heating induce new diffusion effects from the surface to interior. In a direct comparison on sintered Fe-2Ni-0.8C steel processed at 1250°C, the strength was 1.8% higher for microwave sintering, but that came from

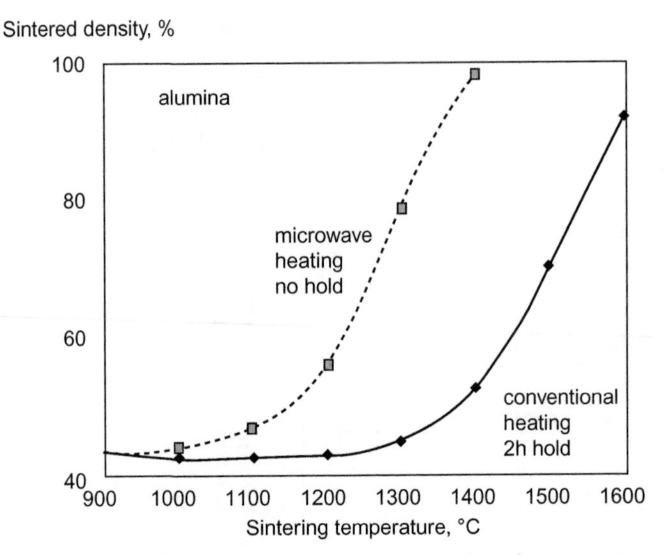

Figure 12.14 Sintered density of alumina versus the measured peak temperature for microwave heating (no hold) and conventional heating (2 h hold) [45].

higher carbon (0.81% versus 0.77% for conventional) due to more carbon loss in slower conventional heating [48].

Some reports on microwave sintering suggest more energy efficiency. However, energy loss from the microwave cavity makes the comparison difficult. In studies reporting higher properties after microwave sintering, usually the difference is from composition or cooling rate differences. For example, after microwave sintering WC-Co compositions are harder, but this is because of rapid cooling from the sintering temperature, not microwave heating. Rapid cooling is possible with conventional furnaces using gas flooding. Since microwave heating suffers a limitation of sintering one compact at a time it is not well matched to industrial sintering rates of 100 kg/h attained in conventional furnaces.

Light—Lasers and Infrared Heating

The high localized heat from a laser beam is employed in welding and surface glazing. Because of the power concentration, lasers provide a means to rapidly heat and sinter a small area. For example, Figure 12.15 plots sintered density for a laser sintered steel powder processed using a 0.4 mm diameter 200 W laser beam [49]. The longer the dwell time the higher the sintered density, although the effect diminishes after 2 ms. Initial sintering is rapid with the intense heating from the laser beam. Infrared heating is similar, except the light is not focused to one spot, but applied over a surface. This

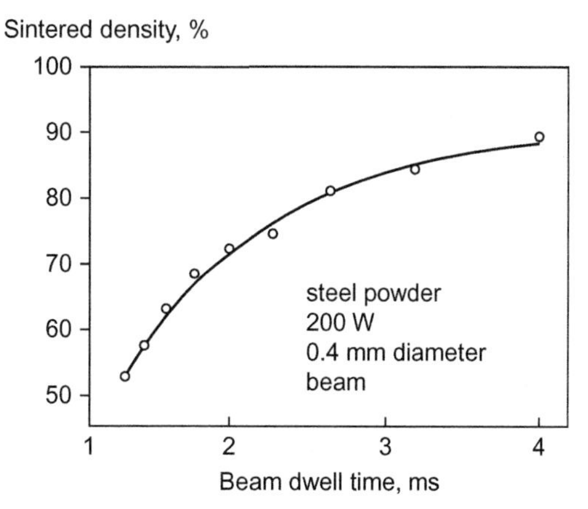

Figure 12.15 Laser sintered steel powder in a 200 W beam of 0.4 mm diameter, plotting sintered density versus beam dwell time [49]. These were small samples with no collateral heating from neighboring beam passes. In practice neighboring passes provide additional heat.

makes it suitable for thin structures, but not thick bodies [50]. Wide area heating units delivering 9 kW/cm^2 are used to sinter surface coatings.

A new use for laser sintering is additive manufacturing, where x-y stage motion is coordinated with laser sintering to build a three-dimensional object, one layer at a time [51]. Additive manufacturing started in the 1980s, initially with paper and plastic, but in the 1990s moved to metals, then ceramics [52]. A schematic of laser additive manufacturing is sketched in Figure 12.16. An image of the desired object is converted into a stack of two-dimensional layers. Powder is placed on the x-y stage and the laser beam (or electron beam) is applied to induce sintering. Heating is usually to the semisolid temperature range similar to supersolidus liquid phase sintering. Laser positioning is coordinated to the projected solid x-y-z coordinate. After each layer is processed, a fresh layer of powder is added and the cycle repeated. The time-temperature-particle size and related factors follow traditional sintering models [53]. The time to build an object is slow, although the local sintering rate is rapid. Beam energy, sweep velocity, and diameter determine the input energy, some of which is reflected off the powder, but the heat capacity of the powder and the sintering temperature determine the build rate.

By proper coordination the sintered object has the desired shape. Unfortunately, the layers are formed in steps, so rounded surfaces take on a step character as shown in Figure 12.17. Laser sintering units range from 8 W to 14,000 W beams with diameters in the 0.4 to 1 mm range. Sintering heating rates exceed 40°C/s. For plastic powder an 8 W unit is effective while for titanium the high power units are required [54,55]. Efforts with zirconia required preheating the powder bed to 800°C (1073 K) prior to laser heating to 1700°C (1973 K). If the laser traverse rate is too fast, then the

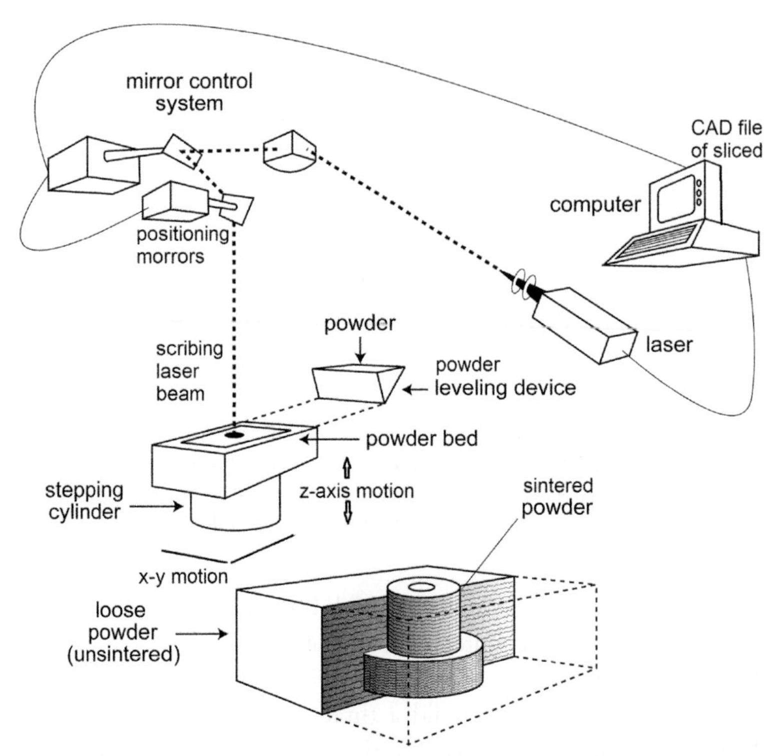

Figure 12.16 Additive manufacturing by laser sintering relies on a computer design file that is sliced to determine the laser rastering over the powder bed. A three-dimensional object is grown by repeated layering and laser rastering. The unsintered powder is removed after the build process.

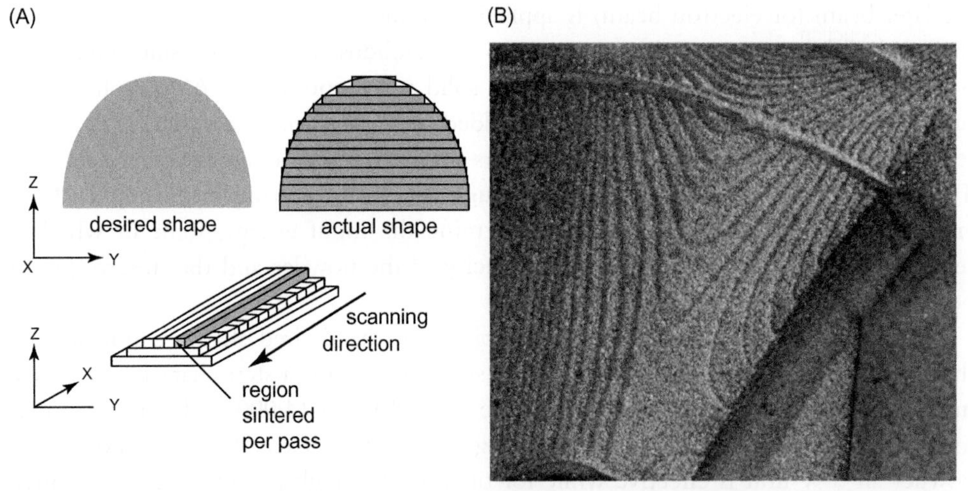

Figure 12.17 Additive manufacturing based on rapid sintering under a laser beam forms stair-step layer approximations to rounded surfaces, (A) illustrates the concept and (B) is a photograph of a curved surface with terraces.

powder fails to sinter and if the traverse rate is too slow resolution is lost. An option is to apply sufficient laser heating to sinter the body and then perform an additional sintering run [49]. Layer thickness is usually between 0.1 to 1 mm. The laser energy is concentrated on the surface so heating is poor below the surface. This results in substantial gradients in the sintered microstructure, density, and dimensions, often requiring top surface overheating to induce sintering on the bottom of the layer.

Induction

Hot pressing has long relied on induction heating. A high frequency alternating current is used to induce eddy currents in graphite tooling to rapidly heat the compact. Rapid heating is also possible without pressure. Demonstrations of nanoscale powder sintering reach heating rates of 15 to 20°C/s [56,57]. With supplemental pressure, slower heating occurs due to the added thermal mass of the tooling. Heating times are a few minutes, but cooling is slower.

Induction sintering surrounds the compact with a conductive coil that carries the alternating current. One design is sketched in Figure 12.18, where the outer copper coil is water cooled to surround a graphite susceptor. Atmosphere control is possible inside the refractory crucible, which is usually made from alumina. Current in the conduction coil creates a magnetic field that induces eddy currents in the susceptor and powder compact. With reversal of the current direction in the coil, the magnetic

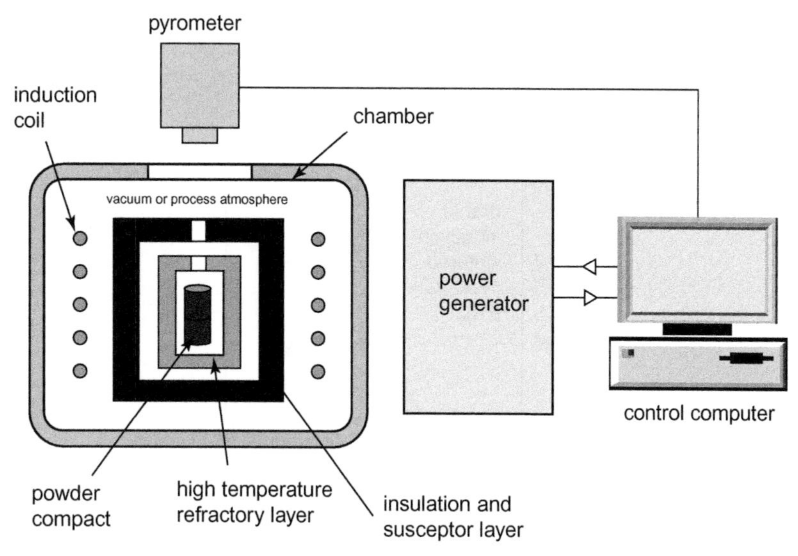

Figure 12.18 Induction heating relies on a water chilled copper coil that delivers a high frequency magnetic field to the powder compact. The cyclic current induces eddy current heating in the powder compact or in the susceptor layer.

field changes polarity, reversing the eddy current direction. Resistance to current passage causes heating.

The inductor coil design is customized to the material, sample size, and inductor frequency. Most trials rely on small cylinders or disks 12 mm in diameter [58]. The induction frequency runs from 50 Hz up to 50 kHz and some efforts have gone to radio frequencies (MHz). Heat transfer in induction heating is upwards of 3000 times better than radiant heating, although it is surface heating and restricted to one component at a time. An outer chamber contains the process atmosphere or vacuum.

The penetration depth limits the compact size to avoid cracking. As the temperature increases the penetration depth decreases. Most demonstrations find best properties at slower heating rates and shorter hold times. In spite of the rapid sintering, the technique is not used much in sintering practice, largely because of low productivity associated with sintering a single compact at a time. On the other hand, induction is widely employed in hot pressing, since heating with graphite tooling is common, allowing for good induction coupling.

A common observation is that sintering densification, properties, and quality are best with a heating rate of near 20°C/s [59]. Figure 12.19 illustrates this optimization for sintering 10 mm diameter urania (UO_2) disks and cylinders [60]. Different heating rates were used to the peak temperature of 1700°C (1973 K), which was held for 5 min. Faster heating cracked even these small compacts.

For materials that tend to outgas during heating, intermediate temperature holds are needed for volatilization of impurities prior to rapid heating. Several powder systems have been demonstrated. For example, for WC-15Co a sintered density of 97% density was attained in 1 min using 50 kHz frequency at 15 kW. Heating was at 20°C/s. When compared to spark sintering at 60 MPa, induction hot pressing gave a higher hardness [61].

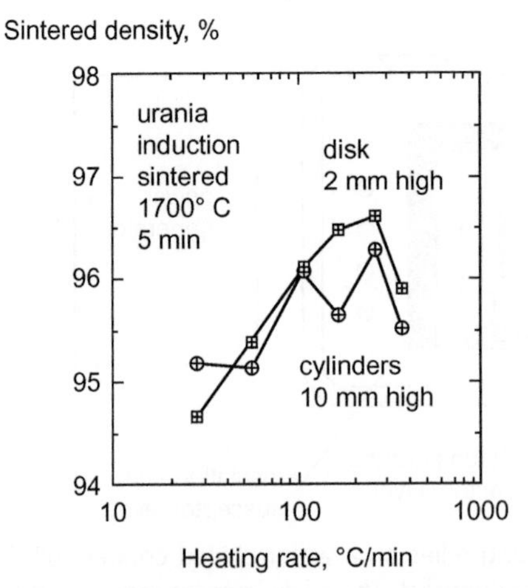

Figure 12.19 Induction sintered urania (UO_2) at 1700°C (1973 K) giving the sintered density for two test geometries after 5 min holds [60]. At heating rates over approximately 300°C/min the density degrades.

SUPPLEMENTAL PRESSURE

As with traditional sintering, supplemental pressure improves particle bonding and densification during rapid heating. A good example of the pressure effect is plotted in Figure 12.20 for zirconia heated at 200°C/min. The plot shows the sintered density using a 5 min consolidation cycle at various pressures for a temperature of 1200°C (1473 K) [33]. Noteworthy is the lack of grain growth during these cycles. These experiments used electrical discharge though graphite tooling surrounding the powder while applying pressure through upper and lower punches, as illustrated in Figure 12.21. This process is common in research laboratories and pressurized electric current sintering devices are sold by several firms.

It is reasonable to expect the combination of rapid heating and supplemental pressure to enhanced densification. Thus, an array of technologies emerges—such as microwave hot pressing, exothermic hot pressing, reactive hot isostatic pressing, spark plasma sintering, induction hot pressing, rapid omnidirectional compaction, and such. This area is an intersection of sintering with forging and related material forming technologies. As such some of the rapid heating and rapid pressurization options revisit old metal forming ideas [62–73]. Exotic approaches are justifications for research funding, but in production these ideas prove difficult to control. Pressurization during rapid heating requires attention to process control to give repeatable densification.

At slower densification rates it is possible to control low strain rate pressurization. For example, exothermic heating to induce rapid sintering is combined with simultaneous

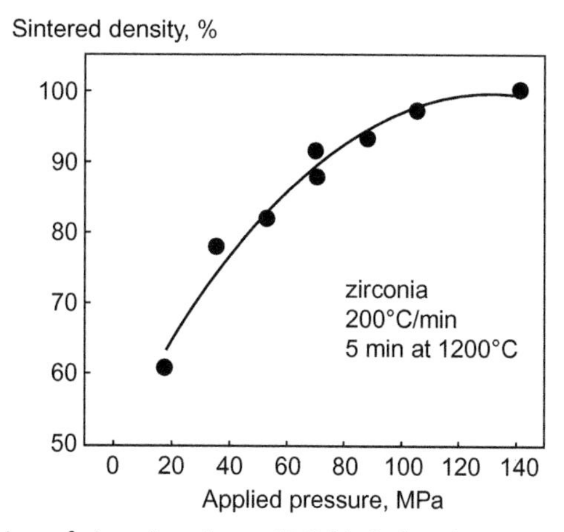

Figure 12.20 Spark sintering of zirconia using a 200°C/min heating rate to 1200°C (1473 K) with a 5 min hold [33]. Sintered density increases with the applied pressure, while the grain size remains relatively constant near 100 nm.

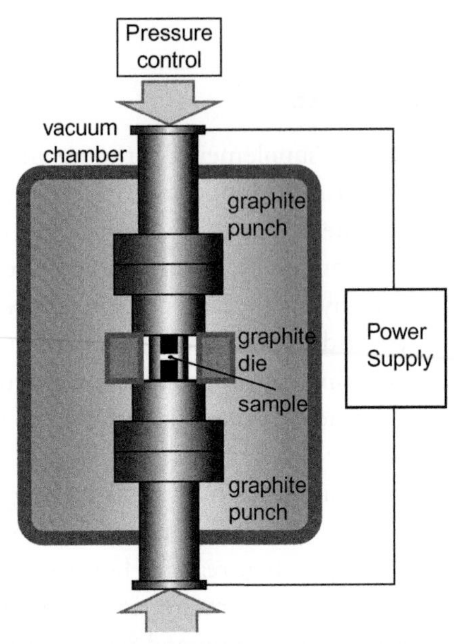

Figure 12.21 Outline of the spark sintering process. The power supply provides both direct current and alternating current to heat the graphite and powder. Pulsed direct current heating is most typical. External force is generated in the graphite punches. For nonconductive powder, current passes only in the tooling, but for a conductive powder the current also passes through the powder.

pressurization. When performed using hot pressing, it is termed exothermic hot pressing and when performed using hot isostatic pressing it is termed reactive hot isostatic pressing. The reaction enthalpy produces heat to the sintering systems. Applications include $MoSi_2$, HfB_2, TiAl, Nb_3Al, TiB_2, and Ni_3Al. The reaction is from mixed powders to form the compound, such as $Mo + 2\ Si \rightarrow MoSi_2 + heat$. Composites are formed by adding reinforcing phases to the reaction species [33,74]. The mixed powders are heated to initiate the reaction and pressurized to densify the product [75].

Pressure applied during rapid sintering aids consolidation, but typically the pressurization pulse needs to delay after the heating pulse. This delay avoids inhomogeneities in the product microstructure. But high strain rates can cause component cracking. The difficulty in controlling and scaling up the rapid heating processes is exacerbated when an external pressure is applied.

PROSPECTS

Rapid heating often provides improved sintering. Several approaches have reached the demonstration stage and show a few characteristic features:

- The components need to be small to respond to rapid energy input.
- The powders need to be small, even nanoscale, to undergo rapid densification.
- Heating rates tend to optimize at about 20°C/s.
- Longer hold times and higher peak temperatures often induce a loss of density.
- Warpage, cracking, and defects are common, especially at higher heating rates.
- Powders must be degassed and polymers removed prior to rapid sintering.
- Most of the rapid heating approaches process one compact at a time.
- Because of cost most of the demonstrations failed to reach production status.
- Equipment cost is often high when measured per kg of sintered material.

Much effort is devoted to developing new sintering schemes and these often return to concepts demonstrated long ago. Rapid processing of a single compact shows significant gains in density, microstructure, and processing times, with shrinkage rates of 1%/s reported in some cases. Such attributes are most useful for metastable materials.

Green powder structures are not very good heat conductors. Consequently, rapid heat delivery induces thermal gradient driven diffusion usually missing in sintering. Rapid heating also generates stresses that activate dislocation flow. But for components over about 10 mm in thickness, the thermal stresses cause damage in the form of cracking or warping. Accordingly, several demonstrations of amazingly fast sintering have not been able to scale to large sections. A few approaches are successful for diamond composites, tellurium thermoelectric compounds, and lead-titanate electronic ceramics. But, for mainstream sintering the rapid heating techniques prove hard to control in production. The one exception is additive manufacturing via laser sintering. This approach is making inroads into limited production situations, such as fabrication of customized dental or medical restorations.

The mechanism of accelerated sintering is uncertain. Most likely impurity roles are amplified in rapid heating. If so, then composition can be tailored to take advantage of the novel heating technique.

Rapid heating opens new pathways to sintering with minimized microstructure coarsening. Sintering one compact at a time is restrictive in terms of cost. Traditional sintering furnaces reach upwards of tons per day in output. So far this level of productivity is not demonstrated using rapid heating schemes. The pending research on rapid heating concepts needs to focus on optimizing the cycles, powders, and impurities. Additionally, proper energy audits are needed to assess property and microstructure gains versus traditional equipment and processing costs.

REFERENCES

[1] P. Vergnon, M. Astier, S.J. Teichner, Initial stage for the sintering of ultrafine particles (TiO_2 and Al_2O_3), in: W.E. Kuhn (Ed.), Fine Particles, The Electrochemical Society, Princeton, NJ, 1974, pp. 299–307.

[2] D.L. Johnson, Ultra rapid sintering, in: G.C. Kuczynski, A.E. Miller, G.A. Sargent (Eds.), Sintering and Heterogeneous Catalysis, Plenum Press, New York, NY, 1984, pp. 243–252.

[3] M. Cologna, B. Rashkova, R. Raj, Flash sintering of nanograin zirconia in less than 5 s at 850 C, J. Amer. Ceram. Soc. 93 (2010) 3556–3559.

[4] M. Cologna, J.S.C. Francis, R. Raj, Field assisted and flash sintering of alumina and its relationship to conductivity and MgO Doping, J. Europ. Ceram. Soc. 31 (2011) 2827–2837.

[5] D. Beruto, R. Botter, A.W. Searcy, Influence of temperature gradients on sintering: experimental tests of a theory, J. Amer. Ceram. Soc. 72 (1989) 232–235.

[6] A.W. Searcy, D. Beruto, Theory and experiments for isothermal and non isothermal sintering, in: D. Taylor (Ed.), Science of Ceramics, vol. 14, The Institute of Ceramics, Stokes-on-Trent, UK, 1988, pp. 1–13.

[7] C.S. Morgan, Observation of dislocations in high temperature sintering, High Temp. High Press. vol. 3 (1971) 317–324.

[8] M.P. Harmer, R.J. Brook, Fast firing – microstructural benefits, J. Brit. Ceram. Soc. 80 (1981) 147–148.

[9] R.M. Young, R. McPherson, Temperature-gradient-driven diffusion in rapid-rate sintering, J. Amer. Ceram. Soc. 72 (1989) 1080–1081.

[10] A.W. Searcy, Theory for sintering in temperature gradients: role of long range mass transport, J. Amer. Ceram. Soc. 70 (1987) C61–C62.

[11] D.J. Chen, M.J. Mayo, Rapid rate sintering of nanocrystalline ZrO_2 – 3 mol.% Y_2O_3, J. Amer. Ceram. Soc. vol. 79 (1996) 906–912.

[12] H. Su, D.L. Johnson, Sintering of alumina in microwave-induced oxygen plasma, J. Amer. Ceram. Soc. 79 (1996) 3199–3210.

[13] M.P. Kassarjian, B.H. Fox, J.V. Biggers, Fast firing of a lead-iron niobate dielectric ceramic, J. Amer. Ceram. Soc. 68 (1985) C140–C141.

[14] R.H.R. Castro, G.J. Pereira, D. Gouvea, Relationship between surface segregation and fast densification of Fe or Mg doped SnO_2 pellets, in: D. Bouvard (Ed.), Proceedings of the Fourth International Conference on Science, Technology and Applications of Sintering, Institut National Polytechnique de Grenoble, Grenoble, France, 2005, pp. 394–397.

[15] Z.A. Munir, D.V. Quach, M. Ohyanagi, Electric current activation of sintering: a review of the pulsed electric current sintering process, J. Amer. Ceram. Soc. 94 (2011) 1–19.

[16] D.M. Hulbert, A. Anders, J. Andersson, E.J. Lavernia, A.K. Mukherjee, A discussion on the absence of plasma in spark plasma sintering, Scripta Mater. 60 (2009) 835–838.

[17] P. Wray, New paradigm prophecy, Ceram. Bull. 92 (3) (2013) 28–33.

[18] S.M. Landin, W.A. Schulze, Rapid sintering of stoichiometric zinc-modified lead magnesium niobate, J. Amer. Ceram. Soc. 73 (1990) 913–918.

[19] C.E.G. Bennett, N.A. McKinnon, L.S. Williams, Sintering in gas discharge, Nature 217 (1968) 1287.

[20] G.E. Tardiff, On the sintering behavior of submicron tungsten-thoria powder blends, Inter. J. Powder Metall. 5 (4) (1969) 29–39.

[21] K. Upadhya, Sintering kinetics of ceramics and composites in the plasma environment, J. Met. 39 (12) (1987) 11–13.

[22] P.C. Kong, Y.C. Lau, E. Pfender, K. Mchenry, W. Wallenhorst, B. Koepke, Sintering of fully stabilized zirconia powders in rf plasmas, in: G.L. Messing, E.R. Fuller, H. Hausner (Eds.), Ceramic Transactions, vol. 1, Amer. Ceramic Society, Westerville, OH, 1987, pp. 939–946.

[23] M. Eriksson, Z. Shen, M. Nygren, Fast densification and deformation of titanium powder, Powder Metall. 48 (2005) 231–236.

[24] P. Vergnon, M. Astier, S.J. Teichner, Sintering of submicron particles of metallic oxides, in: G.C. Kuczynski (Ed.), Sintering and Related Phenomena, Plenum Press, New York, NY, 1973, pp. 301–310.

[25] C. Feng, H. Qiu, J. Guo, D. Yan, W.A. Schulze, Fast firing of nanoscale $ZrO_2 + 2.8$ mol.% Y_2O_3 ceramic powder synthesized by the sol-gel process, J. Mater. Syn. Proc. 3 (1995) 25–29.

[26] C.E. Baumgartner, Fast firing and conventional sintering of lead zirconate titanate ceramic, J. Amer. Ceram. Soc. 71 (1988) C350–C353.

[27] C. Feng, E. Shi, J. Guo, D. Yan, W.A. Schulze, Characterization and fast firing of nanoscale $ZrO_2 + 3$ mol% Y_2O_3 ceramic powder prepared by hydrothermal processing, J. Mater. Syn. Proc. 3 (1995) 31–37.
[28] D.H. Kim, C.H. Kim, Entrapped gas effect in the fast firing of yttria-doped zirconia, J. Amer. Ceram. Soc. 75 (1992) 716–718.
[29] J.B. Holt, D.D. Kingman, G.M. Bianchini, Kinetics of the combustion synthesis of TiB_2, Mater. Sci. Eng. 71 (1985) 321–327.
[30] R.L. Coble, Reactive sintering, in: D. Kolar, S. Pejovnik, M.M. Ristic (Eds.), Sintering – Theory and Practice, Elsevier Scientific, Amsterdam, Netherland, 1982, pp. 145–151.
[31] A. Varma, Combustion synthesis of advanced materials, Chem. Eng. Educ. Winter (2001) 14–21.
[32] A. Sawaoka, S. Saito, Fast sintering of ferrites using high frequency current under pressure, in: Y. Hoshino, S. Iida, M. Sugimoto (Eds.), Ferrites, University Park Press, Baltimore, MD, 1970, pp. 102–104.
[33] Z.A. Munir, U. Anselmi-Tamburini, M. Ohyanagi, The effect of electric field and pressure on the synthesis and consolidation of materials, a review of the spark plasma sintering method, J. Mater. Sci. 41 (2006) 763–777.
[34] K. Feng, Y. Yang, B. Shen, L. Guo, H. He, Rapid sintering of iron powders under action of electric field, Powder Metall. 48 (2005) 203–204.
[35] C.M. Hsu, D.S.H. Wong, S.W. Chen, Generalized phenomenological model for the effect of electromigration on interfacial reaction, J. Appl. Phys. 102 (2007) article 023715.
[36] B. Bernard-Granger, A. Addad, G. Fantozzi, G. Bonnefont, C. Guizard, D. Vernat, Spark plasma sintering of a commercially available granulated zirconia powder: comparison with hot pressing, Acta Mater. 58 (2010) 3390–3399.
[37] G. Bernard-Granger, N. Monchalin, C. Guizard, Comparison of grain size – density trajectory during spark plasma sintering and hot pressing of zirconia, Mater. Lett. 62 (2008) 4555–4558.
[38] L.G. Cordone, W.E. Martinsen, Glow-discharge apparatus for rapid sintering of Al_2O_3, J. Amer. Ceram. Soc. 55 (1972) 380.
[39] E.L. Kemer, D.L. Johnson, Microwave plasma sintering of alumina, Ceram. Bull. 64 (1985) 1132–1136.
[40] S. Sano, K. Oda, Y. Shibasaki, T. Matayoshi, Y. Kayama, Y. Setsuhara, et al., Microwave plasma sintering of low soda alumina, J. Japan Soc. Powder Powder Metall. 41 (1994) 739–741.
[41] J.D. Katz, R.D. Blake, J.J. Petrovic, H. Sheinberg, Microwave sintering of boron carbide, Met. Powder Rept. 43 (1988) 835–837.
[42] J.D. Katz, Microwave sintering of ceramics, Ann. Rev. Mater. Sci. 22 (1992) 153–170.
[43] D.E. Clark, W.H. Sutton, Microwave processing of materials, Ann. Rev. Mater. Sci 26 (1996) 299–331.
[44] Y.V. Bykov, K.I. Rybakov, V.E. Semenov, High-temperature microwave processing of materials, J. Phys. D: Appl. Phys 34 (2001) R55–R75.
[45] D. Agrawal, Microwave sintering of ceramics, composites, and metal powders, in: Z.Z. Fang (Ed.), Sintering of Advanced Materials, Woodhead Publishing, Oxford, UK, 2010, pp. 222–248.
[46] S. Takayama, G. Link, S. Miksch, M. Sato, J. Ichikawa, M. Thumm, Millimeter wave effects on sintering behavior of metal powder compacts, Powder Metall. 49 (2006) 274–280.
[47] G.F. Zu, I.K. Lloyd, Y. Carmel, T. Olorunyolemi, O.C. Wilson, Microwave Sintering of ZnO at Ultra High Heating Rates, J. Mater. Res. 16 (2001) 2850–2858.
[48] M.J. Yang, R.M. German, Comparison of conventional sintering and microwave sintering of two ferrous alloys, Advances in Powder Metallurgy and Particulate Materials – 1999, vol. 1, Metal Powder Industries Federation, Princeton, NJ, 1999, pp. 3.207–3.219.
[49] A. Simchi, F. Petzoldt, H. Pohl, H. Loffler, Direct laser sintering of a low alloy P/M steel, P/M Sci. Tech. Briefs 3 (2001) 5–9.
[50] J.D.K. Rivard, A.S. Sabau, C.A. Blue, D.C. Harper, J.O. Kiggans, Modeling and processing of liquid phase sintered Gamma-TiAl during high density infrared processing, Metall. Mater. Trans. 37A (2006) 1289–1299.

[51] M. Agarwala, D. Bourell, J. Beaman, H. Marcus, J. Barlow, Direct selective laser sintering of metals, Rapid Proto. J. 1 (1995) 26–36.
[52] D.E. Bunnell, D.L. Bourell, H.L. Marcus, Solid freeform fabrication of powders using laser processing, Advances in Powder Metallurgy and Particulate Materials – 1996, Metal Powder Industries Federation, Princeton, NJ, 1996, pp. 15.93–15.106.
[53] D.L. Bourell, J.J. Beaman, Powder material principles applied to additive manufacturing, Materials Processing and Interfaces, vol. 1, Proceedings 141st Meeting the Minerals, Metals, and Materials Society, Warrendale, PA, 2012, pp. 537–544.
[54] T.B. Sercombe, The production of functional aluminum prototypes using selective laser sintering, P/M Sci. Tech. Briefs 3 (6) (2001) 22–25.
[55] F.G. Arcella, F.H. Froes, Producing titanium aerospace components from powder using laser forming, J. Met. (2000, May) 28–30.
[56] H.C. Kim, I.J. Shon, J.K. Yoon, J.M. Doh, Z.A. Munir, Rapid sintering of ultrafine WC-Ni Cermets, Inter. J. Refract. Met. Hard Mater. 24 (2006) 427–431.
[57] H.C. Kim, D.Y. Oh, I.J. Shon, Sintering of Nanophase WC-15 vol.% Co hard metals by rapid sintering process, Inter. J. Refract. Met. Hard Mater. 22 (2004) 197–203.
[58] W. Hermel, G. Leitner, R. Krumphold, Review of induction sintering: fundamentals and applications, Powder Metall. 23 (1980) 130–135.
[59] M. Nakamura, H. Takahashi, Y. Sugaya, Rapid sintering of high strength alloyed steels by induction heating method, J. Japan Soc. Powder Powder Metall. 49 (2002) 534–540.
[60] H.H. Yang, Y.W. Kim, J.H. Kim, D.J. Kim, K.W. Kang, Y.W. Rhee, et al., Pressureless rapid sintering of UO_2 assisted by high-frequency induction heating process, J. Amer. Ceram. Soc. 91 (2008) 3202–3206.
[61] H.C. Kim, I.K. Jeong, I.J. Shon, I.Y. Ko, J.M. Doh, Fabrication of WC-8 Wt.% Co hard materials by two rapid sintering processes, Inter. J. Refract. Met. Hard Mater. 25 (2007) 336–340.
[62] W.D. Jones, Fundamental Principles of Powder Metallurgy, Edward Arnold Publishers, London, UK, 1960.
[63] C.G. Goetzel, Treatise on Powder Metallurgy, vol. 1, Interscience Publishers, New York, NY, 1949, pp. 259–312.
[64] Y. Miyamoto, M. Koizumi, O. Yamada, High-pressure self-combustion sintering for ceramics, J. Amer. Ceram. Soc. 67 (1984) C224–C225.
[65] H.L. Marcus, D.L. Bourell, Z. Eliezer, C. Persad, W. Weldon, High-Energy, High-rate materials processing, J. Met. 39 (12) (1987) 6–10.
[66] S.T. Lin, R.M. German, Mechanical properties of fully densified injection molded carbonyl iron powder, Metall. Trans. 21A (1990) 2531–2538.
[67] Y. Murakoshi, M. Takahashi, K. Hanada, T. Sano, H. Negishi, Dynamic powder compaction method by electromagnetic force, J. Japan Soc. Powder Powder Metall. 48 (2001) 565–570.
[68] B.K. Yen, T. Aizawa, K. Kihara, Reaction synthesis of titanium silicides via self-propagating reaction kinetics, J. Amer. Ceram. Soc. 81 (1998) 1953–1956.
[69] J.H. Lee, N.N. Thadhani, H.A. Grebe, Reaction sintering of shock-compressed Ti + S powder mixtures, Metall. Mater. Trans. 27A (1996) 1749–1759.
[70] I. Sato, A. Hibino, H. Negishi, Hot electromagnetic forming of metal powder compacts and its application to combustion synthesis, J. Japan Soc. Powder Powder Metall. 42 (1995) 283–288.
[71] T. Aizawa, S. Kamenosono, J. Kihara, T. Kato, K. Tanaka, Y. Nakayama, Shock reactive synthesis of TiAl, Intermetallics 3 (1995) 369–379.
[72] I. Song, N.N. Thadhani, Synthesis of Nickel-Aluminum intermetallic compounds by shock-induced chemical reactions, J. Mater. Syn. Proc. 1 (1993) 347–358.
[73] T. Takeuchi, M. Takahashi, K. Ado, N. Tamari, K. Ichikawa, S. Miyamoto, et al., Rapid preparation of lead titanate sputtering target using spark plasma sintering, J. Amer. Ceram. Soc. 84 (2001) 2521–2525.
[74] D.E. Alman, J.A. Hawk, C.P. Dogan, TiAl-Based composites produced by SHS/reactive synthesis techniques, Advances in Powder Metallurgy and Particulate Materials, Metal Powder Industries Federation, Princeton, NJ, 1995, pp. 7.175–7.185.
[75] K. Morsi, The diversity of combustion synthesis processing, A review, J. Mater. Sci. 47 (2012) 68–92.

CHAPTER THIRTEEN

Nanoscale Sintering

OVERVIEW

Nanoscale particles are smaller than 100 nm or 0.1 μm. They are high in surface area and widely used as discrete particles in paints, catalysts, polymer additives, obscuration smokes, inks, pastes, and emulsions. Tremendous quantities are used in white paint (titania) and ultraviolet blocking additives for tires (carbon black). Most of the compositions are pure metals or metal oxides generated by thermal reduction or chemical precipitation. Some of the efforts produce complex compositions, such as WC-Co or Ti(C,N)-WC-Co. Nanoscale particle sintering fosters much hope that new powders will deliver new properties [1].

Early reports on nanoscale particle sintering emerged in the 1970s [2−5]. Subsequently, the research effort intensified to a current publication rate exceeding 10,000 articles each year. A chief focus is on the difficulties in balancing densification and microstructure coarsening. Indeed, nanoscale powders behave in new ways and even sinter at room temperature [6].

Solutions to the densification-coarsening challenge start by manipulating particle agglomeration, usually followed by high pressure compaction, and pressure-assisted sintering [7−11]. Practical difficulties abound; for example nanoscale powders delaminate when subjected to high compaction pressures. Figure 13.1 is a picture of nanoscale copper powder after compaction and sintering. The compact delaminated into several pieces. Further difficulty comes from the sponge-like character of many powders. This causes a large relaxation after pressing, separating the particles on shear planes and these separations grow to become large cracks after sintering.

With attention being given to improved powder synthesis, more nanoscale powder compositions are becoming available to support the study of sintering. The current efforts fall into a number of categories:

- Demonstrations of nanoscale sintering, with focus on density versus time, temperature, or other variables.
- Novel sintering trials, such as microwave heating, plasma heating, spark sintering, or hot isostatic pressing, geared to "seeing what happens."
- Microstructure control in sintering, usually with additives to control coarsening.
- Use of nanoscale particles in the fabrication of sintered microminiature components.

Sintering: From Empirical Observations to Scientific Principles
DOI: http://dx.doi.org/10.1016/B978-0-12-401682-8.00013-6

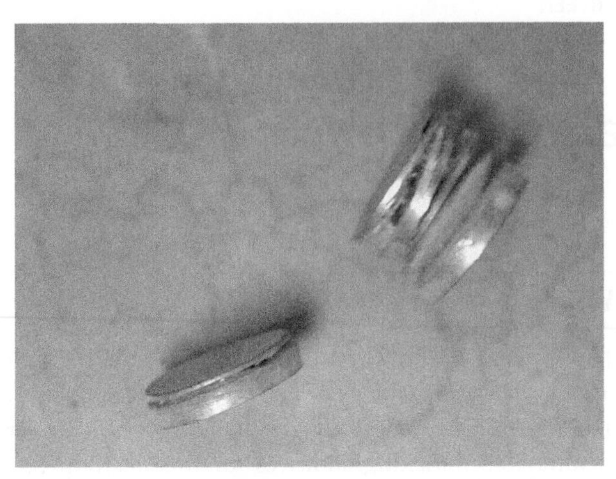

Figure 13.1 Delamination of a 12 mm diameter cylinder pressed and sintered using nanoscale copper powder.

- Flash sintering of thin printed particle films to form complex electrical circuits.
- Supposition about sintering mechanisms based on molecular dynamic simulations.

It is hoped that two of these efforts will lead to significant commercial gains by supporting microminiature device fabrication and new printed electronic circuits. Small features on sintered devices require small particles. For example, the microarray in Figure 13.2 is fabricated by sintering micrometer sized particles. Applications for such devices are found in blood analysis, genetic mapping, and cancer diagnosis. As a reference point, red blood cells range in size from 2 to 7 μm. Component features, such as wall thickness, need to be 20 times the particle size. For a sintered wall thickness of 1 μm this mandates particles smaller than 50 nm. If a medical device is fabricated to operate on the size range of red blood cells, then nanoscale powders are necessary. Many applications are using devices on the microminiature scale, with applications ranging from semiconductor assembly tools to plasma discharge lighting envelopes. Already, significant patents and commercial efforts are being seen in medical dialysis arrays, medical positron emission tomography collimators, micro-drills, heart implant electrodes, minimally invasive biopsy and surgical tools, and consumer electronics. Often the design features are limited by the available powder.

Material properties depend on microstructure, especially for sintered nanoscale materials [1,12]. The size-dependent property changes range from catalytic activity to vapor pressure. An old relation links yield strength to grain size, known as the Hall-Petch relation for yield strength σ_Y and grain size G:

$$\sigma_Y = \sigma_O + \frac{\varphi}{\sqrt{G}} \tag{13.1}$$

Figure 13.2 Microdevices such as depicted here depend on nanoscale particles to form target structures that include fluidic devices, microarrays, blood analysis, and purification devices.

where σ_O is the dislocation flow stress for the material and φ is a constant. Smaller grains greatly improve strength. Similar relations exist for hardness, as illustrated in Figure 13.3 [13]. However, it is not valid to extend the Hall-Petch relation to grain sizes below 10 nm, but in the range plotted here grain size hardening is evident. Thus, a large drive in developing nanoscale sintered sturctures is in the promise of property gains. For example, in the WC-Co system, hardness increases by 20% as the grain size decreases from 0.8 to 0.2 μm. Alternatively, realizing a 20% gain in hardness is possible by alloying without a grain size reduction [14].

In spite of much effort, it has not been possible to fabricate a 100 nm grain size in 100% dense WC-Co. This is because of rapid grain coarsening during sintering, even with sintering times of just 5 min. On the other hand, the addition of a second phase is successful in pinning the microstructure to induce a 50% hardness gain [15]. From a pragmatic point of view, composition development appears to be a more fruitful approach to improving properties than microstructure refinement via nanoscale sintering.

Much speculation has arisen about the improved sintering of nanoscale particles, since the interface area increases dramatically as the particle size decreases. Molecular

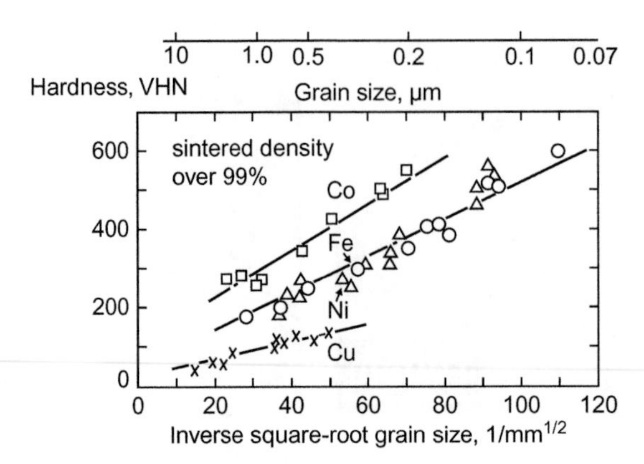

Figure 13.3 Microhardness (VHN) plotted versus the inverse square-root of the sintered grain size for several metals, illustrating agreement with the Hall-Petch relation [13].

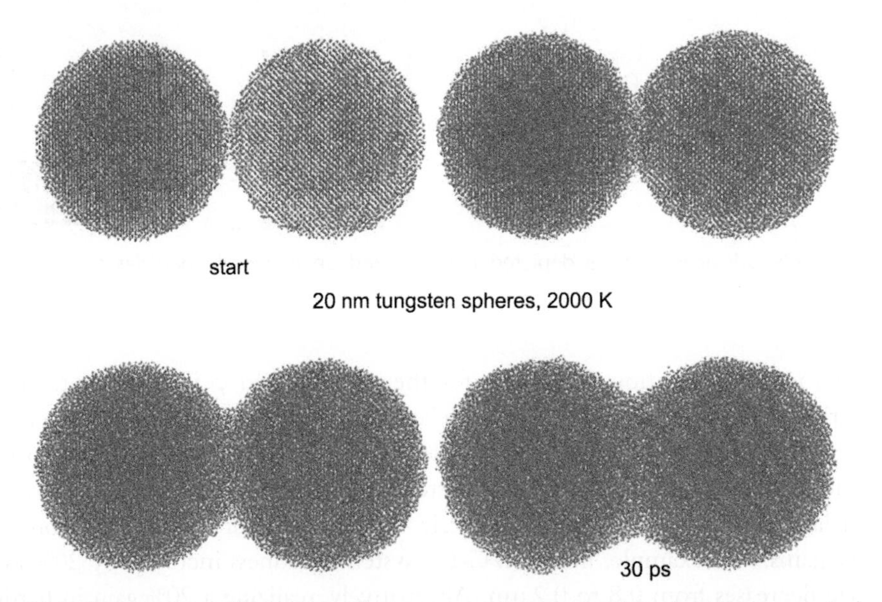

Figure 13.4 Computer graphic images of tungsten nanoscale particle sintering computed using molecular dynamics. The particles are 20 nm in diameter, each consisting of 33,000 atoms, and during heating sinter bond largely by surface diffusion [16].

dynamic simulations illustrate very rapid neck growth rates for nanoscale particles. For example, computed images of 20 nm tungsten particles during sintering are shown in Figure 13.4, and correspond to a total time of $30 \cdot 10^{-9}$ s. Coalescence of two particles requires 50 ps at 2727°C (3000 K) [16]. The neck growth rate and activation energy both agree with surface diffusion controlled sintering, as expected with the high surface area. The molecular dynamics show sintering activity is localized at the particle surface. Accordingly, no new mechanism is evident beyond known processes [17–28].

The sintering stress increases as particle size decreases. Sintering rates respond to these changes, so nanoscale sintering is very rapid, but likewise grain coarsening is rapid too. But problems arise with nanoscale particles [1,8]. For example, particle contamination is higher and proves to be more difficult to remove. For large particles, impurity evaporation is possible prior to sintering because the particle sintering requires a high temperature, but for small particles sintering occurs prior to reaching a temperature that induces evaporation. Accordingly, the contamination could well mask any property gain [13].

The fundamental question pivots around sintering behavior and the determination of any divergence from what happens when larger particles are sintered. It appears that no new mechanism is active in sintering nanoscale particles. Even so, the nanoscale field is filled with much speculation. Keen interest has been generated by the potential to create new properties from smaller particle sizes, as reviewed elsewhere [8,29–31]. The focus here on sintering recognizes that large challenges also exist in synthesis, dispersion, and consolidation.

ROLE OF PARTICLE SIZE

Starting with a small particle size is not necessarily a guarantee of easier sintering. Although the sintering stress increases inversely with the particle size, reaching levels of 100 MPa for 20 nm particles, sintering densification is still elusive in some systems. Figure 13.5 plots the sintered density for 40 nm ZnO after 60 min holds. The density peaks at 97% near 700°C (973 K), never reaches 100%, and declines at higher temperatures [32]. At the same time, the grain size enlarges as the temperature increases, approaching 1 μm by 1000°C (1273 K).

A small particle size inherently increases powder handling difficulties. A first concern is with the new toxicity associated with nanoscale particles [33]. Further, the particles sinter during synthesis, leading to clusters that are difficult to handle. A transmission electron micrograph of nanoscale iron powder after synthesis is given in Figure 13.6. Note the well-developed sinter neck between the particles prior to compaction or sintering. Such thermally induced agglomeration lowers the packing density, often to the 5% range. Figure 13.7 is a transmission electron micrograph showing nanoscale silver particles sintered to form chains. Unfortunately, many nanoscale particles are highly agglomerated. The scanning electron micrograph in Figure 13.8 is an example of extensive agglomeration in what is advertised as nanoscale titanium. Such agglomeration is detrimental to handling, shaping, and sintering [34–42].

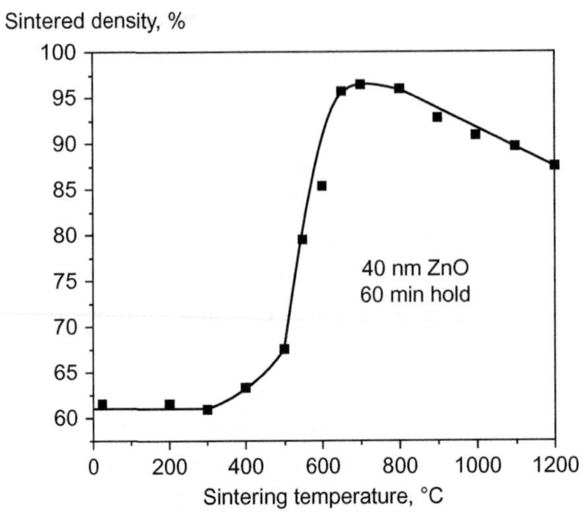

Figure 13.5 Sintered density for 40 nm ZnO powder heated to various temperatures with a 60 min hold, illustrating a peak in density that is less than 100% and swelling at higher temperatures with substantial grain growth [32].

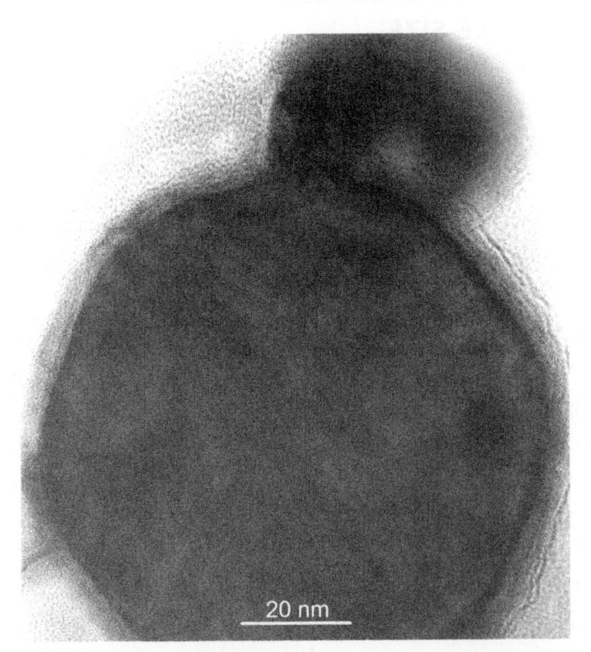

Figure 13.6 Micrograph of as-received nanoscale iron powder. The particles sinter bonded during synthesis.

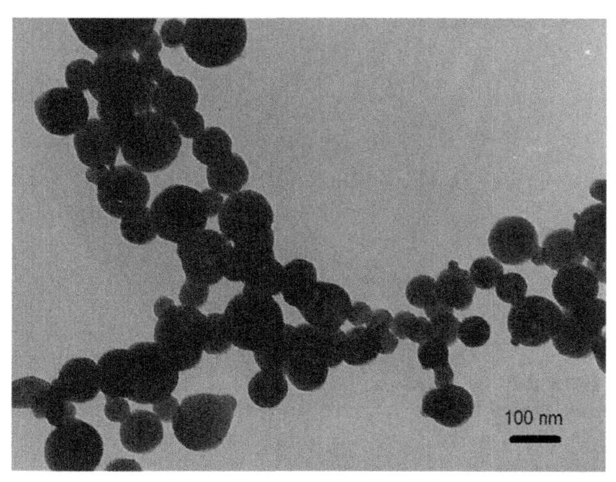

Figure 13.7 Chain structures of sintered nanoscale particles are common as illustrated by this transmission electron micrograph of silver. This level of bonding leads to a low apparent density, much difficulty with handling, and low pressed density.

Figure 13.8 This was sold as nanoscale titanium powder. On examination the grains are indeed in the nanoscale range, but extensive agglomeration makes this sponge powder difficult to process.

It is common to refer to the problem as "sintering during synthesis." One solution is to coat the particles with polymers during synthesis. Once that coating is removed the particles quickly sinter, sometimes at room temperature. Although sintering temperatures fall with smaller particle sizes, other attributes such as polymer pyrolysis are less temperature sensitive. The result can be contamination and formation of unintended phases [5,43].

Compaction proves difficult as particle size decreases, as illustrated in Figure 13.9. This plot traces green density versus compaction pressure for WC-Co powders of

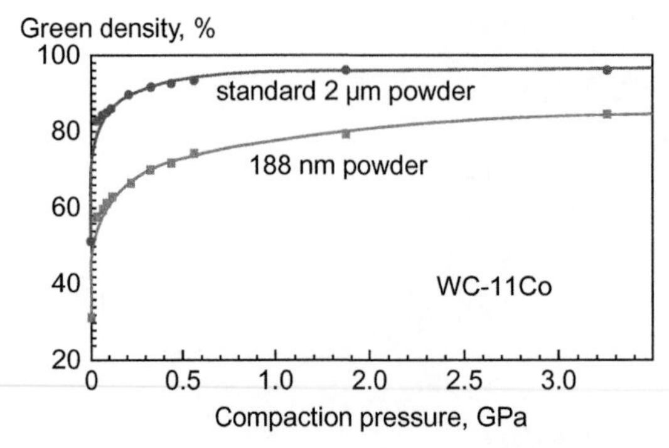

Figure 13.9 Compaction curves for two WC-Co powders differing in particle size by approximately a factor of ten [44]. The smaller powder shows substantial resistance to densification by compaction, leading to compaction pressures in the GPa range in many cases.

2 μm and 188 nm particle size [44]. The smaller powder is more difficult to compact. Such processing challenges lead experts to speculate that nanoscale microstructures are inaccessible by sintering [45]. These calculations assume standard processing cycles, which have proven unsuccessful.

Nanoscale powders provide multiple challenges, and after a few decades the investment community is still awaiting clear successes. In a direct comparison of 30 and 100 nm WC powders, consolidated by vacuum sintering and by hot isostatic pressing, the highest sintered hardness came with the larger initial particle size [46]. After consolidation, the 0.3 to 0.4 μm grain size was as predicted by independent calculations [47]. The conditions which deliver sintered densities over 95% also promoted significant grain coarsening; a conclusion routinely observed across many materials and applications.

SINTERING TEMPERATURE

Early reports on lower temperature sintering observed with nanoscale particles conjectured a melting temperature reduction. Similar conjectures arose earlier, in the 1940s, to explain the sintering of larger particles. Subsequently, both ideas have been dispelled by careful analysis. Measurements conclude that the temperature reduction is small and does not cause lower temperature sintering. Indeed, the temperature reduction is simply an extension of the Herring scaling laws [48].

In nanoscale sintering, grain growth takes place at a rate that is similar to that of standard powders. This means coarsening and sintering go hand in hand, with similar dependencies on sintering temperature. Unfortunately, during sintering the relative grain size change is large, overwhelming the initial particle size. For example, when sintering 30 nm tungsten carbide at 1100°C (1373 K), the grain size is already over 400 nm by the time 90% density is reached [50]. Additives are required to suppress rapid grain growth. The additional drag from a dispersoid slows grain growth, but hinders densification. Additives are employed in several systems, such as ZrO_2, $BaTiO_3$, AlN, Si_3N_4, SiC, and B_4C. For example, yttria reacts with the alumina layer on the surface of aluminum nitride particles to form $Y_3Al_5O_{12}$, and this aids densification, but produces degraded properties due to the introduction of an intergranular phase.

The densification rate changes with particle size. This is formulated here as a sintering temperature change based on particle size:

$$\ln\left(\frac{D_2}{D_1}\right) = \frac{Q}{3R}\left[\frac{1}{T_1} - \frac{1}{T_2}\right] \tag{13.2}$$

where R is the gas constant, T_1 is the sintering temperature for particle size D_1, and T_2 is the sintering temperature for equivalent sintering for particle size D_2. The parameter Q is the activation energy. If sintering is measured by density, then Q is usually the grain boundary diffusion activation energy. This relation explains the sintering temperature changes seen with particle size reduction. Such findings dispel speculation about new sintering mechanisms thought to occur at the nanoscale. Molecular dynamic studies show that surface diffusion and grain boundary diffusion fully explain nanoscale sintering and changes in densification with particle size.

UNCHANGED THERMODYNAMICS

An additional difficulty arises when sintering nanoscale metals, since historically metal processing relied on high temperatures to evaporate contaminants prior to sintering. For nanoscale particles these same heating cycles induce significant microstructural coarsening. The surface area of a nanoscale powder increases the impurity burden, while sintering at lower temperatures traps contaminants prior to evaporation. Large particles delay sintering until high temperature oxide reduction is completed. For example, stainless steels require about 1150°C (1423 K) to remove surface Cr_2O_3, but if particles smaller than 2 μm are used, the lower sintering temperature results in high retained oxygen.

Often thermochemical reactions require high temperatures, independent of the microstructure scale. Hot isostatic pressing and hot pressing densify at lower temperatures with reduced grain growth, but with a high cost. Conflicts between size-dependent events and temperature-dependent events lead to surprises in properties.

TIME-TEMPERATURE-PARTICLE SIZE

From a manufacturing viewpoint, a critical need is to predict the final dimensions and performance of the finished product. This requires analysis of the powder densification and microstructure development using input parameters of particle size, agglomeration level, compaction pressure, heating rate, peak temperature, and hold time. The models are trained using data extracted from experiments. Assumptions are made about how the material parameters change with processing conditions.

Even with detailed data, accurate predictions are taxing. For example, a micromechanical model requires 70 parameters to model nanoscale sintering [50]. Even with that much input information, the model has difficulties and ignores the particle size role in compaction. Experiments show quite the opposite, as illustrated by data from tungsten powders pressed at 160 MPa:

- 46 nm powder gives 41% density
- 110 nm powder gives 47% density
- 490 nm powder gives 50% density.

Accordingly, many critical relations are handled using approximations. A current favorite approach for handling the complex interplay is to apply the master sintering curve [51,52].

Current conceptualizations rely on several constitutive equations individually applied to powder packing, compaction, sintering densification, and other attributes [53–62]. Such approaches are most accurate for average properties, such as component density or mean grain size. They have difficulty predicting attributes such as component warpage. This is an area that needs attention, especially when moving into three-dimensional predictions.

MODELS VERSUS EXPERIMENT

Intense effort has been given to the synthesis and sintering of high value nanoscale materials, especially hard materials such as WC-Co. The promise is improved durability for oil and gas exploration, electronics assembly, and metal-cutting applications. Early efforts with nanoscale WC-Co failed to translate a small starting particle size into a dense, small grain size structure. Processing options covered the whole range of technologies. Even spark sintering has not delivered grain sizes in dense structures smaller than 1 μm [63]. Small grain sizes in dense nanoscale structures are achieved with second phase additions [15,64–68].

As a system, WC-Co is an excellent test of theory. The technical gains from small grain size are industrially attractive—have increased durability, improved strength and hardness, and good toughness. However, there has been an inability to densify nanoscale powders while retaining the desirable microstructure. The nanoscale powders have been available since the 1990s, and sintering has been able to produce full density, but has been unable to deliver on grain size. When sintered to full density, the mean grain size enlarges to several times the particle size. As one example, when 20 to 50 nm particles are sintered to full density at 1477°C (1750 K), the grain size is 0.5 μm [64]. Additives that inhibit grain growth also retard densification [69]. It appears that the sintered hardness-toughness combinations attained using nanoscale powders can be matched using conventional powders. Accordingly, new processing ideas are emerging based on high pressures and rapid heating using pressure-assisted sintering [44].

SOLUTION WINDOWS

Ideas on processing are solving problems for those combinations of factors that deliver dense and small grain size structures [44,70]. This is effectively a solution window similar to sintering maps but based on master sintering curve ideas.

Significant grain growth occurs when traditional sintering cycles are applied to nanoscale powders. For example, in sintering 1 to 2 μm tungsten powder the traditional consolidation cycle relies on compaction at 240 MPa, sintering at 1800 to 2400°C (2073 to 2673 K), with hold times up to 10 h. Such cycles are not appropriate for nanoscale powders. Novel consolidation cycles are needed to preserve nanoscale features, but those cycles must address barriers such as oxide reduction.

Models for the sintering response and microstructure coarsening provide insight into optimized cycles [44,49,52–62,70,71]. A first factor is the difficulty in packing small particles. Nanoscale tungsten packs to a density as low as 5%. A plot of the packing density versus particle size for tungsten powder is given in Figure 13.10, using data for particle sizes from 20 nm to 18 μm. Likewise, models for compaction, sintering, grain growth, and strength evolution provide a means to solve for simultaneous goals, isolating parameters capable of meeting target properties. These rely on constitutive equations for:

1. Compaction density as a function of pressure and particle size.
2. Sintered density as a function of compaction density, temperature, time, and particle size.
3. Grain size as a function of sintered density, particle size, time, and temperature.
4. Sintered strength as a function of sintered density and grain size.

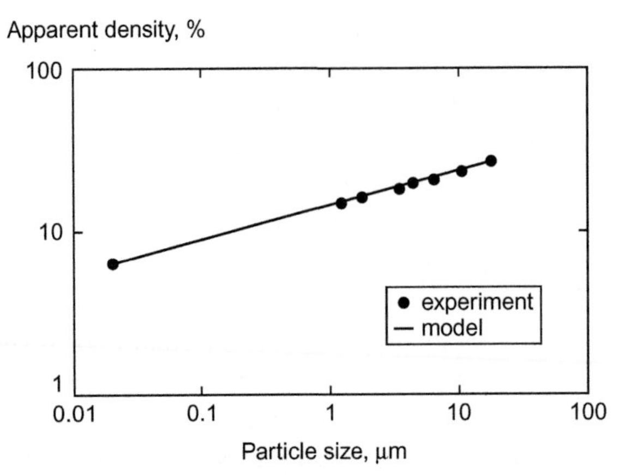

Figure 13.10 Nanoscale particles have considerable surface area and interparticle friction, consequently the packing density significantly decreases. This is a plot of data from tungsten. The constitutive model links apparent density to particle size via a power relation.

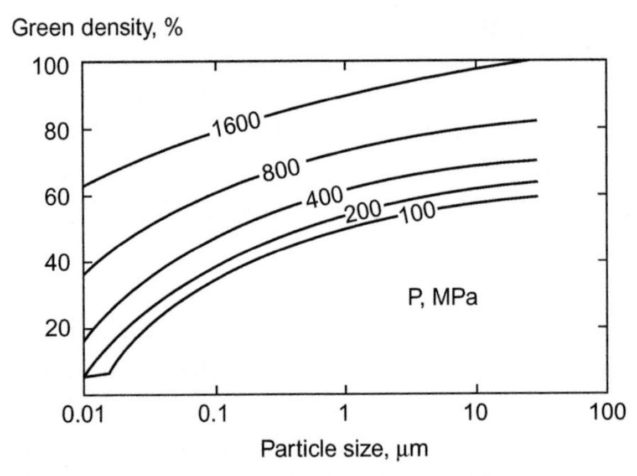

Figure 13.11 Tungsten powder compressibility curves for several pressures and particle sizes, illustrating the improved compaction for larger particles.

Compaction starts at the apparent density giving the green density for a given set of powder characteristics and applied pressure. Figure 13.11 plots the response using five compaction pressures. Larger particles are easier to compress, illustrating a limitation of nanoscale powder. Low green densities result in low sintered densities. This is because of limited densification capacity for any powder [72,73].

Figures 13.12 and 13.13 shows how tungsten powders respond to traditional consolidation cycles, using 500 MPa pressing with 10 h isothermal sintering at temperatures up to 2000°C (2273 K). The first plot gives sintered density and the second plot sintered grain size. Grain size coarsens with densification. The production of a

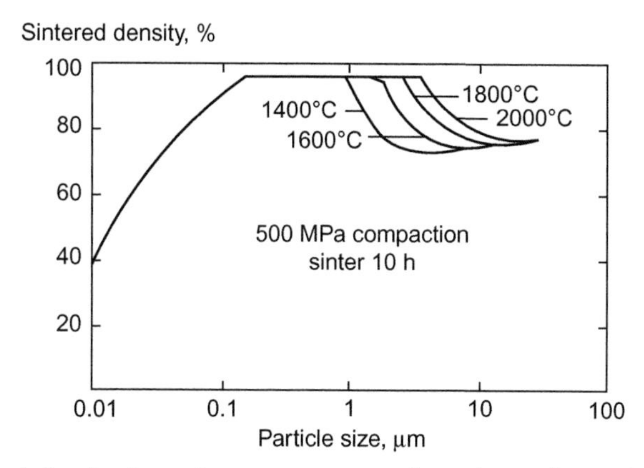

Figure 13.12 Sintered density for various tungsten powders when subjected to classic consolidation cycle of 500 MPa and sintered for 10 h at temperatures from 1400 to 2000°C (1773 to 2273 K).

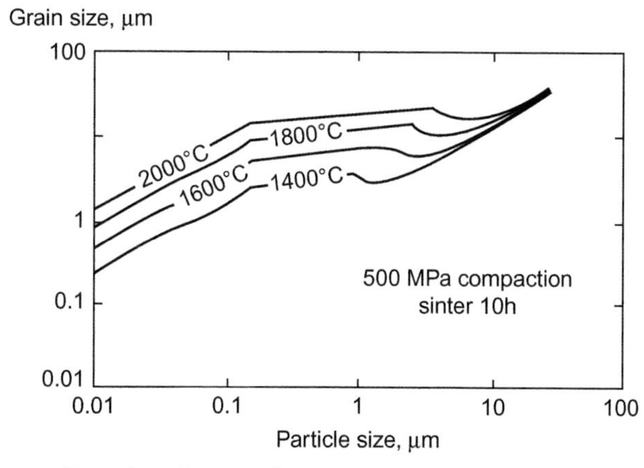

Figure 13.13 Corresponding plot of sintered grain size versus particle size for tungsten powders subjected to pressing at 500 MPa and sintering for 10 h at temperatures from 1400 to 2000°C (1773 to 2273 K).

material with small grain size yet high density is only possible using high compaction pressures, lower sintering temperatures, and short sintering times.

The idea of a solution window is given in Figure 13.14, where sintered grain size is plotted versus sintered density using a cycle of 1000°C (1273 K) for 10 min. For each pressure the solution locus is for particles sizes from 10 nm to 18 μm. The smallest grain size is 10 nm and the largest corresponds to 18 μm powder. Several compaction pressure contours are shown, but only a few result a high density and a grain size below 100 nm. A different view of the solution window is given in Figure 13.15.

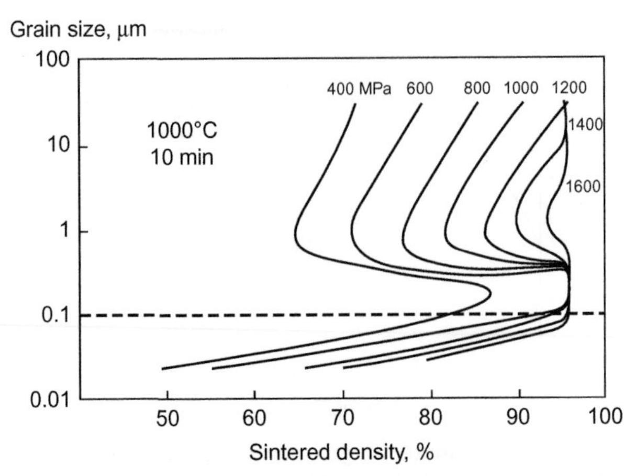

Figure 13.14 Solution curves for nanoscale powder consolidation, plotting sintered grain size versus sintered density for several compaction pressures, using a 10 min at 1000°C (1273 K) sintering cycle. Only a few combinations lead to high sintered density and a 100 nm grain size.

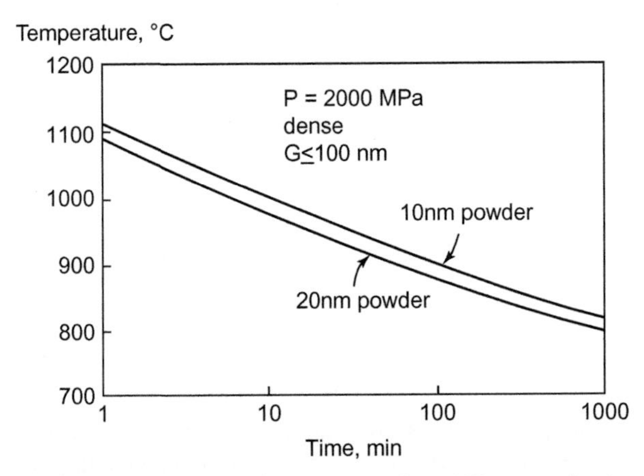

Figure 13.15 Solutions for temperature and time to retain a 100 nm grain size in a sintered structure with at least 95% density. These are for tungsten either 10 or 20 nm initial size compressed at 2 GPa.

For two nanoscale powders the combinations of time and temperature are given for 2 GPa compaction, showing the options for sintering a dense product with a grain size below 100 nm. These cycles contrast with traditional approaches of 240 MPa, 2000°C (2273 K), and 10 h.

From solutions for density and grain size, modeling is possible for performance attributes related to hardness, strength, wear resistance, thermal conductivity, and toughness. In that regard the predictions are for 3 GPa strength under optimal conditions [74]. Such efforts help to identify possible property gains and enable economic analysis of nanoscale powders and their uses [75].

These results explain the great difficulty experienced by early nanoscale efforts using traditional cycles. The processing window for nanoscale powders is significantly different. Higher applied pressures, rapid heating, and short hold times at lower temperatures are required for nanoscale structures [76–81].

TWO-STEP SINTERING

Nanoscale particle sintering involves densification that takes place largely by grain boundary diffusion. Grain growth involves considerable coalescence as well as diffusion across the grain boundaries. Efforts to separate densification from coarsening have encountered difficulties, since both events have similar diffusion processes. Neck growth produces more grain boundary area and that enables more transfer across the boundary. Coarsening is reduced by lower temperatures, such as in pressure-assisted sintering.

A variant applied is to adjust the temperature during sintering, building from the concept of rate controlled sintering [82]. Rate controlled sintering asserts that rapid sintering tends to separate grain boundaries from pores, resulting in lower final density and larger grain size. Control of the temperature is a means to separate grain growth from densification. Related approaches relied on a two-step heating and cyclic heating concepts to densify with minimized coarsening [83,84].

For nanoscale powders, a related approach is to heat to a high temperature and then reduce the temperature to manage densification through slow grain growth. A window of grain size versus density arises that differs from conventional sintering. Figure 13.16 plots an example for 30 nm yttria (Y_2O_3) [85]. Conventional densification is attained using 10°C/min to 1500°C (1773 K) with a 60 min hold, giving 420 nm grain size at full density. A two-step cycle of heating to 1250°C (1523 K) followed by stepping down to 1150°C (1423 K) for 1200 min results in full density at 122 nm grain size, meaning 67 initial particles coalesced to form each final grain. With the two-step cycle, sintering requires quadruple the sintering time, which is a disadvantage from the standpoint of decreased productivity or increased cost. Overall, the approach has seen many trials with nanoscale powders [9,86–94]. The findings vary widely, possibly reflecting sample preparation differences, since the two-step sintering process is sensitive to the homogeneity of the starting structure. It is an approach requiring closer scrutiny to extract the green body conditions that best respond.

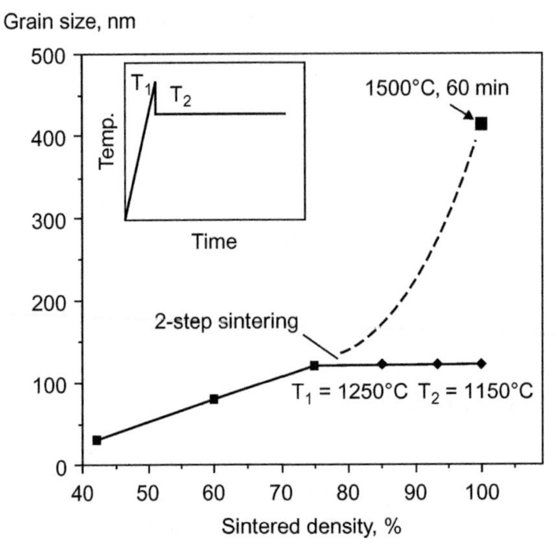

Figure 13.16 Grain size versus sintered density for nanoscale yttria sintered to full density at 1500°C (1773 K) and with two-step sintering. The latter involves a prolonged hold at a reduced temperature as illustrated in the insert [85].

OPPORTUNITIES

Nanoscale particles sinter rapidly due to small diffusion distances and high contact stresses. First surface diffusion dominates, but subsequently grain boundary diffusion induces densification. Grain coarsening is rapid, since conditions that promote grain boundary diffusion typically induce grain growth [47]. Further, grain rotation gives coalescence. The net effect is considerable coarsening while the nanoscale particle structure densifies [34].

Sintered nanoscale materials are different from traditional materials—they are stronger and harder. To date many of the property gains attained in sintered nanoscale materials have been matched by traditional powders.

REFERENCES

[1] K. Lu, Nanoparticulate Materials Synthesis, Characterization, and Processing, Wiley, Hoboken, New Jersey, 2013.

[2] P. Vergnon, M. Astier, S.J. Teichner, Sintering of submicron particles of metallic oxides, in: G.C. Kuczynski (Ed.), Sintering and Related Phenomena, Plenum Press, New York, NY, 1973, pp. 301–310.

[3] K. Kinoshita, K. Routsis, J.A.S. Bett, C.S. Brooks, Changes in the morphology of platinum agglomerates during sintering, Electrochimica Acta 18 (1973) 953–961.

[4] K. Kinoshita, J.A.S. Bett, P. Stonehart, Effects of gas- and liquid-phase environments on the sintering behavior of platinum catalysts, in: G.C. Kuczynski (Ed.), Sintering and Catalysis, Plenum Press, New York, NY, 1975, pp. 117–132.
[5] J. Sautereau, A. Mocellin, Sintering behavior of ultrafine Nb-C and Ta-C powders, J. Mater. Sci. 9 (1974) 761–771.
[6] D. Wakuda, K.S. Kim, K. Suganuma, Room temperature sintering of Ag nanoparticles by drying solvent, Scripta Mater. 59 (2008) 649–652.
[7] J. Langer, M.J. Hoffmann, O. Guillon, Direct comparison between hot pressing and electric field assisted sintering of submicron alumina, Acta Mater. 57 (2009) 5454–5465.
[8] M.J. Mayo, Processing of nanocrystalline ceramics from ultrafine particles, Inter. Mater. Rev. 41 (1996) 85–115.
[9] M. Mazaheri, Z.R. Hesabi, S.K. Sadrnezhaad, Two step sintering of titania nanoceramics assisted by anatase to rutile phase transformation, Scripta Mater. 59 (2008) 139–142.
[10] V.V. Dabhade, T.R. Rama Mohan, P. Ramakrishnan, Dilatometric sintering study of titanium–titanium nitride nano/nanocomposite powders, Powder Metall. 50 (2007) 33–39.
[11] J. Li, Y. Ye, L. Shen, J. Chen, H. Zhou, Densification and grain growth during pressureless sintering of TiO_2 nanoceramics, Mater. Sci. Eng. vol. A390 (2005) 265–270.
[12] A. Munitz, R.J. Fields, Mechanical properties of hot isostatically pressed nanograin iron and iron alloy powders, Powder Metall. 44 (2001) 139–147.
[13] K. Hayashi, H. Etoh, Pressure sintering of iron, cobalt, nickel, and copper ultrafine powders and the crystal size and hardness of the compacts, Materials Trans. Japan Inst. Metals 30 (1989) 925–931.
[14] T.L. Shing, S. Luyckx, I.T. Northrop., I. Wolff, The effect of ruthenium additions on hardness, toughness, and grain size of WC-Co, Inter. J. Refract. Met. Hard Mater. 19 (2001) 41–44.
[15] B. Basu, J.H. Lee, D.Y. Kim, Development of WC-ZrO_2 nanocomposites by spark plasma sintering, J. Amer. Ceram. Soc. 87 (2004) 317–319.
[16] A. Moitra, S. Kim, S.G. Kim, S.J. Park, R.M. German, Atomistic scale study on effect of crystalline misalignment during sintering of nanoscale tungsten powder, in: R.K. Borida, E.A. Olevsky (Eds.), Advances in Sintering Science and Technology, American Ceramic Society, Westerville, OH, 2010, pp. 149–160.
[17] G.Q. Shao, X.L. Duan, J.R. Xie, X.H. Yu, W.F. Zhang, R.Z. Yuan, Sintering of nanocrystalline WC-Co composite powder, Rev. Adv. Mater. Sci. 5 (2003) 281–286.
[18] T. Hawa, M.R. Zachariah, Coalescence kinetics of unequal sized nanoparticles, Aero. Sci. 37 (2006) 1–15.
[19] M.R. Zachariah, M.J. Carrier, E. Blaisten-Barojas, Properties of silicon nanoparticles: a molecular dynamics study, J. Phys. Chem. 100 (1996) 14856–14864.
[20] X. Wang, Z.Z. Fang, H.Y. Sohn, Grain growth during the early stage of sintering of nanosized WC-Co powder, Inter. J. Refract. Met. Hard Mater. 26 (2008) 232–241.
[21] K.E.J. Lehtien, M.R. Zachariah, Energy accumulation in nanoparticle collision and coalescence processes, Aero. Sci. 33 (2002) 357–368.
[22] Y.Q. Wang, R. Smirani, G.G. Ross, F. Schiettekatte, Ordered coalescence of Si nanocrystals in SiO_2, Phys. Rev. 71 (161310 (R)) (2005).
[23] S. Hendy, S.A. Brown, M. Hyslop, Coalescence of nanoscale metal clusters molecular dynamics study, Phys. Rev. B 68 (2003) article 241403.
[24] S. Arcidiacono, N.R. Bieri, D. Poulikakos, C.P. Grigoropoulos, On the coalescence of gold nanoparticles, Inter. J. Multi. Flow 30 (2004) 979–994.
[25] M. Yeadon, J.C. Yang, R.S. Averback, J.W. Bullard, J.M. Gibson, Sintering of silver and copper nanoparticles on (001) copper observed by *In Situ* ultrahigh vacuum transmission electron microscopy, Nano. Mater. 10 (1998) 731–739.
[26] S. Hendy, Coalescence of nanoscale metal clusters: molecular dynamics study, Phys. Rev. B 68 (2003), pp. 241403–1 to 4.
[27] H. Pan, S.H. Ko, C. Grigoropoulos, The solid state neck growth mechanisms in low energy laser sintering of gold nanoparticles, a molecular dynamics simulation study, J. Heat Transfer 130 (2008) paper 092404.

[28] L.J. Lewis, P. Jensen, J.L. Barrat, Melting, freezing, and coalescence of gold nanoclusters, Phys. Rev. B 56 (1997) 2248–2257.
[29] R.A. Andrievski, Review nanocrystalline high melting point compound based materials, J. Mater. Sci. 29 (1994) 614–631.
[30] H. Gleiter, Nanostructured materials basic concepts and microstructure, Acta Mater. 48 (2000) 1–29.
[31] R.A. Andrievski, Review stability of nanostructured materials, J. Mater. Sci. 38 (2003) 1367–1375.
[32] A.P. Hynes, R.H. Doremus, R.W. Siegel, Sintering and characterization of nanophase zinc oxide, J. Amer. Ceram. Soc. 85 (2002) 1979–1987.
[33] A.S. Karakoti, L.L. Hench, S. Seal, The potential toxicity of nanomaterials – the role of surfaces, J. Met. 58 (no. 7) (2006) 77–82.
[34] Z.Z. Fang, H. Wang, in: Z.Z. Fang (Ed.), Sintering of Ultrafine and Nanoscale Particles; Sintering of Advanced Materials, Woodhead Publishing, Oxford, UK, 2010, pp. 434–473.
[35] M. Seipenbusch, P. Toneva, P. Peukert, A.P. Weber, Impact fragmentation of metal nanoparticle agglomerates, Part. Part. Sys. Char. 24 (2007) 193–200.
[36] B.H. Cha, Y.S. Kang, J.S. Lee, Processing of net-shaped Fe-Ni nanomaterials by powder injection molding, J. Japan Soc. Powder Powder Metall. 53 (2006) 769–775.
[37] W. Li, L. Gao, Sintering of Nanocrystalline ZrO_2 (2Y) by Hot Pressing, Mater. Trans. 42 (2001) 1653–1656.
[38] P. Knorr, J.G. Nam, J.S. Lee, Sintering behavior of nanocrystalline gamma Ni-Fe powders, Metall. Mater. Trans. 31A (2000) 503–510.
[39] J.G. Li, X. Sun, Synthesis and sintering behavior of a nanocrystalline alpha alumina powder, Acta Mater. 48 (2000) 3103–3112.
[40] P. Knorr, J.G. Nam, J.S. Lee, Densification and microstructural development of nanocrystalline gamma Ni-Fe powders during sintering, Nano. Mater. 12 (1999) 479–482.
[41] J. Luo, S. Adak, R. Stevens, Microstructure evolution and grain growth in the sintering of 3Y-TZP ceramics, J. Mater. Sci. 33 (1998) 5301–5309.
[42] J.P. Ahn, M.Y. Huh, J.K. Park, Effect of green density on subsequent densification and grain growth of nanophase SnO_2 powder during isothermal sintering, Nano. Mater. 8 (1997) 637–643.
[43] C. Greskovich, J.H. Rosolowski, Sintering of covalent solids, J. Amer. Ceram. Soc. 59 (1976) 336–343.
[44] S.J. Park, J.L. Johnson, R.M. German, Special sintering technologies for nanostructured tungsten carbide, Advances in Powder Metallurgy and Particulate Materials–2006, Part 9, Metal Powder Industries Federation, Princeton, NJ, 2006, pp. 114–122.
[45] K. Hayashi, N. Matsuika, Grain size condition for abnormal grain growth in fine-grained WC-Co hardmetal estimated by numerical calculation based on two kinds of grain size alloy model, J. Adv. Mater. 34 (2002) 38–48.
[46] I. Azcona, A. Ordonez, L. Dominguez, J.M. Sanchez, F. Castro, Hot isostatic pressing of nanosized WC-Co hardmetals, in: P. Rodhammer, H. Wildner (Eds.), Proceedings Fifteenth International Plansee Seminar, 2, Plansee Holding, Reutte, Austria, 2001, pp. 35–49.
[47] J. Kanters, U. Eisele, J. Rodel, Effect of initial grain size on sintering trajectories, Acta Mater. 48 (2000) 1239–1246.
[48] L. Liu, N.H. Loh, B.Y. Tay, S.B. Tor, Y. Murakoshi, R. Maeda, Micro powder injection molding: sintering kinetics of microstructured components, Scripta Mater. 55 (2006) 1103–1106.
[49] H. Wang, Z.Z. Fang, K.S. Hwang, Kinetics of initial coarsening during sintering of nanosized powders, Metall. Mater. Trans. 42A (2011) 3534–3542.
[50] R.S. Iyer, S.M.L. Sastry, Consolidation of nanoparticles–development of a micromechanistic model, Acta Mater. 47 (1999) 3079–3098.
[51] D.Y. Park, S.W. Lee, S.J. Park, Y.S. Kwon, I. Otsuka, Effects of particle sizes on sintering behavior of 316l stainless steel powder, Metall. Mater. Trans. 44A (2013) 1508–1518.
[52] M.G. Bothara, S.V. Atre, S.J. Park, R.M. German, T.S. Sudarshan, R. Radhakrishnan, Sintering behavior of nanocrystalline silicon carbide using a plasma pressure compaction system: master sintering curve analysis, Metall. Mater. Trans. 41A (2010) 3252–3261.

[53] M.I. Alymov, E.I. Maltina, Y.N. Stepanov, Model of initial stage of ultrafine metal powder sintering, Nano. Mater. 4 (1994) 737–742.

[54] G.R. Shaik, W.W. Milligan, Consolidation of nanostructured metal powders by rapid forging: processing, modeling, and subsequent mechanical behavior, Metall. Mater. Trans. 28A (1997) 895–904.

[55] J. Freim, J. McKittrick, Modeling and fabrication of fine grain alumina-zirconia composites produced from nanocrystalline precursors, J. Amer. Ceram. Soc. 81 (1998) 1773–1780.

[56] M.R. Zachariah, M.J. Carrier, Molecular dynamics computation of gas-phase nanoparticle sintering: a comparison with phenomenological models, J. Aero. Sci. 30 (1999) 1139–1151.

[57] H.S. Kim, Modelling strength and ductility of nanocrystalline metallic materials, J. Korean Powder Metall. Inst. 8 (2001) 168–173.

[58] T. Moritz, R. Lenk, J. Adler, M. Zins, Modular micro reaction system including ceramic components, Inter. J. Appl. Ceram. Tech. 2 (2005) 521–528.

[59] D.E. Rosner, S. Yu, MC simulation of aerosol aggregation and simultaneous spheroidization, Amer. Inst. Chem. Eng. J. 47 (2001) 545–561.

[60] S.H. Park, S.N. Rogak, A. One-dimensional, model for coagulation, sintering, and surface growth of aerosol agglomerates, Aero. Sci. Tech. 37 (2003) 947–960.

[61] P. Redanz, R.M. McMeeking, Sintering of a BCC structure of spherical particles of equal and different sizes, Phil. Mag. 83 (2003) 2693–2714.

[62] R.M. German, E. Olevsky, Mapping the compaction and sintering response of tungsten-based materials into the nanoscale size range, Inter. J. Refract. Met. Hard Mater. 23 (2005) 294–300.

[63] C.D. Park, H.C. Kim, I.J. Shon, Z.A. Munir, One step synthesis of dense tungsten carbide-cobalt hard materials, J. Amer. Ceram. Soc. 88 (2002) 2670–2677.

[64] Z. Fang, J.W. Eason, Study of nanostructured WC-Co composites, Inter. J. Refract. Met. Hard Mater. 13 (1995) 297–303.

[65] B.K. Kim, G.H. Ha, G.G. Lee, D.W. Lee, Structure and properties of nanophase WC/Co/VC/TaC hardmetal, Nano. Mater. 9 (1997) 233–236.

[66] M. Hu, S. Chujo, H. Nishikawa, Y. Yamaguchi, T. Okubo, Spontaneous formation of large-area monolayers of well-ordered nanoparticles via a wet-coating process, J. Nano. Res. 6 (2004) 479–487.

[67] R.K. Sadangi, O.A. Veronov, B.H. Kear, WC-Co-Diamond Nano-Composites, Nano. Mater. 12 (1999) 1031–1034.

[68] B.K. Kim, G.H. Ha, D.H. Kwon, Doping method of grain growth inhibitors for strengthening nanophase WC/Co, Materials. Trans. vol. 44 (2003) 111–114.

[69] O. Seo, S. Kang, E.J. Lavernia, Growth inhibition of nano WC particles in WC-Co alloys during liquid phase sintering, Materials Trans. 44 (2003) 2339–2345.

[70] R.M. German, J. Ma, X. Wang, E. Olevsky, Processing model for tungsten powders and extension to the nanoscale size range, Powder Metall. 49 (2006) 19–27.

[71] I. Shimizu, Y. Takei, Thermodynamics of interfacial energy in binary metallic systems: influence of adsorption on dihedral angles, Acta Mater. 53 (2005) 811–821.

[72] J.H. Rosolowski, C. Greskovich, Theory of the dependence of densification on grain growth during intermediate-stage sintering, J. Amer. Ceram. Soc. 58 (1975) 177–182.

[73] C. Greskovich, K.W. Lay, Grain growth in very porous Al_2O_3 compacts, J. Amer. Ceram. Soc. 55 (1972) 142–146.

[74] R.M. German, E. Olevsky, Strength predictions for bulk structures fabricated from nanoscale tungsten powders, Inter. J. Refract. Met. Hard Mater. 23 (2005) 77–84.

[75] J.L. Johnson, Economics of processing nanoscale powders, Inter. J. Powder Metall. 44 (no. 1) (2008) 44–54.

[76] M. Cologna, J.S.C. Francis, R. Raj, Field assisted and flash sintering of alumina and its relationship to conductivity and MgO doping, J. Europ. Ceram. Soc. 31 (2011) 2827–2837.

[77] M. Cologna, B. Rashkova, R. Raj, Flash sintering of nanograin zirconia in less than 5 s at 850 C, J. Amer. Ceram. Soc. 93 (2010) 3556–3559.

[78] H.C. Kim, D.Y. Oh, I.J. Shon, Sintering of nanophase WC-15 vol.% Co hard metals by rapid sintering process, Inter. J. Refract. Met. Hard Mater. vol. 22 (2004) 197–203.

[79] Q. Wei, H.T. Zhang, B.E. Schuster, K.T. Ramesh, R.Z. Valiev, L.J. Kecskes, et al., Microstructure and mechanical properties of super-strong nanocrystalline tungsten processed by high pressure torsion, Acta Mater. 54 (2006) 4079–4089.
[80] D.J. Chen, M.J. Mayo, Rapid rate sintering of nanocrystalline ZrO_2–3 mol.% Y_2O_3, J. Amer. Ceram. Soc. 79 (1996) 906–912.
[81] S.J. Wu, L.C. De Jonghe, M.N. Rahaman, Sintering of nanophase gamma-Al_2O_3 powder, J. Amer. Ceram. Soc. 79 (1996) 2207–2211.
[82] H. Palmour, Rate controlled sintering for ceramics and selected powder metals, in: D.P. Uskokovic, H. Palmour, R.M. Spriggs (Eds.), Science of Sintering, Plenum Press, New York, NY, 1989, pp. 337–356.
[83] M.Y. Chu, L.C. De Jonghe, M.K.L. Lin, F.J.T. Lin, Precoarsening to improve microstructure and sintering of powder compacts, J. Amer. Ceram. Soc. 74 (1991) 2902–2911.
[84] M. Atanasovska, T. Sreckovic, S.M. Radic, A study of cyclic sintering of the 2MgO-2Al_2O_3-5SiO_2 system, Sci. Sintering 25 (1993) 67–70.
[85] I.W. Chen, X.H. Wang, Sintering dense nanocrystalline ceramics without final stage grain growth, Nature 404 (2000) 168–171.
[86] H.D. Kim, B.D. Han, D.S. Park, B.T. Lee, P.F. Becher, Novel two-step sintering process to obtain a bimodal microstructure in silicon nitride, J. Amer. Ceram. Soc. 85 (2002) 245–252.
[87] Y.L. Lee, Y.W. Kim, M. Mitomo, D.Y. Kim, Fabrication of dense nanostructred silicon carbide ceramics through two-step sintering, J. Amer. Ceram. Soc. 86 (2003) 1803–1805.
[88] E.R. Leite, C.A. Paskocimas, E. Longo, C.M. Barrado, M.J. Godinho, J.A. Varela, Two steps sintering of yttria stabilized zirconia, in: R.G. Cornwall, R.M. German, G.L. Messing (Eds.), Proceedings Sintering 2003, Materials Research Institute, Pennsylvania State University, University Park, PA, 2003.
[89] C.J. Wang, C.Y. Huang, Y.C. Wu, Two step sintering of fine alumina-zirconia ceramics, Ceram. Inter. 35 (2009) 1467–1472.
[90] G.J. Wright, J.A. Yeomans, Constrained sintering of yttria stabilized zirconia electrolytes: the influence of two step sintering profiles on microstructure and gas permeance, Inter. J. Appl. Ceram. Tech. 4 (2008) 589–596.
[91] Z.H. Chen, J.T. Li, J.J. Xu, Z.G. Hu, Fabrication of YAG transparent ceramics by two-step sintering, Ceram. Inter. 34 (2008) 1709–1712.
[92] X.H. Wang, P.L. Chen, I.W. Chen, Two step sintering of ceramics with constant grain size, I: Y_2O_3, J. Amer. Ceram. Soc. 89 (2006) 431–437.
[93] J.L. Huang, L.M. Din, H.H. Lu, W.H. Chan, Effects of two-step sintering on the microstructure of Si_3N_4, Ceram. Inter. 22 (1996) 131–136.
[94] X.H. Wang, X.Y. Deng, H.L. Bai, H. Zhou, W.G. Qu, L.T. Li, et al., Two step sintering of ceramics with constant grain size, II: $BaTiO_3$ and Ni-Cu-Zn Ferrite, J. Amer. Ceram. Soc. 89 (2006) 438–443.

CHAPTER FOURTEEN

Computer Models

INTRODUCTION

Computer calculations of sintering are constructed for specific situations. The models range from atomistic atom-atom interactions to models for gas flow in sintering furnaces. Economic simulations help elaborate policy, such as determining which mixture of batch and continuous furnaces provides the lowest cost in spite of possible business fluctuations. Simulations allow for predictions, and are widely used in weather, sporting, finance, politics, logistics, and insurance. Sintering simulations suffer some of the same shortfalls as these other complex systems. Uncertainty in the input knowledge results in low accuracy in the predictions. This chapter introduces the variants while providing expectation management for anyone launching into the area.

Computer simulations exist for all standard sintering processes. Simple simulations make one-dimensional predictions, such as shrinkage versus process parameters of particle size, heating rate, sintering temperature, and hold time. More sophisticated simulations operate in three dimensions to predict final component size, shape, cost, properties, and provide guidance on optimization with respect to target attributes [1].

Unfortunately computer simulations of sintering are good, but not good enough. The problem is not with the simulations, but the large number of unknown or uncontrolled parameters in sintering. For example, the specification of incoming powder at a sintering plant might allow for a 20% variation in particle size. Smaller particles are responsible for more dimensional change in sintering, so a lose incoming material specification allows for sintered dimensional variations. In production the variation is handled by adaptive process control, where operating steps are adjusted to compensate for the variations. On the other hand the simulations assume the powder is always the same. For example, in the production of a sintered stainless steel automotive component, the dimensional change in sintering varies with the powder carbon to oxygen ratio. This ratio is not controlled in production. The maximum values are independently specified at less than 3000 ppm oxygen and less than 300 ppm carbon, but the ratio is not controlled. During sintering the carbon reduces oxygen, promoting better sintering. A high O to C ratio sinters less due to the remaining oxide. In practice, a compaction press is used to adjust the green density to compensate. A simulation trying to predict final dimensions is not accurate if practice is varying to

Sintering: From Empirical Observations to Scientific Principles
DOI: http://dx.doi.org/10.1016/B978-0-12-401682-8.00014-8

compensate for the powder variation. No simulation includes this level of detail, so agreement between simulation and practice is not precise.

Sintering simulations need to accurately predict final size to within 10 μm. Other uses of simulations are in providing guidance toward better production:

- Lower development costs in conceptualization of new processes.
- Avoid mistakes in the planning stage.
- Compress start-up time through the design of experiments.
- Focus on critical problems and knowledge gaps.
- Improve compatibility of equipment and processes.
- Identify process sensitivities.
- Set standards and production tolerances.

Inherently, the greatest value of sintering simulations is ensuring the sintering process is understood.

Simulation Framework

Sintering simulations initially focused on sintering two particles to predict neck size versus time, replicating model experiments from the 1940s. Early simulations considered two-dimensional sintering, effectively the bonding between two wires, by a single diffusion mechanism. The simulations were slow, requiring 10 h of simulation to reproduce 1 h of sintering. Numerical instabilities caused the simulations to lose mass and increase energy. In the 1980s, the simulations added multiple transport mechanisms, multiple sintering stages, and pressure-assisted and liquid phase sintering. Commonly these predicted sintered density versus processing parameters for isothermal sintering of single phase materials. For example, the density of hot isostatic pressed spherical silver for a specific time, temperature, green density, and particle size. The initial software was provided by Ashby [2].

The assumptions of isothermal conditions, ideal spherical powders, and homogeneous green bodies were significant limitations. Sintering occurs during heating to the peak temperature, so isothermal models poorly reflect actual behavior. The particles are often not spherical and green microstructures are not always homogeneous. To dodge the problems, several models elected to simply treat sintering as a viscous flow event. Unfortunately, viscous flow models are not able to handle relevant predictions such as grain size.

Subsequent attention moved to problems in predicting sintered properties, component size and shape, processing costs, furnace operation, and process sensitivities. A wide variety of simulations emerged with hopes of evaluating trade-offs prior to production. Once created, computer simulations are relatively inexpensive and can be repeated to converge on desired attributes.

However, the simulations are hindered a large number of unknown or uncontrolled factors. For example, tabulations of material strength rarely extend to the

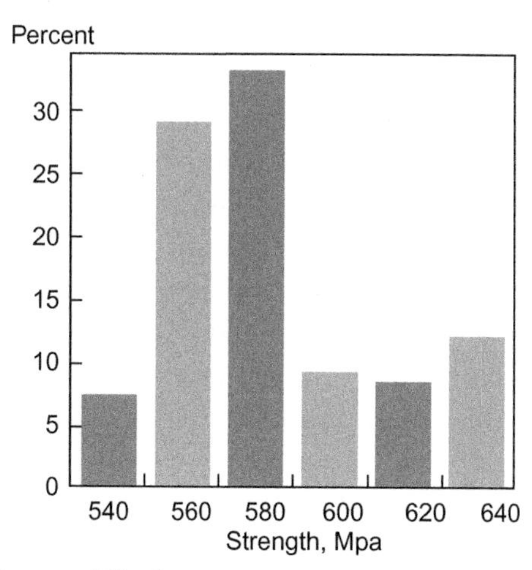

Figure 14.1 One of the inherent difficulties in computer simulation of sintering is the inability to precisely define materials properties. This is a plot of tensile strength for 316L stainless steel showing a typical distribution over several samples. The simulations assume a single value that is not reflective of reality.

sintering temperature. Further, material properties are distributed parameters, not single valued as is usually assumed. An example of the problem is given in Figure 14.1. This is a histogram of tensile strengths from multiple samples for a stainless steel, giving an average 581 MPa with a standard deviation of 20 MPa. Simulations input a single value. To complicate the problem, simple models are assumed for property changes with density or temperature, resulting in significant errors. Consequently, simulations are compromised by approximate data, idealized conditions, and simplified relations. Even so, the prospects are not dismal since commercial simulations are helpful in process set-up, just not final process definition.

Simulation Essentials

If you do not understand the sintering process, then a computer simulation is not meaningful since it requires rules and boundary conditions. Several simulations are available, ranging from atom-level to furnace-level models. These simulations have some common features:

- basic rules, boundary conditions, and goals
- input data
- monitors, error detections, and stability checks
- verification and validation
- experimentation—the key justification for simulations
- output data and output interface.

These elements ensure that the simulation is structured with known rules and assumptions. The input data and boundary conditions use these rules to generate the output.

Early simulations were unstable. Some calculations still fail simple tests for mass and energy conservation. Although the simulation images are attractive, the results are meaningless. Stability checks require systematic changes to the simulation details, such as changing the precision of a variable to establish the sensitivity to the details. Likewise, the simulation needs to avoid metastable conditions. An appropriate test is to examine the prediction versus changes in input variables, establishing sensitivities to variations.

A common problem is the forward time step. It is common to assume a fixed step, say 1 ms, but this creates a systematic error. An appropriate technique is to compute the time advance based on the rate of change. In periods where little change occurs, the time step is large, while smaller steps are used when rapid change occurs.

Material data required for sintering simulations are not single valued. Yet simulations usually assume a single parameter for the input. Consider diffusion data as one illustration. Figure 14.2 plots grain boundary diffusion data for tungsten (in terms of the boundary width times the diffusivity) against the inverse absolute temperature. Note the scatter in measured values. Computer models assume average behavior. However, measurements show differences between large and small angle grain boundaries,

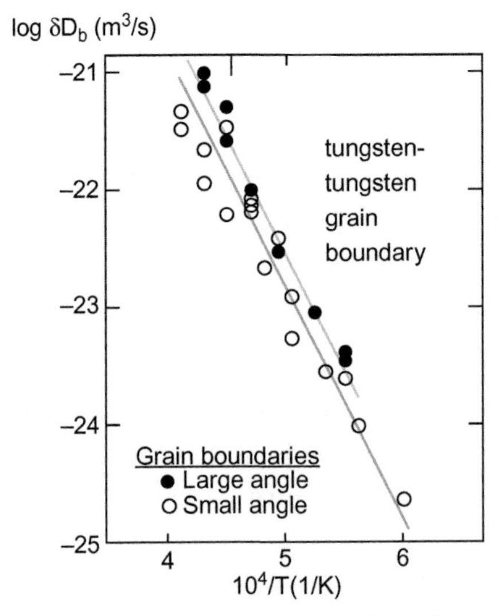

Figure 14.2 Sintering simulations assume average values for diffusion properties. This Arrhenius plot differentiates the large angle and small angle grain boundaries and the diffusion rates, plotted as the diffusivity times the grain boundary width. The variations at any temperature in possible grain boundary diffusion rates are as much as ten-fold, thereby creating difficulty for any assumed value in the simulations.

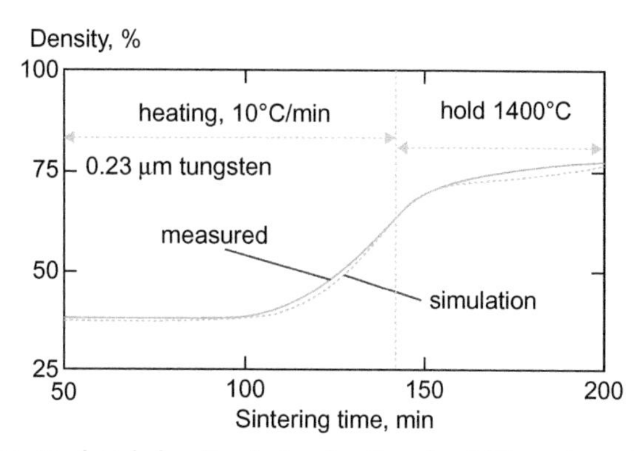

Figure 14.3 Computer simulated density during heating for 0.23 μm tungsten powder heated at 10°C/min to a 1400°C (1673 K) hold. The simulation is compared to the values measured using dilatometry. Many simulations only treat the isothermal portion, failing to include the significant events that occur during heating.

complicated by impurity effects, and measurement errors. These data have a 30% variation at any temperature, contributing to large differences in predicted sintering rates.

As a demonstration of the issue of data quality underpinning simulations, consider predictions for sintering stainless steel powder heated to 1350°C for 60 min in hydrogen. Handbook data for the material properties predicts 92% density. A 10% decrease in the grain boundary diffusion activation energy gives 96% density. This corresponds to a 1.2% difference in sintered dimensions, far larger than a typical tolerance band of ±0.2%. Hence, realistic variations in diffusion data produce sintered dimension variations larger than acceptable in practice.

Sintering simulations moved from two particles neck growth to attacking shape change in complex bodies to address sintering practice. This required inclusion of realistic thermal cycles. An example is plotted in Figure 14.3 for sintering tungsten, showing density versus time during heating and holding. Underlying the calculation is a multiple mechanism sintering model that allows simultaneous changes in neck size, grain size, and transport mechanism. Achieving accurate simulations requires attention to everything, and some of the pending issues include:

- nonspherical particles, agglomerates, and particle size distributions
- nonisothermal heating and cooling
- dopants, impurities, and inhomogeneities
- atmospheres chemistry and atmosphere flow rate
- compaction or shaping to properly include green density gradients
- reactions, melting, transformations, homogenization, and spreading.

With such improvements, the simulations still are limited by the accuracy of the input data.

PROCEDURES

Early sintering simulations relied on simple geometries to calculate neck growth by a single transport mechanism [3–9]. By the late 1970s, multiple mechanism sintering was applied to neck growth, shrinkage, and density [10–14].

There are different dimensionality characteristics levels. The one-dimensional efforts took on the prediction of neck size, shrinkage, surface area, or density for various time-temperature combinations [11–17]. These one-dimensional simulations predicted the average behavior. For example, Figure 14.4 plots sintered density versus applied pressure for 6 μm tungsten powder consolidated at 1500°C (1773 K) with a starting density of 50%. The density is measured at different isothermal hold times ranging from ¼ h to 16 h. Regions are marked to show the major mechanism (yielding, power law creep, grain boundary diffusion, or volume diffusion). There is no information on component warpage or similar attributes related to the different heating cycles.

The next wave of models gave more details, and even predicted particle rotation, pore migration, cracking around reinforcing whiskers, and other axis-symmetric problems [18–23]. Two-dimensional simulations are successful in replicating early sintering studies, such as neck growth between connected spheres as profiled in Figure 14.5. Neck size is further aligned to shrinkage, surface area, and density during sintering.

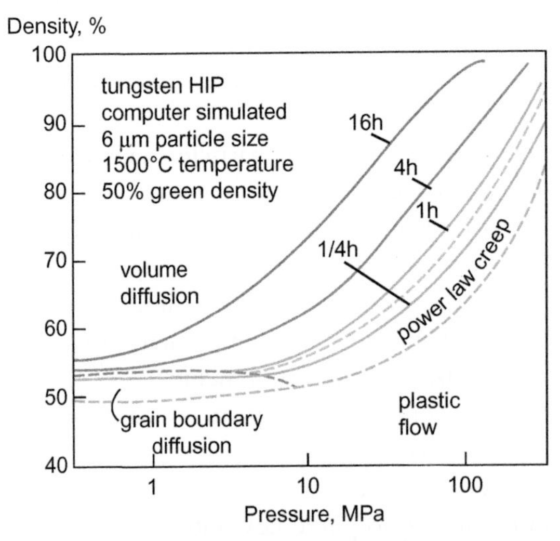

Figure 14.4 Density for 6 μm tungsten powder starting from 50% green density. The calculation is for hot isostatic pressing at 1500°C (1773 K) for times of 0.25, 1, 4, and 16 h over pressures ranging up to 200 MPa. Different mechanisms of grain boundary diffusion, volume diffusion, power law creep, and plastic flow are included in the map; however, at any point there are multiple mechanisms operative.

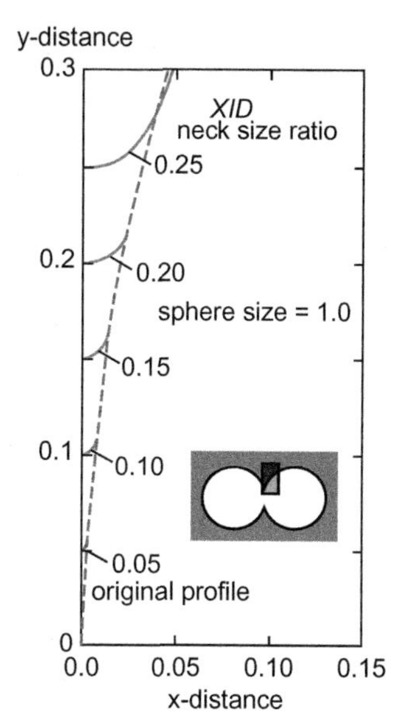

Figure 14.5 Neck profiles computed during sintering by a surface diffusion mechanism. The two sphere geometry is sketched to show the region visualized in the x-y plot.

Answers to questions on component size and shape required three-dimensional simulations [1,24–26]. These are the most challenging and the most rewarding. Even if the results are only approximate, the component level simulations provide striking images. An example is given in Figure 14.6, in which a multiple step thick-thin component is predicted to warp during sintering due to nonuniform heating.

Independent of the dimensionality, a sintering simulation starts with a simple flow chart as diagrammed in Figure 14.7. A problem is defined and the underlying mathematics assembled to construct a model. That model is combined with input data and boundary conditions to execute the simulation. Output from the simulation is used to test technical goals. At the same time the findings are tested against known behavior. If necessary, the conceptual framework is modified and the simulation repeated. Such models range upward from clusters of atoms to furnaces. The latter tend to lean toward economic models.

Construction of sintering simulations requires data. The core simulation is targeted at prediction of a measure such as shrinkage. For such a simulation, the calculation takes input material properties, component specifications, various rules or process models, data on the equipment, and cycle control options. The models are tested for stability over a range of input conditions to assess the sensitivity of the solution obtained to input parameters.

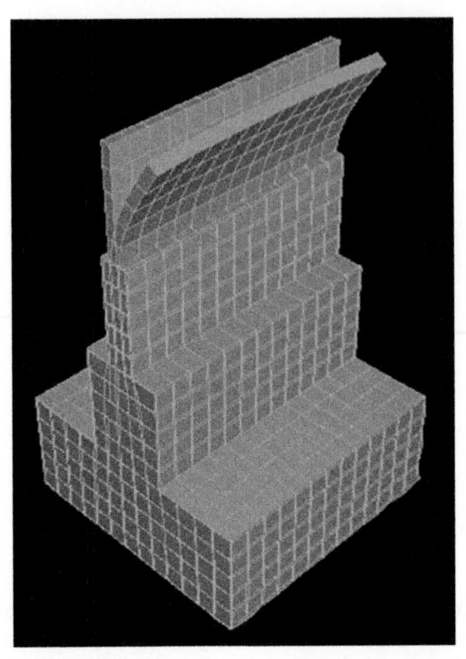

Figure 14.6 A three-dimensional simulation of component warpage during sintering. The finite element grids are evident in the geometry and the distortion is evident in the upper portion.

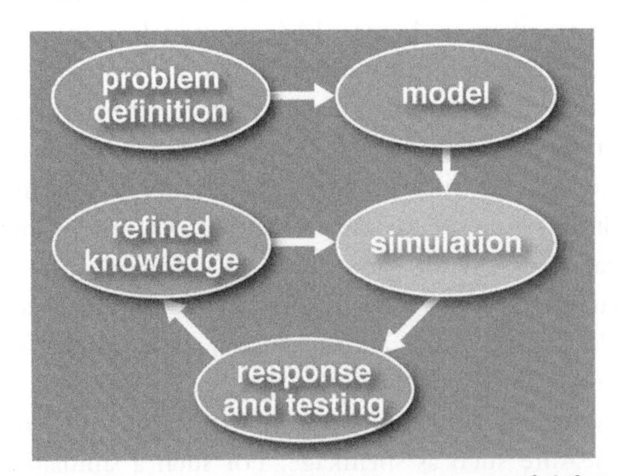

Figure 14.7 The simulation environment is embedded in a process of defining the problem, creating basic mathematical models, and once the simulation is created the output is tested and used to refine basic concepts and knowledge. This process typically repeats until the simulation properly captures the experimental test data.

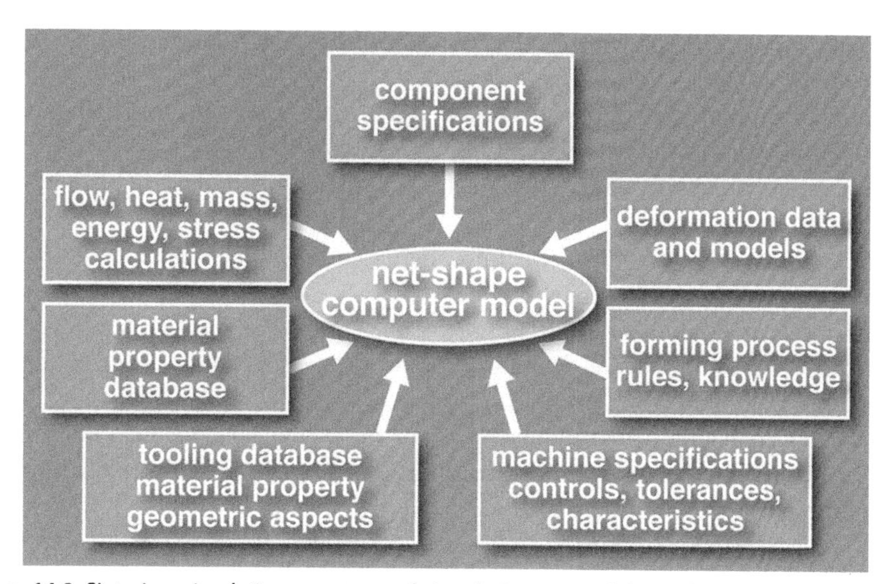

Figure 14.8 Sintering simulations are part of the desire to model net-shape forming processes. This drawing outlines how many different databases impact on the simulation, ranging from component specifications, material data, and tooling factors.

Today, much attention is on prediction of the component shape after sintering; an area of great industrial impact. Initially this was termed the net-shape problem, where knowledge is assembled to predict the final component shape, as outlined in Figure 14.8. The output depends on many factors, ranging from forming machine operation to heat transfer in sintering. This reached semi-commercial software in the late 1990s [27–31]. The inverse problem then became the focus, where a target is specified for the product (size, shape, strength) and the model works to find the process details to meet that target [32–34]. These simulations determined a need for process control to properly understand friction losses, tool wear, heat transfer, phase change, inhomogeneity, and production oscillations. Otherwise, the simulations were inaccurate: Size predictions within 1% were demonstrated, but the needed accuracy was 0.1%. Thus, more knowledge and more complexity are required.

DATA REQUIREMENTS

Depending on the simulation approach, the material data requirements are extensive. As noted earlier, input data include the following types of information:

- material properties–diffusion, plastic flow, elasticity, melting, atomic volume, surface energy

- powder characteristics—particle size distribution, particle shape, packing density, impurities
- microstructure attributes—green density, segregation, pore size distribution, grain size distribution, mixed powder and homogeneity
- component attributes—size, shape, mass, substrate support, green density gradients
- furnace cycle—type of furnace, atmosphere, heating cycle, distance to heating elements, loading, gas flow rate
- special factors—admixed reactants, volatile binders, phase transformation, atmosphere reactions, novel heating situation (microwave, induction, exothermic reaction).

In several simulations, this amounts to 50 or more input parameters. Table 14.1 is an illustration of the data required for densification and grain growth simulations for a tungsten heavy alloy [35–37].

Table 14.1 Example Data Tabulation for Sintering Simulation—Tungsten Heavy Alloy

Material Properties

alloy	88W-8.4Ni-3.6Fe	wt. %
full density	16.80	g/cm^3
volume percent W	68.5	%
liquid formation temperature	1450	°C
W solubility in liquid phase	13.8	%
W self-diffusion pre-exponential	$8\ 10^{-6}$	m^2/s
W self-diffusion activation energy	550	kJ/mol
W diffusion in liquid pre-exponential	1	m^2/s
W diffusion in liquid activation energy	127	kJ/mol
W grain boundary diffusion pre-exponential	$3\ 10^{-13}$	m^3/s
W grain boundary diffusion activation energy	385	kJ/mol
W surface diffusion pre-exponential	0.2	m^2/s
W surface diffusion activation energy	293	kJ/mol
W surface energy	2.8	J/m^2
atomic volume	$1.7\ 10^{-29}$	m^3
activation energy alloy solid state diffusion	241	kJ/mol
pre-exponential alloy solid state diffusion	$2.8\ 10^{-13}$	m^3/s
activation energy alloy liquid state diffusion	105	kJ/mol
pre-exponential alloy liquid state diffusion	$1.1\ 10^{-15}$	m^3/s
room temperature elastic modulus	480	GPa
room temperature yield strength	550	MPa
heat capacity	24	J/(mol °C)
thermal expansion coefficient	$9.27\ 10^{-6}$	1/°C
room temperature thermal conductivity	150	W/(m °C)

(*Continued*)

Table 14.1 (Continued)

Powder Characteristics		
W particle size, D_{10}, D_{50}, D_{90}	2, 6, 10	μm
Ni particle size, D_{10}, D_{50}, D_{90}	3, 10, 24	μm
Fe particle size, D_{10}, D_{50}, D_{90}	2, 6, 10	μm
apparent density	4.1	g/cm^3
Microstructure Attributes		
initial grain size	1.46	μm
solid solubility in liquid	13.8	%
grain growth time exponent	3	
grain growth pre-exponential parameter	$1.1\ 10^3$	$\mu m^3/s$
grain growth activation energy solid state	327	kJ/mol
grain growth activation energy liquid state	105	kJ/mol
Component Attributes		
component diameter	12.7	mm
component height	10	mm
green density	60	%
Furnace Cycle		
heating rate	10	°C/min
hold temperature	1500	°C
hold time	3600	s
furnace type	batch	
atmosphere	hydrogen	
atmosphere thermal conductivity	0.3	W/(m °C)
surface radiation emissivity	0.6	
atmosphere convective coefficient	40	$W/(m^2\ °C)$
component distance from heat source	30	mm
substrate friction coefficient	0.35	

ATOMISTIC CALCULATIONS

The smallest sintering simulation scale involves individual atoms. Although atomistic models are computationally taxing, little information besides atomic interaction parameters are required. The key feature is to assemble a cluster of atoms into spherical particles and to set interaction rules between them. Those rules are the attraction-repulsion energy or force field (force is the derivative of energy versus position). Basic material properties dictate the atomic interactions—elastic modulus, melting temperature, crystal structure, and thermal expansion coefficient. The most common approaches are molecular dynamic or atomistic first-principles modeling. Atomic clusters of thousands of atoms interact and respond to the atomic forces.

Usually two or three atomic clusters are created to represent "particles," where each cluster consists of 10,000 to maybe 50,000 atoms. Energy is input to the system to represent heating and the atoms are allowed to move using small forward time steps. Each atom is described by three position and three velocity indices. Thus, if say 100,000 atoms are involved in the sintering simulation, then considerable computation resource is required for sintering. These calculations are applied to nanoscale particles using supercomputers.

Early applications were sintering coalescence, grain rotation, and additive effects on sintering. The case of nanoscale tungsten sintering is detailed below.

Energetic Approach

The approach applied to sintering simulations is the modified embedded atom method, where the energy for a system of atoms is calculated from the interaction energies relying on the summed pair potentials. The total energy E of a system of atoms is calculated as the sum of the atomic energies E_i:

$$E = \sum_i E_i \tag{14.1}$$

The energy of atom i consists of the embedding energy and the pair potential terms:

$$E_i = F_I(\overline{\rho}_i) + \frac{1}{2}\sum_{j \neq i} \phi_{ij}(r_{ij}) \tag{14.2}$$

The first term is the embedding energy and the second term is the calculated pair potential between neighboring i and j atoms.

The embedding energy is calculated from the sublimation energy (energy to break an atom free of the crystal) [38]. The interaction pair potential is determined in terms of the reference atomic spacing for the material and the packing structure which determines the number of nearest neighbor atoms. Handbook data are used to fit these parameters.

Simulation Procedure

Using the parameters for the atomic interactions, the physical properties are verified, including atomic spacing, elastic modulus, melting temperature, surface energy, and thermal expansion coefficient, all of which provide macroscale insight to atomic scale atomic force profiles [39]. Spherical particles are constructed from the atoms without boundary conditions. The total energy of the each particle is calculated, and the lattice spacing adjusted to relax the structure to its lowest energy. A plot of lattice energy versus atomic spacing is given in Figure 14.9. The handbook value is 0.31585 nm; close to that determined by the simulation.

Figure 14.10 is a computer constructed tungsten particle. The body-centered cubic crystal structure is evident from the atomic positions. Each 10 nm particle consists of 33,700 atoms. Two of these particles are brought into point contact, with a predetermined relative crystal orientation between the particles. Configurations are

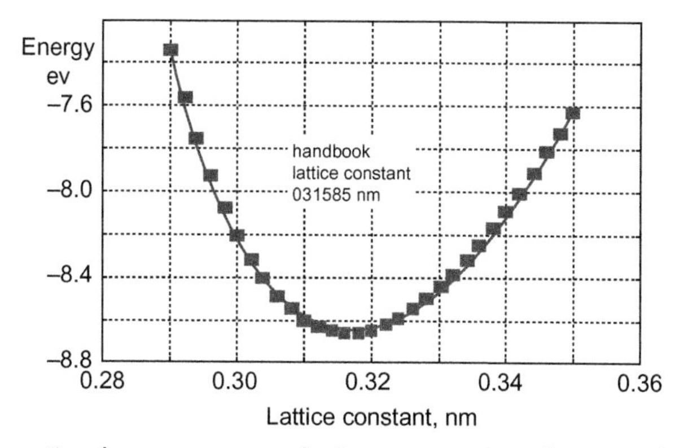

Figure 14.9 Configurational energy versus lattice constant as determined using a molecular dynamic simulation of tungsten particle sintering. The handbook lattice constant agrees with the simulated minimum energy configuration.

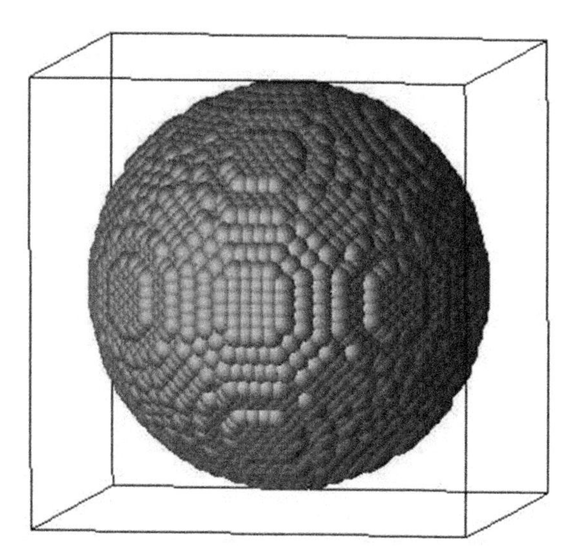

Figure 14.10 Computer simulated nanoscale tungsten particle with the body-centered cubic crystal structure. This is an idealized starting particle consisting of 33,700 atoms as used for sintering simulations.

evaluated at room temperature for 30,000 time steps. Then velocity is added to the atoms, corresponding to heating. Time steps are typically in the 10^{-15} s range to ensure constant interactions between the atoms. Large time steps miss some of the atomic interactions for high velocity atoms. The temperature is increased by increasing the atomic velocities.

For sintering, the particles are energized to a high temperature to induce bonding with an initial separation distance of one atom. Misalignment of the crystal orientations is monitored by the rotation of crystal planes between particles. Once the force vectors are calculated, each atom is allowed to migrate. This produces sintering as evident in the images of Figure 14.11. The particles attract each another while growing a neck. These images correspond to a temperature of 1727°C (2000 K) for progressively longer hold times. Finally, the two particles coalesce into one.

Figure 14.12 illustrates some of the insight that it is possible to derive from atomistic simulations. Superimposed on the two spheres are atomic displacement vectors for each atom that moved during a short time window at 1727°C (2000 K). Most of the motion is near the particle surface, implying that sintering by surface diffusion is dominant. Considering the low temperature and high surface area, surface diffusion neck growth is expected.

Data are extracted from atomic simulations, such as neck size, surface area, or shrinkage versus time, particle size, or temperature for various crystal misorientations. For the case illustrated here, the neck growth rate agrees with surface diffusion

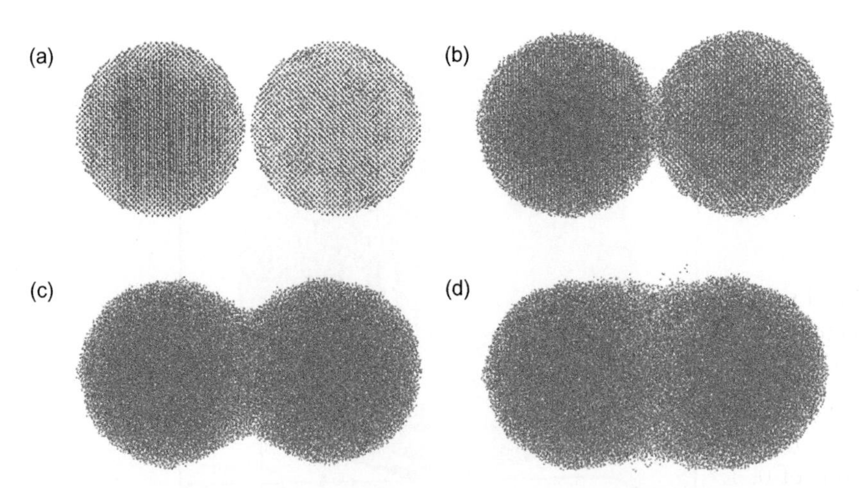

Figure 14.11 A plot of atom positions at three points during computer simulated sintering based on molecular dynamic concepts: a) starting condition just prior to particle contact, b) after initial stage neck growth, c) about three times longer to a point where particle neck growth is advanced and shrinkage is evident, and d) sintered to the point where the two particles bond to form a single ovoid particle.

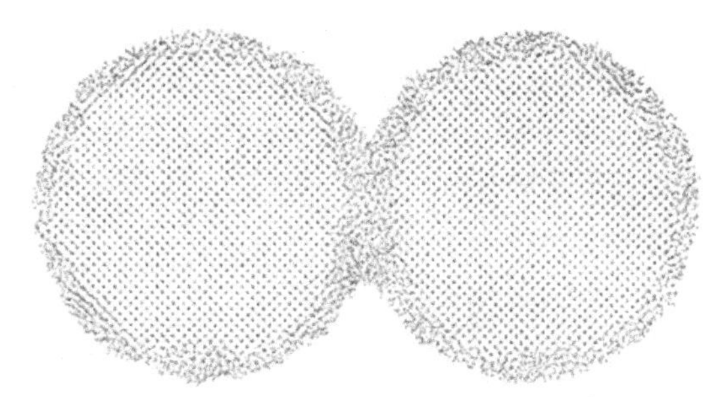

Figure 14.12 Atomic motion vectors captured during an 8 ps time window during the molecular dynamic simulation of sintering at 1727°C (2000 K) for two 10 nm spherical tungsten particles. The atom displacements are evident in the outer surface regions and along the interparticle neck, indicating surface diffusion.

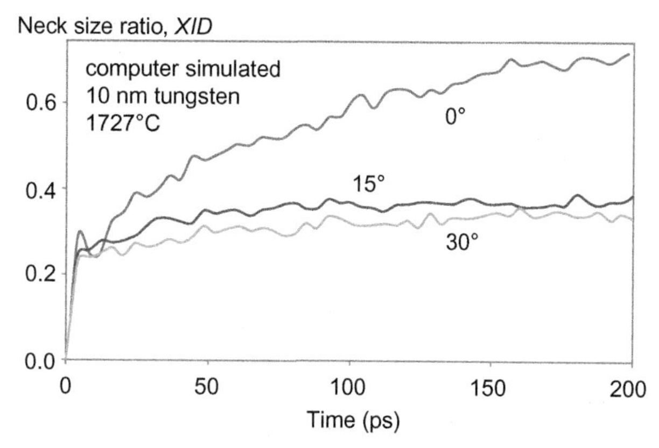

Figure 14.13 Crystal orientation effects (0, 15, and 30° misorientations) on the sintering of two 10 nm spherical tungsten particles at 1727°C (2000 K). The neck size ratio *X/D* is plotted against isothermal hold time, showing that neck growth is faster when the grains are oriented to avoid forming a grain boundary.

controlled sintering. As an example, the activation energy for neck growth at 30° misorientation is calculated to be 269 kJ/mol, and the literature value for tungsten averaged over all crystal orientations is 293 kJ/mol. When the crystal misorientation between particles is 0°, corresponding to aligned crystal planes across the neck and no grain boundary, neck growth is much faster. This is plotted in Figure 14.13 for crystal misorientations of 0, 15, and 30°. At the zero misorientation there is no grain boundary, so neck growth occurs without the impediment of increasing the grain boundary energy.

Similar computer simulations have been reported for several elements, with extensive interest in explaining nanoscale particle agglomeration (sintering) in synthesis, coarsening, and time-dependent behavior at low temperatures. The reports include gold, silver, silicon, and tungsten [40–49]. Lager particle sizes require more atoms and exceed available computational resources. During sintering, data are generated on neck size, shrinkage, surface area, activation energy, and can include impurities, uneven particle sizes, and nonuniform heating cycles.

RESHAPING MODELS

A starting powder structure has excess surface energy and that energy dissipation drives sintering. Given atomic motion to reshape the particles, various simulations provide predictions of the shape change process. In these models, mass transport takes place by any of several mechanisms, such as surface diffusion. Initially each investigator constructed the algorithm, similar to that outlined in Chapter 6. However, now algorithms are available to perform these calculations with user defined rules and starting geometry.

The starting geometry is defined, usually as two equal sized spheres or wires. Often amorphous spheres and viscous flow are assumed to avoid the grain boundary and dihedral angle. The normal motion of a point on the surface is then calculated from the local surface curvature gradient and the transport rate.

$$\frac{\partial N}{\partial t} = B\nabla^2 K \tag{14.3}$$

The form of the equation signifies that the surface normal motion depends on the curvature gradient and the rate parameter B, which for surface diffusion is given as:

$$B = \frac{D_S \gamma n \, \Omega^2}{R\,T} \tag{14.4}$$

where D_S is the surface diffusion coefficient at absolute temperature T, γ is the solid-vapor surface energy, n is the population of mobile atoms per unit area, Ω is the atomic volume, and R is the gas constant. The flux of atoms on the surface, and by implication the volume of material arriving at the neck, depends on the surface curvature gradient. Starting from a contact point, time is advanced in small steps. The normal motion of the surface is calculated at all spots, and the surface is repositioned to reflect the motion during the next time step. Recalculation loops, with surface motion, give snapshots of the way that the surface moves over time. Early problems arose with cumulative numerical errors, leading to loss of mass and artifacts such as undercutting in the necks.

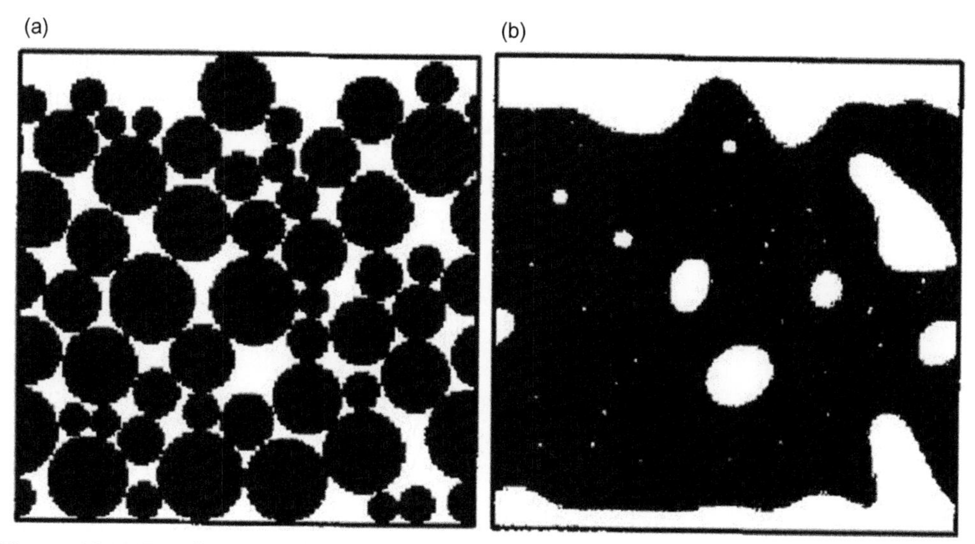

Figure 14.14 Two-dimensional simulation of pore reshaping during sintering: a) is the starting geometry and b) is the sintered geometry with a few large pores [54].

Refinements increased the simulation accuracy, and allowed for sphere-plate, wire-wire, sphere-sphere, and uneven sphere sintering [4,50,51]. At times the results differed from experimental findings by as much as a factor of ten. Over time, more stable computers allowed the simulations to be revisited and added factors to improve the predictions [7,51–53]. A favorite calculation is coalescence—two spheres in point contact eventually fuse into a single large sphere.

If multiple particles of differing size and packing coordination are included in the assembly, then, pore coarsening is predicted. A representation is provided in Figure 14.14, based on a two-dimensional simulation [54]. The circles are initially in point contact, and late in the simulated sintering the pore count has decreased, but the few remaining pores have enlarged. Other simulations include multiple mechanism sintering with grain boundary migration and viscous flow sintering [51–58]. Figure 14.15 plots the calculated velocity field during viscous flow sintering of two spheres at a neck size ratio $X/D = 0.5$. Because of mass conservation, the sphere shrinks in diameter to provide mass for the neck.

Surface reshaping simulations advanced with software constructed for surface motion toward reduced energy and reduced curvature gradient morphologies. The Surface Evolver software by Brakke [http://www.susqu.edu/brakke/evolver/evolver.html, updated 25 August 2013] allows calculation of the shaping events based on input geometries, material properties, and boundary conditions. The predictions include terminal grain shapes and grain size distributions [59,60].

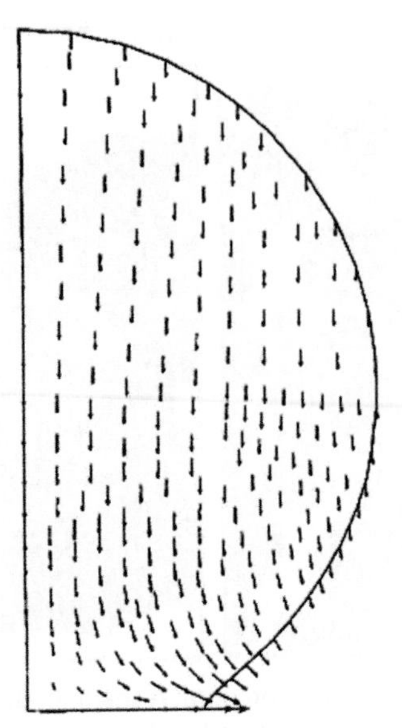

Figure 14.15 Velocity vectors overlaid on a calculated two sphere viscous flow sintering neck, indicating the magnitude and direction of mass flow during neck growth.

PHYSICAL EVENT MODELS

A more reassuring simulation approach is to simply calculate the contributions from the several events during sintering. Computers are required to step time forward through the complex array of mass flow processes and the evolving microstructure. This approach is used for one-dimensional predictions of density, targeted at sintering and hot isostatic pressing. When the pressure is set to zero the simulation is for sintering (what is sometimes called pressureless sintering or free sintering). Early models assumed isothermal conditions, spherical powders, and monolithic materials such as silver.

These simulations relied on multiple mechanism solutions to simulate neck growth and build time-dependent combined processes. Using classic models for sintering, such as linear shrinkage $\Delta L/L_O$, a rate equation is assumed, such as the following:

$$\left(\frac{\Delta L}{L_O}\right)^N = \frac{B\,t}{G^M R\,T}\exp\left[-\frac{Q}{R\,T}\right] \tag{14.5}$$

This form includes N and M as mechanism exponents (for example $N=3$ and $M=4$ for grain boundary diffusion), isothermal hold time t, absolute temperature T, with $R=$ gas constant, $G=$ average grain size, $Q=$ activation energy, and $B=$ collection of geometric and material constants such as surface energy, vibrational frequency, and atomic volume. From the first derivative based on time, the incremental change in component length during sintering is obtained. Using a small time step (where temperature might also be a function of cumulative time during heating) the shrinkage is incremented to calculate a new shrinkage using the instantaneous rate of change and a small time step Δt:

$$\left[\frac{\Delta L}{L_O}\right]_{new} = \left[\frac{\Delta L}{L_O}\right]_{old} + \Delta t \frac{d\left(\frac{\Delta L}{L_O}\right)}{dt} \tag{14.6}$$

Time is likewise advanced, such that the total time is the summation of the number of cycles through time steps Δt. Computers are useful in combining the incremental changes from all possible sintering processes. This approach allows non-isothermal heating, and evolving material and geometric parameters, for example grain size or phase transformation. Inherently such simulations are underpinned by sintering theory, so the only adjustable parameter is the forward time step, Δt. Techniques for sensing rapid change to make smaller time steps ensure stable calculations.

If each mechanism can be evaluated for the rate of shrinkage, densification, surface area change, or neck growth, then the combination of mechanisms (given the geometry, temperature, and material properties) contribute to the overall rate. Summation of the individual contributions over all mechanisms gives the instantaneous total rate, and integration over time gives the cumulative total change in density, shrinkage, neck size, surface area, or other parameter. Figure 14.16 is an example simulation output for

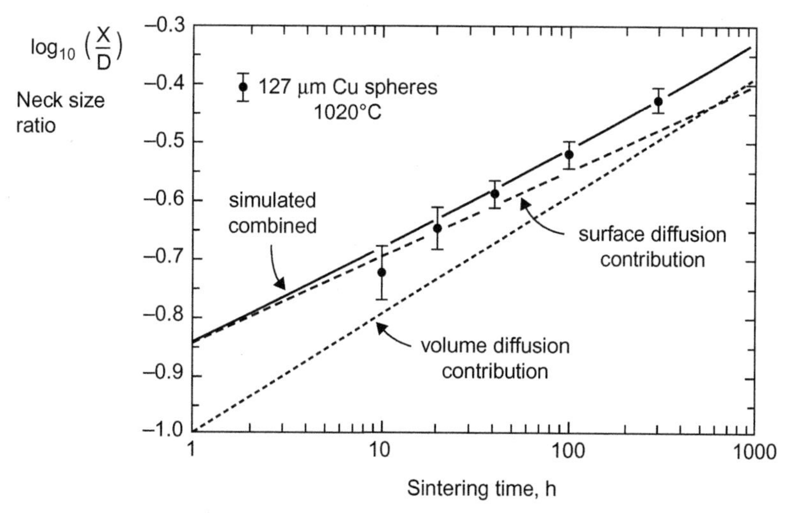

Figure 14.16 Calculated neck size ratio (*X/D*) during sintering for 127 μm copper at 1020°C (1293 K) [12]. The experimental points are compared to three sintering simulations – one is for volume diffusion, one is for surface diffusion, and one is for combined simultaneous surface and volume diffusion.

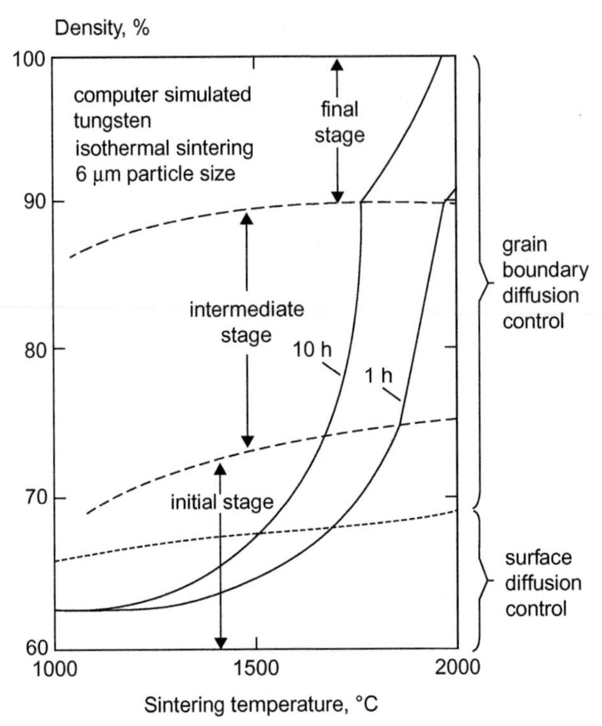

Figure 14.17 Computer simulated density versus temperature for hold times of 1 or 10 h for the sintering of 6 μm tungsten powder. The three stages of sintering are indicated and the shift in dominance from surface diffusion to grain boundary diffusion is also traced.

neck growth between 127 μm copper spheres [12]. The neck size ratio (*X/D*) observed during sintering is compared with the calculated size for pure surface diffusion, pure volume diffusion, and the two acting in combination. The combined mechanism best fits the experimental results.

The most successful models were constructed by Ashby [61]. He added pressure into a sintering calculation to include hot isostatic pressing. The resulting model gives density versus sintering temperature for isothermal sintering, but has the ability to vary several independent parameters. Figure 14.17 plots the simulation results for 6 μm tungsten held for 1 or 10 h at temperatures from 1000 to 2000°C (1273 to 2273 K). In this simulation, the sintered density is dominated by grain boundary diffusion, although other mass transport processes are active. Surface diffusion is dominant when there is little densification.

The application of pressure adds mechanisms to the sintering problem, including creep and plastic flow events. This leads to the hot isostatic pressing plot of density versus time, temperature, or pressure. Figure 14.18 maps full density solutions for Ti_3Al consolidation starting at a green density of 62% with the goal of reaching 100%

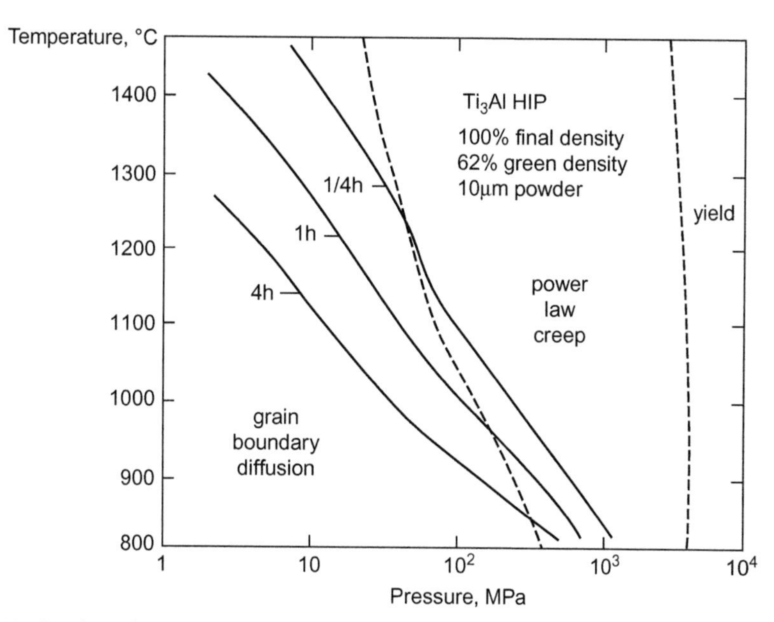

Figure 14.18 Combined pressure, temperature, and time map for the hot isostatic pressing of 10 μm Ti_3Al starting at a green density of 62%. This map shows the combinations of the three parameters required to attain full density. A lower pressures densification is dominated by grain boundary diffusion, but at higher pressures power law creep dominates densification.

density. The various combinations of temperature, pressure, and time required to achieve densification are given by the three curves. Densification at the lower pressures is dominated by grain boundary diffusion. Subsequent efforts added heating, stoichiometry, polymorphic materials, and particle size distributions [15,62–65]. Recent efforts have extended modeling to include laser sintering in additive manufacturing and other industrially relevant topics.

Computer simulations based on physical event models are appealing since the approximations are few and behavior is directly tied to measured material parameters. This same attribute is a limitation, since new materials have poorly defined properties, such as surface or grain boundary diffusion rates. Also sintering situations are complex problems, and require considerable care to properly include all of the relevant factors. For example, in modeling the sintering dimensional change in a mixed powder alloy of Fe-2Cu-0.8C, several factors had to be included [32]; internal porosity in the iron powder, particle sizes, heating rate and peak temperature, hold time, and atmosphere composition. Dimensional dilation was predicted as illustrated in Figure 14.19. The net dimensional change is small, amounting to 2% expansion at the peak temperature.

Supersolidus liquid phase sintering includes composition and temperature as factors that determine the system viscosity [66]. Densification depends on the microstructure.

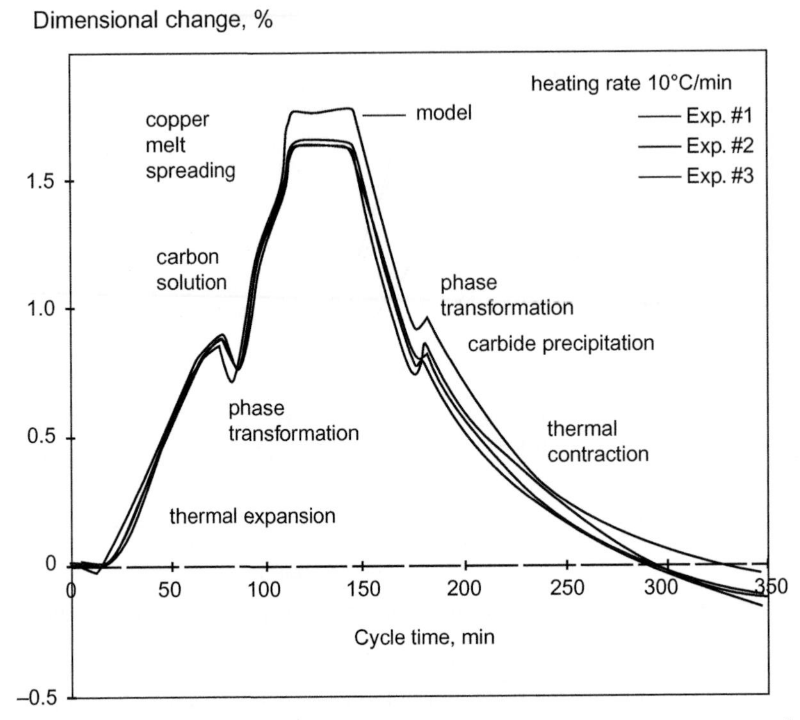

Figure 14.19 Computer simulated dimensional change for a powder mixture corresponding to the composition Fe-2Cu-0.8C (wt.%) [32]. Three dilatometer experiments are plotted to show the measurement variation along with the model behavior. Some of the key events during heating such as carbon dissolution and polymorphic phase changes are labeled on the plot.

Early in sintering the grain size is small with a high grain boundary area not covered by liquid, since there is insufficient liquid to film over all grain boundaries. Densification is slow. However, grain growth leads to a critical condition where the grain boundary area declines, allowing the liquid volume to finally coat grain boundaries to soften the structure. Densification is rapid at this point, as is evident in Figure 14.20. This plot of density versus temperature corresponds to a 10°C/min heating rate, comparing measured and simulated results [67]. During heating, both microstructure coarsening and liquid spreading lead to a complicated viscosity behavior that determines when densification occurs. Further, the model predicts that overheating results in excessive softening and component distortion. The result is a sintering window—reflecting combinations of temperature, time, heating rate, and particle size that produce densification without distortion.

The physical models predict average sintered properties, such as density, reasonably accurately. Fluctuations in materials, compositions, and powder characteristics give

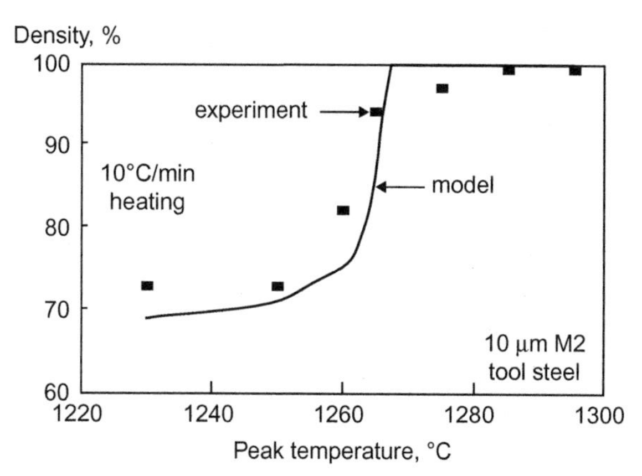

Figure 14.20 A comparison density for both experiment and model during the supersolidus sintering of 10 μm tool steel powder heated at 10°C/min to various peak temperatures [67].

large output variations. To truly provide predictive models requires a large body of input data, more than is currently justified. Thus, in large part manufacturing operates in an adaptive process control mode, in which time, temperature, pressure, or other independent variables are used to compensate for variations in input materials.

MONTE CARLO METHODS

Monte Carlo simulations rely on rules (such as energy reduction) and random probability events to mimic atomic motion. This is equivalent to literally rolling the dice to decide what happens. It is a favorite technique for modeling microstructure evolution during sintering and is well positioned to track grain and pore behavior. Unfortunately, most Monte Carlo simulations are not useful since they rely on coarse simulation elements (their typical scale is much larger than atoms and even larger than grains) and are often only two-dimensional. They reflect how oil droplets floating on water coalesce over time. Three-dimensional models have been developed more recently [17,68].

To start with, a microstructure is constructed consisting of a large number of discrete points in space. These are termed pixels in two-dimensional simulations and voxels in three-dimensional simulations. Each point is assigned a flavor, color, chemistry, or orientation—some means to create granularity in the initial microstructure. The starting point might be input by the operator. A few models nucleate the structure using random coordinates, and then grow around those nuclei—the first nucleus is "1" and the second is "2" and so on. After nucleating, then particles, grains, or pores are expanded around each nucleus, usually with uniform radial growth, until all the space is filled. This gives the starting randomized microstructure.

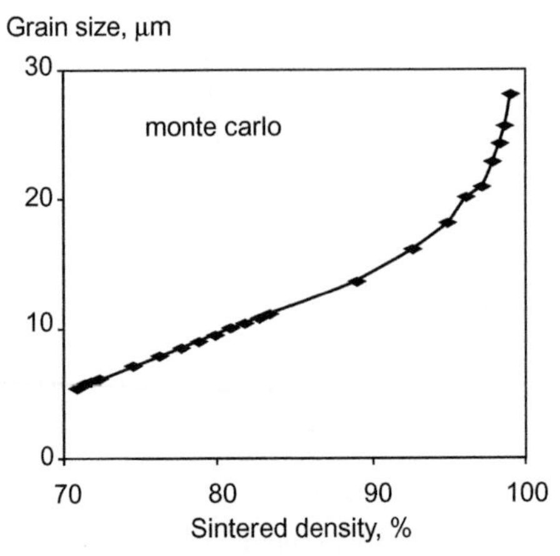

Figure 14.21 Grain size versus sintered density for a Monte Carlo simulation of sintering [29]. The ability to extract such data during a computer experiment is one attraction of the approach.

Random events and energy reduction rules then determine the evolution of the structure and its final stability [69]. A random number is generated, from which a candidate point is selected and tested versus a random possible move. Evolution of the structure depends on simple rules; in a two-dimensional simulation a point joins a neighboring group if the energy goes down. If the points are alike, they tend to agglomerate or bond together, and that leads to a decline in interface energy.

Monte Carlo methods are a favorite for microstructural coarsening studies, such as grain growth during liquid phase sintering [70], particle coalescence by surface diffusion [71], percolation and conductivity [72], and pore-boundary interactions [73]. As an example, Figure 14.21 plots grain size versus density during sintering, giving behavior similar to that reported in several experiments [29].

Another gain from Monte Carlo simulations is the isolation of generalized rules on the microstructure evolution during sintering. Figure 14.22 is an example of the grain edge length distribution determined for a microstructure by a Monte Carlo simulation [74]. This provides insight that is not easily gained in traditional laboratory experiments. Although not yet pinned to actual materials, progress in this field is promising and some of the simulations are developing into commercial packages.

CONTINUUM MODELS AND FINITE ELEMENT ANALYSIS

The evolution of continuum models goes hand in hand with the use of finite element analysis. In these, a structure is subdivided into many small elements.

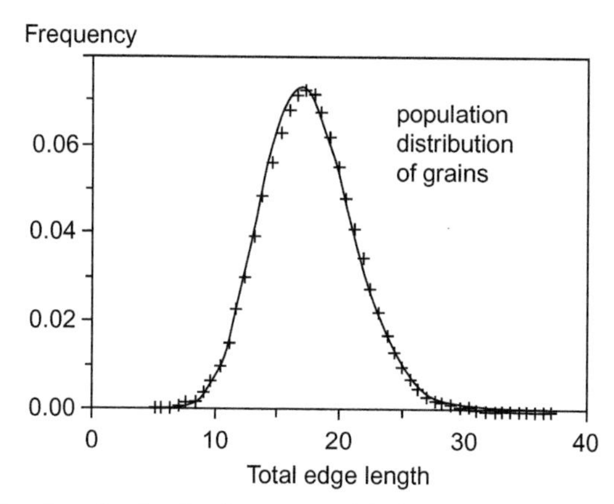

Figure 14.22 Grain edge length distribution as modeled using Monte Carlo techniques [74]. The ability to extract detailed data of this sort from the model is an advantage over laboratory experiments.

The element size is small compared to the component scale, but large compared to the atomic scale. In the simulation, force, stress, heat, or other effects are propagated though the body from element to element. The finite element method solves descriptive equations through approximation functions applied to each element. To properly implement the finite element approach requires descriptive equations that establish the behavior inside each element. Continuum models are a means to move from complicated relations to equations that relate state variables. So instead of including diffusion rates, the continuum approach relies on stress-strain-strain rate relations.

The combination of constituent equations and finite elements is a powerful means for predicting final sintered component size and shape for complex geometries. The green body properties need to be known if simulations are to be accurate. Commercial software packages exist for performing the calculations using three-dimensional thermo-mechanical processes which are adaptable to sintering problems; some examples include liquid phase sintering densification, distortion, microstructure evolution, and warpage [21,27,33,34,75–79]. A common approach is to ignore diffusion rates and treat the solid as a viscous system.

Finite element calculations rely on constitutive equations, but not measured material properties. Simulation of the final component size and shape, properties, and defects are routinely performed using personal computers. The simulations are fast for axisymmetric shapes, but it is possible to use complicated three-dimensional shapes. A three-dimensional simulation provides a detailed dimensional description of the sintering behavior for a given geometry. Material flows according to constitutive equations that derive from diffusional creep laws. The phenomenological models rely on the determination of a sintering stress that pulls the particles together and viscous flow rules related to the material, porosity, grain size, and temperature. The three-dimensional stress state is then related to the deformation strain rates, or shrinkage. The boundary conditions invoke mass conservation, momentum conservation, and energy relations. It is in the constitutive equations that much disagreement arises regarding how to approximate particle-particle sinter bonding (strengthening), thermal softening, densification, grain growth, pore rounding, pore growth, and other microscopic events. Usually the models assume the form of a strain rate tensor as follows:

$$\varepsilon'_{ij} = \frac{1}{2\eta}\sigma'_{ij} + \frac{1}{3\kappa}(\sigma_M - \sigma_S)\delta_{ij} \tag{14.7}$$

where the first term is the shear viscosity η for the material during sintering and the derivative of the stress tensor σ_{ij} and the second term is the bulk viscosity κ and the difference between the hydrostatic stress σ_M and the sintering stress σ_S arising from surface energy, with δ_{ij} being 1 if $i = j$, otherwise 0. It is in the determination of each of these parameters that the models vary by a factor of two. In some cases σ_S is simply set as a constant (not very realistic) and in others it is calculated from the fractional density (better), and in the most involved models it is determined from particle size, surface energy, porosity, grain size, and detailed factors.

In a finite element calculation the strain rates are small. The body is not granular, but simply a continuum. This way the first term describes the distortional behavior of the continuum by relating the deviatoric stress and rate through the shear viscosity. The second term describes the shrinkage by relating applied hydrostatic stress (such as in hot isostatic pressing) and sintering stress to the volumetric strain rate.

Conservation of mass provides the link between the volumetric strain rate and the densification rate. Several factors need to be included in the analysis, such as gravity, substrate friction, nonuniform heating, and the density gradients arising from the powder forming process. Density calculations rely on knowing or assuming the input density gradients. If a homogeneous green body is assumed, then the resulting simulated shape is inaccurate.

Input response equations are required to simulate sintering via finite element analysis. These can be as simple as Hooke's law, in which stress is proportional to strain for elastic deformation. Viscosity is a function of composition, temperature, density, and grain size and this proves to be a dominant factor. Such information is rarely known, so many assumptions are required. Usually the relations are simplistic, yet produce reasonable results. Additional sensitivity comes from impurities, requiring empirical treatments. Although there is much error in the assumed behavior, fortunately many of the errors seem to offset one another, so as a consequence the results are surprisingly reasonable.

Initiation of a simulation requires definition of the boundary and initial conditions. Initial conditions are input for density in each element. Symmetric boundaries apply if there is rotational symmetry. A major concern is friction with the supporting substrate. Free boundaries describe the outer surface.

The governing equations and the component geometry are combined with the constitutive models, and boundary and initial conditions, and other parameters to simulate the shape and size evolution during sintering. The sintering cycle is also entered. As the temperature changes with time, variables in the governing equations and constitutive models change.

To illustrate the use of such simulations, a T-shaped liquid phase sintering geometry is shown in Figure 14.23a with initial meshing on the body. The starting height is 30 mm and during sintering it shrinks to about 19 mm. The simulated shape after sintering is given in Figure 14.23b, in which slumping and edge rounding are evident. For comparison, the actual shape after sintering is pictured in Figure 14.23c. The simulation is reasonable except for the corners, where the rounding was not captured by the simulation. This simulation required relations for sintering stress, bulk viscosity, shear viscosity, and grain growth. Independent experiments had to be run to provide estimates for these relations. More accuracy in the simulations implies more experimental work, to the point where it is easier to optimize via experiment than by simulation. Experience shows that about five finite element simulations are required to make the calculations converge with known behavior.

One use of these simulations is in the prediction of ways to optimize the powder forming operation to deliver target component features. The flowchart in Figure 14.24 outlines how the optimization is approached. The geometric model and processing conditions are input. The core activity is the simulation, which is driven by an overall design of experiments, focused on attaining target attributes. The simulation is repeated over and over to converge to an optimal design—possibly in terms of forming tool motion, powder selection, or sintering cycle.

DISCRETE ELEMENT METHOD

Discrete element techniques are related to finite element analysis. They differ in that there is no mesh in the former; each particle is an element. Sometimes this approach is termed mesh-less modeling. It is most applicable to simulations where the focus is on particles, since it does not handle coarsening, changes in grain size, and disappearance of grains.

In discrete element analysis, the system equilibrium is determined between particles in terms of mechanical, chemical, capillary, gravity, or other forces. Each particle is treated as a single element to assess interactions between particles. Such an approach allows for assemblies of thousands of particles and it is preferred for treating powder

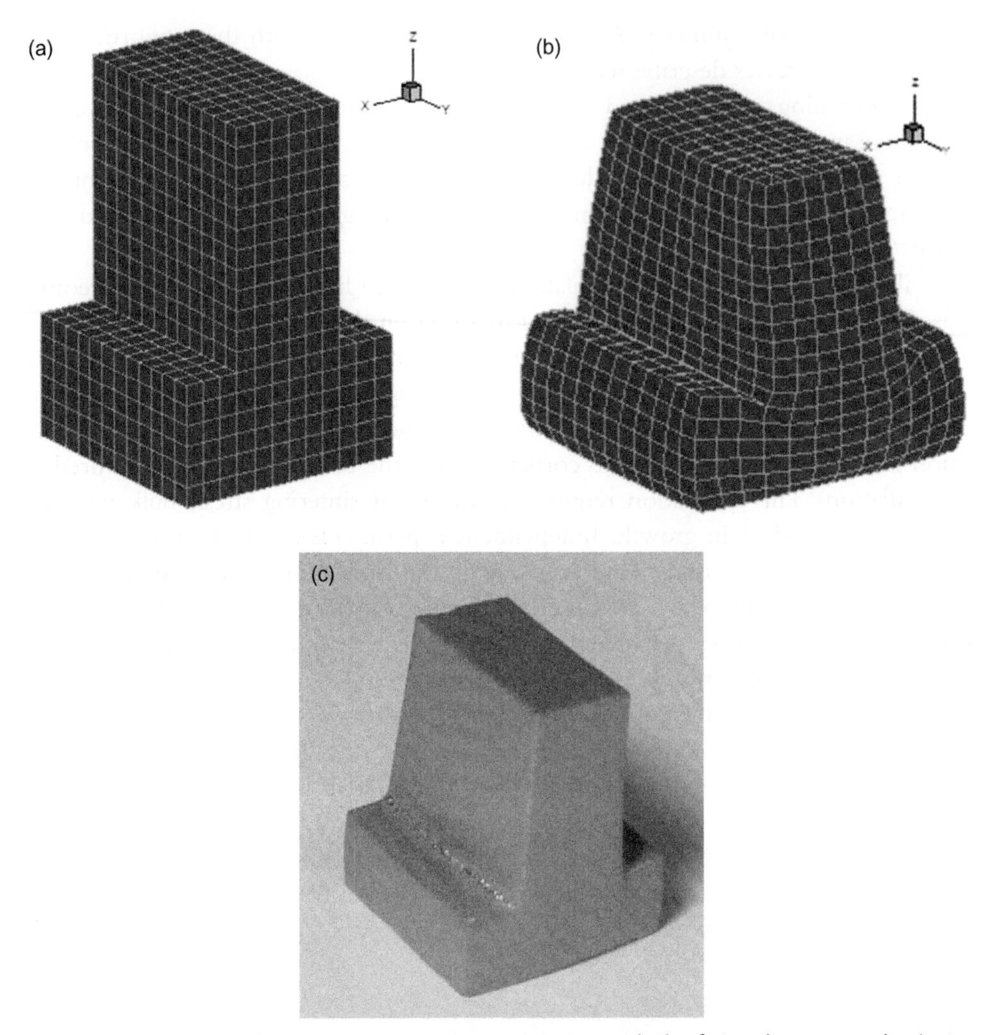

Figure 14.23 An inverted T test geometry prior to sintering with the finite element mesh: a) gives the initial geometry for the 30 mm high sample, b) gives the predicted shape after sintering where the sample is now 19 mm high, and c) shows the physical sample for comparison.

flow, container filling, and compaction and other situations involving particle arrays [80–84]. Figure 14.25 is a case where a 70% large powder is mixed with 30% small powder with a particle size ratio of 3:1. This simulation provides information on the large-small, small-small, and large-large particle coordination. Subsequent neck growth depends on the neighbor size. As a consequence the simulation provides input on particle neighbors for neck growth simulations, but is less useful for sintering. Other solutions include analysis of substrate friction [85], densification induced

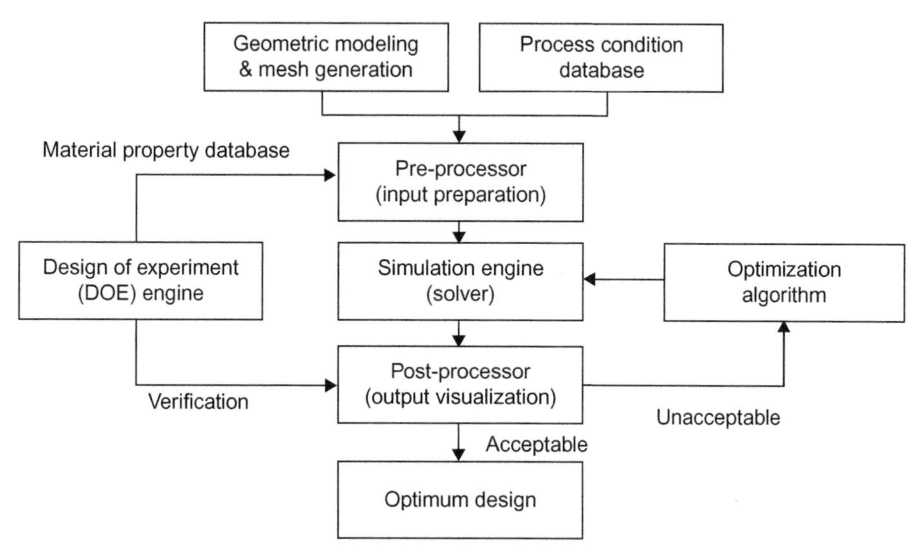

Figure 14.24 Optimization relies on a sequence of decisions and determination of interactions as outlined in this schematic. This is an active area for sintering simulations.

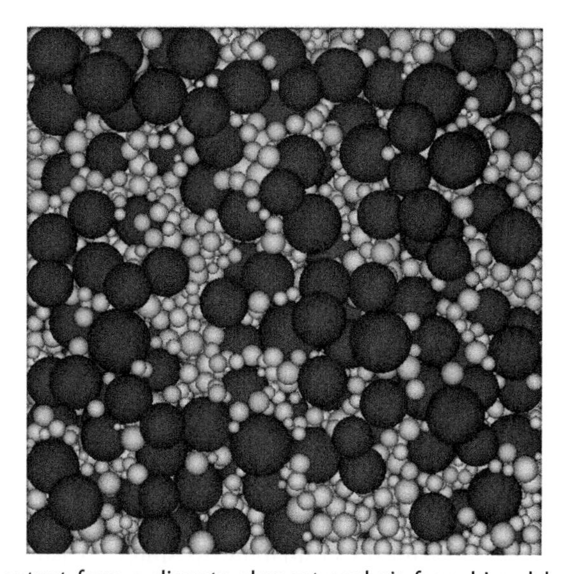

Figure 14.25 Visual output from a discrete element analysis for a bimodal mixture of 70% large powder and 30% small powder with a 3:1 size ratio. The simulation allows for statistical extraction of the contact frequency for large-large, large-small, and small-small pairs. Such information is then useful in sintering neck growth simulations.

strain [86], and anisotropic shrinkage [87]. Unfortunately, many of the early simulations were two-dimensional and failed to provide insight. Recent simulations have added coarsening to the calculations [88].

MASTER SINTERING CURVE

Sintering models have a characteristic body of mathematical equations. These include terms based on independent parameters of time, temperature, pressure and include the evolving density and microstructure. A certain work of sintering (time and temperature) is embedded in these terms [89]. The integral work of sintering then links this work to the sintering response, such as shrinkage, density, strength, grain size, or distortion [90]. A master curve concept arises via a separation of variables, giving a combined stage and mechanism model applicable to sintering, as well as debinding, coarsening, distortion, creep, and component failure. For example, in describing sintered density, the following rearranged densification model is applied:

$$\frac{R\,G^N}{3f\,\Omega\,\Gamma\,D_O}\,df = \frac{1}{T}\,exp\left(-\frac{Q}{R\,T}\right)dt \tag{14.8}$$

where G is the grain size, T is the absolute temperature, t is time, R is the gas constant, f is the fractional density, Ω is the atomic volume, $N=3$ is the grain growth exponent, Q is the apparent activation energy, D_O is the diffusion pre-exponential frequency factor, and Γ is a combination of geometric terms. These are all related to the constitutive equations encountered in approaches such as finite element analysis models. Similar forms are possible for shrinkage, grain size, distortion, and other metrics applied to sintering.

Separation of variables enables calculation of a work of sintering parameter Θ, that includes only time t and absolute temperature T,

$$\Theta(t,\ T) = \int_0^t \frac{1}{T}\,exp\left(-\frac{Q}{R\,T}\right)dt \tag{14.9}$$

In this form the only unknown is the apparent activation energy Q. Often a first estimate of Q is possible from grain boundary diffusion or even melting temperature. However, since multiple mechanisms contribute to sintering (surface diffusion, grain boundary diffusion, plastic flow, and so on), it is appropriate to solve for a best fit activation energy. This is performed by measuring grain size or density using experiments with intentional variations in heating rate, peak temperature, and hold time. Parameters such as density approach an asymptotic value. Other parameters are

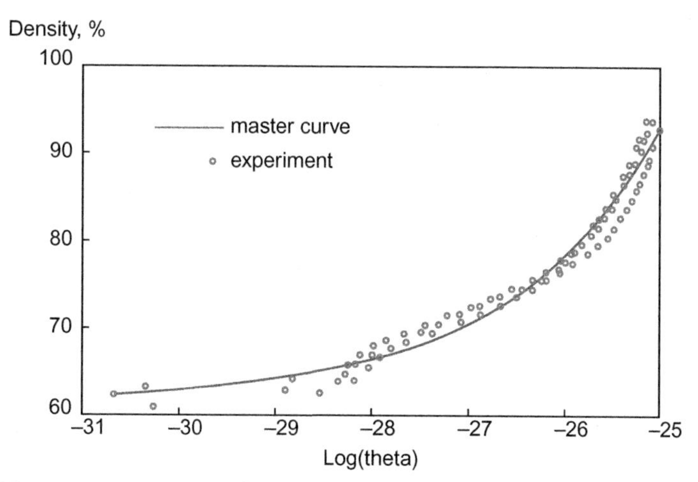

Figure 14.26 Master sintering curve for sintering stainless steel powder [91]. The experiments involve different heating rates, peak temperatures, and hold times. The resulting master curve combines the variables into the integral work of sintering (theta).

unlimited, such as grain size, so there is a difference in the way they are treated. Note that the master curve route has considerable advantages over other simulation routes since it is based on experiments, but has only one key unknown material parameter.

As an example of the master sintering curve concept, Figure 14.26 plots the sintered density for a stainless steel [91]. Here the final density is capped at 100%. Once the master curve is known, then different time-temperature combinations are explored without the need for further experiments. For example, if a slower heating rate is employed then the consequence can be predicted, leading to potentially a longer hold to attain the same degree of sintering. Additional treatments have included debinding, role of particle size, and even determination of the means to form nanoscale structures to full density. The latter is performed by creating two master curves, one for densification and one for grain growth, and then solving for the intersection to identify processing options [92].

NEURAL NETWORKS

One approach to modeling sintering in the context of manufacturing is to rely on a neural network. The intent is optimized control by including the input powder variation and a host of parameters that might fluctuate. Training a neural network depends on a body of knowledge where powder attributes, processing conditions, and final product properties are monitored [93].

A neural network model implies some form of artificial analogy to biological nervous systems. Once trained, it is a means to link input variables (say particle size) to the output variables (say strength), without knowing the detailed linkage or how sintering time, temperature, and similar variables enter into a model. In other words the model is a nonlinear "black box" trained using experimental data [94–95]. It is effective when the variables remain within the range used for training the model.

Two types of neural network models exist:

- Those that predict the behavior of a given powder or processing cycle.
- Those that advise on the processing or powder characteristics appropriate for certain target properties [96].

The simulation seeks a representative mathematical linkage of variables without paying close attention to physical realities. The fact that sintering involves many independently adjustable parameters makes the approach helpful in controlling industrial processes.

SUMMARY COMMENTS

Accurate sintering predictions require a large body of information ranging from powder details to thermal cycle details, so the predictions are very sensitive. Small errors in the material properties, especially activation energies for diffusion and the role of crystal orientation or impurity segregation, lead to considerable variations in the predicted behavior. Accurate simulations require a wide breadth of knowledge that is only approximated to date. Although the simulations appear close to experiment, the translation into levels useful in industry is a major difficulty. Industry holds to a narrower sintered dimensional variation than can be supported by the existing simulations.

Although commercial software packages are available, simulations are not widely employed in sintering. This is due to the inherent process variation. Indeed, adaptive process control has long been adopted in industry as a means of continuously adjusting production to meet tight tolerances. In other words, industry has learned to live with the variation that is a barrier to accurate simulations.

Computer models are sensitive to a balance between speed and accuracy. Early models converged to reasonable predictions for solid state sintering neck growth. However, the real challenge is to predict final dimensions with a production level accuracy. Components tend to distort during sintering, so controlling this is a problem of great relevance. There are two approaches to distortion. One is to control the process conditions to obtain as uniform density as possible after powder compaction. The other is to calculate and anticipate distortion in designing the forming process and tooling. The latter approach is preferred by industry and remains a target for the sintering simulations.

Optimization is an area where sintering simulations are of great value. Simulation predictions are linked to control parameters and then used to identify situations insensitive to normal process variations. The fundamental need is to predict robust conditions that stabilize component attributes. This lowers production cost and improves process yield. The master sintering curve approach is effective in this regard since it provides predictions using spreadsheet calculations [97–100].

REFERENCES

[1] S.H. Chung, Y.S. Kwon, S.J. Park, R.M. German, Modeling and simulation of press and sinter powder metallurgy, in: D.U. Furrer, S.L. Semiatin (Eds.), Metals Process Simulation, ASM Handbook Volume 33B, ASM International, Materials Park, OH, 2010, pp. 323–334.
[2] M.F. Ashby, Operating Manual for HIP 6.0, Engineering Department Cambridge University, Cambridge, UK, 1990.
[3] K.E. Easterling, A.R. Tholen, Computer-simulated models of the sintering of metal powders, Z. Metall. 61 (1970) 928–934.
[4] F.A. Nichols, Theory of sintering of wires by surface diffusion, Acta Metall. 16 (1968) 103–113.
[5] K. Breitkreutz, D. Amthor, Monte-carlo-simulation des sinterns durch volumen-und oberflachen-diffusion, Metall 29 (1975) 990–993.
[6] G.J. Cosgrove, J.A. Strozier, L.L. Seigle, An approximate analytical model for the late-stage sintering of an array of rods by viscous flow, J. Appl. Phys. 47 (1976) 1258–1264.
[7] R.M. German, J.F. Lathrop, Simulation of spherical powder sintering by surface diffusion, J. Mater. Sci. 13 (1978) 921–929.
[8] J.W. Ross, W.A. Miller, G.C. Weatherly, Dynamic computer simulation of sintering by volume diffusion, Z. Metall. 73 (1982) 391–398.
[9] P.W. Voorhees, M.E. Glicksman, Ostwald ripening during liquid phase sintering- effect of volume fraction on coarsening kinetics, Metall. Trans. 15A (1984) 1081–1088.
[10] P. Bross, H.E. Exner, Computer simulation of sintering processes, Acta Metall. 27 (1979) 1013–1020.
[11] N. Rosenzweig, M. Narkis, Dimensional variations of two spherical polymeric particles during sintering, Poly. Sci. Eng. 21 (1981) 582–585.
[12] K.S. Hwang, R.M. German, Analysis of initial stage sintering by computer simulation, in: G.C. Kuczynski, A.E. Miller, G.A. Sargent (Eds.), Sintering and Heterogeneous Catalysis, Plenum Press, New York, NY, 1984, pp. 35–47.
[13] T.M. Hare, Statistics of early sintering and rearrangement by computer simulation, in: G.C. Kuczynski (Ed.), Sintering Processes, Plenum Press, New York, NY, 1980, pp. 77–93.
[14] C.M. Sierra, D. Lee, Modeling of shrinkage during sintering of injection molded powder metal compacts, Powder Metall. Inter. 20 (5) (1988) 28–33.
[15] S.V. Nair, J.K. Tien, Densification mechanism maps for hot isostatic pressing (HIP) of unequal sized particles, Metall. Trans. 18A (1987) 97–107.
[16] R.M. German, E.A. Olevsky, Predictions of tungsten alloy coarsening during sintering, Proceedings of the International Conference on Tungsten, Refractory and Hardmaterials VIII, Metal Powder Industries Federation, Princeton, NJ, 2011, pp. 6.9–6.23.
[17] A. Luque, J. Aldazabal, J.M. Martinex-Esnaola, A. Martin-Meizoso, J.G. Sevillano, R.S. Farr, Geometrical Monte Carlo model of liquid phase sintering, Math. Comp. Simul. 80 (2010) 1469–1486.
[18] A. Nissen, L.L. Jaktlind, R. Tegman, T. Garvare, Rapid computerized modelling of the final shape of HIPed axisymmetric containers, in: R.J. Schaefer, M. Linzer (Eds.), Hot Isostatic Pressing Theory and Applications, ASM International, Materials Park, OH, 1991, pp. 55–61.

[19] H. Djohari, J.I. Martinez-Herrera, J.J. Derby, Transport mechanisms and densification during sintering: I, viscous flow versus vacancy diffusion, Chem. Eng. Sci. 64 (2009) 3799–3809.
[20] R.S. Farr, A. Luque, M.J. Izzard, M. Van Ginkel, Liquid phase sintering of two roughened ice crystals in sucrose solution: a comparison of theory and simulation, Comp. Mater. Sci. 44 (2009) 1135–1141.
[21] R. Huang, P. Pan, A further report on finite element analysis of sintering deformation using densification data – error estimation and constrained sintering, J. Europ. Ceram. Soc. 28 (2008) 1933–1939.
[22] O. Guillon, R.K. Bordia, C.L. Martin, Sintering of thin films / constrained sintering, in: Z.Z. Fang (Ed.), Sintering of Advanced Materials, Woodhead Publishing, Oxford, UK, 2010, pp. 415–433.
[23] J. Rathel, M. Herrmann, W. Beckert, Temperature distribution for electrically conductive and non conductive materials during field assisted sintering (FAST), J. Europ. Ceram. Soc. 29 (2009) 1419–1425.
[24] X. Kong, T. Barriere, J.C. Gelin, C. Quinard, Sintering of powder injection molded 316L stainless steel: experimental investigation and simulation, Inter. J. Powder Metall. 46 (3) (2010) 61–72.
[25] S. Kim, S. Ahn, S.J. Park, S.V. Atre, R.M. German, Integrated simulation of mold filling (Powder-Binder Separation), debinding, and sintering in powder injection molding, Advances in Powder Metallurgy and Particulate Materials – 2008, Part 1, Metal Powder Industries Federation, Princeton, NJ, 2008, pp. 76–86.
[26] M. Shimizu, H. Nomura, H. Matsubara, S.G. Shin, Analysis of porous structures of crystalline particles with anisotropic surface energy by sintering simulation, J. Japan Soc. Powder Powder Metall. 55 (2008) 3–9.
[27] S. Kucherenko, J. Pan, J.A. Yeomans, A combined finite element and finite difference scheme for computer simulation of microstructure evolution and its application to pore-boundary separation during sintering, Comp. Mater. Sci. 18 (2000) 76–92.
[28] V. Tikare, E.A. Olevsky, M.V. Braginsky, Combined macro-meso scale modeling of sintering. Part II, mesoscale simulations, in: A. Zavaliangos, A. Laptev (Eds.), Recent Developments in Computer Modeling of Powder Metallurgy Processes, ISO Press, Ohmsha, Sweden, 2001, pp. 94–104.
[29] E. Olevsky, V. Tikare, Combined macro-meso scale modeling of sintering. Part I: continuum approach, in: A. Zavaliangos, A. Laptev (Eds.), Recent Developments in Computer Modeling of Powder Metallurgy Processes, ISO Press, Ohmsha, Sweden, 2001, pp. 85–93.
[30] Y.S. Kwon, S.H. Chung, C. Binet, R. Zhang, R.S. Engel, N.J. Salamon, et al., Application of optimization technique in the powder compaction and sintering processes, Advances in Powder Metallurgy and Particulate Materials – 2002, Metal Powder Industries Federation, Princeton, NJ, 20029.131–9.146
[31] A. Petersson, J. Agren, Numerical simulation of shape changes during cemented carbide sintering, in: F.D.S. Marquis (Ed.), Powder Materials Current Research and Industrial Practices III, The Minerals, Metals and Materials Society, Warrendale, PA, 2003, pp. 65–75.
[32] R. Raman, T.F. Zahrah, T.J. Weaver, R.M. German, Predicting dimensional change during sintering of FC-0208 parts, Advances in Powder Metallurgy and Particulate Materials – 1999, vol. 1, Metal Powder Industries Federation, Princeton, NJ, 1999, pp. 3.115–3.122.
[33] T. Kraft, Determination of the optimum tool geometry of a cutting insert by finite element simulation of compaction and sintering, Advances in Powder Metallurgy and Particulate Materials – 2003, Part 4, Metal Powder Industries Federation, Princeton, NJ, 2003, pp. 120–126.
[34] T. Kraft, H. Riedel, Numerical simulation of solid state sintering: model and applications, J. Europ. Ceram. Soc. 24 (2004) 345–361.
[35] S.J. Park, S.H. Chung, J.M. Martin, J.L. Johnson, R.M. German, Master sintering curve for densification derived from a constitutive equation with consideration of grain growth, application to tungsten heavy alloys, Metall. Mater. Trans. 39A (2008) 2941–2948.
[36] J. Park, J.M. Martin, J.F. Guo, J.L. Johnson, R.M. German, Densification behavior of tungsten heavy alloy based on master sintering curve concept, Metall. Mater. Trans. 37A (2006) 2837–2848.
[37] D.C. Blaine, S.J. Park, P. Suri, R.M. German, Application of work of sintering concepts in powder metallurgy, Metall. Mater. Trans. 37A (2006) 2827–2835.

[38] M.I. Baskes, Modified embedded-atom potentials for cubic materials and impurities, Phys. Rev. B 46 (1992) 2727–2742.
[39] M.S. Daw, S.M. Foiles, M.I. Baskes, The embedded atom method: a review of theory and applications, Mater. Sci. Rept. 9 (1993) 251–310.
[40] A. Moitra, S. Kim, S.G. Kim, S.J. Park, R.M. German, Investigation on sintering mechanism of nanoscale powder based on atomistic simulation, Acta Mater. 58 (2010) 3939–3951.
[41] M.R. Zachariah, M.J. Carrier, Molecular dynamics computation of gas-phase nanoparticle sintering: a comparison with phenomenological models, J. Aero. Sci. 30 (1999) 1139–1151.
[42] T. Hawa, M.R. Zachariah, Coalescence kinetics of unequal sized nanoparticles, J. Aero. Sci. 37 (2006) 1–15.
[43] W. Chen, A. Pechenik, S.J. Dapkunas, G.J. Piermarini, S.G. Malghan, Novel equipment for the study of the compaction of fine powders, J. Amer. Ceram. Soc. 77 (1994) 1005–1010.
[44] K.E. Harris, V.V. Singh, A.H. King, Grain rotation in thin films of gold, Acta Mater. 46 (1998) 2623–2633.
[45] L.J. Lewis, P. Jensen, J.L. Barrat, Melting, freezing, and coalescence of gold nanoclusters, Phys. Rev. B 56 (1997) 2248–2257.
[46] P. Zeng, S. Zajac, P.C. Clapp, J.A. Rifkin, Nanoparticle sintering simulations, Mater. Sci. Eng. A2 (1998) 301–306.
[47] H.Y. Kim, S.H. Lee, H.G. Kim, J.H. Ryu, H.M. Lee, Molecular dynamic simulation of coalescence between silver and palladium clusters, Mater. Trans. 48 (2007) 455–459.
[48] T. Hawa, M.R. Zachariah, Molecular dynamics simulation and continuum modeling of straight chain aggregate sintering: development of a phenomenological scaling law, Phys. Rev. B 76 (2007) 0541091–9.
[49] J. Houze, S. Kim, S.G. Kim, S.J. Park, R.M. German, The effect of Fe atoms on the adsorption of a W atom on W(100) surface, J. App. Phys. 103 (2008) 106103.
[50] F.A. Nichols, Coalescence of two spheres by surface diffusion, J. Appl. Phys. 37 (1966) 2805–2808.
[51] R.S. Berry, J. Bernholc, P. Salamon, The disappearance of grain boundaries in sintering, J. Appl. Phys. Lett. 56 (1991) 595–597.
[52] F. Amar, J. Bernholc, R.S. Berry, J. Jellinek, P. Salamon, The shapes of first-stage sinters, J. Appl. Phys. 15 (1989) 3219–3225.
[53] P. Basa, J.C. Schon, R.S. Berry, J. Bernholc, J. Jellinek, P. Salamon, Shapes of wetted solids and sinters, Phys. Rev. B 43 (1991) 8113–8122.
[54] J.W. Bullard, Digital image based models of two dimensional microstructural evolution by surface diffusion and vapor transport, J. Appl. Phys. 81 (1997) 159–168.
[55] A. Jagota, P.R. Dawson, Micromechanical modeling of powder compacts I. Unit problems for sintering and traction induced deformation, Acta Metall. 36 (1988) 2551–2561.
[56] J. Bernholc, P. Salamon, R.S. Berry, Annealing of fine powders: initial shapes and grain boundary motion, in: P. Jena, B.K. Rao, S.N. Kahanna (Eds.), Physics and Chemistry of Small Clusters, Plenum Press, New York, NY, 1987, pp. 43–48.
[57] A. Jagota, P.R. Dawson, Simulation of the viscous sintering of two particles, J. Amer. Ceram. Soc. 73 (1990) 173–177.
[58] R.S. Garabedian, J.J. Helble, A model for the viscous coalescence of amorphous particles, J. Coll. Inter. Sci. 234 (2001) 248–260.
[59] F. Wakai, N. Enomoto, H. Ogawa, Three-dimensional microstructural evolution in ideal grain growth - general statistics, Acta Mater. 48 (2000) 1297–1311.
[60] J. Fikes, S.J. Park, R.M. German, Equilibrium states of liquid, solid, and vapor and the configurations for copper, tungsten, and pores in liquid phase sintering, Metall. Mater. Trans. 42B (2011) 202–209.
[61] M.F. Ashby, A. First, Report on sintering diagrams, Acta Metall. 22 (1974) 275–289.
[62] R. Laag, W.A. Kaysser, R. Maurer, G. Petzow, The influence of stoichiometry on the prediction of HIP parameters for intermetallic prealloyed Ni-Al powder, in: R.J. Schaefer, M. Linzer (Eds.), Hot Isostatic Pressing Theory and Applications, ASM International, Materials Park, OH, 1991, pp. 101–113.

[63] B.K. Lograsso, D.A. Koss, Densification of titanium powder during hot isostatic pressing, Metall. Trans. 19A (1988) 1767–1773.
[64] C. Schuh., P. Noel, D.C. Dunand, Enhanced densification of metal powders by transformation mismatch plasticity, Acta Mater. 48 (2000) 1639–1653.
[65] B.B. Panigrahi, M.M. Godkhindi, Sintering of titanium effect of particle size, Inter. J. Powder Metall. 42 (2) (2006) 35–42.
[66] R.M. German, Supersolidus liquid-phase sintering of prealloyed powders, Metall. Mater. Trans. 28A (1997) 1553–1567.
[67] R.M. German, Computer model for the sintering densification of injection molded M2 tool steel, Inter. J. Powder Metall. 35 (4) (1999) 57–67.
[68] K. Mori, H. Matsubara, N. Noguchi, Micro macro simulation of sintering process by coupling Monte Carlo and finite element methods, Inter. J. Mech. Sci. 46 (2004) 841–854.
[69] P.L. Liu, S.T. Lin, The K value distribution of liquid phase sintered microstructures, Mater. Trans. 43 (2002) 2115–2119.
[70] J. Aldazabal, A. Martin-Meizoso, J.M. Martinez-Esnaola, Simulation of liquid phase sintering using the Monte Carlo method, Mater. Sci. Eng. A365 (2004) 151–155.
[71] M.K. Akhtar, G.G. Lipscomb, S.E. Pratsinis, Monte Carlo simulation of particle coagulation and sintering, Aerso. Sci. Tech. 21 (1994) 83–93.
[72] A. Sur, J.L. Lebowitz, J. Marro, M.H. Kalos, S. Kirkpatrick, Monte Carlo studies of percolation phenomena for a simple cubic lattice, J. Stat. Phys. 15 (1976) 345–353.
[73] I.W. Chen, G.N. Hassold, D.J. Srolovitz, Computer simulation of final stage sintering: II, influence of initial pore size, J. Amer. Ceram. Soc. 73 (1990) 2865–2872.
[74] S. Kumar, S.K. Kurtz, Monte Carlo study of angular and edge length distributions in a three dimensional poisson-voronoi tessellation, Mater. Char. 34 (1995) 15–27.
[75] P.E. Mchugh, H. Riedel, A liquid phase sintering model: application to Si_3N_4 and WC-Co, Acta Mater. 45 (1997) 2995–3003.
[76] E.A. Olevsky, R.M. German, A. Upadhyaya, Effect of gravity on dimensional change during sintering - II. Shape distortion, Acta Mater. 48 (2000) 1167–1180.
[77] D. Blaine, S.H. Chung, S.J. Park, P. Suri, R.M. German, Finite element simulation of sintering shrinkage and distortion in large PIM parts, P/M Sci. Tech. Briefs 6 (2) (2004) 13–18.
[78] S.J. Park, S.H. Chung, J.L. Johnson, R.M. German, Finite element simulation of liquid phase sintering with tungsten heavy alloys, Mater. Trans. 47 (2006) 2745–2752.
[79] E.A. Olevsky, A.L. Maximenko, J.H. Arterberry, V. Tikare, Sintering of multilayer powder composites: distortion and damage control, Advances in Powder Metallurgy and Particulate Materials – 2002, Metal Powder Industries Federation, Princeton, NJ, 2002, pp. 9.49–9.59.
[80] I.C. Sinka, A.C.F. Cocks, Evaluating the flow behavior of powders for die fill performance, Powder Metall. 52 (2009) 8–11.
[81] F. Tsumori, K. Hayakawa, Simulation of powder behavior based on discrete model taking account of adhesion force between particles (second report) – three-dimensional simulations, J. Japan Soc. Powder Powder Metall. 53 (2006) 565–570.
[82] A. Petersson, J. Agren, Rearrangement and pore size evolution during WC-Co sintering below the eutectic temperature, Acta Mater. 53 (2005) 1673–1683.
[83] M. Gan, N. Gopinathan, X. Jai, R.A. Williams, Predicting packing characteristics of particles of arbitrary shapes, Kona 22 (2004) 82–93.
[84] B. Henrich, A. Wonisch, T. Kraft, M. Moseler, H. Riedel, Simulations of the influence of rearrangement during sintering, Acta Mater. 55 (2007) 753–762.
[85] C.L. Martin, R.K. Bordia, The effect of a substrate on the sintering of constrained films, Acta Mater. 57 (2009) 549–558.
[86] K. Mori, M. Ohashi, K. Osakada, Simulation of microscopic shrinkage behavior in sintering of powder compact, Inter. J. Mech. Sci. 40 (1998) 989–999.
[87] A. Wonisch, O. Guillon, T. Kraft, M. Moseler, H. Riedel, J. Rodel, Stress induced anisotropy of sintering alumina, discrete element modeling and experiments, Acta Mater. 55 (2007) 5187–5199.

[88] C.L. Martin, L.C.R. Schneider, L. Olmos, D. Bouvard, Discrete element modeling of metallic powder sintering, Scripta Mater. 55 (2006) 425–428.

[89] H. Su, D.L. Johnson, Master sintering curve: a practical approach to sintering, J. Amer. Ceram. Soc. 79 (1996) 3211–3217.

[90] D.C. Blaine, S.J. Park, R.M. German, Linerarization of master sintering curve, J. Amer. Ceram. Soc. 92 (2009) 1400–1409.

[91] R. Bollina, S.J. Park, R.M. German, Master sintering curve concepts applied to full density supersolidus liquid phase sintering of 316L stainless steel powder, Powder Metall. 53 (2010) 20–26.

[92] R.M. German, E. Olevsky, Mapping the compaction and sintering response of tungsten-based materials into the nanoscale size range, Inter. J. Refract. Met. Hard Mater. 23 (2005) 294–300.

[93] L.N. Smith, A Knowledge Based System for Powder Metallurgy Technology, Professional Engineering Publishing, London, UK, 2003.

[94] H. Hofmann, Characterization of the sintering behavior of commercial alumina powder with a neural network, in: R.M. German, G.L. Messing, R.G. Cornwall (Eds.), Sintering Technology, Marcel Dekker, New York, NY, 1996, pp. 301–308.

[95] D. Drndarevic, B. Reljin, Accuracy modelling of powder metallurgy process using backpropagation neural networks, Powder Metall. 43 (2000) 25–29.

[96] L.N. Smith, R.M. German, M.L. Smith, A neural network approach for solution of the inverse problem for selection of powder metallurgy materials, J. Mater. Proc. Tech. 120 (2002) 419–425.

[97] J.C. Lasalle, S.K. Das, B. Snow, B. Chernyavsky, M. Goldenbert, D. Blaine, et al., Injection molding and sintering of large components, Advances in Powder Metallurgy and Particulate Materials – 2003, Part 8, Metal Powder Industries Federation, Princeton, NJ, 2003, pp. 199–204.

[98] S.J. Park, S.H. Chung, D. Blaine, P. Suri, R.M. German, Master sintering curve construction software and its applications, Advances in Powder Metallurgy and Particulate Materials – 2004, Part 1, Metal Powder Industries Federation, Princeton, NJ, 2004, pp. 13–24.

[99] D.C. Blaine, S.J. Park, R.M. German, J. Lasalle, H. Nandi, Verifying the master sintering curve on an industrial furnace, Advances in Powder Metallurgy and Particulate Materials – 2005, Metal Powder Industries Federation, Princeton, NJ, 2005, pp. 1.13–1.19

[100] M.W. Reiterer, K.G. Ewsuk, An analysis of four different approaches to predict and control sintering, J. Amer. Ceram. Soc. 92 (2009) 1419–1427.

Sintering Practice

CRITICAL METRICS

Questions abound when setting up fresh sintering cycles. The first is how to establish a best sintering practice? Usually the critical technical metrics are composition control and dimensional control. Right behind the technical criteria are issues related to cost.

The first concern is whether sintering is capable of delivering the desired material properties in the specified component. Engineering properties improve as the degree of sintering increases, but then decline due to over-sintering. Microstructural damage arises from too high a temperature or too long a time. Cycle optimization depends on the required properties. The ideal cycle for fatigue strength is not meaningful for filtration applications, meaning that there is no single "best" way to sinter a material, so practice varies as a result.

A limiting aspect of sintering is cost. Cost goes up as time, temperature, and pressure increase. For example, once a sintering temperature is identified, then the construction materials for the furnace follow. Higher temperature materials invariably cost more. This makes sense. If a material is resistant to high temperatures, then fabrication of that material is more difficult. Process atmospheres also limit the options. Practical limitations arise as a sintering cycle is specified with significant cost implications.

Dimensional Control

Component production is geared to meeting engineering specifications, largely based on component size. Dimensional variations arise in sintering from green mass variations and sintering cycle variations. Effectively, sintering identifies and amplifies any forming step variations. Moreover, sintering is unable to correct for variations. The usual situation is for sintering to amplify part-to-part differences. The typical required coefficient of variation (standard deviation divided by mean size) is 0.2% for most sintered products, but can reach 0.02% in special situations.

Forming steps have natural variations in temperature, pressure, mass, tool temperature, and tooling friction. Although the green dimensions appear uniform, the green mass is distributed. Figure 15.1 plots the mass variation for an automotive main

Sintering: From Empirical Observations to Scientific Principles
DOI: http://dx.doi.org/10.1016/B978-0-12-401682-8.00015-X

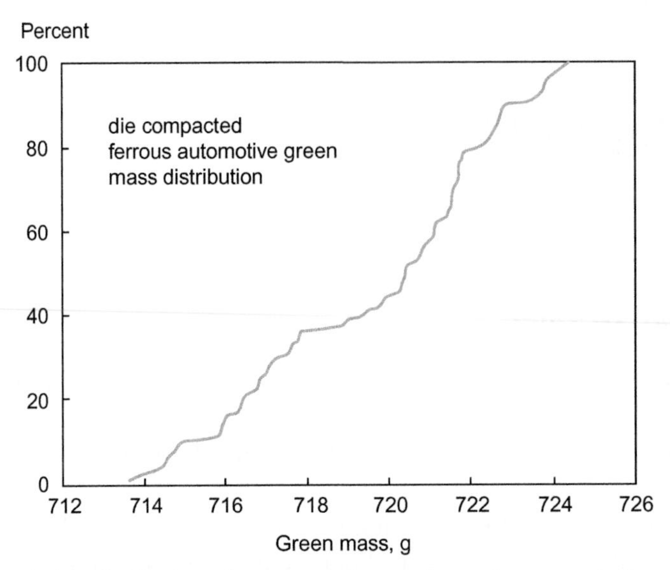

Figure 15.1 Cumulative distribution of component mass for a die compacted main bearing cap. Although the components had uniform green dimensions due to the pressing operation, the mass variation results in sintered dimensional variation.

Table 15.1 Green Mass Variation in Die Pressed Bearing Cap

mean mass	719.6 g
maximum mass	724.4 g
minimum mass	713.6 g
median mass	720.4 g
standard deviation	3.0 g
coefficient of variation	0.41%

bearing cap and summary statistics are given in Table 15.1. The coefficient of variation (standard deviation divided by mean size) is 0.4%, so the dimensional scatter after sintering is expected to be about 0.13%. It is the mass variation in forming that induces the sintered size variation [1]. With proper controls, sintering delivers uniform finial dimensions if the green body is uniform. When dimensional control problems arise, attention to the shaping step is more productive than altering the sintering step. Sintering is the "messenger" not the "cause" for this dimensional scatter, and you should not shoot the messenger.

Dimensional change is a concern in all sintering operations. At the one extreme, a low sintering temperature will minimize dimensional change, as illustrated for a Fe-2Ni-0.5C alloy in Figure 15.2. With a 7.2 g/cm^3 or 90% of theoretical density, zero dimensional change is possible by sintering at 1140°C (1413 K) for 30 min. Such a

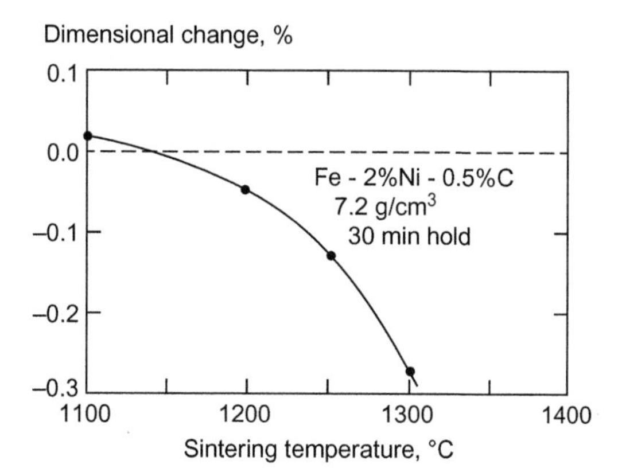

Figure 15.2 Sintering dimensional change from the compaction die size for a Fe-2Ni-0.5C powder composition sintered at various temperatures for 20 min. The data show zero dimensional change for 1140°C (1413 K) sintering.

combination of conditions is ideal for automotive components, because the specified final dimensions are used to design the tooling. Similar low dimensional change cycles are encountered in the production of ceramic casting cores and filters.

Sintering shrinkage ranges upwards from 0% to as much as 25% for liquid phase sintered materials, leading to significant control issues. For binder-assisted forming techniques such as slip casting or injection molding, sintering shrinkage is often 15%. These changes lead to difficulty in maintaining precise final dimensions.

The green density, sintering shrinkage, and sintered density are related. As green density increases or as final density decreases there is less shrinkage. Low density regions undergo more sintering shrinkage. Due to green density gradients, die pressed components warp during sintering. Alternatively, isotropic green microstructures, such as those formed by injection molding or cold isostatic pressing, exhibit more uniform sintering shrinkage. For this reason, die compaction technologies are used with lower sintering temperatures to minimize dimensional change, thereby reducing distortion. On the other hand, hydrostatic forming approaches allow higher temperature sintering without loss of dimensional control.

One cause of distortion in die pressed powders traces to the anisotropic pores and green density gradients. The pores are flattened in the pressing direction if the particles deform during compaction. These pores spheroidize during sintering; a flat pore produces dimensional change that depends on its orientation with respect to the pressing direction [2]. Swelling might occur in the pressing direction while shrinkage occurs in the perpendicular direction. Further, tool friction induces density gradients in the green body. Regions of high density at the pressed faces shrink less compared with regions away from the compaction tools. Differential dimensional change with

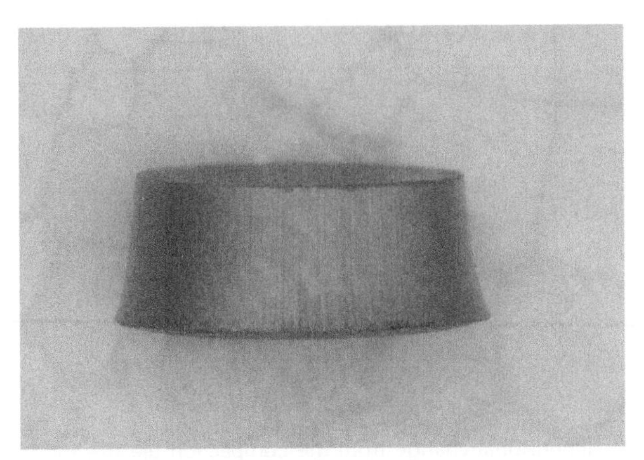

Figure 15.3 An example of distortion in sintering where an elephant foot develops due to gravitational force in the vertical direction and substrate friction in the horizontal direction.

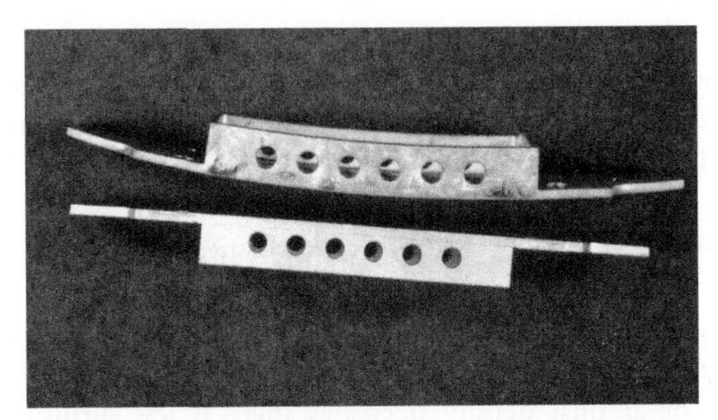

Figure 15.4 Two microelectronic packages after sintering. The lower version was heated uniformly, but the upper version experienced nonuniform heating and warped toward the heat source.

location causes warpage. Gravity also contributes to anisotropic dimensional change [3,4]. This effect is accentuated by a large component height. As the component mass increases, friction with the substrate constricts thermal expansion and sintering shrinkage. In severe cases an "elephant foot" forms due to component mass and substrate friction. This is depicted in Figure 15.3. Initially the component shape was a right circular cylinder.

Sintering furnace temperature gradients influence final dimensions. Gradients encountered during heating are especially problematic. Figure 15.4 is a photograph of two versions of the same electronic package after sintering at the same temperature. Curvature occurred in one case due to non-uniform heating.

Practical steps can be taken to control dimensional changes during sintering to ensure uniformity—repeatable and uniform green bodies can be subjected to uniform heating. Two key dimensional control parameters are sintering temperature and time. Dimensional change depends exponentially on temperature but exhibits a subdued response to time. Temperature, green density, and component mass must all be uniform to attain dimensional control in sintering.

Composition Control

Three factors are dominant with respect to sintered composition—temperature, substrate, and atmosphere. Evaporative loss during sintering is only a factor in a few systems. Vapor pressure depends on the composition. At high temperatures, preferential evaporation will change the component composition. For example, in a stainless steel, chromium has the highest vapor pressure of the components. A high sintering temperature preferentially evaporates chromium, resulting in reduced corrosion resistance. This is mitigated by the process atmosphere which helps suppress evaporation. Evaporation problems are assessed through vapor pressures using Table 15.2; this table gives the boiling temperature and temperature for a vapor pressure of 10^{-3} Pa (the practical upper limit in vacuum systems). Evaporation problems are seen in sintering materials containing zinc, barium, lead, manganese, tin, indium, silver, gallium, and aluminum.

Compounds can differ in their volatility when compared to the constituent elements, as illustrated in Table 15.3. The table is organized to show the temperature which corresponds to a vapor pressure of 1 Pa (10^{-5} atmospheres), corresponding to measurable weight loss during sintering. As an example, a substantial difference is evident for molybdenum compared to its oxide. Molybdenum is volatile in the presence of oxygen. In contrast, aluminum has the opposite behavior and becomes stable when oxidized.

Sintering is performed by resting the powder compacts on substrates. The substrates might have conformal shapes to help maintain shape. The alternative is to simply pile the compacts together. Unfortunately, contacting components sinter bond together if they are in contact during sintering, as illustrated by the titanium gears pictured in Figure 15.5.

Sintering substrates are fabricated from high temperature materials, such as stainless steel, molybdenum, silica, graphite, zirconia, or alumina. Improper selection results in contamination. For example, the bar chart in Figure 15.6 plots how different substrates change strength and oxygen content for sintered titanium [5,6]. The comparison includes both virgin and baked version of zirconia and yttria. The higher oxygen content from the substrate aids strength but degrades corrosion resistance. The substrate must remain rigid at the sintering temperature without contaminating the compact. Table 15.4 lists the maximum used temperature, atmosphere, and uses for common substrates [7]. They are categorized by primary composition. Alumina-based supports

Table 15.2 Evaporation Characteristics for Some Elements

Material	Boiling Temperature, °C	Temperature for Vapor Pressure of 10^{-3} Pa, °C	Vapor Pressure at 1120 °C, Pa
Ag	2212	730	8
Al	2467	1140	$2 \cdot 10^{-1}$
Au	2807	1050	$9 \cdot 10^{-3}$
B	3658	1630	$8 \cdot 10^{-8}$
Ba	1637	450	$2 \cdot 10^{-3}$
Co	2870	1180	$4 \cdot 10^{-4}$
Cr	2672	1070	$6 \cdot 10^{-3}$
Cu	2567	930	$1 \cdot 10^{-1}$
Fe	2750	1120	$2 \cdot 10^{-3}$
Ga	2403	730	6
Ge	2817	1030	$1 \cdot 10^{-2}$
In	2080	650	$3 \cdot 10^{1}$
Ir	4130	1930	$5 \cdot 10^{-11}$
Mn	1962	720	$2 \cdot 10^{1}$
Mo	4612	1930	$7 \cdot 10^{-11}$
Nb	4742	2130	$6 \cdot 10^{-13}$
Nd	3027	970	$6 \cdot 10^{-2}$
Ni	2732	1180	$5 \cdot 10^{-4}$
Pb	1740	460	$8 \cdot 10^{-2}$
Pd	3140	1100	$3 \cdot 10^{-3}$
Pt	3827	1600	$6 \cdot 10^{-9}$
Si	2355	1230	$5 \cdot 10^{-5}$
Sn	2270	800	$2 \cdot 10^{-1}$
Ta	5425	2380	nil
Ti	3286	1310	$1 \cdot 10^{-5}$
U	3745	1560	$2 \cdot 10^{-7}$
V	3377	1230	$5 \cdot 10^{-7}$
W	5657	2550	nil
Y	3338	1430	$9 \cdot 10^{-4}$
Zn	906	200	$6 \cdot 10^{5}$
Zr	4650	1860	$1 \cdot 10^{-9}$

are the most popular, followed by graphite, silica, and ferrous or nickel alloys. Some materials are unstable in certain atmospheres. For example, nitrides (AlN, BN, and Si_3N_4) are unstable in oxidizing atmospheres.

Defect Avoidance

At low sintering temperatures, before significant bonding, a powder compact has a relatively low strength. The fracture toughness for loose powder is about 0.02 $MPa\sqrt{m}$ [8]. The stresses associated with compact shrinkage during sintering are

Table 15.3 Approximate Temperature for a Vapor Pressure of 1 Pa (10^{-5} atmosphere)

Material	Temperature, °C
Ag	1050
Al	1000
Al_2O_3	2000
Au	1470
B	1360
Ba	720
BaO	1550
C	2680
Ca	600
CaO	2050
Co	1650
Cr	1200
Cr_2O_3	1700
Cu	1270
Fe	450
Mg	440
MgO	1800
Mo	2530
MoO_3	620
Na	290
NaCl	660
Ni	1510
Pt	2090
Si	1340
SiO_2	1750
Sr	540
Ta	3070
Ti	1550
Y	1650
Zn	340
Zr	2000

only a few MPa [9,10]. Cracking arises during sintering because of the low fracture toughness. Sintering cracks are caused by gravity, thermal gradients, polymer burnout, or vibration. Because of the low starting fracture toughness, small flaws grow, even defects as small as 10 μm enlarge during heating. The photograph in Figure 15.7 is of a crack that was not evident prior to sintering. It formed early in the heating cycle due to differential densification between thick and thin regions.

As sintering progresses the interparticle bonds grow to increase strength. Thermal softening or liquid formation subsequently reduces the compact's strength. Usually the most susceptible point is during early heating. For example, in a vacuum furnace,

Figure 15.5 A cluster of titanium gears that bonded together because of close proximity during sintering.

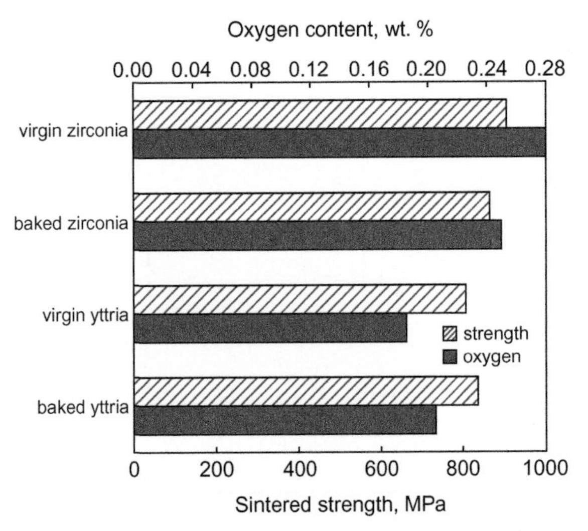

Figure 15.6 Data on the strength and oxygen content after sintering for titanium, showing four substrate conditions—zirconia and yttria, virgin material and previously baked material.

radiant heating is poor until about 500°C (773 K), so a compact might experience rapid heating near that temperature even when the furnace cycle is programmed for a modest heating rate. Creep occurs at high temperatures, usually leading to distortion but not cracking. Cantilevered sections slump and large compacts buckle or warp, but usually there is no cracking from high temperature deformation. As the component mass decreases, these high temperature problems are less evident. Also, slow heating

Table 15.4 Sintering Substrates and Common Applications

Material	Maximum Use Temperature, °C	Atmospheres*	Example Sintering Materials
alumina (Al_2O_3)	1500	R, C	most metals, oxides, glass,
	1800	O, I, V	intermetallics, no carbides
boron carbide (B_4C)	2000	R, I, V	borides, carbides
boron nitride (BN)	1800	R, I, V	Ta, AlN, nitrides
calcia (CaO)	2400	O, I	Fe, Pt
fecraly (Fe-Cr-Al-Y)	1300	O, R, I	steels, intermetallics
graphite (C)	2000	V	Au, Ag, Cu, Zn, WC, Cr_2O_3,
	2500	I, V, C	SiC, TaC, carbides
Ni alloys (Ni-Cr)	1300	R, I, V, C, D	Ni, Fe, stainless
iron (Fe)	1200	R, I, V, D	Fe, polymers, Cu
magnesia (MgO)	1600	V	Ag, Au, Co, Cu, Fe, Ni, MgO,
	2000	O, I	no carbides
molybdenum (Mo)	2100	R, I, V	steel, rare earths, SiC
silica (SiO_2)	1000	I, D	Ag, Al, Au, Cu, Zn, Ni
	1200	O, V	
silicon nitride (Si_3N_4)	1600	I, V	many metals
tantalum (Ta)	2000	I, V	Ta, TaB_2, avoid oxides
tungsten (W)	2800	R, I, V	Mo, W, $MoSi_2$, $TiSi_2$
zirconia (ZrO_2)	2500	O, I	Au, Co, Cr, Fe, Ni, Ti

*atmosphere key
O = oxidizing, R = reducing, I = inert,
V = vacuum, C = carburizing, D = decarburizing.

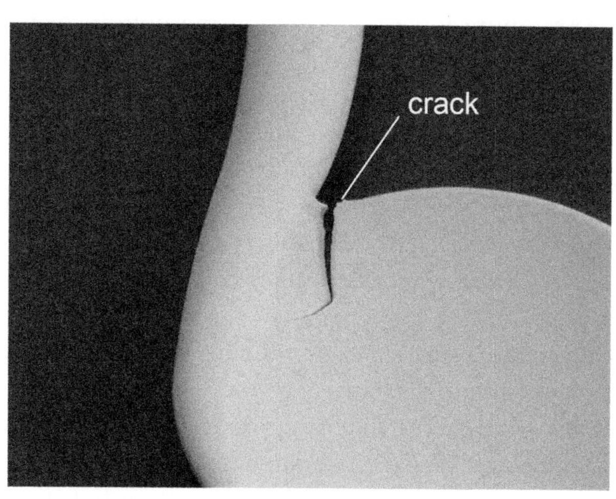

Figure 15.7 Crack formed during sintering of an alumina surgical spoon. The long crack indicates it formed early prior to significant strengthening, and the location corresponds to a junction of thick and thin sections.

induces sufficient sinter bonding prior to shrinkage to offset damage from thermal stresses or shrinkage [11].

The key point about damage control is that patience in the form of slower heating is the first step in curing the problem, of course assuming the defects were not generated in the forming step prior to sintering.

CONTROL PARAMETERS

Practical sintering cycles are concerned with the delivery of proper composition and dimensions. Green body homogeneity is critical with respect to both goals. The homogeneity of mixed powders impacts chemical reactions and homogenization. Complex mixed powder chemistries are common, in which the selection of small particles, especially for the minor phase, gives the most uniform initial distribution. Likewise, green body density gradients contribute to dimensional control issues. Regions of low density and high density produce different shrinkage or swelling responses.

The density gradients in the green body usually arise from the forming step. If the component is fabricated oversize, the density gradients and component warpage can be ignored. Of course the compensation is in the added cost of post-sintering machining, grinding, or sizing to correct the dimensions.

Gravitational forces induce distortion during sintering. The problem is especially evident for large components. A result of gravity is differential dimensional change depending on height; bulging might occur near the component bottom. As the component height increases, more force is exerted on the bottom of it, so friction with the sintering support constricts dimensional change at the base. Factors that increase compact strength during heating, such as a slow heating rate, help reduce distortion. One trick is to use conformal setters that guide the compact during sintering. For example, tall structures are placed at an angle with respect to the horizontal to reduce friction, and thin structures are placed on setters that allow shrinkage up to a limit corresponding to the final size. Figure 15.8 shows a thin walled stainless steel component sitting on the conformal setter after sintering. The stainless steel shape shrank onto the ceramic support.

Temperature gradients in a furnace cause product variation. Temperature gradients during heating are especially problematic, since they distort the otherwise weak component. Hotter regions give earlier sintering than cold regions. Local differences in thermal expansion, sintering shrinkage, and strength contribute to warpage. Usually a component warps toward the heat. One trick for offsetting hot spots is to sinter in a high thermal conductivity atmosphere of hydrogen or helium or to encapsulate the components to distribute heat.

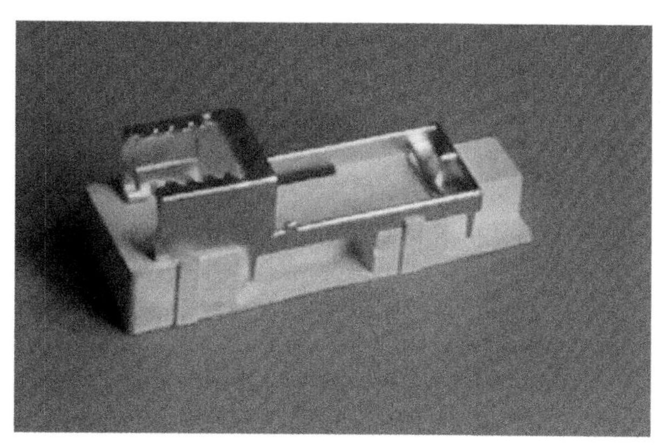

Figure 15.8 Picture of a stainless steel sintered component sitting on a conformal alumina setter after sintering.

ATMOSPHERE

The sintering atmosphere surrounds the powder compact. Several options are in use, ranging from air to hydrogen. The sintering atmosphere decision balances cost against the technical objectives of the process [12–14]. Options for atmospheres include the following:

- hydrogen—good heat transport, costly, explosive
- nitrogen—less expensive, not used with polymers, no reduction role, generally neutral but can form nitrides
- nitrogen-hydrogen—usually 4 to 10% hydrogen, mixture of costs, reactivity, and reduction roles
- argon—neutral, no oxidation, no reduction, and not good for heat transport, costly
- helium—neutral, no oxidation, no reduction, excellent heat transport, costly
- air—no significant cost, oxidizing, lower thermal conductivity
- natural gas—usually partially combusted, variable quality, low cost
- dissociated ammonia—similar to nitrogen-hydrogen in merits, toxic if not properly dissociated—1000°C (1273 K) catalyzed reaction, $2\ NH_3 \rightarrow 3\ H_2 + N_2$
- vacuum—absence of atmosphere, some residual vapor reactions possible, not effective at low temperatures due to lack of heat conduction.

Reactions

Oxides, nitrides, and carbides have specific chemical needs if they are to remain stable during sintering. They are sintered in atmospheres with oxygen, nitrogen, or carbon. Metals at high temperatures form oxides unless protected, and this is done by sintering in hydrogen (reducing), nitrogen (usually neutral) atmospheres, or under vacuum.

The atmosphere plays a role in the removal of binders or lubricants. The polymers must be extracted at low temperatures to avoid undesired contamination. For example, steels become harder as the carbon content increases. Since polymers are mostly carbon they harden, and possibly embrittle, ferrous alloys if not properly extracted during heating. Carbon is added or subtracted from the component during sintering by adjustments to the atmospheric composition by controlling the levels of oxygen, water, methane, carbon monoxide, and carbon dioxide [15]. Neutral atmospheres are ineffective, so doping with oxygen (to form CO or CO_2), hydrogen (to form CH_4), or water (to form CO and CH_4) is effective, while argon, helium, or nitrogen fail to contribute to these reactions.

The cost of the sintering atmosphere is a consideration. High purity reducing atmospheres, such as hydrogen, produce repeatable sintered properties, but are expensive. Alternatively, low cost sintering is possible using a mixture of natural gas and air, but this gives less repeatable properties. A common option for cost is to mix 4 to 10% hydrogen with nitrogen.

The vapor phase surrounding the compact influences sintering in several possible ways:

- Polymers used in powder forming are removed at low temperatures to avoid contamination; the proper atmosphere accelerates polymer removal.
- Many materials require precise carbon control during sintering and the atmosphere is selected to obtain the desired final carbon level.
- Nitride ceramics are unstable at high temperatures, so to preserve the desired phase the sintering atmosphere must be adjusted to stabilize the nitride during sintering.
- Alumina can be sintered in air, but to obtain optical transparency without trapping insoluble vapor in the pores requires vacuum or hydrogen sintering.
- Many materials prove unstable with high vapor pressures during sintering, so a protective atmosphere is selected to inhibit volatilization and loss of composition.
- Ionic materials are modified in sintering rate by the generation of desired ionic vacancies through reactions with the atmosphere.

Sintering atmospheres resort to combinations, such as low vacuum doped with hydrogen or nitrogen and hydrogen. Some compositions are listed in Table 15.5. Endothermic and exothermic atmospheres based on natural gas have high contents of nitrogen and hydrogen, as well as water, carbon monoxide, and carbon dioxide. An endothermic atmosphere results from catalytic conversion of natural gas and air to form a mixture of hydrogen, nitrogen, and carbon monoxide. Exothermic atmospheres are a partially combusted mixture of air and natural gas. They are composed of approximately 6.5 parts of air and one part natural gas and prove useful in sintering

Table 15.5 Compositions of Synthetic Sintering Atmospheres

Constituent	Endothermic	Exothermic	Dissociated Ammonia	Nitrogen Based
% N_2	39	70 to 98	25	75 to 97
% H_2	39	2 to 20	75	2 to 20
% H_2O	0.8	2.5	0.004	0.001
% CO	21	2 to 10	–	–
% CO_2	0.2	1 to 6	–	–
% CH_4	0.5	<0.5	0	0
ppm O_2	10 to 150	10 to 150	10 to 35	5
dew point, °C*	−16 to 10	25 to 45	−50 to −30	−75 to −50

*dew point is the temperature for moisture condensation.

Table 15.6 Properties of Common Sintering Atmosphere Gases at 527°C (800 K)

Gas	Density, kg/m^3	Heat Capacity, kJ/(kg K)	Viscosity, μPa s	Thermal Conductivity, mW/(m K)
air	0.435	1.10	37	57
carbon dioxide	0.661	1.17	34	55
carbon monoxide	0.421	1.14	34	56
helium	0.061	5.19	38	304
hydrogen	0.030	14.7	17	378
nitrogen	0.421	1.12	35	55
oxygen	0.481	1.05	42	59
steam	0.274	2.15	28	59

common materials. Because of their high oxidation potential, they are not useful for sintering materials containing easily oxidized components such as chromium, titanium, or zinc. A small percentage of unconverted methane usually remains and this increases the carbon content of the sintered body.

Some attributes of sintering atmospheres are given in Table 15.6. A high thermal conductivity, evident in hydrogen and helium, helps to control the temperature and ensure uniform sintering. Heating is dependent on heat conduction via the atmosphere and a gas with high conductivity is beneficial. At temperatures over about 500°C (773 K) radiation dominates heat transfer, so generally vacuum sintering is not recommended for lower temperature sintering.

Oxides have an equilibrium partial pressure of oxygen at high temperatures, so even neutral atmospheres result in some oxide reduction. However, this reduction is small in comparison with atmospheres containing hydrogen or carbon monoxide. On the other hand, several materials react with hydrogen to form brittle hydrides, or with CO to form oxides or carbides. Tantalum, titanium, zirconium, rare earths, and niobium are problematic in this respect, so they are sintered in vacuum or inert gas.

Nitrogen is neutral in many situations. It is an active agent in sintering aluminum, aluminum nitride, silicon nitride, and similar materials. Since it forms the majority of air, large scale sintering operations generate the nitrogen on-site using refrigeration or chemical absorption techniques.

Hydrogen provides reduction if the water vapor is swept away by proper gas flow. Liquid hydrogen supplies introduce gas containing 1 ppm oxygen and 8 ppm water. However, moisture lowers the ability to reduce oxides and makes the atmosphere decarburizing.

Vacuum sintering involves heating in a sealed chamber with a pumping mechanism to continuously extract the evolved vapors. The pressure in the sintering chamber might be between 10^{-4} and 10^{-7} bar (10 and 0.01 Pa), but the composition of the atmosphere is often unqualified. Evacuation is achieved using pumps, ranging from rotary pistons to turbomolecular and ion pumps. Although vacuum is a neutral atmosphere, chemical reactions still occur. For example, a low oxygen partial pressure coupled with the continual extraction of vapors in the vacuum chamber leads to oxide reduction for many compounds.

Most materials can be sintered under vacuum, but this may not necessarily give the best properties. Vacuum sintering is preferred for reactive materials (titanium, tantalum, and beryllium), high temperature materials (tool steels, molybdenum, and titanium carbide), hydriding elements (uranium, $Fe_{14}Nd_2B$, or zirconium), and corrosion resistant materials (stainless steels). Unfortunately some elements evaporate preferentially (Zn, Mg, Li, Ca, Ag, Mn, and Al) and prove difficult to sinter under vacuum. A slight inert gas pressure inhibits evaporation. One variant of this process is to sinter in vacuum until the pores are closed, then shift to an inert atmosphere late in the sintering cycle. This ensures the pores are empty, but reduces surface evaporative losses.

Often vacuum furnaces employ graphite heating elements, insulation, or support hardware. During heating, residual oxygen or moisture evolving from the powder compact reacts with this graphite to form carbon monoxide. The reduction potential of carbon monoxide is large at high temperatures; at a level near 100 ppm it is useful for sintering stainless steels, tool steels, and superalloys. Figure 15.9 plots the relation between carbon monoxide partial pressure and temperature for the vacuum reduction of chromium oxides, such as are found in stainless steel [16]. The lower the carbon monoxide partial pressure, the lower the temperature required to convert Cr_2O_3 into Cr. This reaction is accelerated by the intentional addition of graphite to the compact.

Oxygen Control

Oxygen is detrimental to metal sintering. A high oxygen partial pressure drives oxidation. Reduction is the opposite reaction where oxides are broken into their

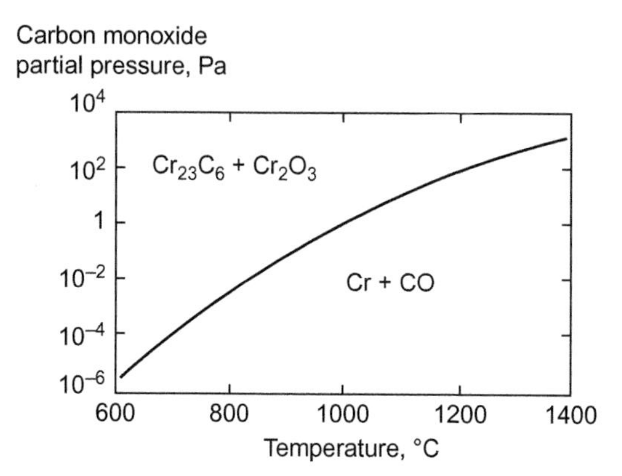

Figure 15.9 Carbon monoxide is an effective reducing agent at high temperatures. Plotted here is the reduction of chromium (below the curved line) for low particle pressures of CO at higher temperatures.

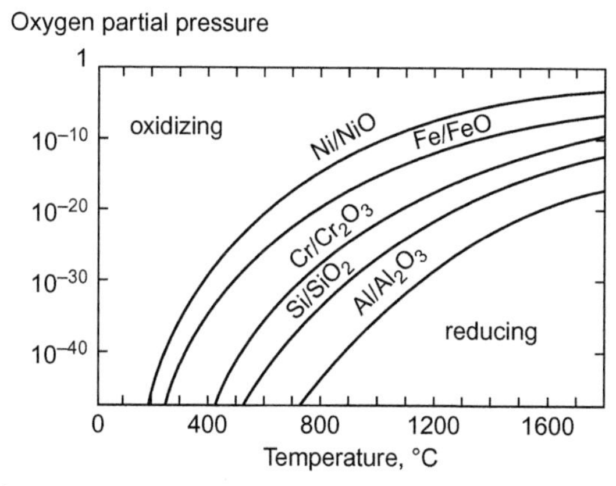

Figure 15.10 Plots of the oxygen partial pressure (such as in vacuum sintering) required to reduce oxides of nickel, iron, chromium, silicon and aluminum. For any temperature the condition above the line is oxidizing and below the line is reducing.

constituent components using hydrogen or carbon monoxide. The propensity for oxidation versus reduction is determined by the temperature and atmosphere. Example oxidation-reduction conditions are illustrated for Ni, Fe, Cr, Si, and Al and their oxides in Figure 15.10. The oxygen partial pressure is plotted versus temperature to indicate the sintering atmosphere conditions; reduction for each species is associated with conditions below the respective lines. For example, chromium sintering at 1300°C (1573 K) requires an oxygen pressure below 10^{-16} atmospheres (10^{-11} Pa) to avoid oxidation, but chromia (Cr_2O_3) requires an oxygen partial pressure higher than this.

Table 15.7 Ranking of Compound Stabilities at 1000°C (1273 K); Most Stable to Least Stable

Oxides	Sulfides	Nitrides	Carbides	Borides
CaO	CeS	TiN	HfC	HfB_2
BeO	CaS	HfN	ZrC	ZrB_2
MgO	MgS	ZrN	TiC	TiB_2
Li_2O	Na_2S	Th_3N_4	TaC	TaB_2
Al_2O_3	MnS	AlN	NbC	MoB
TiO_2	ZnS	UN	ThC_2	NbB_2
SiO_2	Cu_2S	TaN	UC	VB_2
V_2O_3	FeS	NbN	V_2C	UB_2
NbO	MoS_2	VN	Be_2C	CrB_2
Ta_2O_5	WS_2	Mg_3N_2	SiC	Fe_2B
MnO	PbS	Si_3N_4	CaC_2	MnB
Cr_2O_3	Ag_2S	Cr_2N	WC	MgB_2
Na_2O	H_2S	Fe_4N	MoC	WB
ZnO	PtS	Mo_2N	Cr_3C_2	Ni_3B
FeO	IrS		Al_4C_3	
P_2O_5	Cu_2S		Fe_3C	
K_2O			Ni_3C	
CoO				
SO_2				
NiO				
PbO				
Cu_2O				

Reducing conditions for one species can be oxidizing for another. For example, ferrous oxides are more easily reduced than chromium oxide. Table 15.7 ranks the relative stability of several oxides at 1000°C (1273 K). Included are lists for carbides, nitrides, borides, and sulfides. Certain metals prove very difficult to sinter in the presence of oxides, notably Al, Ba, Ca, Cd, In, Mg, Pb, Sn, Sr, and Zn. Oxide reduction must be attained prior to reaching the sintering temperature.

Carbon monoxide and hydrogen react with oxides to form carbon dioxide or water vapor. Accordingly the partial pressure ratios of CO_2 to CO or H_2O to H_2 are atmosphere control parameters. A high partial pressure of carbon dioxide or water leads to oxidation. Figure 15.11 shows plots of the CO/CO_2 partial pressure ratio over a range of temperatures for the same five systems shown earlier. Vacuum sintering with graphite heating elements generates a strong reduction potential since oxygen reacts with the graphite to form carbon monoxide. A water-hydrogen atmosphere is similar, as shown in Figure 15.12 using a plot of hydrogen to water pressure versus temperature.

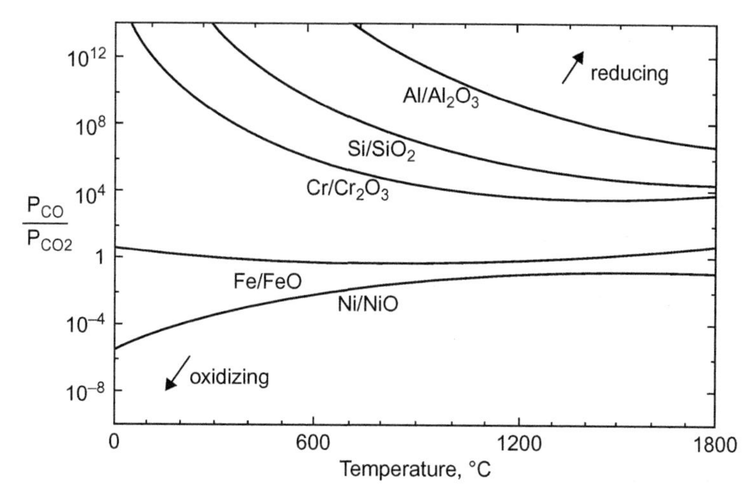

Figure 15.11 The relation between carbon monoxide and carbon dioxide partial pressures and temperature to demark the oxidizing and reducing conditions for five metals.

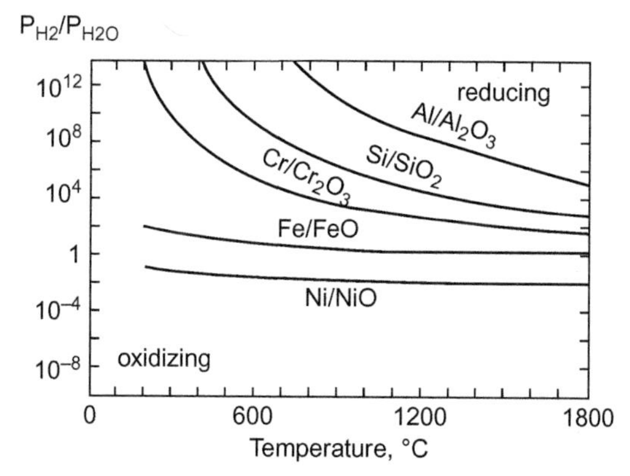

Figure 15.12 The relation between hydrogen (reducing) and steam (oxidizing) partial pressures and temperature to demark the oxidizing and reducing conditions for five metals.

Carbon Control

Related concerns arise regarding carbon control during sintering. Carburization is desirable for carbides, but an excess of carbon degrades the properties of the product. The reactions depend on the carbon source; such as methane (CH_4), carbon monoxide (CO), or carbon dioxide (CO_2). As with the oxidation-reduction reactions, carburization-decarburization depends on temperature and pressure.

Some metals do not form carbides, or temperature increases make them unstable, but other metals form carbides that are stable at elevated temperatures (Zr, Ti, Si, Cr,

U, W, Mn, Th, and Ta). Carbon control during sintering is important to cermets, cemented carbides, tool steels, and ceramics such as silicon carbide. One annoying carbon source arises from decomposing polymers used as lubricants and binders, so special provision is required to thoroughly remove them during heating. As an aside, simple hydrocarbons, such as paraffin wax, polyethylene, and stearic acid tend to contaminate the least, while polymers with ring structures, such as polystyrene, tend to form graphite contaminants.

Five main species (CO, CO_2, H_2, H_2O, and CH_4), consisting of three components (carbon, hydrogen, and oxygen), give atmospheric control through two controlling reactions:

$$CO_2 + H_2 \leftrightarrows CO + H_2O$$
$$CH_4 + H_2O \leftrightarrows CO + 3H_2$$

Sintering atmosphere control is gained by monitoring these species.

The atmosphere changes during sintering. The compacts carry contaminants into the furnace that change the atmosphere. It is most desirable to sweep the contaminants away using a flowing atmosphere. Hydrocarbon species which decompose to form carbon soot particles make control difficult. Soot formation requires more atmosphere flow and possibly more oxidizing conditions. Thus, the atmosphere composition, atmosphere flow rate, sintering temperature, powder compact, and contaminants interact during sintering. Accordingly, analysis of the atmosphere requires sensors coupled to control systems to continuously adjust quality.

The cost of the atmosphere is always a consideration in sintering. On a relative basis, air is the lowest in cost, and atmospheres based on natural gas are relatively inexpensive. Hydrogen is the most expensive of the reducing atmospheres, giving a relative cost ranking as follows:

hydrogen = 1.0
nitrogen-based = 0.6
dissociated ammonia = 0.4
endothermic gas = 0.2
exothermic gas = 0.1

Although vacuum has no direct gas expense, the equipment and its operation do make vacuum sintering an expensive alternative.

Density Changes

The atmosphere influences sintering densification. Some examples are as follows:

- inert gases trapped in pores hinder full densification
- coarsening of the gas-filled pores causes desintering
- gas reactions generate vapor species that collect in pores and cause swelling.

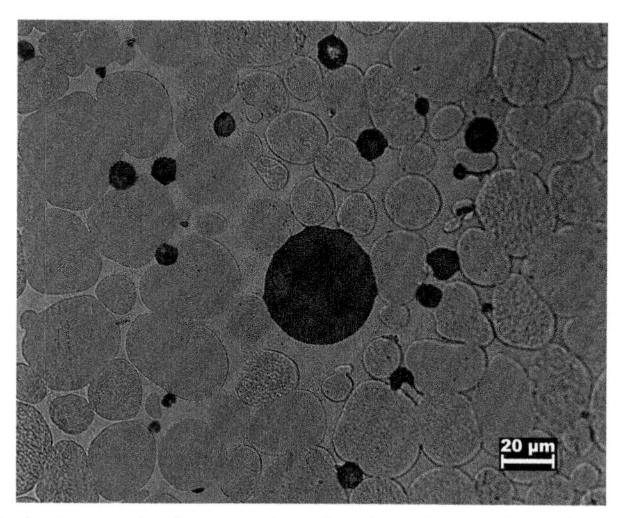

Figure 15.13 Optical micrograph after sintering, showing a large spherical pore indicative of gas reactions impeding sintering.

Even when sintering under vacuum, delayed reactions result in bloating, swelling, blistering, and other deleterious changes.

One straightforward means of assessing problems with the atmosphere is via microscopy. Gas reactions form spherical pores that grow over time. As the pore size enlarges, the pressure in the pore decreases, leading to component swelling. Optical microscopy of cross-sectioned samples gives evidence of stable gas pores, such as imaged in Figure 15.13. Large, spherical pores are clear indications of atmospheric reactions in sintering. In the image shown here, the pores in the green body were under 1 μm, but the coarsened pores are up to 40 μm, accompanied by a loss of sintered density.

EQUIPMENT

A production sintering furnace is expensive, so selection requires assessment of the technical and economic objectives. The sintering furnace provides time-temperature control while containing the atmosphere. There are two basic designs—batch and continuous.

Figure 15.14 is a picture of a batch furnace. In the sintering process, it is loaded with the green components, sealed, purged to establish the desired atmosphere, and heated to temperature. A batch furnace is flexible and only runs when required.

Figure 15.14 An example batch sintering furnace, in this case approximately 250 kg or product is sintered in each run. This furnace is capable of hydrogen, argon, nitrogen, vacuum, or any combination of these atmospheres (including partial pressure of hydrogen).

Options exist for special cycles (vacuum or pressure-assisted) and small production quantities. In batch sintering treatments (especially those involved in vacuum sintering) the furnace walls are kept cold via reflective heat shields around the working zone. An alternative design is a hot wall where the heating elements are located outside a retort that contains the sintering material and protective atmosphere. The components are loaded from the front, top, or bottom, using elevators, hoists, or lift-trucks.

A continuous furnace operates by controlling the position of the powder compact in a preheated multiple zone furnace. Figure 15.15 is a photograph showing the layout. The components move on a conveyor, such as a belt, tray, lifting beam, or roller device. The conveyor limits the furnace operating temperature. For temperatures below about 1150°C (1423 K) a stainless steel mesh belt is successful. Ceramic belts increase the possible working temperature to about 1300°C (1573 K) and refractory metal or graphite belts can reach 2200°C (2472 K). Alternatively, pusher mechanisms use trays that are stoked one against another and can operate at up to 2200°C (2473 K).

There are at least three stages in production sintering. The heating stage, or the first zone in a continuous furnace, initiates compact heating, removes lubricants, binders, and contaminants from the pores, and starts gas reactions with the powder. Outgassing products contaminate the atmosphere, so gas flow is adjusted to sweep the

Figure 15.15 The layout of a continuous sintering furnace where the work traverses from the left to the right, passing through several heated and cooled zones over several hours to exit as a sintered component.

contaminants away from the hot zone. The next stage or zone is the high heat region, where the desired temperature-time combination is maintained. Cooling occurs in the last stage where the compact is subjected to a high gas flow. It is possible to restore the desired carbon, oxygen, or nitrogen levels during cooling. Stratification of the process atmosphere is possible to tailor chemical reactions during sintering. If necessary, the furnace heating elements are located externally to the atmosphere heat radiating through a furnace muffle.

The optimal furnace type and dimensions depend on the intended production quantity, material, operating costs, type of atmosphere, and post-sintering cooling rate. Production furnaces exist for loads ranging from less than 1 kg to 250,000 kg. Routine production sintering handles about 100 kg/h, but large tunnel furnaces run at up to 2000 kg/h.

The majority of technical sintering furnaces rely on electrical heating, although natural gas is also in common use due to its lower operating cost. The heating elements depend on the desired temperature, atmosphere, and application. A wide variety of designs arise. The most common heating elements are made from molybdenum, silicon carbide, nickel-chromium, tungsten, molybdenum disilicide, or graphite. Each has advantages and limitations. Thus, custom design furnaces are created to fit the specific needs of each application.

As noted earlier, sintering has several process monitors. It is expected that the furnace will include these in its control system. Thermocouples are most commonly used for measuring temperature. Optical and infrared pyrometers are employed at higher temperatures.

CYCLES

Production sintering cycles consist of a sequence of heat and hold stages, intended to remove polymers, contaminants, and soak the bodies at the peak temperature, and control the cooling pathway. Often the furnace limits the thermal cycle; for example the support trays might not be able to accommodate rapid heating. The higher the sintering temperature and the more reactive the atmosphere or test material, the fewer are the furnace options which exist. The limitations include melting temperature, reactivity with the atmosphere, and stability under repeated heating cycles. Chemical reactions with the components being sintered are not acceptable, but support trays also react with the heating elements. As the temperature increases, the furnace furniture progressively increases in cost and decreases in durability. High temperature sintering, over about 2000°C (2273 K) is difficult.

Temperature gradients during sintering are another difficulty. Especially problematic are gradients during heating, since heat is delivered to the outside envelope of the furnace. Components near the center see a lagging thermal cycle compared to components near the edge. Such thermal gradients contribute to distortion and position sensitive sintered dimensions. The typical cure is to slow the heating down to minimize the temperature gradients.

COSTS

Large components require slower heating and large furnaces, thereby mandating a higher expense for sintering. As a first estimate, the cost of sintering (atmosphere, energy, and maintenance) is between $0.10 and $1.00 per kg. Many factors, such as furnace cost, labor rates, gas usage, energy costs, depreciation schedules, maximum temperature, and component size are involved in cost calculations [17]. In ferrous powder metallurgy, this cost accounts for 20 to 30% of the total component fabrication expense of sintering. Large scale oxide ceramic sintering, such as brick production has about the lowest sintering cost. On the other hand, for specialty sintering of materials such as titanium, the cost can reach upwards to $7/kg.

In a cost analysis, the space occupied by the furnace is a factor. Further, the level of automation determines the labor costs. The location of the sintering plant impacts labor and environmental costs. High-value sintered products, such as electronic circuits requiring high skill levels, operate in urban areas. On the other hand, low-value products, such as construction bricks, are sintered in more remote areas. There is a progressive migration of the sintering industries to low cost labor markets.

EXAMPLES

With a million publications on sintering, much detailed information is available on sintering cycles for different applications. This section highlights a few examples to show the variety of cycles available for some popular materials.

Alumina

Alumina (Al_2O_3) is an ionic compound sensitive to oxygen ion mobility during sintering. It is routinely sintered from submicrometer powders to near full density in air or hydrogen. Alumina is sintered to a translucent condition using temperatures of 1600 to 1850°C (1873 to 2123 K) in hydrogen. Densification is achieved by grain boundary diffusion control, but surface diffusion is a major factor during heating. The sintering response is sensitive to changes in particle size, time, temperature, green density, additives, and heating rate [18–20]. Small quantities (0.1%) of MgO or NiO aid densification by retarding grain growth. Additionally, impurities such as carbon or calcia retard densification.

Many high temperature devices rely on silica additions to the alumina to lower the sintering temperature. Sintered alumina is used in insulator, electronic, lighting, medical, furnace furniture, and wear applications. For most of these, the final mechanical properties are not a dominant concern. As purity increases the sintering difficulty increases. When employed in high temperature applications, sintering additives are minimized to improve creep resistance. High purity alumina is usually 99.5 to 99.8% pure, with strength in the 350 to 550 MPa range. An example microstructure after sintering is shown in Figure 15.16. It is the basis for many technical ceramics, especially in electronics.

Aluminum

Aluminum powders are highly compressible, so green densities of 90% are common. Sintering is focused on bonding these particles, not on densification. Because of the oxide films on the particles, sintering is performed using alloys that form a liquid phase, under atmospheric conditions that prevent further oxidation [21–24]. Alloying relies on Zn (up to 10%), Mg (up to 3%), Si (up to 1%), and Cu (up to 5%) additions as mixed

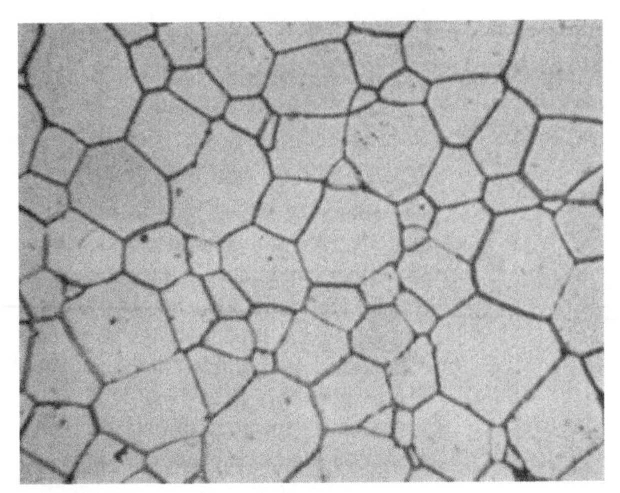

Figure 15.16 Etched microstructure of sintered alumina, showing essentially pore free material as required for optical translucency for lighting applications.

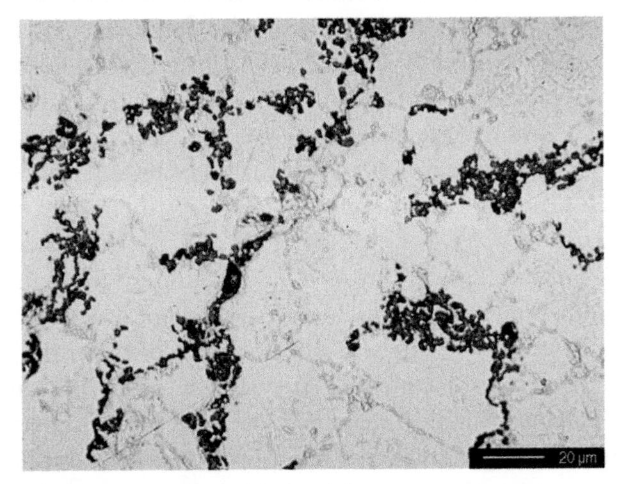

Figure 15.17 Sintered aluminum, with residuals of the oxygen scavenging reactions between the particles.

powders. Copper and silicon form eutectic liquids that give rapid sintering by 650°C (923 K). A typical time at temperature would be 20 min and the most popular atmospheres are nitrogen or nitrogen-hydrogen with low water contents. As shown in Figure 15.17, the grain boundaries show evidence of the reactions with the additives.

Lubricant is added to the powder prior to compaction and is burned out during heating. The sintering dimensional change depends on the green density and alloying ingredients, and is between 2% shrinkage and 1% swelling. Strengthening in aluminum is achieved via post-sintering heat treatments. Typical strengths are in the 220 to 300 MPa range with 2 to 7% fracture elongation. Sintered aluminum is used for mechanical components, such as automotive gears and levers.

Early sintered aluminum alloy was known as sintered aluminum product (SAP). It intentionally contained from 7 to 14 vol.% alumina as an oxide dispersion. The oxide is stable at elevated temperatures and provides creep resistance. Recent variants have introduced intermetallic precipitates, such as $FeNiAl_9$. Because of their high temperature strength, these powders require pressure-assisted consolidation in the 400°C (673 K) range. The product has a strength of 500 MPa with 5% fracture elongation.

Brass and Bronze

Brass and bronze are copper alloys which have been produced by sintering for nearly a hundred years, but our understanding of the mechanism involved is more recent [25–27]. Brass contains 10 to 35% Zn and is typically sintered from particles in the 40 μm size range. Lithium stearate is a common additive for compaction. It decomposes during heating to the sintering temperature. For better mechanical or electrical properties, it is common to sinter to 95% density at 870 to 900°C (1173 K) for about 30 min. Dimensional change is typically about 1% linear shrinkage, giving a sintered yield strength of 80 MPa, a tensile strength of 220 MPa, 20% elongation to failure, and electrical conductivity equal to 35% of pure copper. For better properties the compact is repressed and resintered.

Bronze is a copper alloy containing about 10% Sn. Unlike brass, sintered bronze is often fabricated for controlled porosity for bearings and filters. These are fabricated from mixed powders (Cu + Sn) or prealloyed powders. Additives such as lead, graphite, iron, or phosphorous are used to modify friction attributes. The prealloyed powder is sintered in the solid state, but the mixed powder is heated over the melting temperature of tin to induce a transient liquid phase. Typical peak temperatures are in the 800 to 870°C (1073 to 1143 K) range for less than 30 min. High temperatures induce densification, otherwise shrinkage is about 1%. Atmospheres are nitrogen-based, with additions of carbon monoxide or hydrogen. Since the final product is porous, the mechanical properties after sintering are relatively low, with tensile strengths in the 100 MPa range.

Cemented Carbide

Cemented carbides are based on tungsten carbide and other additions (TaC, VC, TiC as examples) with cobalt used to form a liquid phase [28–30]. The cementing phase might involve other transition metals. The mainstay compositions are based on WC-Co, mostly with 6 to 12 wt.% Co. The powders are in the micrometer size range and are usually highly milled prior to sintering; the milling provides stored energy.

Figure 15.18 Microstructure of liquid phase sintered cemented carbide, consisting of angular WC grains in a matrix (previously liquid) of cobalt.

Densification occurs during heating prior to liquid formation, due the stored strain energy and defects introduced during milling.

The tungsten carbide crystal structure is hexagonal close-packed with an anisotropic surface energy. This causes grain reshaping on heating as the carbide grains polygonize. Additions of VC, TiC, or TaC accelerate grain shape changes, leading to smaller sintered grains and higher strengths. Slow heating is used to remove lubricants. The final heating rate is about 5 to 10°C/min. Sintering is performed under vacuum. The peak temperature depends on the composition, being higher with lower cobalt levels. A typical peak temperature is about 1400°C (1673 K) with a 60 min hold. The furnace heating elements and support trays are made from graphite to prevent carbon loss. If carbon is lost, the product is brittle or experiences exaggerated grain growth. Carbon control is tricky and critical to attaining desired hardness, strength, wear resistance, and fracture toughness combinations.

Late in the sintering cycle, after pore closure, pressure is introduced into the furnace to collapse any pores. This gives a pore free microstructure consisting of angular carbide grains in a solidified cobalt-rich matrix, as shown in Figure 15.18. The elimination of residual pores has a considerable benefit for toughness and fracture resistance. A WC-10Co composition should exhibit a transverse fracture strength of 3500 MPa.

Copper

Pure copper sinters by grain boundary, surface, and volume diffusion, with the details being dependent on the particle size, temperature, and green density. Most applications

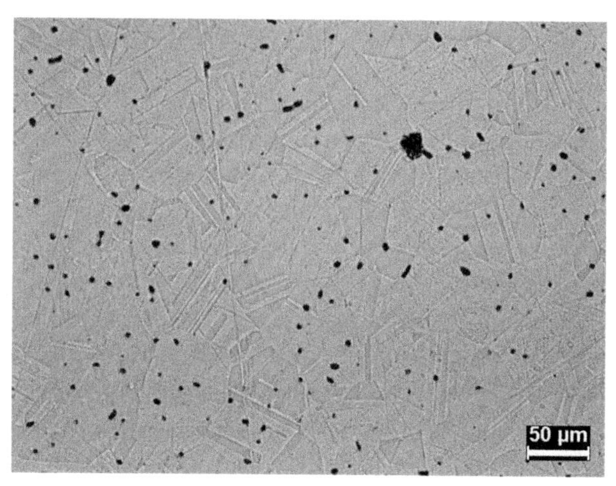

Figure 15.19 Sintered copper with some residual pores, but conductivity properties about 90% of full dense copper.

are in electrical and thermal components, such as heat spreaders. Die pressing relies on particles sizes closer to 100 μm, but injection molding relies on particles sizes closer to 10 μm. Components sintered from copper are fabricated in the 1000°C range for times of 30 min. The usual sintering atmosphere is hydrogen or hydrogen-nitrogen. In the unalloyed condition, the sintered product is relatively weak, with 100 MPa strength and residual pores as evident in Figure 15.19. At this point the thermal and electrical conductivities are near 90% that of wrought copper. Too high a green density causes swelling, since oxygen trapped in the pores reacts with the hydrogen to produce insoluble water vapor. Additions of Fe, Cr, Al or other oxide formers are effective in removing the oxygen contamination effect.

Nickel silver is an alloy (with no silver) consisting of 65Cu-18Ni-17Zn. Typical sintering is in the 1000°C range for 30 min using nitrogen-hydrogen. For a 150 μm powder, the sintering shrinkage is about 1.5%, product strength about 300 MPa with 12% fracture elongation. This material is used in hardware, jewelry, and camera components.

Diamond

Diamond is hard and sets the record for thermal conductivity. Sintering diamond is a difficult task, in part because the material is metastable [31]. Originally, diamond was mined, but subsequently artificial diamond powder arose, and since the 1980s it has been sintered to make polycrystalline diamond. The applications for sintered diamond are in tools for oil well drilling, wire drawing, metal cutting, and stone cutting.

Additionally, the high thermal conductivity is useful for heat dissipation in high performance computers.

To sinter polycrystalline diamond, a broad particle size distribution is prepared. The mixed powders are liquid phase sintered in carbide anvil presses with tantalum encapsulation. Typical pressures are 5 to 6 GPa range with peak temperatures of 1500°C (1773 K) or higher. The sintering cycles last about 15 min to avoid decomposition of the diamond. During this cycle a liquid forms from added transition meals (Ni, Co, Fe, or various blends). Sintering occurs via solution-reprecipitation where the larger diamond grains bond using material from the small diamonds. As a consequence, the final product is dense polycrystalline diamond with sinter necks determined by the dihedral angle. In turn, the dihedral angle depends on the liquid phase composition. If no liquid phase is used, then temperatures over 2300°C (1573 K) at 6 GPa are required to directly sinter diamond.

Polycrystalline cubic boron nitride is fabricated in a similar manner, and this is mainly used for high speed cutting and milling of ferrous alloys; situations where diamond tends to react with the work piece.

The consolidation conditions have long been known for both diamond and cubic boron nitride, but the design of the consolidation devices is proprietary. Anvil and belt presses are most typical, constructed from cemented carbides. These are typically 10 m tall with an output of a few kg per hour. Thus, thousands of these devices are installed around the world. It is estimated this application accounts for $1 billion in product per year.

Iron and Steel

About one billion kilograms of ferrous powder are used to form sintered components each year, with about 80% used in automobiles. The applications range from shock absorbers to fuel injectors. High pressure compaction to 800 MPa is used to press the powders (typically mixed elemental powders since they are softer than prealloyed powder) to 85 to 90% density. The intent is to attain as much densification outside the sintering furnace, allowing tool design to deliver complex features with minimum post-sintering dimensional adjustments. There are hundreds of compositions, tailored to balance cost and properties; some common alloys are Fe-2Cu-0.8C, Fe-2Ni-0.5C, and Fe-2Ni-0.5Mo-0.2Mn-0.5C.

Ferrous sintering is performed in continuous furnaces, usually relying on a woven stainless steel wire belt for conveyance. Over about six hours, the traveling belt conveys the compacts through a sequence of heat zones. In facilities dedicated to automotive production up to 150 continuous furnaces run in parallel. The sintering furnace provides stepwise heating and cooling to induce proper sintering and heat treatment [32]. The compact is kept at a peak temperature for long enough to allow uniform heating without inducing dimensional change. Microstructural coarsening is avoided by the short time at the peak temperature, although there is no clear guide for selecting the best temperature [33–35].

In the first stage of continuous heating, lubricants are removed using a temperature near 550°C (823 K). Subsequently, the compact is heated to at least 1120°C (1393 K). Higher temperatures require more expensive furnaces. Iron is a polymorph that has carbon solubility only when transformed to the face-centered cubic high temperature austenite phase. The level of carbon determines the mechanical properties, so sintering at low temperatures is not successful because the graphite remains undissolved.

Oxidation at high temperatures is prevented by the use of an atmosphere of nitrogen and hydrogen (90% N_2 and 10% H_2 is typical) [36]. Nitrogen is almost neutral with respect to iron (nitrogen is an interstitial strengthener similar to carbon) while hydrogen removes residual oxygen.

After reaching the peak temperature, the compact is cooled and removed from the furnace. Heat treatments are included in the cooling zone to manipulate final properties; a process termed sinter-hardening. One important aspect of sintering is the control of the compact surface. A reducing atmosphere using some hydrogen removes oxides, but an improper atmosphere results in discoloration, loss of carbon, and even warpage.

A typical production rate in steel sintering is about 350 kg/h, depending on the furnace width, loading on the belt, and belt speed. Depending on the furnace design, the cost of sintering ranges from about $0.20 to $1 per kg. Globally, sintered iron and steel account for over $6 billion in products, while consuming about 10^9 kg of powder each year.

Rare Earth Magnets

The highest performance permanent magnets are based on the $Fe_{14}Nd_2B$ compound. This composition provides exceptional magnetic energy, useful in electric motors, power hand tools, stereo speakers, headphones, and electronic fuel injectors. The powders are fabricated from a rapidly solidified melt using ingot casting or strip casting. Alloys contain a broad range of rare earth (Nd, Dy and Pr are common), transition metal (Co and Cu), and carbon additions. The composition is tailored to give magnetic properties in the form of a high coercive force, high energy product, high Curie temperature, or high remanence [37].

Powder is generated in a sequence of milling, crushing, and hydrogen pulverization steps, ending with a submicrometer particle size. To align crystals the powder is compacted in a strong magnetic field. Excess rare earth components provides for liquid phase sintering. Although the ideal magnet cotains 26 wt.% Nd, in practice about 30 to 33 wt.% is used for best sintering. Heating is performed in a graphite vacuum

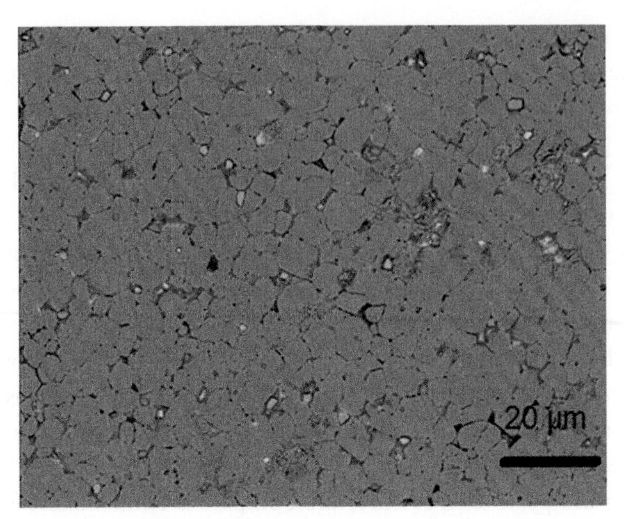

Figure 15.20 Sintered rare earth permanent magnet microstructure, consisting of primary $Fe_{14}Nd_2B$ grains, solidified rare earth rich liquid (white), and black residual pores.

furnace to about 1100°C (1373 K) for up to 300 min, giving a sintered density of 95%. The scanning electron micrograph in Figure 15.20 shows the primary grains, black pores, and small amounts of solidified liquid phase (white). Considerable grain growth occurs, giving final grain sizes near 10 to 15 μm. A post-sintering heat treatment and magnetic alignment step provides the desired properties. Due to a high reactivity the magnets are coated, plated, or painted. The final product is fairly expensive due to the high rare earth content, resulting in price fluctuations from $20 to $100 per kg. Globally the sales of sintered rare earth permanent magnet products are valued at about $12 billion per year.

Silicon Carbide

Silicon carbide is a difficult material to sinter, reflecting the stability of the material at high temperatures. The resistance to high temperature industrial environments inherently adds to the processing difficulty in forming engineering components. Covalent bonding in the compound inhibits diffusion, making it necessary to use sintering additives to form a liquid phase [38]. The high grain boundary energy makes pore elimination difficult and promotes rapid grain growth. Practical sintering cycles are aimed at densification with controlled grain growth, often with sintering aids of alumina, boron, or oxides of titanium and yttria. These form liquid phases at the sintering temperature.

In one approach, SiC powder with a specific surface area of 10 m^2/g is combined with 0.5% boron and 1.5 to 3.5% carbon. Boron reduces surface diffusion at low

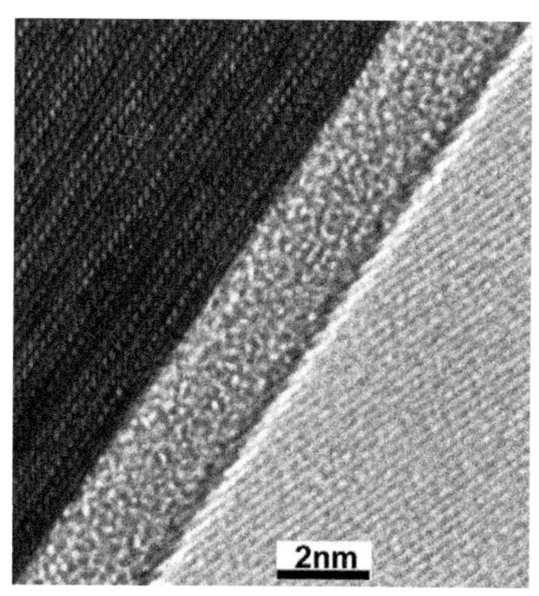

Figure 15.21 A high resolution transmission electron micrograph of the amorphous grain boundary film between SiC grains after liquid phase sintering [39].

temperatures and carbon prevents silicon evaporation. An oxygen content below 0.2 wt.% is required for sintering densification. A sintering temperature of 2100°C (2373 K) gives a high sintered density with minimal grain growth. Additives are commonly used to form amorphous grain boundary phases, giving liquid phase sintering but with a high viscosity glass. Figure 15.21 is a high resolution transmission electron micrograph showing the grain boundary film [39].

Typical sintered strengths are in the 500 MPa range and this strength is retained up to 1200°C. Sintering is performed in a graphite element furnace. Initial heating to 1700°C (1973 K) is in vacuum, but argon is used at higher temperatures to reduce evaporation. The carbon monoxide partial pressure is maintained as low as possible to ensure densification.

Applications are predominantly as abrasives, but include wear components, semiconductor processing equipment, high temperature semiconductors, heating elements, igniters for combustion devices, and diesel engine components. Not all of these uses rely on sintering, but overall the material accounts for about $1 billion in commercial product each year.

Silicon Nitride

Silicon nitride is a covalent ceramic that remains strong at high temperatures, making it difficult to sinter. For this same reason the sintered component provides attractive

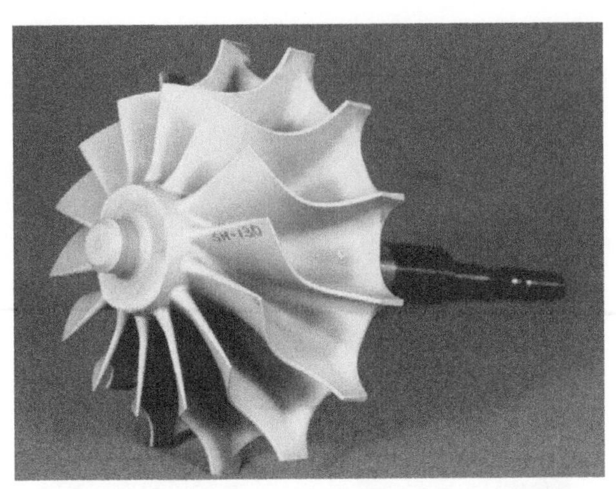

Figure 15.22 A sintered silicon nitride truck turbocharger shows the detailed features possible by sintering.

high temperature properties which are useful in turbochargers, bearings, cutting tools, and armor. Figure 15.22 is a turbocharger after sintering. Additives to Si_3N_4 form a viscous liquid at the sintering temperature that enables densification [40]. Target processing is 1600°C (1873 K) for 300 min, usually in a batch furnace but continuous sintering is possible [41]. Liquid phase sintering relies on magnesia, yttria, or alumina additions. Accordingly the final composition might be termed SiAlON or SiAlYON. Sintering is performed in boron nitride crucibles in an embedding layer of silicon nitride powder. The furnace elements are graphite and the process atmosphere is nitrogen. If sintering reaches 1800°C (2073 K), mass loss becomes a problem. Because silicon nitride is unstable at high temperatures, it is difficult to attain full density without pressure, such as in hot isostatic pressing. Final strengths are in the 700 to 800 MPa range.

Silicon nitride is also formed by a reaction between silicon powder and nitrogen, termed reaction bonded silicon nitride. In reaction bonding, silicon powder with a particle size below 40 μm is shaped to a density of 2.7 g/cm^3, corresponding to a fractional density of 0.735. Iron oxide additions aid the conversion of silicon into silicon nitride. The compact is nitrided in an atmosphere of 4% hydrogen and 96% nitrogen. Since the nitriding reaction is exothermic, temperature and atmosphere control are critical, especially in response to the effects of differing furnace loads. The gas pressure during heating is maintained slightly over atmospheric pressure (0.12 MPa). The expanded volume due to the nitridation of silicon progressively fills the pore space, so that the final product is near full density with no significant dimensional change. A feedback control system replenishes the atmosphere as nitrogen combines to form the nitride. The temperature is slowly increased to 1400°C to sustain a constant nitridation rate. After reaction bonding, the strength of the resulting material is in the 350 to 400 MPa range.

Stainless Steel

Thousands of stainless steels are used, but sintering technologies focus on just a few—the 300 series (304L, 316L), 630 (17-4 PH), and the 400 series (440 and 410). The 300 series is widely used to form filters, watch cases, eyeglass hinges, orthodontic brackets, and the 603 grade is a favorite for surgical tools, aircraft components, and cellular telephone and computer buttons, hinges, and sockets. These alloys are selected for applications sensitive to corrosion [42].

Green bodies are formed several ways:

- by die compaction using 100 μm powders pressed at 500 MPa to about 85% density for flat shapes
- injection molding 10 μm powder to 65% density for complex, small shapes
- cold isostatic pressing 100 μm irregular powder to form tubular products
- hot isostatic pressing 150 μm spherical powder to form large, dense bodies.

For the die compaction option, the peak sintering temperature is near 1250°C (1523 K) for 20 min giving from 0.5 to 2.5% shrinkage. As illustrated by the microstructure shown in Figure 15.23, the residual porosity makes an excellent filter. For

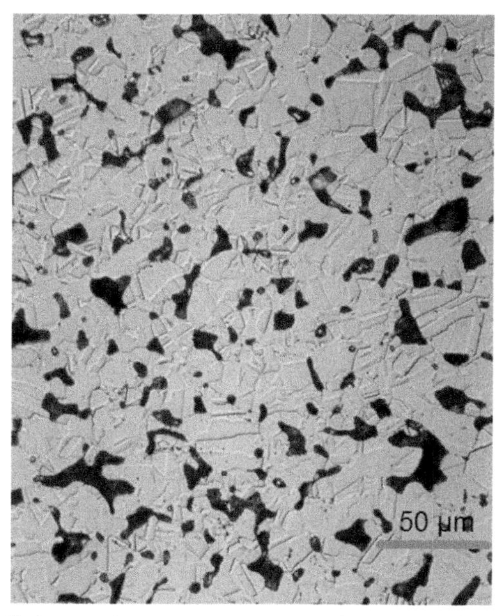

Figure 15.23 The controlled pore size structure associated with a sintered stainless steel filter, the black regions are pores.

the injection molds, smaller particle sizes are sintered to near full density at 1350°C (1623 K) in 60 min with 15% shrinkage [43]. Similar to cast stainless steel, a small amount of porosity remains (2 to 5%). Cold isostatic pressing is a favorite for forming filters, so they are sintered to generate controlled pore sizes and porosity. Hot isostatic pressing is a favorite for large, one-of-a-kind, petrochemical pipes and valves. The sintered mechanical properties are competitive, but residual porosity degrades corrosion resistance. Mass transport during sintering is a combination of grain boundary diffusion and dislocation climb. Boron-doped stainless steels sinter to full density by supersolidus liquid phase sintering.

Atmospheric control during sintering has a major impact on corrosion resistance. Carbon, nitrogen, and oxygen are detrimental, since these elements form chromium compounds to prevent passivation. Corrosion resistance greatly degrades at more than 0.4% retained nitrogen, due to the formation of chromium nitrides. A typical cycle involves heating in hydrogen to temperatures at which any organic additives decompose and holding until decomposition is complete. Subsequent heating takes place at approximately 10°C/min to the final sintering temperature. A hold at approximately 1000°C (1273 K) helps to eliminate contaminants. Vacuum sintering is a viable option, as long as the pressure is not so low that chromium evaporation occurs.

Sintered stainless steels are capable of a wide range of final properties, depending on the composition and microstructure. For precipitation hardenable alloys, yield strengths of 1100 MPa with 12% elongation are possible. Alternatively, for the 300 series austenitic stainless steels, sintered yield strength is 250 to 450 MPa with good ductility (30% or more). The sintered metal powder injection molded applications, including surgical tools, cellular telephone and computer components, is approaching the $1 billion per year range, and the filter, automotive, and industrial uses for sintered stainless steel by other fabrication routes is smaller, at about $200 million per year.

Titanium

Sintered titanium was demonstrated in the 1950s and 1960s. All titanium is formed as a sponge powder, so milling that powder to enable press-sinter processing was an early option. Sintered titanium is available in two primary compositions—pure titanium and Ti-6Al-4V. The alloy Ti-6-4 accounts for the vast majority of applications in electronics, aerospace, medical, and petrochemical uses. The three main forming routes are [44—47]:

- press-sintering to form filters with intentional porosity and this includes cold isostatic pressing for tubular filters
- injection molding and sintering to form components for watches, cameras, and surgical tools

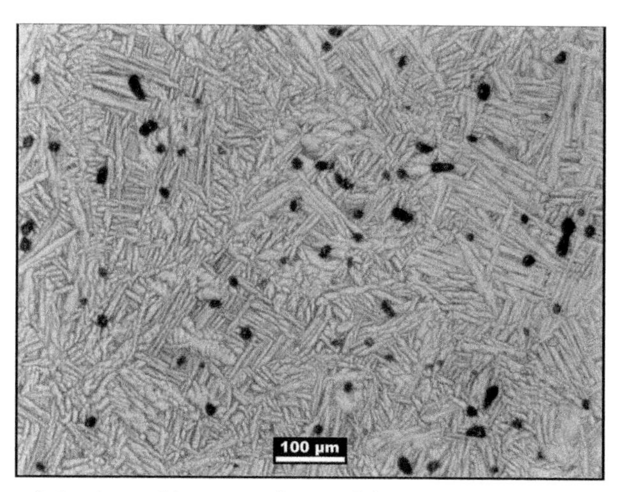

Figure 15.24 Sintered titanium with a two phase solid structure and residual pores that require pressure-assisted sintering for full densification.

- hot isostatic pressing to form tubes, landing gear, sputter targets, and medical implants.

Titanium is sensitive to oxygen, so considerable effort is required to minimize contamination. Sintering is in the 1250 to 1400°C (1523 to 1673 K) range and times of 120 to 240 min are typical, almost always in vacuum [14]. Everything is important to successful titanium sintering and considerable attention must be paid to setters, vacuum levels, furnace design, heating ramps, as well as peak temperature and hold time. It is fairly easy to over-sinter titanium, so cycles are customized to the powder and furnace. Typically, the best sintered properties arise with a dual-phase sintered microstructure, as seen in Figure 15.24. The residual pores are detrimental in some situations and are removed using hot isostatic pressing.

Sintered titanium components are expensive, largely since quality raw powders can cost from $150 to $225 per kg. But note that for injection molded components, the sale price averages $8000 per kg. The annual market for sintered titanium is comparatively small, estimated at $250 million globally.

Tool Steels

Tool steels are a hybrid between a cemented carbide and steel. The sintered microstructure consists of a dispersion of refractory carbides in a steel matrix. Two sintering approaches are applied to tool steels—supersolidus liquid phase sintering and pressure-assisted sintering [48,49]. Both start with prealloyed powder containing up to 1.5% carbon and 15% refractory metal (Cr, W, Mo, or V) that forms the carbide dispersion.

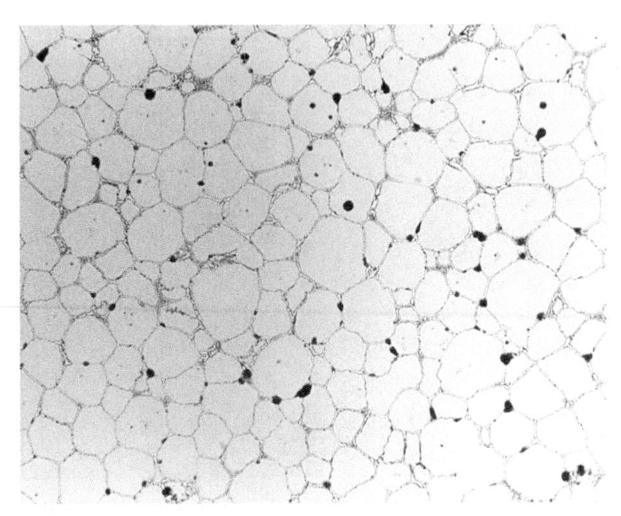

Figure 15.25 Supersolidus liquid phase sintered tool steel, with carbides dispersed on the grain boundaries and a small residual porosity.

Supersolidus liquid phase sintering densifies the powder over a narrow temperature range where a carbide-metal eutectic forms. The liquid allows rapid densification, resulting in small levels of porosity and carbides on the grain boundaries, as evident in Figure 15.25. Sintering is best performed in vacuum and typical temperatures are near 1250°C. Precise temperature control is required to form sufficient liquid for densification while avoiding carbide coarsening. The sintered properties are sensitive to residual porosity and carbide grain size. An alternative approach is via pressure-assisted sintering starting with vacuum sintering and ending with inert gas pressurization, thereby collapsing residual porosity. In both approaches, low oxygen levels are required to prevent decarburization.

Hot isostatic pressing of tool steels is typically performed in the 800 to 1100°C (1073 to 1373 K) range using pressures near 100 MPa. Large compacts of 1000 kg are densified and then hot worked into full density ingots. Hot isostatic pressed tool steels are widely used for the construction of high quality metal working tools.

Tungsten Heavy Alloy

The high density associated with tungsten-based composites leads to the term heavy alloy. They are used in self-winding watches, projectiles, radiation shields, cellular telephones, and computer heat sinks. Heavy alloys are liquid phase sintered using mixed elemental powders, reaching full density at temperatures near 1500°C and times of 30 min [50–52]. The sintering temperature varies with composition, but the most common are W-Ni-Fe or W-Ni-Cu alloys. The tungsten content is usually in the 83 to 98% window. Other examples include tungsten with Ni-Mn, Co-Ni, Ni-Fe, or Mo-Ni-Fe. These systems sinter rapidly because the solid is soluble in the liquid. Sintering starts in the solid state during heating, and rapidly progresses when the first liquid forms. Coarsening accompanies sintering due to the coalescence and dissolution of the small solid grains into the liquid followed by reprecipitation on the larger grains. The mean grain size increases with the cube root of the time. There is a concomitant grain shape accommodation event that eliminates pores and forms a rigid solid skeleton.

These alloys are sensitive to oxygen and carbon contamination, so sintering is usually carried out in a hydrogen atmosphere (to remove oxygen) doped with water (to remove carbon). Initial heating is done in hydrogen at 10°C/min to 1000°C (1273 K), with a 5°C/min ramp to 1500°C (1773 K), with a 30 min hold. In this latter cycle, the moisture content is progressively increased to suppress swelling. Sintering shrinkage is about 16%.

To reduce the unfavorable effects of hydrogen on mechanical properties, cooling is performed in argon, or a post-sintering heat treatment is applied in nitrogen or argon. The final product is fully dense with tensile strengths near 900 MPa and 25% elongation to fracture. The high absolute density of the tungsten alloys (typically 17 to 19 g/cm^3) contributes to distortion, most notably when the liquid phase forms. Several microgravity liquid phase sintering experiments have relied on tungsten heavy alloys to accentuate the presence or absence of gravity in small samples. When gravity is removed, the pores agglomerate, since there is no buoyancy force, leading to enormous central cavities in an otherwise dense shell.

Zirconia

Zirconium oxide is widely used in watches, cutlery, heart pacemakers, scissors, and golf clubs. Zirconia is unique since it is a tough ceramic, with a combination of hardness, strength, fracture toughness, wear resistance, and chemical inertness. Stabilizers such as yttria are added to control the microstructure and for toughening. Since the melting temperature is very high and diffusion rates are low, small particles are required to promote sintering densification [53,54]. Particles sizes in the 20 to 30 nm range are available, but agglomeration is a problem. When properly deagglomerated, nanoscale zirconia sinters to full density at 1100°C in 60 min, with a final grain size of 0.2 μm. However, without deagglomeration, the same powder only sinters to 95% density at 1500°C in 240 min. A sintered microstructure is shown in Figure 15.26. Sintered strength ranges from 400 to 900 MPa, depending on grain size, additive, and density.

It is best to heat this material rapidly to avoid low temperature surface diffusion coarsening. Accordingly, fast firing in a microwave is one option. Alternatively, sintering

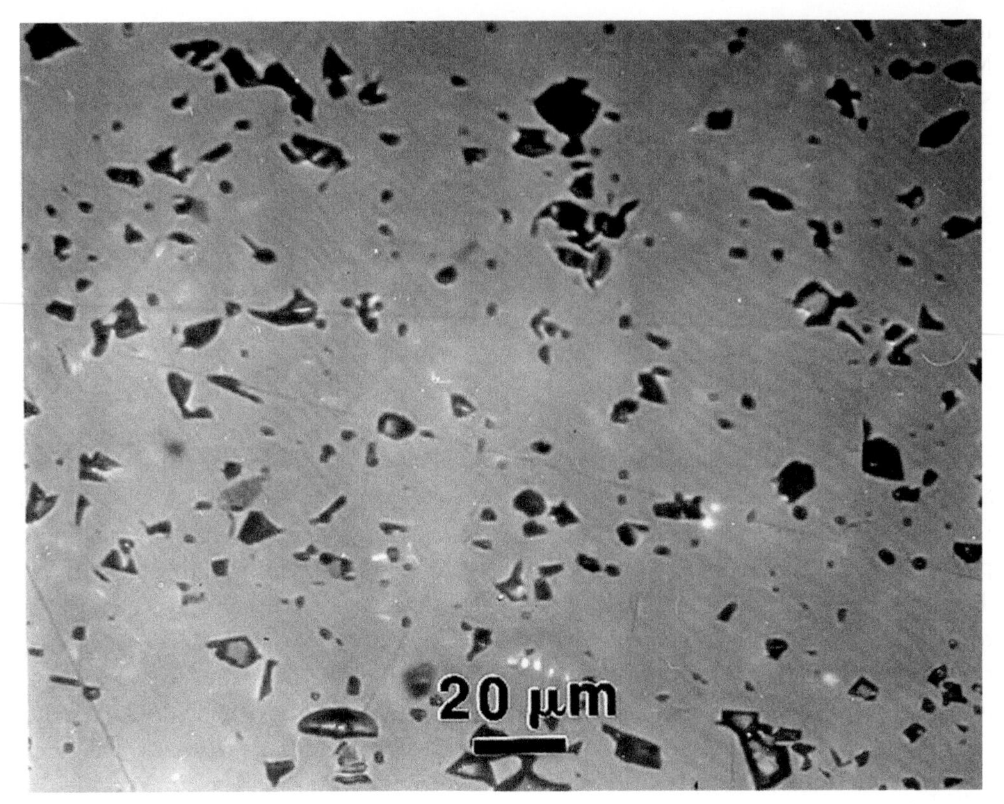

Figure 15.26 Sintered microstructure for zirconia with a reinforcing phase to improve toughness.

additives are used to increase densification; MgO reduces surface diffusion to assist in high temperature densification. However, MgO lowers the pore mobility and hinders pore motion during grain growth.

SUMMARY

Earlier it was stated that sintering is a "messenger" not the "cause" for defects, composition shifts, and other issues. To emphasize this point, Figure 15.27 is a picture of a sintered stainless steel pump housing. Cracks are marked on this structure. In production, several experiments were conducted with different sintering conditions, but the cracks did not go away. In such situations, an experimental design is employed which includes intentional variations in peak temperature, heating rate, hold time, furnace loading, and component location in the furnace. After much frustration,

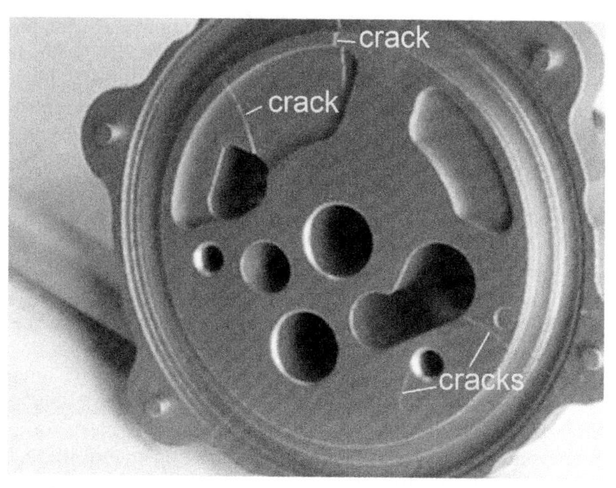

Figure 15.27 A sintered stainless steel pump housing with several cracks (marked) after sintering. This is a case where the cracks opened during sintering, but inherently sintering did not cause cracking.

upon careful inspection these cracks were identified as molding hairline cracks in the green body, caused by the forming tool. They were not sintering cracks. This is an example of the defect being amplified by sintering, but sintering was not the cause, and sintering was not able to heal the defect. Defects observed after sintering often have root causes in steps prior to sintering.

Sintering as a technology contributes to probably $100 billion in global commercial impact. Underlying this practice is considerable empirical study. Compilations of industrial sintering cycles provide much assistance in facing practical sintering problems, and several starting point references are included below [1–67]. There is an obvious abundance of information. This chapter touches on some of the more common materials and cycles, but makes no attempt to recount the extensive body of prior reports. In spite of the progress, still the intersection between sintering science, materials science, mechanical engineering, and commercial development remains rich with opportunities for new discoveries. Good hunting.

REFERENCES

[1] R.M. German, Green body homogeneity effects on sintered tolerances, Powder Metall. 47 (2004) 157–160.

[2] M.Y. Nazmy, M.S. Abdel-Azim, Investigation of the dimensional changes in soft metal powder compacts, Powder Metall. 17 (1974) 13–20.

[3] F.V. Lenel, H.H. Hausner, O.V. Roman, G.S. Ansell, The influence of gravity in sintering, Trans. TMS-AIME 227 (1963) 640–644.

[4] J.A. Alvarado-Contreras, E.A. Olevsky, R.M. German, Modeling of gravity induced shape distortions during sintering of cylindrical specimens, Mech. Res. Comm. 50 (2013) 8–11.

[5] M. Gauthier, E. Baril, High temperature interaction of titanium with ceramic materials, Advances in Powder Metallurgy and Particulate Materials – 2005, Metal Powder Industries Federation, Princeton, NJ, 2005, pp. 7.127–7.140.
[6] T. Uematsu, Y. Itoh, K. Sato, H. Miura, Effects of substrate for sintering on the mechanical properties of injection molded Ti-6Al-4V alloy, J. Japan Soc. Powder Powder Metall. 53 (2006) 755–759.
[7] I. Smid, G. Aggarwal, Powder injection molding of niobium, Mater. Sci. Forum 475 (2005) 711–716.
[8] R.C. Chiu, M.J. Cima, Drying of granular ceramic films: II, drying stress and saturation uniformity, J. Amer. Ceram. Soc. 76 (1993) 2769–2777.
[9] A.G. Evans, Considerations of inhomogeneity effects in sintering, J. Amer. Ceram. Soc. 65 (1982) 497–501.
[10] G.A. Shoales, R.M. German, In situ strength evolution during the sintering of bronze powders, Metall. Mater. Trans. 29A (1998) 1257–1263.
[11] W.A. Kaysser, A. Lenhart, Optimization of densification of ZnO during sintering, Powder Metall. Inter. 13 (1981) 126–128.
[12] R.M. German, Sintering Theory and Practice, Wiley-Interscience, New York, NY, 1996.
[13] C. Blias, Atmosphere sintering, in: Z.Z. Fang (Ed.), Sintering of Advanced Materials, Woodhead Publishing, Oxford, UK, 2010, pp. 168–188.
[14] D.F. Heaney, Vacuum sintering, in: Z.Z. Fang (Ed.), Sintering of Advanced Materials, Woodhead Publishing, Oxford, UK, 2010, pp. 189–221.
[15] R. Speaker, R. Osterreich, S. Kazi, J. Buonassisi, Sintering atmosphere analysis: selection and use of analyzers to monitor and control sintering atmospheres, Advances in Powder Metallurgy and Particulate Materials – 2003, Part 5, Metal Powder Industries Federation, Princeton, NJ, 2003, pp. 16–31.
[16] D.P. Duncavage, C.W.P. Finn, Debinding and sintering of metal injection molded 316L stainless steel, Advances in Powder Metallurgy and Particulate Materials – 1993, vol. 5, Metal Powder Industries Federation, Princeton, NJ, 1993, pp. 91–103.
[17] S.K. Smith, High temperature sintering cost analysis, Inter. J. Powder Metall. 40 (2004) 54–56.
[18] G. Bernard-Granger, C. Guizard, A. Addad, Influence of co-doping on the sintering path and on the optical properties of a submicronic alumina material, J. Amer. Ceram. Soc. 91 (2008) 1703–1706.
[19] G. Bernard-Granger, C. Guizard, Influence of MgO and TiO_2 doping on the sintering path and on the optical properties of a submicrometer alumina material, Scripta Mater. 56 (2007) 983–986.
[20] X. Mao, S. Shimai, M. Dong, S. Wang, Gelcasting and pressureless sintering of translucent alumina ceramics, J. Amer. Ceram. Soc. 91 (2008) 1700–1702.
[21] M. Qian, G.B. Schaffer, Sintering of aluminum and its alloys, in: Z.Z. Fang (Ed.), Sintering of Advanced Materials, Woodhead Publishing, Oxford, UK, 2010, pp. 291–323.
[22] A. Arockiasamy, R.M. German, P. Wang, W. Morgan, S.J. Park, Sintering behavior of Al-6061 powder produced by rapid solidification process, Powder Metall. 54 (2011) 354–359.
[23] H. Asgharzadeh, A. Simchi, Supersolidus liquid phase sintering of Al 6061/SiC metal matrix composites, Powder Metall. 52 (2009) 28–35.
[24] G.J. Kipouros, W.F. Caley, D.P. Bishop, On the advantages of using powder metallurgy in new light metal alloy design, Metall. Mater. Trans. 37A (2006) 3429–3436.
[25] S. Gheorghe, C. Teisanu, I. Ciupitu, Considerations regarding sintering of the copper based alloys with low tin content, in: D. Bouvard (Ed.), Proceedings of the Fourth International Conference on Science, Technology and Applications of Sintering, Institut National Polytechnique de Grenoble, Grenoble, France, 2005, pp. 421–424.
[26] C. Menapace, P. Costa, A. Molinari, Influence of Mg on sintering of 90/10 elemental bronze, Powder Metall. 48 (2005) 171–178.
[27] C. Menapace, M. Zadra, A. Molinari, C. Messner, P. Costa, Study of microstructural transformations and dimensional variations during liquid phase sintering of 10% tin bronzes produced with different copper powders, Powder Metall. 45 (2002) 67–74.

[28] J. Soares, L.F. Malheiros, J. Sacramento, M.A. Valente, F.J. Oliveira, Microstructure and properties of submicrometer carbides obtained by conventional sintering, J. Amer. Ceram. Soc. 94 (2011) 84–91.
[29] G.S. Upadhyaya, Materials science of cemented carbides – an overview, Mater. Des. 22 (2001) 483–489.
[30] S. Luckx, The hardness of tungsten carbide – cobalt hardmetal, in: R. Riedel (Ed.), Handbook of Ceramic Hard Materials, vol. 2, Wiley-VCH, Weinheim, Germany, 2000, pp. 946–964.
[31] J.D. Belnap, Sintering of ultrahard materials, in: Z.Z. Fang (Ed.), Sintering of Advanced Materials, Woodhead Publishing, Oxford, UK, 2010, pp. 389–414.
[32] M.C. Thomason, Sintering furnace cooling method study, Advances in Powder Metallurgy and Particulate Materials – 2000, Metal Powder Industries Federation, Princeton, NJ, 2000, pp. 5.103–5.108.
[33] H. Danninger, G. Jangg, B. Weiss, R. Stickler, Microstructure and mechanical properties of sintered iron Part I: basic considerations and review of literature, Powder Metall. Inter. 25 (3) (1993) 111–117.
[34] C. Lall, Principles and applications of high temperature sintering, Rev. Powder Metall. Part. Mater. 1 (1993) 75–107.
[35] R.M. German, Powder Metallurgy of Iron and Steel, Wiley-Interscience, New York, NY, 1998.
[36] D. Garg, K.R. Berger, D.J. Bowe, J.G. Marsden, Effects of various nitrogen based atmospheres on sintering of carbon steel components, in: R.M. German, G.L. Messing, R.G. Cornwall (Eds.), Sintering Science and Technology, The Pennsylvania State University, State College, PA, 2000, pp. 1–8.
[37] J. Ormerod, The physical metallurgy and processing of sintered rare earth permanent magnets, J. Less Common Met. 111 (1985) 49–69.
[38] W. Dressler, R. Riedel, Progress in silicon-based non-oxide structural ceramics, Inter. J. Refract. Met. Hard Mater. 15 (1997) 13–47.
[39] Y.W. Kim, Y.S. Chun, T. Nishimura, M. Mitomo, Y.H. Lee, High Temperature strength of silicon carbide ceramics sintered with rare earth oxide and aluminum nitride, Acta Mater. 55 (2007) 727–736.
[40] F.L. Riley, Silicon nitride and related materials, J. Amer. Ceram. Soc. 83 (2000) 245–265.
[41] D.E. Wittmer, C.W. Miller, Comparison of continuous sintering to batch sintering of Si_3N_4, Ceram. Bull. 70 (1991) 1519–1527.
[42] E. Klar, P.K. Samal, Powder Metallurgy Stainless Steels, ASM International, Materials Park, OH, 2007.
[43] S. Krug, S. Zachmann, Influence of sintering conditions and furnace technology on chemical and mechanical properties of injection moulded 316L, Powder Inj. Mould. Inter. 3 (4) (2009) 66–71.
[44] H. Wang, Z.Z. Fang, P. Sun, A critical review of the mechanical properties of powder metallurgy titanium, Inter J. Powder Metall. 46 (5) (2010) 45–57.
[45] M. Qian, G.B. Schaffer, C.J. Bettles, Sintering of titanium and its alloys, in: Z.Z. Fang (Ed.), Sintering of Advanced Materials, Woodhead Publishing, Oxford, UK, 2010, pp. 324–355.
[46] I.M. Robertson, G.B. Schaffer, Review of densification of titanium based powder systems in press and sinter processing, Powder Metall. 53 (2010) 146–162.
[47] R.M. German, Status of metal powder injection molding of titanium, Inter. J. Powder Metall. 46 (5) (2010) 11–17.
[48] A. Bose, W.B. Eisen, Hot Consolidation of Powders and Particulates, Metal Powder Industries Federation, Princeton, NJ, 2003.
[49] R.M. German, Supersolidus liquid phase sintering Part I: process review, Inter. J. Powder Metall. 26 (1990) 23–34.
[50] J.L. Johnson, Sintering of refractory metals, in: Z.Z. Fang (Ed.), Sintering of Advanced Materials, Woodhead Publishing, Oxford, UK, 2010, pp. 356–388.
[51] V. Srikanth, G.S. Upadhyaya, Sintered heavy alloys – a review, Inter. J. Refract. Met. Hard Mater. 5 (1986) 49–54.

[52] A. Belhadjhamida, R.M. German, Tungsten and tungsten alloys by powder metallurgy – a status review, in: A. Crowson, E.S. Chen (Eds.), Tungsten and Tungsten Alloys, The Minerals, Metals and Materials Society, Warrendale, PA, 1991, pp. 3–18.
[53] M.J. Mayo, Processing of nanocrystalline ceramics from ultrafine particles, Inter. Mater. Rev. 41 (1996) 85–115.
[54] R.A. Andrievski, Review nanocrystalline high melting point compound based materials, J. Mater. Sci. 29 (1994) 614–631.
[55] F. Thummler, W. Thomma, The sintering process, Metall. Rev. 12 (1967) 69–108.
[56] M.B. Waldron, B.L. Daniell, Sintering, Heyden, London, UK, 1978.
[57] M.N. Rahaman, Ceramic Processing and Sintering, Marcel Dekker, New York, NY, 1995.
[58] S.J.L. Kang, Sintering Densification, Grain Growth, and Microstructure, Elsevier Butterworth-Heinemann, Oxford, United Kingdom, 2005.
[59] Z.Z. Fang, Sintering of Advanced Materials, Woodhead Publishing, Oxford, UK, 2010.
[60] R. Orru, R. Licheri, A.M. Locci, A. Cincotti, G. Cao, Consolidation synthesis of materials by electric current activated assisted sintering, Mater. Sci. Eng. R63 (2009) 127–287.
[61] R.M. German, P. Suri, S.J. Park, Review liquid phase sintering, J. Mater. Sci. 44 (2009) 1–39.
[62] J.R. Blackford, Sintering and microstructure of ice, a review, J. Phys. D: Appl. Phys. 40 (2007) R355–R385.
[63] B. Uhrenius, J. Agren, S. Haglund, On the sintering of cemented carbides, in: R.M. German, G.L. Messing, R.G. Cornwall (Eds.), Sintering Technology, Marcel Dekker, New York, NY, 1996, pp. 129–139.
[64] E.M. Rabinovich, Preparation of glass by sintering, J. Mater. Sci. 20 (1985) 4259–4297.
[65] B.A. James, Liquid phase sintering in ferrous powder metallurgy, Powder Metall. 28 (1985) 121–130.

CHAPTER SIXTEEN

Future Prospects for Sintering

LINKAGES

Invention is a leading indicator of technical growth. In sintering much of the growth comes from advances in powder shaping technologies. As one example, Figure 16.1 plots the commercial growth in powder injection molding from its 1986 start, reaching $1.5 billion in sales by 2012—effectively a 22% per year growth rate. This is an outstanding application of sintering. The current rush is into additive manufacturing based on inkjet or laser construction of powder layers to form three-dimensional sintered objects [1,2]. Other ideas include shaping using novel centrifugal, electrophoretic, freezing, eddy current, and magnetic approaches.

Sintering abounds with new ideas, including a wide array of concepts [3–26]:

- sintered nitride-carbide, high surface area electrodes for super-capacitors
- sintered transparent, tough aluminum oxy-nitride armor
- aligned grain growth for anisotropic microstructures using seeded gels during sintering
- fullerene nanotube sintering to form stiff, strong, light structural materials
- porous scaffolds for tissue ingrowth formed by sintering titanium foams
- high temperature ceramics for plasma vapor containment in high efficiency lighting
- hard material composites of diamond bonded in a silicon carbide matrix
- freeform aerospace components based on sintering via a computer controlled lasers
- silicon circuits formed by ink jet printing and sintering nanoscale silicon powder
- bone replacements made by additive laser sintering using a computer image file
- sintered silver nanoscale particles used to replace lead solder in electronics
- carbon-copper composites for thermal management applications
- polycrystalline diamond bonded to tungsten carbide substrates for rock drilling
- sintered phosphors such as CdSe and CdS for advanced lighting.

In 1960 Jones [27] said:

> *"... the development of the theories of sintering ... is a fascinating story of a phenomenon which at first appeared mysterious, then temporarily simple, and eventually unexpectedly complex."*

Sintering: From Empirical Observations to Scientific Principles
DOI: http://dx.doi.org/10.1016/B978-0-12-401682-8.00016-1

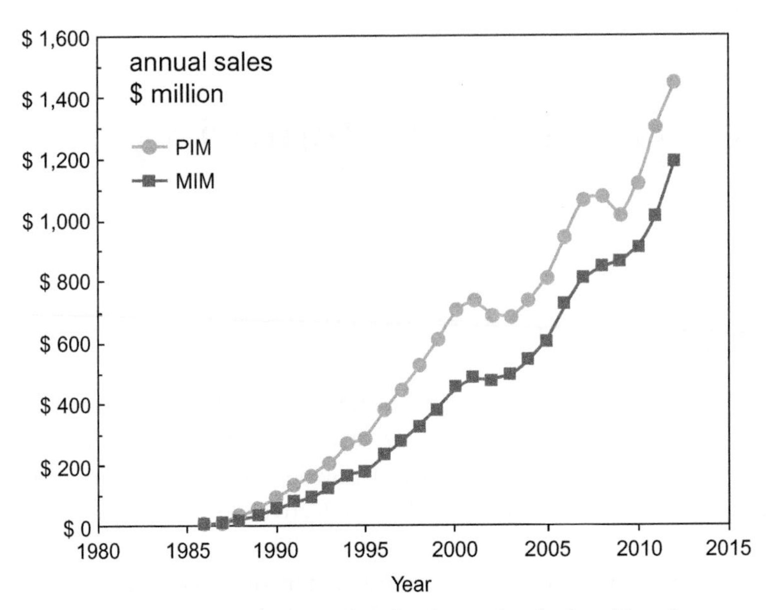

Figure 16.1 An indication of how new powder shaping technologies drive the expanded use of sintering is evident in powder injection molding. This plot shows sales growth from 1986 to 2012, averaging 22% per year.

From the examples cited above, sintering is clearly moving toward complicated chemistries, complex shapes, high performance levels, and tighter tolerances. In that spirit, this chapter is constructed to show where significant change is emerging in sintering, the materials, processes, and applications.

NEW MATERIALS

Sintering excels when the material is difficult to form by other means. This frequently involves unstable or metastable compositions, or unique powder mixtures. A good example would be the cemented carbides. The WC phase is unstable at high temperatures, but liquid phase sintering with Co enables densification and the creation of tough, hard, and wear resistant composites. Many additions are possible to change grain size, hardness, or other properties [28]. Similar comments apply to other sintered compositions, such as high performance magnets, and improved thermoelectrics. Success is assured when alternative manufacturing processes are not possible. Below are some of the materials being promoted for the future.

Light metals: Light metals are evolving, with new ventures focused on titanium, aluminum, and magnesium. Each effort involves sintering targeted at structural applications. If the cost falls sufficiently, the benefit from lower weight structural automotive components is massive, creating a substantial opportunity for press-sinter processing. Accordingly, much attention is directed toward determining the alloys, compaction and sintering cycles, and means to optimize mechanical properties [29–33].

Foamed materials: Foamed polymers are widely used for insulation, ranging from coffee cups to construction insulation boards. Today, foamed materials are emerging in ceramics and metals. A variety of approaches rely on sintering and the use of selective pore forming agents, intentional gas forming inclusions, and use of polymer foam templates and particle infiltration. Distention is now possible to 5% of theoretical density. As inventions arise, applications are found for sintered foams in heat pipes, electrolytes, fuel cells, crash protection barriers, filters, tissue affixation structures, and furnace hardware [34–40].

Composites: Composite structures are ideal for sintering. The constituent powders are mixed prior to sintering to given the desired ratio of amount, size, and spacing to custom design the final microstructure and properties. Some new composites are based on mimicking biological structures such as tissue scaffolds with intentional porosity [40–45]. This leads to customized functionality—such mixtures of magnetic and nonmagnetic phases. A wide variety of materials and property combinations are possible.

Hard and metastable materials: Hard materials have long been formed by using sintering. Critical relations are now recognized regarding the links between microstructure, composition, processing, and properties [46,47]. Added to the carbides and other cermets, the sintered superhard materials are the most recent additions. For example, silicon and diamond are sintered to form diamond bonded with silicon carbide. Polycrystalline diamond is pressure sintered, and the sister compound cubic boron nitride is processed in similar cycles. Inexpensive direct current presses (spark sintering) are a key to the consolidation of these hard materials. By 2012 there were more than 9000 such presses in use for diamonds and diamond composites.

Low temperature materials: Although sintering tends to excel with materials difficult to fabrication using traditional approaches, such as casting, an exciting option arises with nanoscale powders and low temperature firing [48]. For example, nanoscale silver is able to sinter at temperatures as low as 100°C (373 K) and is now used to replace solder in electronic circuit assembly or to build quantum dot circuits [25,48–54]. Considering the range of applications, the advent of sintering for low temperature electronic components is potentially the biggest growth field, enabled by nanoscale particle availability.

Brittle materials: Brittle ceramics, and especially high value ceramics such as superconductors, rely on sintering. Early intermetallic compound superconductors relied on pressure-assisted sintering, and as ceramic superconductors emerged, sintering was the optimal approach for their manufacture [53–56]. As mankind awaits room temperature superconductors, sintering remains poised to be the consolidation route. The age of structural ceramics is slow in arriving. So far attention has been focused on silicon carbide, silicon nitride, zirconia, and ceramic composites of oxides, nitrides, borides, silicides, and carbides. These are sintered materials, so when mankind finally turns to ceramics for everyday use in automobiles, garden tools, and home appliances, sintering will be the core process for their manufacture.

Insoluble materials: Composites consisting of mixed but insoluble phases are another target for sintering. Materials such as copper-chromium, invar-silver, molybdenum-copper, or tungsten carbide-silver cannot be made via fusion metallurgy, but have long been produced for electrical applications using sintering. New variants include ceramic phases dispersed in metallic matrices or metallic phases dispersed in ceramic matrices, for use in electrical and wear components. Tool steels containing dispersed carbides are formed using pressure-assisted sintering, and the ideas include a host of composites strengthened by intermetallic, silicon carbide, or carbon nanotube reinforcements [22,57–62]. Indeed, the wide range of possible combinations provides a new playground to custom design material combinations.

High temperature materials: Refractory materials, both ceramic and metallic, are sintered to avoid the difficulty of reaching high melting temperatures [63,64]. Alloys of tungsten, rhenium, molybdenum, tantalum, niobium, iridium, and ruthenium are important in high temperature rockets, lighting, furnaces, welding, and energy systems. Tantalum is the mainstay of high reliability capacitors. Magnesia is a mainstay for glass and steel refractories. As energy generation pushes to new efficiencies, such as via fusion reactors, sintered high temperature materials remain a key to service success. To offset the high sintering temperatures, a natural inclination is to move to nanoscale powders, which introduce other sintering difficulties, such as the difficulty in evaporating contaminants.

Multiple function materials: The use of sintering to form laminated electronic structures is an idea widely used in computer and electrical devices. Much opportunity arises for damage during sintering due to differential strains, so this topic area is steeped in mathematical analysis [12,65–76]. The number and type of applications is enormous, ensuring many new materials in the future now that basic guidelines for sintering these materials are isolated.

Controlled porosity materials: A few examples of controlled porosity have already been mentioned in this chapter. Sintering is easily able to customize the materials for energy absorption, insulation, catalyst supports, capacitors, battery electrodes, filters, heat pipes, and tissue ingrowth devices. The key is to select a proper particle size and to sinter to the desired combination of porosity, pore size, and permeability. In some mixed powder systems, swelling is common and the product is distended. Sacrificial pores are possible via additives, sometimes called space holders, which evaporate during sintering. With the advent of small powders, the sacrificial pore former enables separation of the pore size from the particle size. In early porous materials the pore size, porosity, and permeability were restricted by the narrow array of powder sizes. Such limitations are offset by innovations in pore forming agents [35–38,77–82].

NEW APPLICATIONS

Electronic systems are a massive opportunity for sintering. A new application for sintering relies on printed powders to form electronic circuits. This started with solder replacements, where a paste containing tin and copper powders reacts via transient liquid phase sintering. Solder containing lead is commonly used to join electronic components, but there are environmental concerns about its use with respect to worker and facility contamination. Although lead-free solders are known, often relying on silver, an alternative approach relies on a sintered polymer-powder mixture. Nanoparticle silver is mixed with a polymer and the mixture is screen printed to form a conductive circuit that sinters at a low temperature to replace solder. Since the sintering temperature depends on the particle size, nanoscale powders allow sintering below the polymer burnout temperature. The product is an electrical connection with intermixed polymer.

Silicon semiconductor devices initially were discrete diodes and transistors. As circuit complexity increased, integration of multiple devices on a single silicon chip resulted in the integrated circuit. A higher density of computing required closer proximity of components and smaller line widths between components, achieved using lithography. Now circuits include sintered silicon as a new fabrication option. This significantly changes the world of electronics, displacing lithography with faster print-sinter approaches. One variant places a circuit on a slender tube to allow temperature control during cauterization inside the human body. Variants relying on small particles, printed patterns, and low temperature sintering arise in flexible circuits, radio frequency identification tags, near-field cellular circuits, and in-store displays. A sintered device in use is for storing identification information, such as authentication for wine or designer luggage. Radio frequency identification tags rely on printed and sintered silicon inks. Unlike lithography, thin circuit patterns are produced at low cost using analogs to photocopiers or newspaper printing presses. Similar to newspaper production, production of printed-sintered silicon circuits will reach 30 billion units per year as prices fall. Applications are already found in retail check-out, product inventory, package identification, transit fares, toll cards, luggage identification, and security passkeys, which will quickly drive production levels extremely high.

The idea of component assembly prior to sintering has some positive benefits. The intent is to avoid welding or other bonding processes. Sinter bonding combines two components into a single, multiple material device. Applications include electronic packaging, in which solder is replaced by a small particle that sinters at a low temperature. Firing cycles of 180 to 250°C (453 to 523 K) are possible using nanoscale silver or copper particles. The combination of dissimilar materials is an area of current research – examples include zirconia and stainless steel, electronic alloys and stainless steel, and cemented carbide and stainless steel. For complicated shapes or small structures, the sinter bonding route offers higher quality, lower cost, and improved productivity.

Structural color is another new area for sintering. Most color is determined by phonon interaction with a surface. Depending on the electronic state, some of the incoming "white" light is absorbed and the remainder is reflected. The energy and intensity of the reflected light determines color. Pigments provide a means to tailor the reflected light to give red, blue, green, or pink. However, the color of a butterfly's wing is determined by a repeated surface pattern that interacts with light. Many opportunities arises in display technologies. When nanoscale particles are packed and partially sintered, the periodic pore array creates a color [83–87]. Structural color is controlled by the particle size, packing, and sintering. Early targets are in cellular telephone or other displays which are readable in outdoor sunlight.

Small particles are used to form structures with controlled porosity. Initially, conventional powders were lightly pressed and sintered to attain pore sizes from 1 μm and above, with porosity levels near 15 to 40%. The addition of pore forming agents, such as a polymer, carbonate, ice, or salt (NaCl) allows lower densities and controlled pore sizes. An option popular in sintering aluminum is to add titanium hydride to the aluminum powder. During sintering, the hydride decomposes, releasing hydrogen to swell the aluminum. Final aluminum porosities are in the 90% range. Foamed metal is useful for tissue scaffolds, where pore sizes of 100 μm allows tissue ingrowth in structures tailored to about 40% density. With smaller particles, the sintering is rapid and porosities of 96% have been demonstrated. Combinations of small and easily sintered particles, low forming pressures, and pore forming agents can produce very low densities. The low density foams are like sponges, with an ability to undergo 50% compressive strain yet return to their initial size when the stress is released. Envisioned uses are in automobiles for crash energy absorption, as well as solar energy for light absorption, and as supports for gas reaction catalysts.

Diamond particles can be synthesized from graphite using high temperatures and pressures. It is now possible to sinter diamond into polycrystalline objects using high pressure sintering. The sintered microstructure is akin to that seen in cemented carbides. During the high temperature sintering cycle, the small grains dissolve preferentially into the liquid phase, usually cobalt, and reprecipitate at the diamond-diamond bonds. The result is a rigid diamond skeleton in a material that has good toughness. The durability is very desirable and should expand interest in high pressure sintering.

There is significant opportunity for sintered materials that combine attributes. Consider the hybrid automobile. The silicon control system undergoes significant thermal stress as it switches electric loads between motors, engine, batteries, and regenerative brakes. Sintered heat dissipation plates are used to ensure that the silicon control system does not overheat. New concepts in heat dissipation rely on sintered silicon carbide with internal heat pipe features. This gives exceptional heat dissipation without causing thermal fatigue. However, sintering the gradient microstructure is a significant challenge—the structure has internal channels (sinter bonding) yet the structure must remain leak tight. The most demanding applications are in military vehicles, for example where a vehicle engine is also used to drive a portable power generator, such as for mobile radar units.

NEW PROCESSES

A host of new sintering processes have been mentioned. The field is full of invention, often with an emphasis on faster sintering. A traditional industrial sintering cycle takes about a day. For continuous sintering, the cycles span from 6 hours (dental orthodontic brackets) to a week (large abrasive wheels).

The old adage is to find a problem and solve it. Unfortunately many new sintering concepts seem to start with a novel process and seek a problem. Without understanding the basic problems in sintering, new approaches often are curious, but not significant. A good example comes with electric fields. They induce rapid sintering for flat, thin structures. If the geometry is complex, then non-uniform heating results in warpage and property gradients. Rapid sintering has a long history, but due to component distortion it is not widely used. The innovative rapid sintering processes include a variety of heating approaches, including solar, microwave, induction, pulsed intense light, laser, magnetic pulse, capacitance discharge, exothermic reactions, spark discharges, and plasma. It is expected that these innovations will continue and eventually attack the critical questions of precision and cost.

In a sense, some of the new sintering ideas are revitalized versions of earlier concepts. Electric discharge sintering arose between 1890 and 1910. In 1955, Lenel [88] applied direct current heating to titanium and other materials and Goetzel and De Marchi [89] revisited the idea in 1971. However, outside of diamond composite consolidation, spark sintering remains largely a research topic. It is successful for making simple disk shapes from a wide range of conductive metals, but it has not graduated to difficult shapes, although about 10% of all sintering studies now involve spark sintering.

For nonconductive ceramics, the current passes through the graphite tooling and the densification and microstructure are simply the same as obtained with rapid hot pressing [90,91]. In spite of much early uncertainty, there is no plasma generated during spark sintering, even so it is often misnamed spark plasma sintering [92].

On the other hand, laser sintering, especially for rapid prototyping, is beneficial for tool-less manufacturing. However, laser sintering is slow compared to traditional sintering. For example, sintered steel requires 18 MW/kg for press-sinter fabrication, so a 50 W laser sintering facility gives only 10 g/h or slightly more than 1 cm^3/h at 100% efficiency. A laser sintering facility is expensive, costing about the same as a continuous furnace able to sinter 50 to 100 kg/h. The high capital cost causes laser sintering to be only used for high value small components, such as dental crowns and bridges, but not bulk production. Likewise, sintering in a home microwave is exciting, but nonuniform heating and concomitant distortion make it a doubtful application for production. Although smaller grain sizes are possible from nanoscale powders with microwave heating, the approach has few applications [93]. Many approaches characteristically are solutions looking for a problem.

Innovation abounds in sintering, but one becomes skeptical since only a few innovations move beyond the laboratory. The reverse situation relies on identification of problems to guide development of new sintering concepts. A few persistent problems include:

- **cost**—although always a concern, processing cost is less an issue than furnace capital expense; the need is for fast, lower cost production furnaces with no penalty in sintered dimensional uniformity.
- **dimensions**—dimensional adjustment after sintering by machining, grinding, or coining adds to the cost; much gain comes from improved sintered dimension control, most likely via uniform heating and improved control of factors causing dimensional variation, negating interest in rapid heating.
- **metastable**—short sintering cycles allow consolidation of unstable materials, and a science base is needed to identify additives to modify sintering for these materials.

CONCLUDING REMARKS

This chapter touches on the exciting future prospects for sintering. At the same time, that excitement must be tempered by some reality checks. As the investment community says—"you always hear the good news first." This is very true with new sintering developments. As with so many things, economics are a dominant consideration. The new materials, applications, and production practices all face cost-based litmus tests. Significant technological advances do occur that pass this test. Some recent examples are double-wide continuous walking beam furnaces sintering 5 million cell phone components per day, batch iron ore induration furnaces reaching to 20,000 tons per day, and batch vacuum furnaces with integrated plasma polymer burnout capabilities.

In reviewing past forecasts, a few reminders emerge. Computer modeling was going to change production, enabling faster convergence to optimized cycles, more precise components, and fewer process adjustments. Unfortunately, computer modeling has shown that the underlying understanding of sintered materials production is weak. Simulation is not an impediment; rather the failure to understand all variables and their interactions during sintering is the barrier [94,95]. Basic materials science has yet to isolate several important relations, such as impurity effects on grain boundary diffusion, surface emissivity changes during densification, and *in situ* material property changes during sintering, and the interaction of pores and grain boundaries. Thus, sintered dimension predictions are scattered. The simulations are not able to give accurate dimensional control relevant to production (say 10 μm). Even so, simulation is a favorite topic at sintering conferences.

Every few years, distinguished researchers provides a sense of where sintering is moving [95–101]. These forward-looking views focus on models and how they are improving. Better models help explain empirical observations, but observation and the empirical aspect of sintering still are ahead of theory [102]. We are seeing significant advances via improved powders, uniform powders, and smaller powders, but the largest gains come from improved green body homogeneity. A critical need is models relevant to industrial sintering and the attributes important to sintering practice—density, size, distortion, grain size, and dimensional uniformity. Cost minimization is a topic rarely included in sintering simulations.

It is unclear wheter different heating technologies fundamentally impact sintering. It is a topic worthy of careful comparative research. Meanwhile, the complexity increases—as evident in the material chemistry, shapes, property demands, and dimensional precision.

REFERENCES

[1] L.I. Kivalo, V.V. Skorokhod, N.F. Grigorenko, Effect of nickel on sintering processes in the Ti-Fe system. Part I, Powder Metall. Metal Ceram. 22 (1983) 543–546.

[2] D.L. Bourell, J.J. Beaman, Powder material principles applied to additive manufacturing, in: Materials Processing and Interfaces, vol. 1, Proceedings 141st Meeting the Minerals, Metals, and Materials Society, Warrendale, PA, 2012, pp. 537–544.

[3] T.B. Sercombe, G.B. Schaffer, Selective laser sintering of aluminum, Advances in Powder Metallurgy and Particulate Materials – 2003, Part 3, Metal Powder Industries Federation, Princeton, NJ, 2003, pp. 148–155.

[4] J.W. McCauley, N.D. Corbin, Phase relations and reaction sintering of transparent cubic aluminum oxynitride spinel (AlON), J. Amer. Ceram. Soc. 62 (1979) 476–479.

[5] D.T. Colbert, J. Zhang, S.M. Mcclure, P. Nikolaev, Z. Chen, J.H. Hafner, et al., Growth and sintering of fullerene nanotubes, Science 266 (1994) 1218–1222.
[6] H.Y. Suzuki, M. Fukuda, H. Kuroki, Colloidal compaction of fine metallic powders under high speed centrifugal force – compaction and sintering of high speed steel, J. Japan Soc. Powder Powder Metall. 50 (2003) 856–864.
[7] L.E. McCandlish, B.H. Kear, B.K. Kim, Processing and properties of nanostructured WC-Co, Nano. Mater. 1 (1992) 119–124.
[8] W.L. Li, K. Lu, J.Y. Walz, Freeze casting of porous materials: review of critical factors in microstructure evolution, Inter. Mater. Rev. 57 (2012) 37–60.
[9] B. Levenfeld, A. Varez, J.M. Torralba, Effect of residual carbon on the sintering process of M2 high speed steel parts obtained by a modified metal injection molding process, Metall. Mater. Trans. 33A (2002) 1843–1851.
[10] S. Tsurekawa, K. Harada, T. Sasaki, T. Matsuzaki, T. Watanabe, Magnetic sintering of ferromagnetic metal powder compacts, Mater. Trans. Japan Inst. Met. 41 (2000) 991–999.
[11] H.H. Yang, Y.W. Kim, J.H. Kim, D.J. Kim, K.W. Kang, Y.W. Rhee, et al., Pressureless rapid sintering of UO_2 assisted by high-frequency induction heating process, J. Amer. Ceram. Soc. 91 (2008) 3202–3206.
[12] M. Eriksson, M. Radwan, Z. Shen, Spark plasma sintering of WC, cemented carbide, and functional graded materials, Inter. J. Refract. Met. Hard Mater. 36 (2013) 31–37.
[13] J.S. Lee, J.C. Yun, J.P. Choi, G.Y. Lee, Consolidation of iron nanopowder by nanopowder agglomerate sintering at elevated temperature, J. Korean Powder Metall. Inst. 20 (2013) 1–6.
[14] S.A. Deshpande, T. Bhatia, H. Xu, N.P. Padture, A.L. Ortiz, F.L. Cumbrera, Microstructural evolution in liquid phase sintered SiC: Part II, effects of planar defects and seeds in the starting powder, J. Amer. Ceram. Soc. 84 (2001) 1585–1590.
[15] J.M.M. Rheme, J. Carron, L. Weber, Thermal conductivity of aluminum matrix composites reinforced with mixtures of diamond and SiC particles, Scripta Mater. 58 (2008) 393–396.
[16] T. Hawa, M.R. Zachariah, Molecular dynamics simulation and continuum modeling of straight chain aggregate sintering: development of a phenomenological scaling law, Phys. Rev. B 76 (2007) 054109 pp. 1–9.
[17] J.M. Williams, A. Adewunmi, R.M. Schek, C.L. Flanagan, P.H. Krebsbach, S.E. Feinberg, et al., Bone tissue engineering using polycaprolactone scaffolds fabricated via selective laser sintering, Biomater 26 (2005) 4817–4827.
[18] A. Simchi, F. Petzoldt, H. Pohl, H. Loffler, Direct laser sintering of a low alloy P/M steel, P/M Sci. Tech. Briefs 3 (2001) 5–9.
[19] C.T. Campbell, S.C. Parker, D.E. Starr, The effect of size-dependent nanoparticle energetics on catalyst sintering, Science 298 (2002) 811–814.
[20] X. Shi, K. Su, R.R. Varshney, Y. Wang, D.A. Wang, Sintered microsphere scaffolds for tissue engineering, Pharm. Res. 28 (2011) 1224–1228.
[21] G.C. Wei, Transparent ceramics for lighting, J. Europ. Ceram. Soc. 29 (2009) 237–244.
[22] Z.F. Zu, Y.B. Choi, K. Matsugi, D.C. Li, G. Sasaki, Mechanical and thermal properties of vapor grown carbon fiber reinforced aluminum matrix composites by plasma sintering, Mater. Trans. 51 (2010) 510–515.
[23] J. Yan, G. Zou, X. Wang, F. Mu, H. Bai, B. Wu, et al., Characterization of low temperature bonding with Cu nanoparticles for electronic packaging applications, in: Proceedings Materials Science and Technology Conference, Columbus, OH, 2011, pp. 1526–1531.
[24] E. Herderick, Additive manufacturing of metals: a review, in: Proceedings Materials Science and Technology Conference, Columbus, OH, 2011, pp. 1413–1425.
[25] H.A. Colorado, S.R. Dhage, J.M. Yang, H.T. Hahn, Intense pulsed light sintering technique for nanomaterials, in: Materials Processing and Interfaces, vol. 1, Proceedings 141st Meeting the Minerals, Metals, and Materials Society, Warrendale, PA, 2012, pp. 577–584.
[26] M. Yuki, H. Yamaoka, Y. Nakanishi, A. Kawasaki, R. Watanabe, Experimental study of laser sintering process of ceramics, in: Y. Bando, K. Kosuge (Eds.), Proceedings of 1993 Powder Metallurgy World Congress, Part 2, Japan Society of Powder and Powder Metallurgy, Kyoto, Japan, 1993, pp. 939–942.

[27] W.D. Jones, Fundamental Principles of Powder Metallurgy, Edward Arnold Publishers, London, UK, 1960.
[28] J.D. Belnap, Sintering of ultrahard materials, in: Z.Z. Fang (Ed.), Sintering of Advanced Materials, Woodhead Publishing, Oxford, UK, 2010, pp. 389–414.
[29] M. Qian, G.B. Schaffer, Sintering of aluminum and its alloys, in: Z.Z. Fang (Ed.), Sintering of Advanced Materials, Woodhead Publishing, Oxford, UK, 2010, pp. 291–323.
[30] H. Wang, Z.Z. Fang, P. Sun, A critical review of the mechanical properties of powder metallurgy titanium, Inter J. Powder Metall. 46 (5) (2010) 45–57.
[31] I.M. Robertson, G.B. Schaffer, Review of densification of titanium based powder systems in press and sinter processing, Powder Metall. 53 (2010) 146–162.
[32] C.H. Caceres, Economical and environmental factors in light alloys automotive applications, Metall. Mater. Trans. 38A (2007) 1649–1662.
[33] R.M. German, Status of metal powder injection molding of titanium, Inter. J. Powder Metall. 46 (5) (2010) 11–17.
[34] S. Barg, C. Soltmann, M. Andrade, D. Koch, G. Grathwohl, Cellular ceramics by direct foaming of emulsified ceramic powder suspensions, J. Amer. Ceram. Soc. 91 (2008) 2823–2829.
[35] T. Shimizu, K. Matsuzaki, K. Kikuchi, N. Kanetake, Space holder method to produce high porosity metal foam using gelation of water base binder, J. Japan Soc. Powder Powder Metall. 55 (2008) 770–775.
[36] U.T. Gonzenbach, A.R. Studart, E. Tervoort, L.J. Gauckler, Macroporous ceramics from particle stabilized wet foams, J. Amer. Ceram. Soc. 90 (2007) 16–22.
[37] N. Sakuari, J. Takekawa, Shape recovery characteristics of NiTi foams fabricated by a vacuum process applied to a slurry, Mater. Trans. 47 (2006) 558–563.
[38] L. Tuchinskiy, Novel fabrication technology for metal foams, J. Adv. Mater. 37 (3) (2005) 60–65.
[39] I. Thijs, J. Luyten, S. Mullens, Producing ceramic foams with hollow spheres, J. Amer. Ceram. Soc. 87 (2004) 170–172.
[40] B. Levine, A new era in porous metals: applications in orthopaedics, Adv. Eng. Mater. 10 (2008) 788–792.
[41] J.L. Jonson, Opportunities for PM processing of metal matrix composites, Inter. J. Powder Metall. 47 (2) (2011) 19–28.
[42] K. Nishiyaku, S. Matsuzaki, S. Tanaka, Liquid infiltration property of micro porous stainless steel produced by powder space holder method, Powder Inj. Mould. Inter. 2 (3) (2008) 60–63.
[43] V. Friederici, A. Bruinink, P. Imgrund, S. Seefried, Getting the powder mix right for design of bone implants, Met. Powder Rept. (2010) 14–16.
[44] N. Salk, B. Troger, Precision manufacturing of micro ceramic injection molded implants, Advances in Powder Metallurgy and Particulate Materials - 2010, Metal Powder Industries Federation, Princeton, NJ, 2010, pp. 4.11–4.17.
[45] K. Doi, T. Matsushita, T. Kokubo, S. Fjibayashi, M. Takemoto, T. Nakamura, et al., Mechanical properties of porous titanium and its alloys fabricated by powder sintering for medical use, in: P. Rodhammer (Ed.), Proceedings of the Seventeenth Plansee Seminar, vol. 1, Plansee Group, Reutte, Austria, 2009, pp. GT10.1–GT10.8.
[46] S. Luckx, The hardness of tungsten carbide – cobalt hardmetal, in: R. Riedel (Ed.), Handbook of Ceramic Hard Materials, vol. 2, Wiley-VCH, Weinheim, Germany, 2000, pp. 946–964.
[47] Z.J. Lin, J.Z. Zhang, B.S. Li, L.P. Wang, H.K. Mao, R.J. Hemley, et al., Superhard diamond/tungsten carbide nanocomposites, Appl. Phys. Lett. 98 (2011), paper 121914, 3 pages
[48] K. Lu, Nanoparticulate Materials Synthesis, Characterization, and Processing, Wiley, Hoboken, New Jersey, 2013.
[49] S. Jang, H. Cho, Y. Lee, D. Kim, Atmospheric effects on the thermally induced sintering of nanoparticulate gold films, J. Mater. Sci. 47 (2012) 5134–5140.
[50] N. Marjanovic, J. Hammerschmidt, J. Perelaer, S. Farnsworth, I. Wawson, M. Kus, et al., Inkjet printing and low temperature sinteirng of CuO and CdS as functional electronic layers and Schottky diodes, J. Mater. Chem. 21 (2011) 13634–13639.
[51] H. Alarifi, A. Hu, M. Yavuz, N. Zhou, Silver nanoparticle paste for low temperature bonding of copper, J. Elect. Mater. 40 (2011) 1394–1402.

[52] A. Hu, J.Y. Guo, H. Alarifi, G. Patane, Y. Zhou, G. Compagnini, et al., Low temperature sintering of Ag nanoparticles for flexible electronics packaging, Appl. Phys. Lett. 97 (2010) paper 153117.

[53] S.M. Salamone, L.C. Stearns, R.K. Bordia, M.P. Harmer, Effect of rigid inclusions on the densification and constitutive parameters of liquid phase sintered $YBa_2Cu_3O_{6+x}$ powder compacts, J. Amer. Ceram. Soc. 86 (2003) 883–892.

[54] Z. Ma, Y. Liu, J. Huo, Influence of ball milled amorphous B powders on the sintering process and superconductive properties of MgB_2, Supercond. Sci. Tech. 22 (2009) article 125006, 5 pages.

[55] K. Shinohara, T. Futatsumori, H. Ikeda, Effect of hot press method on the critical current density of MgB_2 bulk samples, Phys. C 468 (2008) 1369–1371.

[56] Y.C. Liu, Q.Z. Shi, Q. Zhao, Z.Q. Ma, Kinetics analysis for the sintering of bulk MgB_2 superconductor, J. Mater. Sci.: Mater. Elect. 18 (2007) 855–861.

[57] H. Chang, T.P. Tang, K.T. Huang, F.C. Tai, Effects of sintering process and heat treatments on microstructures and mechanical properties of VANADIS 4 tool steel added with TiC powders, Powder Metall. 54 (2011) 507–512.

[58] W. Acchar, A.M. Segadaes, Properties of sintered alumina reinforced with niobium carbide, Inter. J. Refract. Met. Hard Mater. 27 (2009) 427–430.

[59] A. Upddhyaya, S. Balaji, Sintered intermetallic reinforced 434L ferric stainless steel composite, Metall. Mater. Trans. 40A (2009) 673–683.

[60] C. Padmavathi, A. Upadhyaya, Densification, microstructure, and properties of supersolidus liquid phase sintered 6711 Al – SiC metal matrix composite, Sci. Sintering 42 (2010) 363–382.

[61] H. Asgharzadeh, A. Simchi, Supersolidus liquid phase sintering of Al 6061/SiC metal matrix composites, Powder Metall. 52 (2009) 28–35.

[62] S. Gustafsson, L.K.L. Falk, E. Liden, E. Carstrom, Pressureless sintered Al_2O_3-SiC nanocomposites, Ceram. Inter. 34 (2008) 1609–1615.

[63] J.L. Johnson, Sintering of refractory metals, in: Z.Z. Fang (Ed.), Sintering of Advanced Materials, Woodhead Publishing, Oxford, UK, 2010, pp. 356–388.

[64] R.G. Pileggi, A.R. Studart, M.D.M. Innocentini, V.C. Pandolfelli, High performance refractory castables, Ceram. Bull. 81 (6) (2002) 37–42.

[65] J.N. Calata, A. Matthys, G.Q. Lu, Constrained-film sintering of cordierite glass-ceramic on silicon substrate, J. Mater. Res. 13 (1998) 2334–2341.

[66] J. Guille, A. Bettinelli, J.C. Bernier, Sintering behavior of tungsten powders co-fired with alumina or conventionally sintered, Powder Metall. Inter. 20 (5) (1988) 26–28.

[67] C. Pascal, A. Thomazic, A.A. Zdziobek, J.M. Chaix, Co-sintering and microstructural characterization of steel/cobalt base alloy biomaterials, J. Mater. Sci. 47 (2012) 1875–1886.

[68] A. Simchi, F. Petzoldt, Cosintering of powder injection molding parts made from ultrafine WC-Co and 316 L stainless steel powders for fabrication of novel composite structures, Metall. Mater. Trans. 41A (2010) 233–241.

[69] H. He, Y. Li, P. Liu, J. Zhang, Design with a skin-core structure by metal co-injection moulding, Powder Inj. Mould. Inter. 4 (1) (2010) 50–54.

[70] H. Ye, X.Y. Liu, H. Hong, Fabrication of metal matrix composites by metal injection molding – a review, J. Mater. Proc. Tech. 200 (2008) 12–24.

[71] A. Baumann, M. Brieseck, S. Hohn, T. Moritz, R. Lenk, Development of multi-component powder injection moulding of steel-ceramic compounds using green tapes for inmould label process, Powder Inj. Mould. Inter. 2 (1) (2008) 55–58.

[72] Y. Boonyongmaneerat, C.A. Schuh, Contributions to the interfacial adhesion in co-sintered bilayers, Metall. Mater. Trans. 37A (2006) 1435–1442.

[73] J.C. Chang, J.H. Jean, Camber development during the Co-firing of bi-layer glass-based dielectric laminate, J. Amer. Ceram. Soc. 88 (2005) 1165–1170.

[74] T. Harikou, Y. Itoh, K. Satoh, H. Miura, Joining of stainless steels (SUS316L) and hard materials by insert injection molding, J. Japan Soc. Powder Powder Metall. 51 (2004) 31–36.

[75] D.F. Heaney, P. Suri, R.M. German, Defect-Free sintering of two material powder injection molded components: Part I, experimental investigations, J. Mater. Sci. 38 (2003) 4869–4874.

[76] O. Guillon, R.K. Bordia, C.L. Martin, Sintering of thin films/constrained sintering, in: Z.Z. Fang (Ed.), Sintering of Advanced Materials, Woodhead Publishing, Oxford, UK, 2010, pp. 415–433.

[77] K. Nishiyabu, S. Matsuzaki, S. Tanaka, Dimensional accuracy of micro-porous metal injection and extrusion molded components produced by powder space holder method, J. Japan Soc. Powder Powder Metall. 53 (2006) 776–781.

[78] S.H. Li, J.R. De Wijn, P. Layrolle, K. De Groot, Novel method to manufacture porous hydroxyapatite by dual phase mixing, J. Amer. Ceram. Soc. 86 (2003) 65–72.

[79] T.S. Kim., I.C. Kang, T. Goto, B.T. Lee, Fabrication of continuously porous alumina body by fibrous monlithic and sintering process, Mater. Trans. 44 (2003) 1851–1856.

[80] U.T. Gonzenbach, A.R. Studart, E. Tervoort, L.J. Gauckler, Macroporous ceramics from particle stabilized wet foams, J. Amer. Ceram. Soc. 90 (l) (2007) 16–22.

[81] T. Shimizu, K. Matsuzaki, Y. Ohara, Process of porous titanium using a space holder, J. Japan Soc. Powder Powder Metall. 53 (2006) 36–41.

[82] N. Sakurai, J. Takekawa, Fabrication of foamed titanium by vacuum process using slurry, J. Japan Soc. Powder Powder Metall. 50 (2003) 1025–1039.

[83] H. Ghiradella, Light and color on the wing: structural colors in butterflies and moths, Appl. Opt. 30 (1991) 3492–3500.

[84] J. Silver, R. Withnall, T.G. Ireland, G.R. Fern, Novel nanostructured phosphor materials cast from natural morpho butterfly scales, J. Mod. Opt. 52 (2005) 999–1007.

[85] O.L.J. Pursiainen, J.J. Baumberg, H. Winkler, B. Viel, P. Spahn, T. Ruhl, Nanoparticle tuned structural color from polymer opals, Opt. Exp. 15 (2007) 9553–9561.

[86] W. Zhang, D. Zhang, T. Fan, J. Ding, J. Gu, Q. Guo, et al., Biomimetic zinc oxide replica with structural color using butterfly (*Ideopsis similis*) wings as templates, Bioinsp. Biomim. 1 (2006) 89–95.

[87] A. Kimura, K. Yagi, F. Suzuki, Development of colored pearly lustre pigment: I – synthesis of colored mica, coated with titanium dioxide and electroless plated nickel, J. Japan Soc. Powder Powder Metall. 44 (1997) 387–392.

[88] F.V. Lenel, Resistance sintering under pressure, Trans. TMS-AIME 203 (1955) 158–167.

[89] C.G. Goetzel, V.S. De Marchi, Electrically activated pressure sintering (spark sintering) of titanium powders, Powder Metall. Inter. 3 (1971), pp. 80–87 and 134–136

[90] J. Langer, M.J. Hoffmann, O. Guillon, Direct comparison between hot pressing and electric field assisted sintering of submicron alumina, Acta Mater. 57 (2009) 5454–5465.

[91] J. Langer, M.J. Hoffmann, O. Guillon, Electric Field-assisted sintering in comparison with the hot pressing of yttria stabilized zirconia, J. Amer. Ceram. Soc. 94 (2010) 24–31.

[92] D.M. Hulbert, A. Anders, J. Andersson, E.J. Lavernia, A.K. Mukherjee, A discussion on the absence of plasma in spark plasma sintering, Scripta Mater. 60 (2009) 835–838.

[93] J.L. Johnson, Economics of processing nanoscale powders, Inter. J. Powder Metall. 44 (1) (2008) 44–54.

[94] R.M. German, Phenomenological observations and the prospects for predictive computer simulations, Advances in Powder Metallurgy and Particulate Materials - 2012, Part 1, Metal Powder Industries Federation, Princeton, NJ, 2012, pp. 32–42.

[95] S.J. Park, S. Ahn, T.G. Kang, S.T. Chung, Y.S. Kwon, S.H. Chung, et al., A review of computer simulations in powder injection molding, Inter. J. Powder Metall. 46 (3) (2010) 37–46.

[96] M.M. Ristic, Science of Sintering and Its Future, International Team for Science of Sintering, Beograd, Yugoslavia, 1975.

[97] R.J. Brook, W.H. Tuan, L.A. Xue, Critical issues and future directions in sintering science, in: G.L. Messing, E.R. Fuller, H.H. Hausner (Eds.), Ceramic Transactions, vol. 1, American Ceramic Society, Westerville, OH, 1987, pp. 811–823.

[98] T. Honda, Trend and future for fatigue characteristics of sintered ferrous products, J. Japan Soc. Powder Powder Metall. 44 (1997) 475–482.

[99] A. Fujiki, Present state and future prospects of powder metallurgy parts for automotive applications, Mater. Chem. Phys. 67 (2001) 298–306.
[100] K. Brookes, Some tribulation on the way to a nano future for hardmetals, Met. Powder Rept. 60 (2005) 24–30.
[101] G.S. Upadhyaya, Future directions in sintering research, Sci. Sintering 43 (2011) 3–8.
[102] R.M. German, History of sintering: empirical phase, Powder Metall. 6 (2013) 117–123.

INDEX

Note: Page numbers followed by "*f*" and "*t*" refer to figures and tables, respectively.

D

E

F

G

Q

R

Printed and bound by CPI Group (UK) Ltd, Croydon, CR0 4YY
08/05/2025
01864900-0006